WITHDRAWN
FROM STOCK

Clay fills

Proceedings of the Conference held at the Institution of Civil Engineers, 14-15 November 1978

The Institution of Civil Engineers, London, 1979

Conference sponsored by the British Geotechnical Society, the British National Committee on Large Dams, the Institution of Highway Engineers and the Institution of Civil Engineers

Organizing committee: P. R. Vaughan (Chairman), J. A. Charles, K. W. Cole, I. F. Symons and C. E. J. Wood

Published by the Institution of Civil Engineers, P.O. Box 101, 26-34 Old Street, London EC1P 1JH

First published 1979

ISBN: 0 7277 0069 3

Produced and distributed by Thomas Telford Ltd, P.O. Box 101, 26-34 Old Street, London EC1P 1JH

Made and printed in Great Britain by The Burlington Press (Cambridge) Ltd, Foxton, Royston, Herts

CONTENTS

Opening address

A.M. MUIR WOOD, MA, FEng, FICE, President 1977-78, Institution of Civil Engineers

This is a timely occasion to review the state of knowledge in the safe disposal of clay fills; not only has there been a great increase in knowledge since the Aberfan disaster in 1966 but also much information well known at that time has become much more widely appreciated.

This Conference is yet more timely in the use of clay fills, recognizing the merit in saving resources in easy-to-use fills and in the areas of land necessary for their extraction. It is good to find many aspects of individual experience in the wider use of what has in the past been designated unsuitable material.

There are special problems concerning the behaviour of fills by comparison with soils in situ. The latter may, by site investigation, be related to a known history of deposition and subsequent physical and chemical alteration by geological events. Understanding all these processes should provide a more or less reliable knowledge of continuity and variability. In the world of fills the circumstances are much different. The source, admixture, treatment and workmanship all depend on the fallibility of man which introduces uncertainties with possible serious implications not found with in situ soils.

A number of new concepts are required in dealing with clay fills; many of these are brought out in the papers. Another which may be of importance in dealing with the composite situation of poorly compacted and well compacted fills concerns definition of brittleness in terms of strain to complement brittleness in stress terms.

This is an occasion on which a particular strength can be shown, under the auspices of the Institution of Civil Engineers, in bringing together theory and practice for discussion in an informed manner.

R. A. ABEYESEKERA, Graduate Instructor,
C. W. LOVELL, Professor of Civil Engineering, and
L. E. WOOD, Associate Professor of Civil Engineering,
Purdue University

Stress-deformation and strength characteristics of a compacted shale

When shales are encountered in road cuts, economic and environmental considerations usually dictate that they be used in adjoining embankments. However, unless special precautions are taken, the stability of a shale embankment can deteriorate with time on account of the nondurable nature of most shales. Previous investigators have developed various tests to classify shales with respect to their hardness, durability and degradability. This paper presents the results of a laboratory investigation to define the effective stress strength parameters of mechanically hard but nondurable shale pieces of a lower Mississippian age formation (New Providence) from the State of Indiana, U.S.A. The initial gradation of crushed shale aggregate, the molding water content, the compaction pressure, and the pre-shear consolidation pressure were adopted as the independent test variables. Triaxial specimens formed by kneading compaction were back pressure saturated under a low effective confining pressure, consolidated to the desired isotropic effective stress, and sheared undrained at a constant rate of deformation. The compaction characteristics, the volume changes during saturation and consolidation, and the undrained shearing response including pore water pressure changes were studied. The effective stress strength parameters evaluated at maximum deviator stress were found to be essentially independent of the initial conditions, except for loose uncompacted aggregate. The compaction characteristics, the volume change characteristics, the induced pore water pressure changes, and the consolidated undrained strength were found to be greatly dependent on the initial conditions.

INTRODUCTION

1. Between December 1971 and January 1972, a major slope failure occurred in an embankment on I-74 near St. Leon in Dearborn County, Indiana, U.S.A., as reported by Wood, Lovell and Deo (ref. 1). It was constructed in 1961 and the fill material was obtained from intermittent beds of limestone and shale, and weathered soil principally from the shale. The harder limestone and the weaker shale were present in about equal amounts and their distribution within the embankment was random. The embankment was built as a rockfill in lifts up to 3 foot thick. Large settlements preceded the actual slide, due to the degradation of the shale in the service environment.

2. The embankment problems along I-74, an interstate highway, motivated research and development activities both directly by the Indiana State Highway Commission (ISHC) and through the Joint Highway Research Project (JHRP) at Purdue University. These efforts were directed to set design and construction guidelines for shale embankments and fall into four principal categories; (a) shale classification, (b) study of compaction and degradation characteristics of shales, (c) storage and retrieval of existing data on Indiana shales, and (d) definition of shear strength parameters for the long-term stability of compacted saturated shales.

3. The results of studies on categories (a), (b) and (c) have been published in Deo (ref. 2), Chapman (ref. 3), Bailey (ref. 4) and van Zyl (ref. 5). The results of studies on category (d) are reported in Abeyesekera (ref. 6) and summarized herein.

4. New Providence shale was selected for testing as it has a relatively high hardness with a relatively low durability. It is such shales that present most of the field problems because they are difficult to break down during compaction and yet degrade in the service environment.

DESCRIPTION OF NEW PROVIDENCE SHALE

Geology

5. The shale used in the study was sampled from the New Providence formation at an elevation of 575 feet along a road excavation on I-265 in Floyd County, south central Indiana, U.S.A. The geologic description of the shale as reported by the ISHC is as follows:

Era	- Paleozoic
Period	- Carboniferous
Epoch	- Mississippian
Series	- Valmeyeran (Osage)
Group	- Borden
Formation	- New Providence

The age of the formation is around 241 to 261 million years, and its pre-consolidation pressure is estimated to be about 2600 psi.

Mineralogy

6. The mineralogical composition of the test shale was found by X-ray diffraction analysis to be predominantly quartz, kaolinite and illite, with traces of chlorite and vermiculite likely to have been present. Montmorillonite was absent.

Indiana State Highway Commission tests

7. Table 1 shows the results of routine tests performed by the ISHC on the test shale. The shale classification tests were developed by Deo (ref. 2). The CBR test was performed on minus No. 4 material compacted at optimum moisture content using standard AASHTO effort.

Table 1. Summary of ISHC test results on New Providence shale (test shale)

Laboratory No. (ISHC) 75-55731			
General Physical Description			
Colour: Light Gray	Hardness: Medium	Fissility: Massive	
Shale Classification		**Soil Classification**	
Degradability:	Soil-like	Textural	Silty-clay
Slaking Index		AASHTO	A-4 (10)
- 1 cycle	3%	Plastics Limit	21%
- 5 cycles	48%	Liquid Limit	31%
Slake Durability Index		Plasticity Index	10%
Dry- 200 revs.	91%	Sand Size	2%
- 500 revs.	76%	Silt Size	67%
Wet- 200 revs.	71%	Clay Size	21%
- 500 revs.	41%	Colloid Size	10%
Fissility Number	26%	**Moisture-Density**	
Physical Properties		Effort	Std. AASHTO
Nat. Wet Dens. (pcf)	155	Maximum Size	No. 4
Nat. Dry Dens. (pcf)	145	Max. Wet Dens. (pcf)	135
Nat. Moist. Cont.	6.1%	Max. Dry Dens. (pcf)	120
Sp. Gr.	2.78	Opt. Moist. Cont.	12.1%
pH	6.5	**CBR Test**	
Shrinkage Limit	21.2%	As-compacted	10.2%
Lineal Shrinkage	1.9%	After Soaking	1.1%
Loss on Ignition	4.0%	Average Swell	2.8%

Scleroscope hardness

8. The scleroscope hardness H measured perpendicular to the bedding planes of shale pieces was found by Bailey (ref. 4) to be a function of the particle moisture content w% according to the linear regression equation,

$$H = 32.1 - 2.4\ w$$

with a Pearson's R^2 correlation coefficient of 0.43. Moisture contents ranged from 0 to 3.6% The scatter in hardness values for oven dried (105°C) specimens ranged from 27.5 to 40.5. At the highest moisture content of 3.6%, the scatter was much reduced and ranged from 23.5 to 24.5. Bailey attributed the wide scatter at zero moisture content to the presence of small hairline cracks, chipping and flaking in the test specimen.

Point load strength

9. Bailey (ref. 4) found that the point load strength at zero moisture content also showed a wide scatter ranging from 490 to 1980 psi due to hairline cracks, chipping and flaking on oven drying. The relationship between the point load strength and the moisture content of the test shale showed too much scatter within the moisture content range of 0 to 2.46% for any statistical trends to be verified. However, for four other shales, the point load strength decreased more or less linearly with increase in moisture content of the test specimen.

Collapse test

10. This test might be used to classify shales according to their hardness and durability. It is a one-dimensional compression test on aggregate at natural moisture content, followed by wetting. The results of such a test on a specimen composed of shale aggregate passing the 3/8 inch sieve and retained on the No. 4 sieve is shown in Figs. 1 and 2. Under a vertical stress of 20 psi the axial compression of the specimen amounted to 1.1%. When water was introduced, the axial strain increased dramatically to 12.8% in 7 minutes, and less rapidly to 17.5% in one day. The collapse that took place on wetting was mainly due to the slaking degradation of the shale. In the absence of any load acting on the specimen the rate of slaking degradation is very much reduced.

TESTING PROCEDURE

11. Consolidated undrained triaxial compression tests with pore water pressure measurements were performed on nearly 50 cylindrical specimens, 4.0 inch diameter and 8.5 inch high, compacted by means of a California type kneading compactor. The specimens were saturated under a back pressure, isotropically consolidated to the desired effective stress, and sheared undrained at a constant rate of deformation up to an axial strain pf 20%.

Independent variables and levels

12. The initial gradation of the shale aggregate, the molding water content, the compaction pressure, and the pre-shear consolidation pressure were adopted as the test variables.

13. Six initial gradations (Table 2) were used. Gradation No. 1 corresponds to the exponential function

$$P = 100(d/D)^{0.5}$$

where P is the percentage finer than a given diameter d, and D is the top size. The molding water added ranged from 5% to 8% of the weight of the shale aggregate at natural moisture content. The compaction pressures used were 200, 100 and 50 psi. Two specimens were formed by light tamping with a standard Proctor hammer (weighing 5.5 lb) in a dry condition but at natural moisture content. The consolidation pressures used ranged from 0 to 40 psi with intermediate values of 1, 5, 10 and 20 psi.

Table 2. Initial gradations used

Size Fraction	Weight of Size Fraction (gm) 1	2	3	4	5	6
P 3/4" R 1/2"	95	50	50	50	-	-
P 1/2" R 3/8"	52	150	150	150	250	200
P 3/8" R No. 4	103	100	100	100	250	100
P No.4 R No.8	73	100	100	200	-	100
P No.8 R No.16	52	-	100	-	-	100
P No.16 R No.30	37	100	-	-	-	-
P No.30 R No.50	26	-	-	-	-	-
P No.50 R No.100	19	-	-	-	-	-
P No.100 R No.200	30	-	-	-	-	-
P No.200	13	-	-	-	-	-
Total Weight (gm)	500	500	500	500	500	500

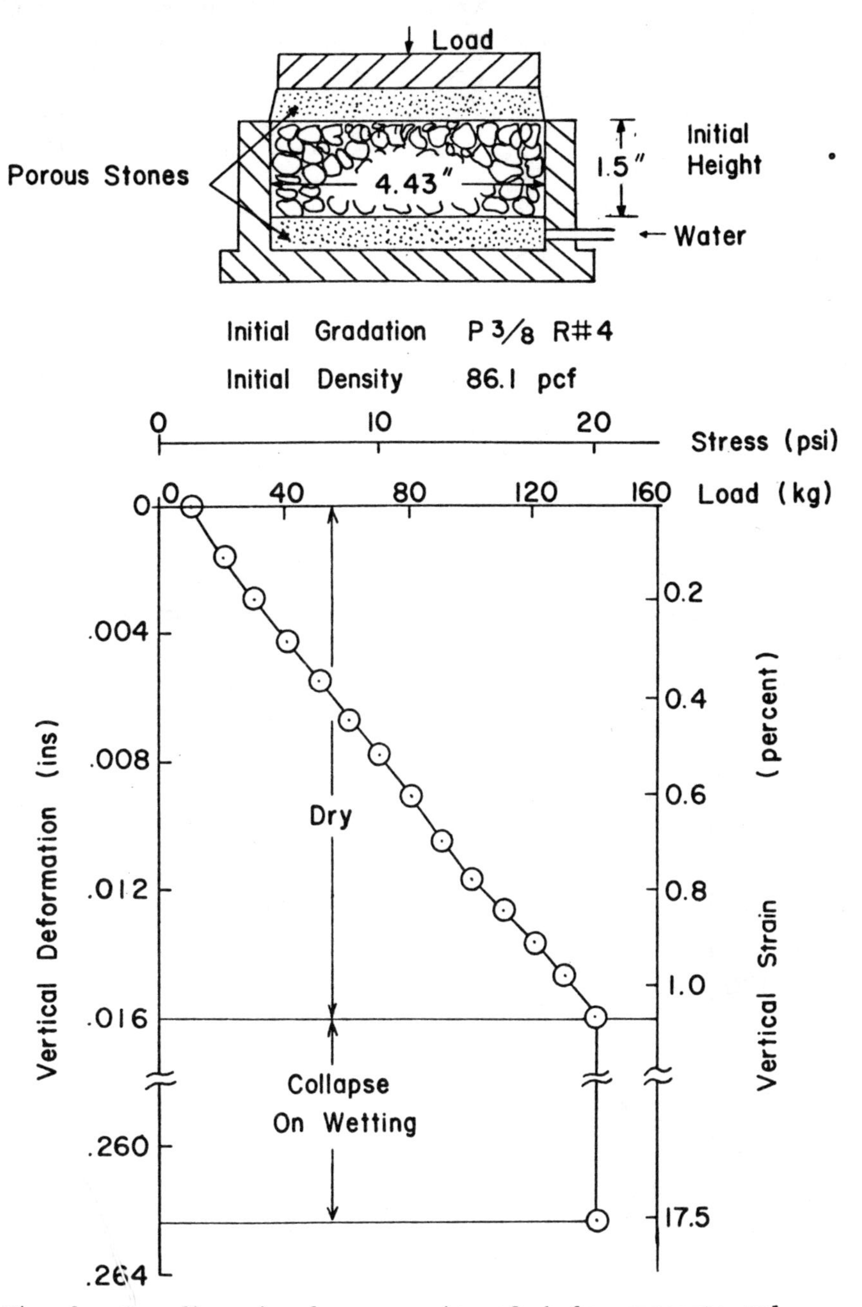

Fig. 1. One-dimensional compression of shale aggregate and collapse after wetting

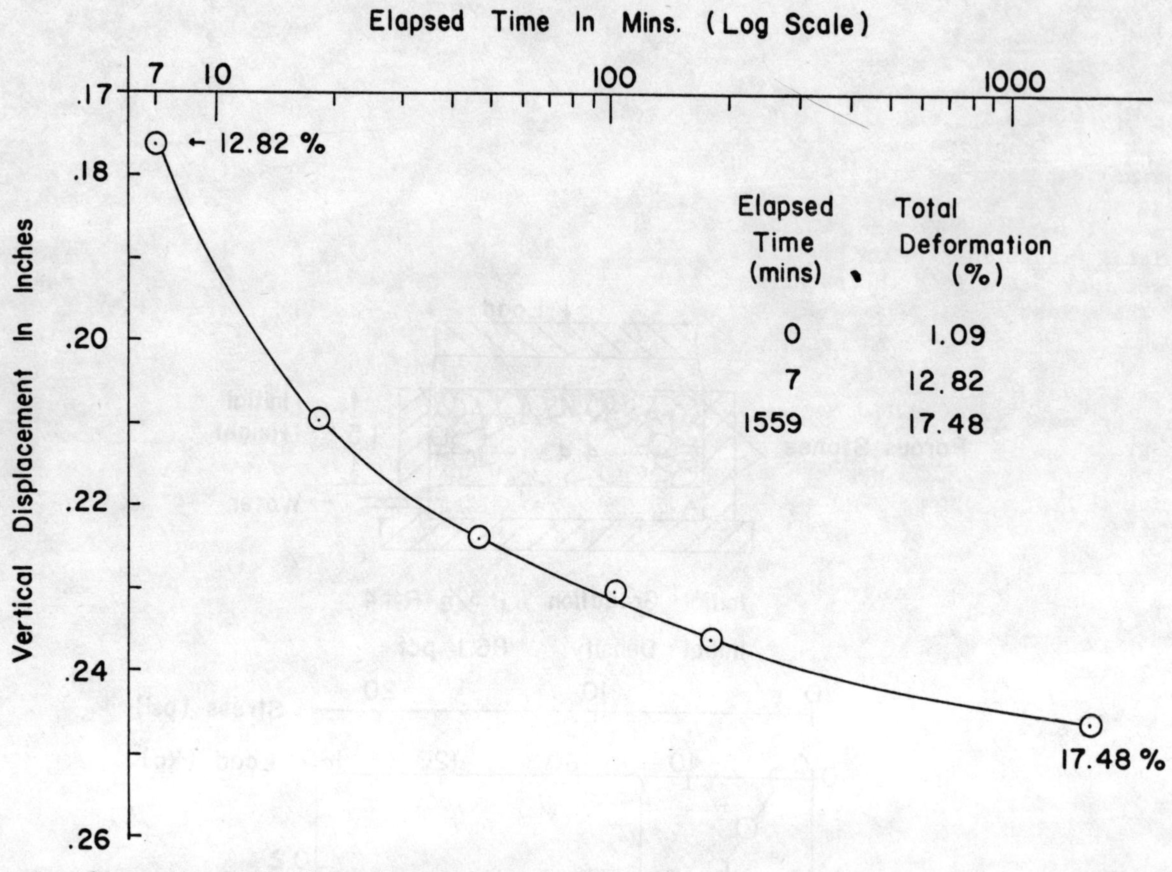

Fig. 2. Collapse of shale aggregate under load as a function of log time after wetting

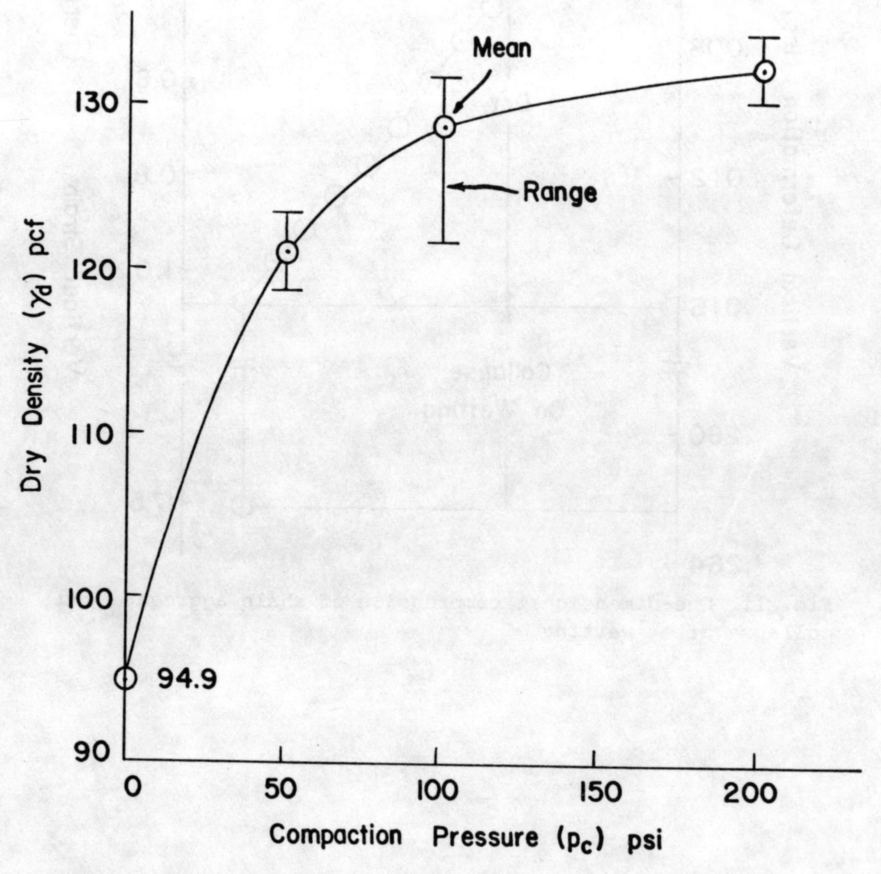

Fig. 3. Variation of the range and average dry density with compaction pressure for various gradations and molding water contents

Formation of compacted specimens

14. The specimens were formed using crushed shale aggregate at natural moisture content with a top size of 3/4 inch. The crushing and sorting into different size fractions was spread out over the testing program duration and it was found that the aggregate moisture content decreased from 4.0% to 1.1% over a period of thirteen months. The initial gradations were artificially prepared by weighing and mixing the various size fractions. The weight of each lift was kept constant at 500 gm.

15. For each test a total of nine 500 gm parcels of shale aggregate, conforming to one of the chosen gradations, were prepared and placed in small polythene bags. One parcel was used to determine the average moisture content of the aggregate. The remaining parcels were used for compacting the specimens.

16. The required amount of water to be added to the shale aggregate was measured and poured into the polythene bag in stages. The water was distributed by rotating and shaking the bag and by imparting a gentle molding action to the mixture by squeezing the bag. The contents of each bag were then completely emptied into a split steel compaction mold, previously set up on the rotating base of the kneading compactor. Great care was taken to prevent loss of moisture during these operations.

17. After spreading the mixture with a large steel spatula, each lift was subjected to 30 tamps over a period of one minute at the foot pressure previously set by means of an air pressure regulator. Successive lifts were similarly placed and compacted and the top of the specimen was trimmed and levelled off while the specimen was still in the mold.

18. The compacted specimen was allowed to remain in the mold, which was sealed top and bottom after compaction, for a period of 24 hours. Thereafter, the bolts holding the two halves of the mold were unscrewed and the specimen was carefully extruded, weighed and measured.

Triaxial testing

19. After enclosing the specimen in a 0.01 inch thick rubber sleeve, it was set up in the triaxial cell with top and bottom porous stones. Filter paper discs were placed between the specimen and the porous stones to facilitate the removal of the specimen after the test, and to prevent the porous stones from getting clogged by fine clay particles during drainage. Stretched O-rings were placed to provide a water-tight seal between the sleeve and the loading caps, the cylindrical surfaces of which were smeared with a silicone vacuum grease. De-aired distilled water was used as the cell fluid.

20. The specimen was back pressure saturated using the procedure described by Lowe and Johnson (ref. 7). During the initial stages, a vacuum was applied to the top of the specimen and de-aired distilled water was allowed to enter through its bottom. The specimen was left to saturate under a back pressure in excess of 50 psi until the inflow of water ceased.

21. The following procedure was used to check whether the specimen was fully saturated. The specimen was prevented from expanding by keeping the cell water lines closed. The pressure in the back pressure lines was increased keeping the drainage lines closed, and the depression of the water level in the burette (due to system flexibility) was observed. The drainage valves to the specimen were then opened and the water level in the burette observed.

22. When the specimen is fully saturated there will be no significant change in the water level. An instantaneous drop in the water level due to the compression of air will take place when the specimen is not saturated.

23. When the structure of the specimen is very stiff, Skempton's pore pressure parameter B is much less than unity, even if the specimen is fully saturated, as reported by Wissa (ref. 8). This criterion was therefore not used for compacted shale. A B parameter of around 0.95 was observed for the previously saturated compacted shale.

24. After saturation had been completed under a low confining pressure, the cell pressure was increased in a single increment and the specimen allowed to drain against the back pressure. No air was found to come out of the specimen during drainage.

25. The saturated consolidated specimen was then sheared undrained at a constant rate of deformation of approximately 0.01 inches per minute up to an axial strain of 20%. The axial load was measured by means of a load cell, the pore water pressure changes by means of a transducer, and the axial deformation by means of a dial gauge. The weight of the sample was determined after the test.

RESULTS AND DISCUSSION

26. The compaction characteristics, the volume change characteristics during saturation and consolidation, and the undrained shearing response were found to be greatly dependent on the independent test variables and levels. On the other hand, the effective stress strength parameters c' and ϕ' were found to be essentially independent of the initial conditions for compacted specimens having dry densities ranging from 120 to 130 pcf. Specimens that were not compacted, but tested in a loose condition at a dry density of around 95 pcf, had lower strength parameters. The parameters were evaluated at maximum deviator stress.

Compaction characteristics

27. Fig. 3 shows the range and the average dry densities obtained for kneading compaction pressures of 200, 100 and 50 psi. The range in dry density for a given compaction pressure is a function of the initial gradation, the molding water content, and the compaction pressure itself.

28. Although the influence of the variables on

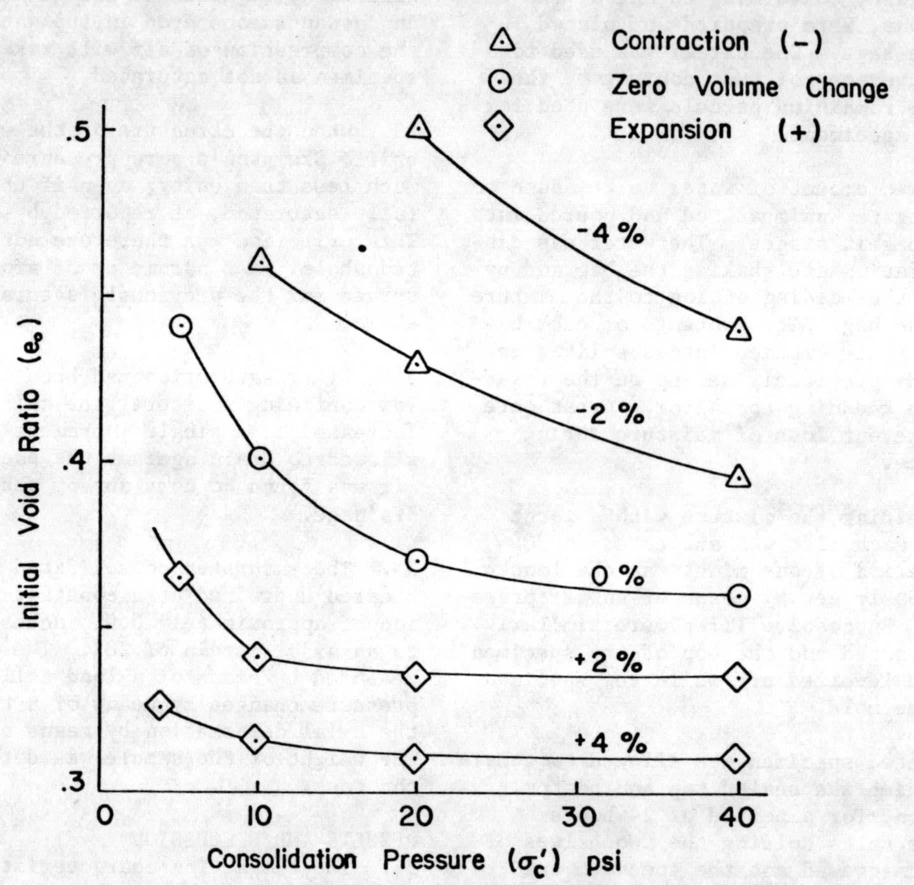

Fig. 4. Volume change tendency of compacted shale after soaking as a function of the initial void ratio and consolidation pressure

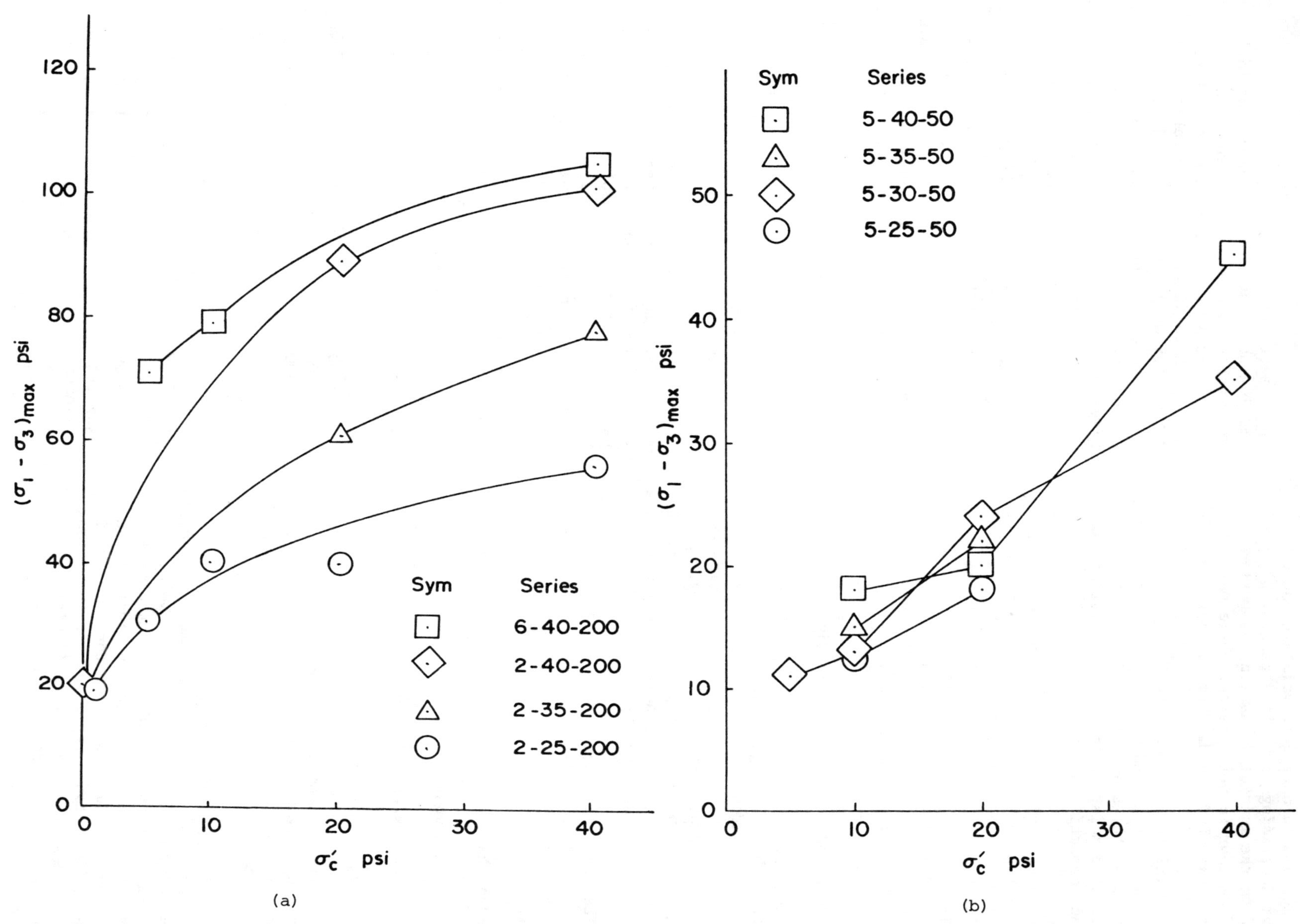

Fig. 5. Relationship between undrained strength and consolidation pressure for specimens compacted at (a) 200 psi, and (b) 50 psi

the compacted dry density was not studied on a statistical basis, the following trends were observed.

1. The dry density tended to increase with increase in aggregate moisture content, or decreasing particle strength.
2. When the initial gradation was varied from poor to well graded, there was a tendency for the density to increase.

29. The influence of molding water on the dry density appears to depend on the slaking characteristics of the shale, the initial gradation, the particle moisture content, the curing time, and the compaction pressure.

30. When the curing time is relatively long, the shale aggregate will absorb some or all of the molding water and the strength of the shale pieces will tend to decrease. This decrease in strength will make the aggregate more susceptible to break down during compaction and the resulting density will tend to increase. Further, there will be less water present between the shale pieces and the expulstion of air during compaction will tend to be easier than when the voids between the pieces contain more molding water.

31. When the curing time is relatively short, the strength of the aggregate will not be altered as much. Further, the molding water will tend to remain on the surfaces of the particles and the likelihood of pore water and pore air pressures building up during compaction will tend to increase. The result of these effects will be a lowering of the dry density.

32. The compaction results reported in this paper were obtained using a short curing time of around 5 minutes, to simulate common field practice.

33. When the average dry density is plotted against the logarithm of the compaction pressure a linear relationship is obtained. The relationship between dry density and compaction pressure also fits the hyperbolic equation

$$(\gamma_d - \gamma_o) = p_c/(a + b\ p_c)$$

where p_c is the compaction pressure, γ_d is the dry density obtained under the compaction pressure, and γ_o is the dry density prior to compaction (ref. 6).

Volume change characteristics

34. The partially saturated specimens of compacted New Providence shale changed in volume when saturated and consolidated. The net change in volume for a given consolidation pressure and initial void ratio is shown in contour form in Fig. 4. The contours represent average values for specimens having different initial degrees of saturation and structures compacted at various compaction pressures.

35. The current practice of compacting a fill lift to a specified dry density and moisture content range, irrespective of its location in an embankment, is from a compaction control standpoint the most feasible. Varying the as-compacted condition of the fill to achieve zero volume change on saturation (under a given confining pressure) is theoretically possible (Fig. 4), but would in general be impractical.

Shear strength

36. The shear strength of specimens highly prestressed by compaction increases in a non-linear manner and at a decreasing rate with consolidation pressure (Fig. 5a). The relationship is linear for slightly prestressed specimens (Fig. 5b). These figures also show that the influence of molding water on the shear strength at a given consolidation pressure is very significant at high compaction pressures (Fig. 5a), and marginal at low compaction pressures (Fig. 5b).

Effective stress strength parameters.

37. The effective stress strength parameters c' and ϕ' were evaluated at maximum deviator stress $(\sigma_1-\sigma_3)_{max}$ from the slope and intercept of the straight line fitted through the data points representating the conditions at failure as shown in Figs. 7 and 8. The following relationship was used

$$q_f = c'\cos\phi' + p_f'\sin\phi'$$

where $q_f = (\sigma_1-\sigma_3)_f/2$ and $p_f' = (\sigma_1'+\sigma_3')_f/2$, are the shear stress and the normal stress on a 45° plane to the plane of major or minor principal stress, respectively.

38. Table 3 shows the values of c' and ϕ' obtained for several series of tests having different initial gradation, molding water content, compaction pressure and consolidation pressure. Each row represents a series of tests having the same compaction history but tested under different consolidation pressures.

Table 3. Effective stress strength parameters evaluated at maximum deviator stress from consolidated undrained triaxial compression tests on soaked compacted New Providence shale

Series G-W_a-p_c No gm psi	Consolidation Pressures σ_c' psi	Strength Parameters c' psi	ϕ degrees
2-25-200	1 5 10 20 40	1.75	30.5
2-35-200	20 40	2.29	29.1
2-40-200	0 20 40	1.16	30.6
6-40-200	5 10 40	1.15	30.0
1-25-100	0 5 10	1.74	30.6
1-40-100	0 5 10 40	1.74	30.3
5-25-100	20 40	1.73	30.0
6-25-100	0 5 10 40	1.15	30.2
6-35-100	0 10 40	1.70	28.2
5-25-50	10 20	1.15	30.2
5-30-50	5 10 20 40	0.59	33.4
5-35-50	10 20	1.16	30.8
5-40-50	10 20 40	1.75	31.3
5-0-0	10 20	0	24.6

G initial gradation used (see Table 1)
W_a molding water added (gm/500 gm)
p_c kneading compaction pressure (psi)

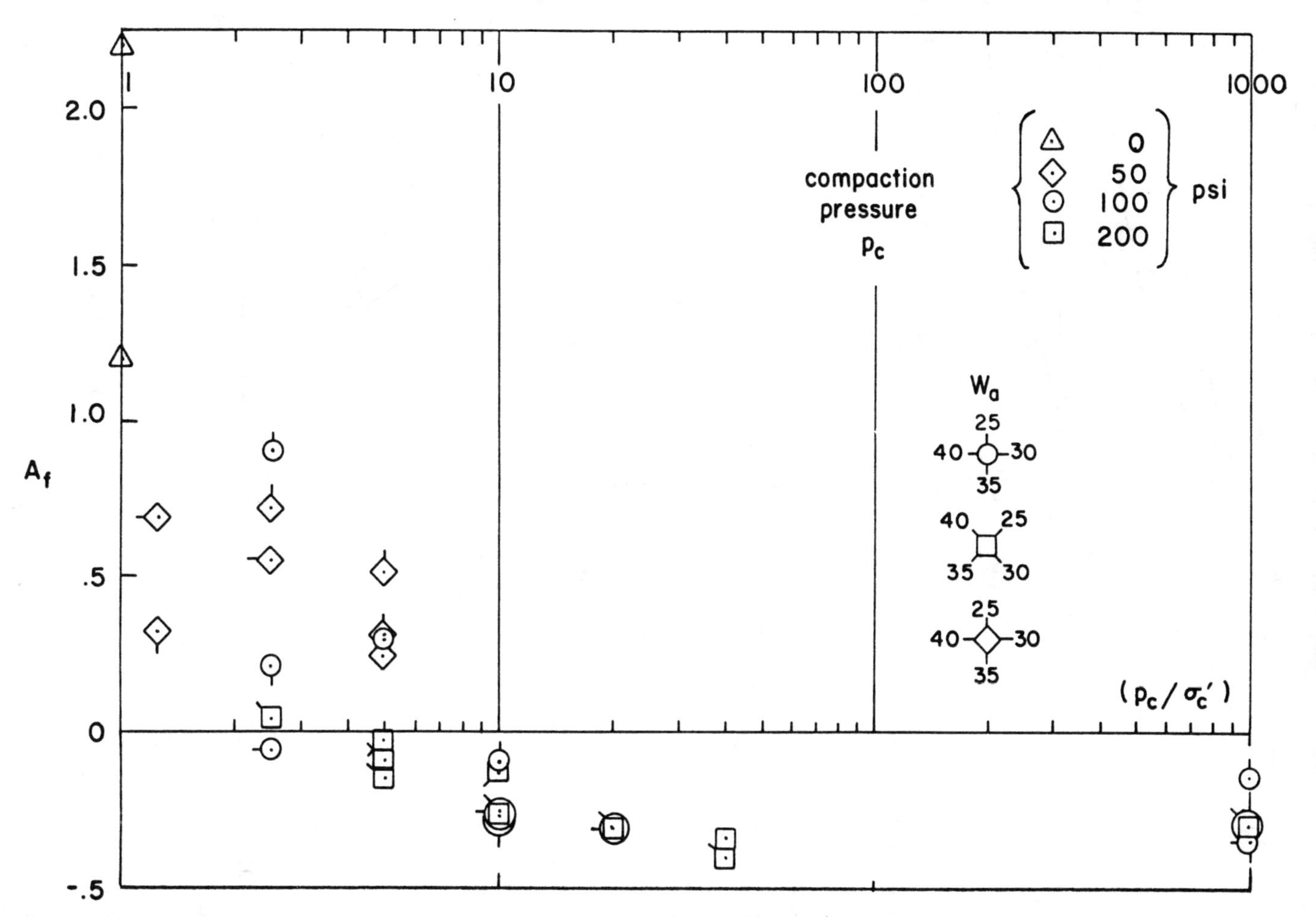

Fig. 6. Variation of A_f with log (p_c/σ_c') for compacted specimens having different initial gradations and molding water contents

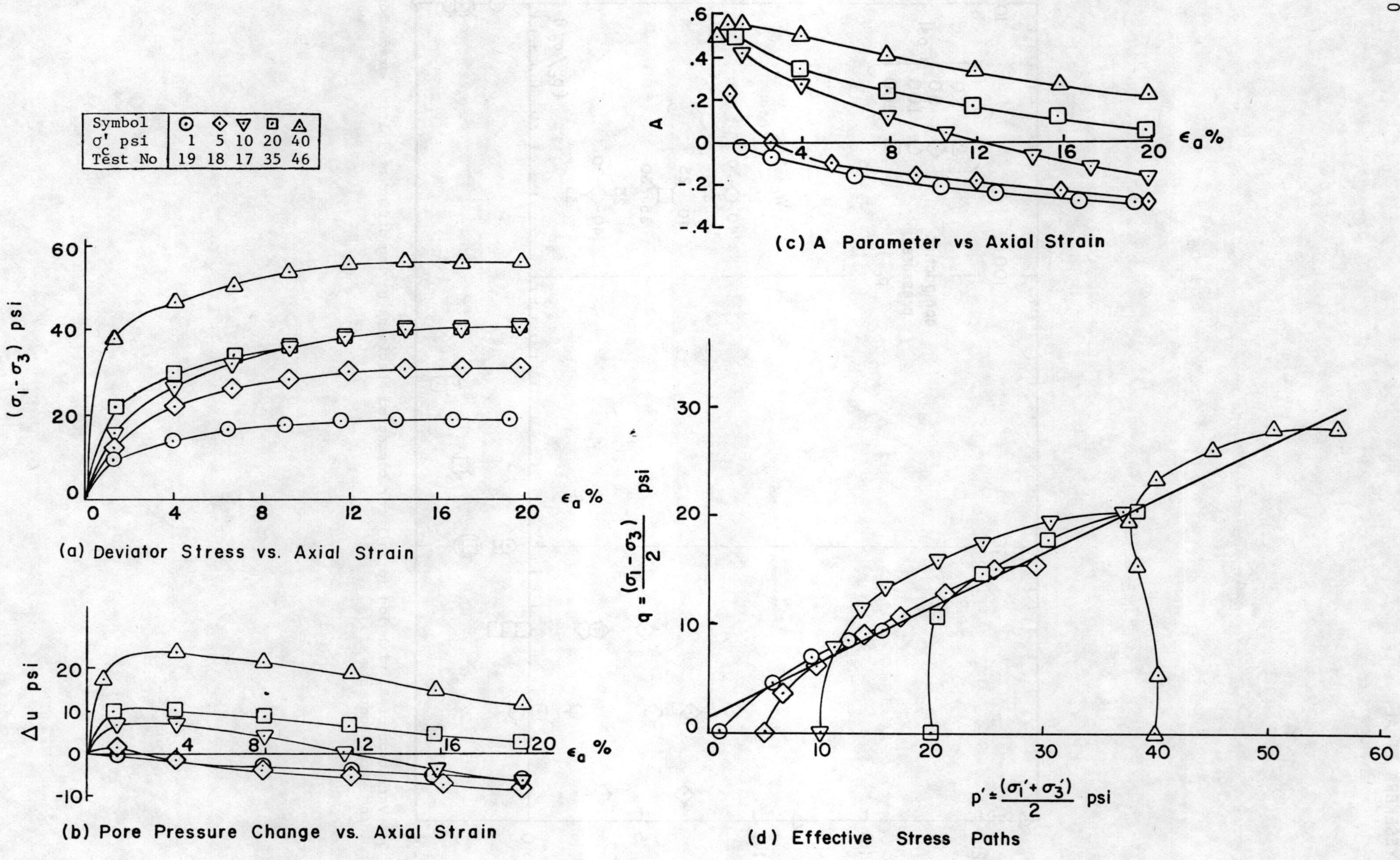

Fig. 7. Undrained shearing response of highly prestressed specimens in series (2-25-200) after saturation and isotropic consolidation

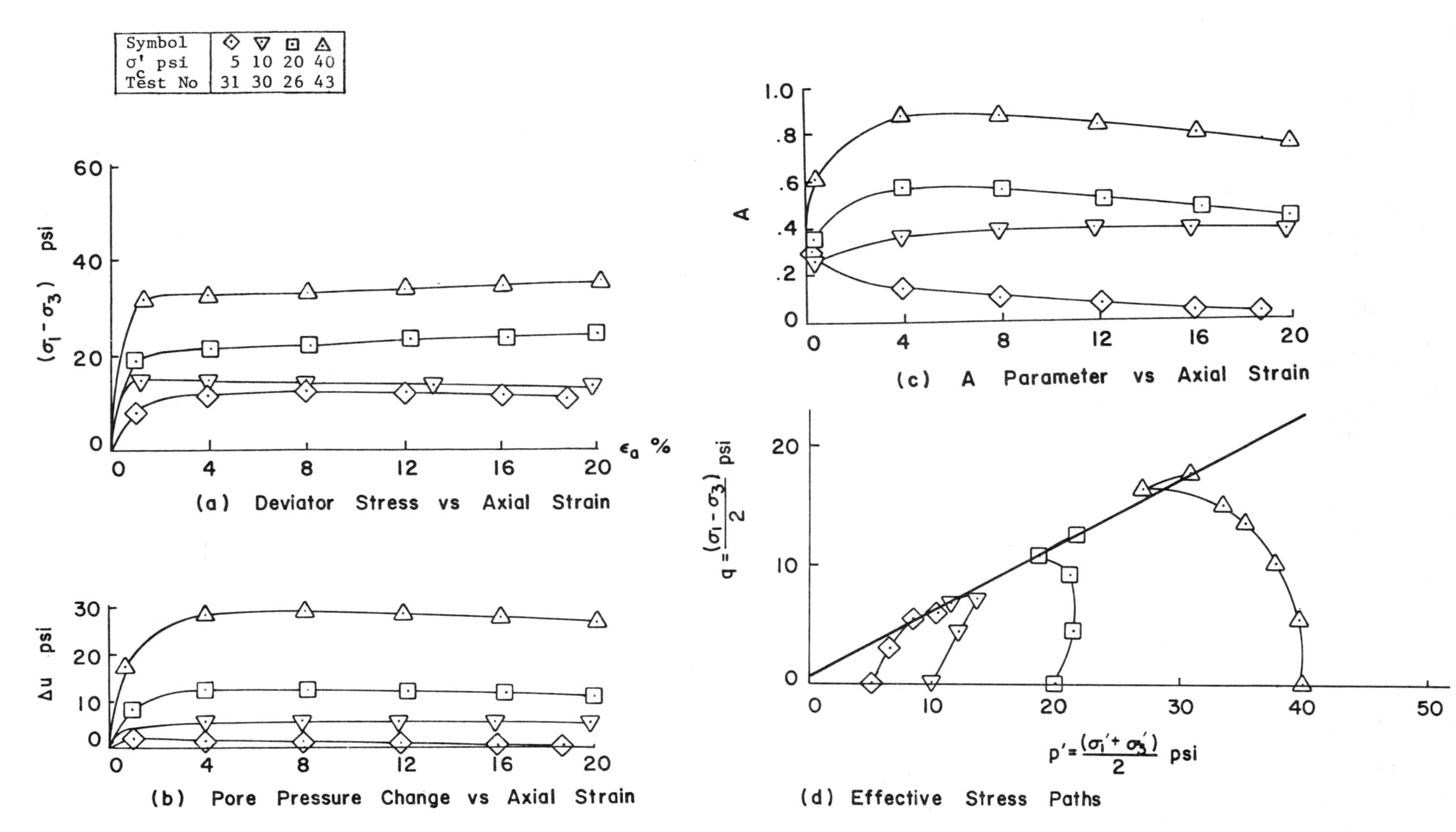

Fig. 8. Undrained shearing response of slightly prestressed specimens in series (5-30-50) after saturation and isotropic consolidation

39. Although the strength parameters are not the same for the several series of tests, they are very similar, except for the uncompacted loose specimens in Series (5-0-0). This experimental result shows that the parameters may be evaluated from a limited number of tests where the compacted densities and consolidation pressures are representative of the field conditions.

Undrained shearing response

40. The undrained shearing response of compacted New Providence shale was found to be greatly influenced by the compaction history (initial gradation, molding water content, and compaction pressure) and the pre-shear consolidation pressure. Figs. 7 and 8 show detailed results for two series of tests having the following initial conditions.

Test Variables		Series 2-25-200	Series 5-30-50
Initial Gradation	(No)	2	5
Molding Water	(gm/500 gm)	25	30
Compaction Pressure	(psi)	200	50
Consolidation Pressure	(psi)	1,5,10, 20,40	5,10, 20,40

41. Series (2-25-200). The deviator stress increases non-linearly and at a decreasing rate up to an axial strain of 20%, and the shape of the curves are well represented by the hyperbolic equation proposed by Kondner (ref. 9)

$$(\sigma_1 - \sigma_3) = \varepsilon/(\alpha + \beta\,\varepsilon)$$

where ε is the axial strain, and α and β are experimental parameters. The pore water pressure changes are initially positive and they tend to become negative at higher strains due to dilatation tendencies, particularly at the lower consolidation pressures. All the stress paths display features typical of highly overconsolidated clays. The principal stress ratio reached a peak value at relatively low strains of the order of 2% to 4%, and decreased thereafter to a residual value at 20% strain.

42. Series (5-30-50). The deviator stress reaches its maximum value at relatively low strains of the order of 2% to 4%, and remains constant up to 20% strain. The pore water pressure changes are always positive due to the tendency for the specimens to decrease in volume. The stress paths of specimens tested under low consolidation pressures display features typical of slightly overconsolidated clays, whereas those tested under the higher consolidation pressures are similar to normally consolidated clays. The principal stress ratio attained its maximum value at about 4% strain, and remained essentially constant thereafter up to 20% strain.

43. Series (5-0-0). This series of tests was carried out on specimens of shale aggregate lightly tamped under the weight of a 5.5 lb Proctor hammer. The deviator stress reached a peak value at a relatively low strain of the order of 2% and dropped to a residual value at 20% strain. The pore water pressure changes were exceedingly large (positive) and continued to increase after the deviator stress reached its maximum value. These changes were due to the the degradation of the shale aggregate under the action of the shear stresses. The stress paths were very similar to those of loose sands and sensitive clays which have a metastable structure. The principal stress ratio increased continuously up to 20% strain.

Pore water pressures

44. The changes in pore water pressure are initially positive due to the tendency for the spemens to compress under the axial load. The magnitude of the positive excess pore water pressure was found to vary linearly with the consolidation pressure as follows:

$$(\Delta u)_{max} = 0.875\ \sigma_c', \text{ for } p_c = 0 \text{ psi and}$$

$$(\Delta u)_{max} = 0.625\ \sigma_c', \text{ for } p_c = 50, 100 \text{ and } 200 \text{ psi}$$

where p_c is the compaction pressure.

45. During subsequent stages of loading, the larger shear strains induced in the specimen have a significant effect on its structure. Specimens that are loosely packed tend to become dense (compress), whereas specimens that are densely packed tend to become loose (dilate). The volume change tendencies are reflected by the development of positive and negative pore water pressure changes depending on whether the specimen is tending to compress or dilate, respectively.

46. The pore water pressure change at failure $(\Delta u)_f$ was found to increase with increasing consolidation pressure and decreasing molding water content. The variation of $(\Delta u)_f$ may be represented by an equation of the form,

$$(\Delta u)_f = -\,a + b\ \sigma_c', \text{ for } \sigma_c' = 5 \text{ to } 40 \text{ psi}$$

where a and b are positive numbers. The parameter a is greatly dependent on the compaction history of the specimen, whereas the parameter b is essentially independent of this compaction history. The magnitude of a was found to be zero for specimens compacted at 50 psi, and to decrease with decreasing molding water content. The magnitude of b was found to be close to 0.625 for compacted specimens.

47. The pore water pressure change represented by

$$(\Delta u)_d = (\Delta u)_{max} - (\Delta u)_f$$

is due to the dilatation tendency of the specimen when strained beyond the point where $(\Delta u)_{max}$ is reached. For a given compaction history, the variation of $(\Delta u)_d$ with consolidation pressure is relatively small, as can be seen from the shapes of the pore water pressure change vs axial strain curves in Figs. 7 and 8. However, for a given consolidation pressure, the variation of $(\Delta u)_d$ with compaction history is very significant. For specimens compacted at 50 psi $(\Delta u)_d$ was found to be zero, and for specimens compacted at 200 psi it was found to increase with increasing molding water content. In general, $(\Delta u)_d$ = a, where the parameter 'a' decreases

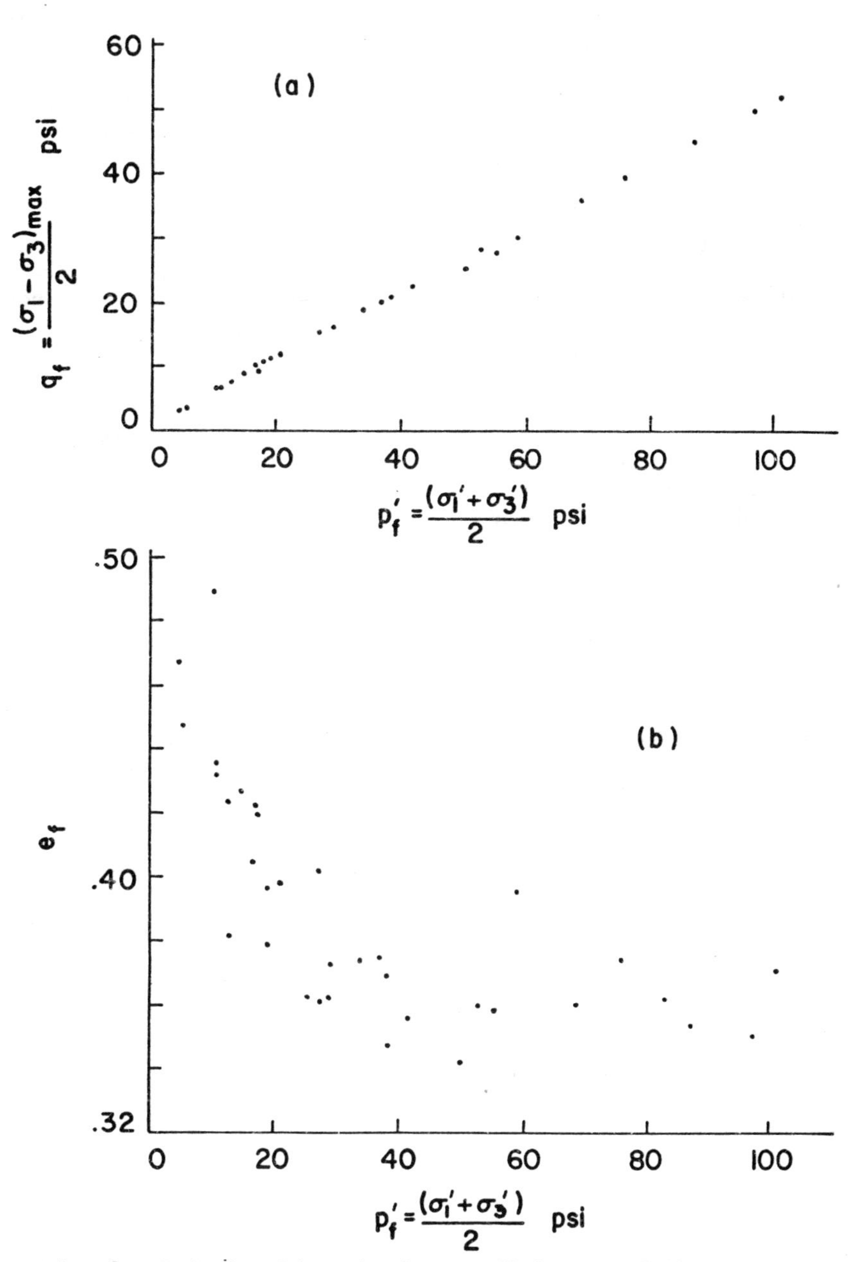

Fig. 9. Relationships showing conditions at failure, (a) shear stress vs mean effective normal stress, and (b) average void ratio vs mean effective normal stress

with decreasing compaction pressure and decreasing molding water content.

Skempton's A parameter

48. Skempton's A parameter varies with axial strain as shown in Figs. 7b and 8b. The maximum value of this parameter, designated A_{max}, is reached at relatively low strains during which the specimens tend to compress. The magnitude of A_{max} was found to increase with increasing consolidation pressure, decreasing molding water content, and decreasing compaction pressure.

49. The value of the A parameter at failure, designated A_f, is shown in Fig. 6 as a function of the ratio of the compaction pressure to the consolidation pressure (p_c/σ_c'), which is analagous to the overconsolidation ratio, OCR. Whereas, the OCR (defined in terms of effective stresses) and A_f are uniquely related for a saturated remolded clay, the relationship between the ratio (p_c/σ_c') and A_f is not unique, for samples compacted with different initial gradations and molding water contents. However, the general trend in the variation of A_f with (p_c/σ_c') is similar to the variation of A_f with OCR. For the shale tested, A_f ranged from 2.2 for uncompacted aggregate to around -0.4 for specimens compacted at 200 psi. It was also found that for a given value of (p_c/σ_c'), the higher values of A_f were observed at the lower molding water contents.

SUMMARY AND CONCLUSIONS

50. The compaction characteristics, the volume changes during saturation and consolidation, and the undrained shearing response including pore water pressure changes were studied for New Providence shale.

51. The compaction and volume change characteristics, the induced pore water pressures and the consolidated undrained strength were found to be greatly dependent on the initial conditions. The effective stress strength parameters c' and ϕ' evaluated at maximum deviator stress were found to be essentially independent of the initial conditions, except for loose uncompacted aggregate.

ACKNOWLEDGEMENTS

52. The U. S. Federal Highway Administration and the Indiana State Highway Commission sponsored the research through the Joint Highway Research Project of Purdue University.

REFERENCES

1. WOOD L.E., LOVELL C.W. and DEO P. Building embankments with shales. Joint Highway Research Project Report No. 73-16, Purdue University, Lafayette, Indiana, U.S.A., 1973, August, 32.
2. DEO P. Shales as embankment materials. Ph.D. Thesis, Purdue University, Lafayette, Indiana, U.S.A., 1972, December, 202.
3. CHAPMAN D.R. Shale classification tests and systems: a comparative study. MSCE Thesis, Purdue University, Lafayette, Indiana, U.S.A., 1975, May, 90.
4. BAILEY M.J. Shale degradation and other parameters related to the construction of compacted embankments. MSCE Thesis, Purdue University, Lafayette, Indiana, U.S.A., 1976, August, 230.
5. van Zyl D.J.A. Storage, retrieval and analysis of Indiana shale data. Joint Highway Research Project Report No. JHRP-77-11, Purdue University, Lafayette, Indiana, U.S.A., 1977, 140.
6. ABEYESEKERA R.A. Stress deformation and strength characteristics of a compacted shale. Ph.D. Thesis, Purdue University, Lafayette, Indiana, U.S.A., 1978, May, 420.
7. LOWE J. and JOHNSON T.C. Use of back pressure to increase degree of saturation of triaxial test specimens. Proceedings of the American Society of Civil Engineers, Research Conference on Shear Strength of Cohesive Soils, Boulder, Colorado, U.S.A., 1960, June, 819-836.
8. WISSA A.E. Pore pressure measurement in saturated stiff soils. Journal of the Soil Mechanics and Foundations Division, American Society of Civil Engineers, 1969, Vol. 95, No. SM4, July, 1063-1073.
9. KONDNER R.L. Hyperbolic stress-strain response: cohesive soils. Journal of the Soil Mechanics and Foundations Division, American Society of Civil Engineers, 1963, Vol. 89, No. SM 1, February, 115-143.

M. M. H. AL-SHAIKH-ALI, MSc, BScTech, MICE, MIHE, AIArb, Senior Soils Engineer, North Western Road Construction Unit, Cheshire County Council Sub-Unit

The behaviour of Cheshire Basin lodgement till in motorway construction

The complex structure of the glacial till in the Cheshire Basin presents interesting engineering problems in the Motorway Industry at both design and construction stages.

The variable fabric, macro-structure, fissures, laminations, sand pockets and lenses are the products of its geological history and formation during the advances and retreats of the ice sheets.

The behaviour of the lodgement till as a granular material offers a unique challenge to Civil Engineers to assess its past stress history and to some extent its suitability as fill.

The excavation, storage, placement and placement control in Motorway construction is also discussed.

INTRODUCTION

The variability of the lodgement till in Cheshire has attracted special attention from Engineers in almost every aspect of Geotechnical Engineering.

Each locality in the Cheshire basin offers different engineering problems from even the adjacent one, due to variable ground conditions.

Its use in Motorway construction requires the knowledge of its stress history, fabric, structure and its insitu stresses in order to assess any possible problems at the design stage and to overcome them during the construction of the Motorway.

The factors which affect the compactibility of the lodgement till during the construction of a Motorway are natural and placement moisture content, plasticity index, shear strength (insitu and remoulded), construction pore water pressure and the stress ratio factor "k" and, of course, the compacting plant.

Chester Southerly By Pass, which was designed and constructed to allow for possible conversion to "Motorway standard", will be used here as a typical site for which the performance of the lodgement till as a fill material is discussed.

SOIL CONDITIONS

The lodgement till in the Cheshire basin, which is locally known as "Boulder Clay", is variable in fabric, texture, strength and index properties, Figs.1-5.

It is generally sandy, with 14-45% clay particles < 2μ and specific gravity 2.65-2.73. Stones and cobbles of various sizes (from 1mm to 100mm) are often found with occasional erratic boulders of up to 1.5 metres in diameter.

Weathered and unweathered igneous rocks are present which indicate that they were imported by the movement of the glacier from the north and north west.

Irish Sea erratics and sea shells are also found which indicate advances and retreats of the ice sheets from the north west.

The till also varies from very stiff to generally soft and occasionally very soft. Lacustrine clay and laminated clay are common in the Cheshire basin.

Fissures are generally pronounced in the till on the west side of Chester city (ref.1), but are not excluded from the east and north east of the Cheshire basin. Figs.1,2 and 3 show the fissures, particle size distribution and the index properties. The fissures were thought to be the product of desiccation during the ice retreat, and the roots, which are also found in these fissures at greater depths, suggest that trees, or possibly forest, covered most of the Cheshire basin (refs.7,11 & 14).

The mineralogy of the till is, with the exception of its erratics, local and analogous to the parent under-lying Triassic rock series.

PRECONSOLIDATION PRESSURE AND THE OVER-CONSOLIDATION RATIO (OCR) OF THE LODGEMENT TILL

The lodgement till in the Cheshire basin was deposited during the advances and retreats of the ice sheets and was covered by hundreds of metres of ice. Boulton and Paul (ref.4) propounded their theory that the weight of the ice has little effect on the effective overburden pressure but, of course, erosion has contributed to its present stress condition.

The advances and retreats of the glaciers have obliterated its past history and as a result of this Casagrande's technique (ref.6) for comput-

Fig. 1. Natural fissuring at the Chester Southerly By Pass

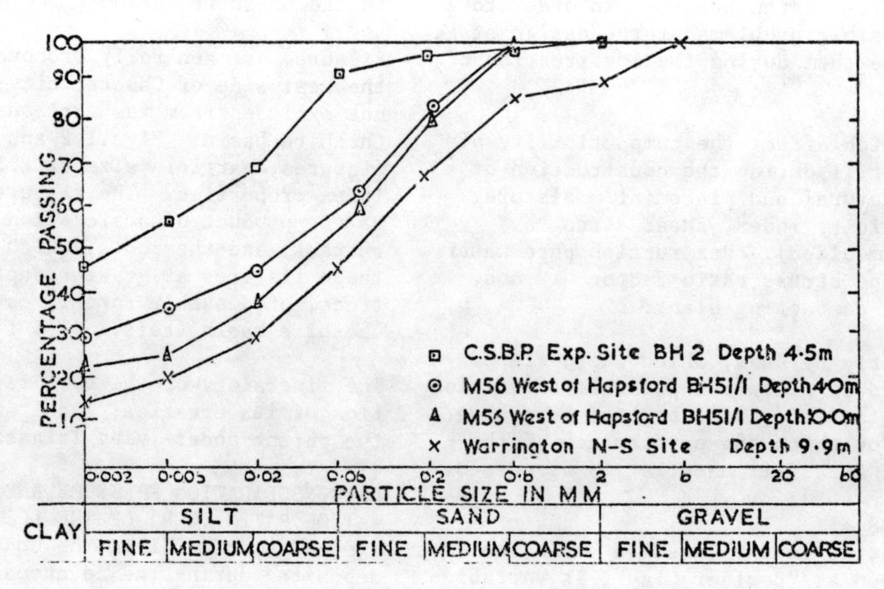

Fig. 2. Particle size distributions of Cheshire tills

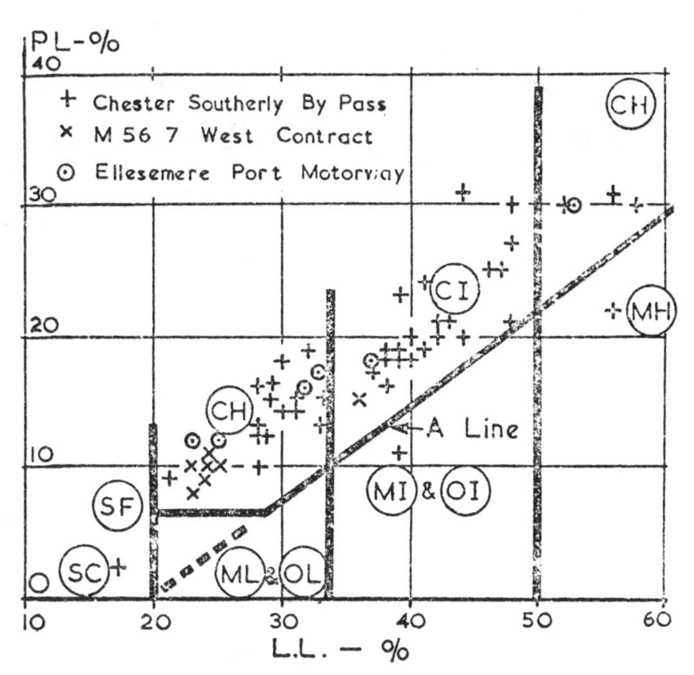

Fig. 3. Plasticity index vs liquid limit

Table 1. Maximum depth of cracks 750 mm
Maximum width of cracks 25-75 mm

DEPTH (m)	M/C	L.L.	P.L.	P.I.
Trial Hole No.1				
0.1	8.9	44	19	25
0.25	16.17	38	20	18
0.40	16.44	38	17	21
0.75	15.16	33	20	13
Trial Hole No.2				
G.L.	6.77	30	19	11
0.18	9.59	35	18	15
0.35	13.77	35	21	14
0.70	12.66	31	18	13

Table 2

Position	Dry Density kg/m^3	Moisture Content %	Air Voids %	Plastic Limits
1	1789	16.72	3.82	20
2	1740	18.88	2.70	
3	1706	20.84	1.26	
4	1750	17.63	4.33	
5	1712	19.62	3.00	19
6	1788	16.80	3.73	
7	1717	20.08	1.92	19
8	1712	20.43	1.61	20
9	1744	18.88	2.48	
10	1750	19.78	0.56	19
Average	1741	18.97	2.541	
Range	83	4.12	3.77	

ing its OCR was found to be, in most cases, unsuitable.

Janbu's approach (ref.15) for sandy and granular material seems to be suitable for computing the OCR of the lodgement till and the results of several tests confirm that the lodgement till in the Cheshire basin is normally to lightly over-consolidated. The OCR profile varies from 3-4 at the surface, decreasing to unity at about 6.0m below ground level.

THE ENGINEERING PROPERTIES

The lodgement till in the Cheshire basin has always been considered sandy and behaving as granular and permeable.

The rapid rate of consolidation is thought to be due to the fissures, root holes and its macro-structure and fabric.

The undrained cohesion is generally erratic but increases with depth and its value lies between 30 and 400kN/m^2, occasionally falling outside these limits.

In certain locations where the lodgement till is sandwiched between layers of saturated sand with sub-artesian or artesian pressures, the strength usually decreases with depth due to a softening effect.

The coefficients of permeability, consolidation and volume compressibility (k, m_v, c_v) are also variable and the average values for each are:

$k = 10^{-6} - 10^{8}$ cm/sec

$m_v = 0.15 - 0.4$ m^2/MN

$c_v = 2 - 10$ m^2/year.

The angle of the effective shear resistance, Ø', ranges between 20-29° and occasionally values for Ø' fall outside these limits. Pore water pressure parameter "A_f" ranges from 0.19 to 0.5.

The lodgement till has been used as a founding material for spread footings and piles. For spread footings a net safe bearing pressure in excess of 350kN/m^2 has been used.

Due to its granular and permeable nature, pore water dissipation has been found to be fairly fast and because of this the writer is of the opinion that the long term working safe bearing pressure could be raised relying on the increase in shear strength after dissipation of the pore water pressure during construction.

SUITABILITY OF THE LODGEMENT TILL AS A FILL MATERIAL

As mentioned earlier, lodgement till is the product of its geological history, which has resulted in the formation of sand pockets, and sand layers and lenses, perched water tables and sub-artesian or artesian pressures.

The frequency of each item mentioned above is variable on each site, and occasionally all items can be found on one site. Obviously each of these items is a potential source of problems.

The 1976 DTp Specification defines cohesive soils for the purpose of compaction to Clause 608, as having a moisture content not less than the value of the plastic limit minus 4. Its upper limit for compaction is usually defined by each contract depending on whether or not there is a surplus of suitable material. In many Motorway contracts in Cheshire the limit has been 1.2 x P.L.

On Chester Southerly By Pass the limitation was that: "the moisture content should be less than 20% irrespective of the plastic limit, and shear strength."

This approach has the distinct advantage of enabling the Engineer to reject the lacustrine and laminated clays both of which have high plastic limits and moisture contents and are very difficult to handle during embankment construction.

Generally, typical lodgement till in the Cheshire basin has a plasticity index between 18 and 25, plastic limits 16-35 and natural moisture content below 20%.

Intermixing

In cut, as mentioned earlier, sand layers are frequently found. The sand, usually medium to fine, may be, on its own, suitable, i.e. the moisture content may be about its optimum, but when it is mixed with suitable till the product will, more often than not, be unsuitable.

It has now been established that Intermixing will reduce the plastic limit and increase the moisture content of the mixed material without actually adding water to the level of rendering this mixture unsuitable.

Jones (ref.8) produced graphs confirming this, and similar graphs have been produced with Cheshire till, see Fig.4 (A and B).

Jones (ref.8) outlined the factors which may affect the suitability of the till as follows:

1. Intermixing.
2. Water table variation during the construction of a Motorway.
3. Swelling due to stress relief.
4. Mechanical disturbance.
5. Atmospheric agencies (wind and rain).

Intermixing in Motorway construction in Cheshire is undesirable and is generally avoided.

Swelling and shrinkage

It has already been mentioned that the till is lightly over-consolidated and its OCR may reach 4. The stress release due to this is not very pronounced, but it will be exacerbated if the till comes into contact with water and the problem will be clouded with the swelling which is mainly due to the expansive minerals in the till. Mineralogical tests using Keeling

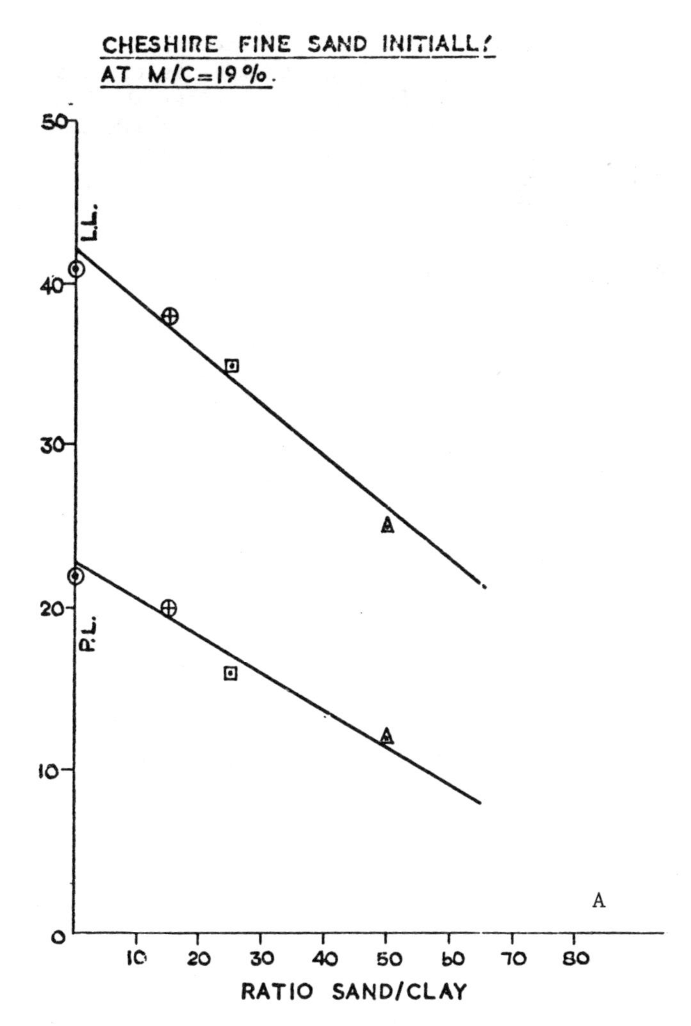

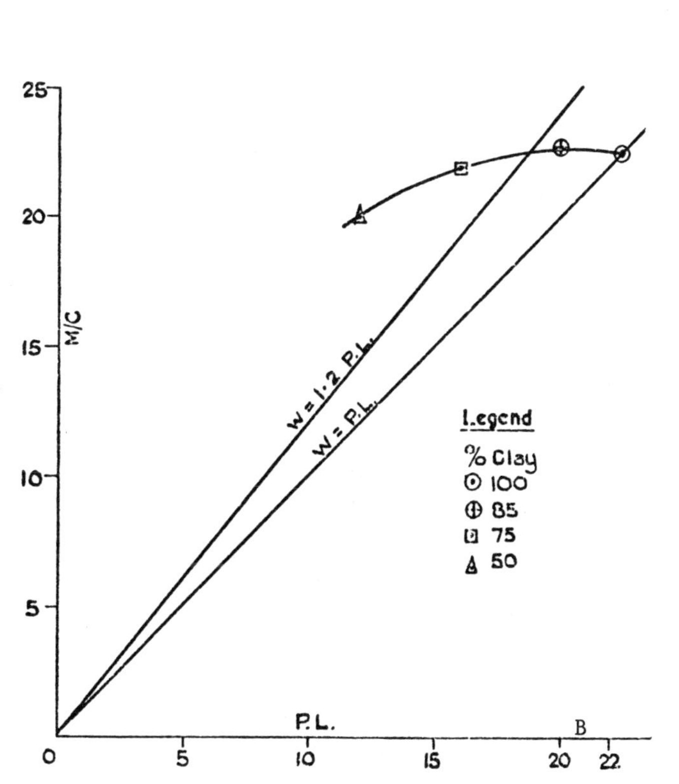

Fig. 4 A and B. Relationship between LL, PL, M/C and the ratio sand/clay mixture

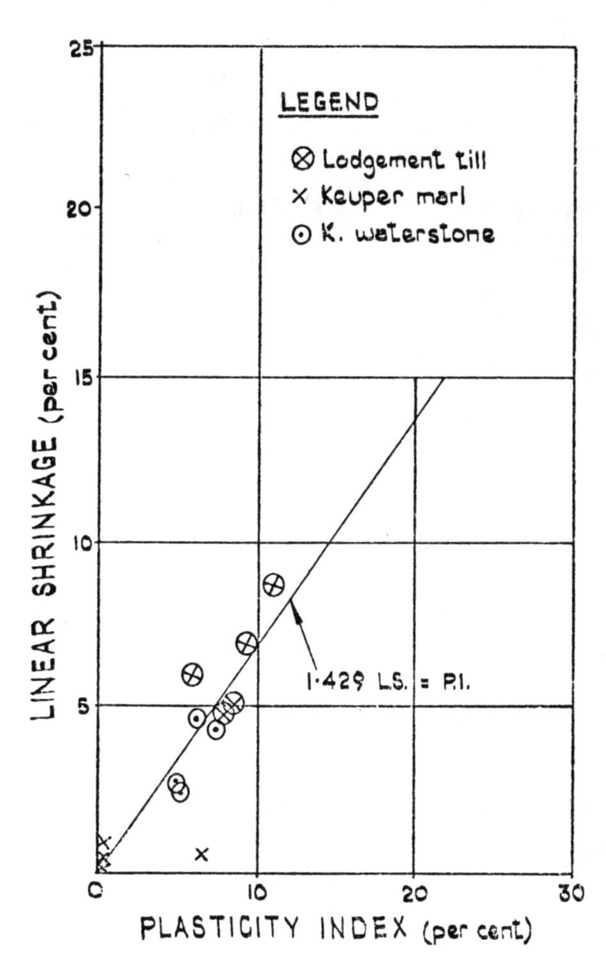

Fig. 5. Linear shrinkage vs plasticity index

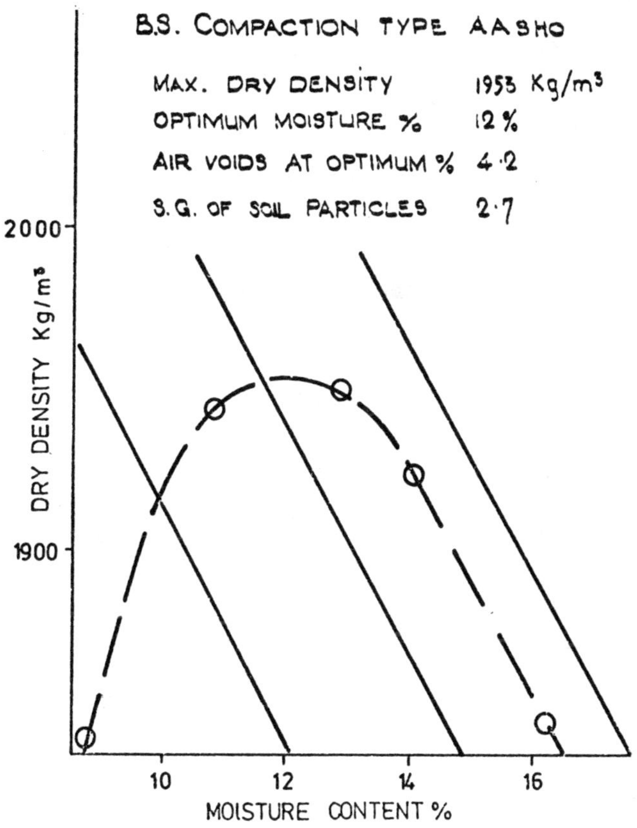

Fig. 6. Typical BS compaction results of Cheshire tills

(ref.9) indicated that the till has moisture adsorption (MA) in the range of (2.3-3.0) (well-ordered to disordered Kaolinite) with clay mineral in the order of 65%. This suggests that expansive minerals may be present.

The expansive minerals are often the reason for shrinkage, particularly in dry seasons. For instance, at an embankment of an accommodation bridge at the Chester Southerly By Pass site during the long hot summer of 1976, the till fill suffered severe cracking, up to 75mm in width and more than 750mm deep. The moisture content of the fill was slightly less than 20% and that of the cracked layers as low as 12.66% at 0.70m BGL, see Table 1. The linear shrinkage of the Cheshire till and of the parent rock of Keuper Marl and Keuper Waterstone is shown in Fig.5.

It is of interest to note that the maximum value for the linear shrinkage for lodgement till is about 8% which is just over that for the argillaceous Keuper Marl. A linear relationship of the plasticity index and linear shrinkage is suggested to be $LS = 1.429I_p$ (see Fig.5). This, perhaps, explains the magnitude of the cracks mentioned earlier, which were remedied in one part by filling the cracks with non-plastic, well compacted, fines of sub-base material prior to placing of the pavement, and in the other part of the accommodation bridge by removing and recompacting the desiccated materials. Both methods proved to be satisfactory.

Mechanical disturbance

Storing and double handling is not recommended and is not generally used in Cheshire because of the breaking down of the aggregation and diagenetic bonds, releasing of the recoverable strain energy, if any, and reduction in shear strength due to weathering (ref.2).

Preventive measures from inclement weather

Preventive measures from rain are normally taken in road construction when lodgement till is used as fill, one of which is to grade over the surface of the embankment at the end of the working day leaving a protective layer until the pavement construction is about to commence.

COMPACTION OF THE LODGEMENT TILL

Typical dry density/moisture content relationship obtained from BS compaction test - AASHO - is shown in Fig.6 for reference purposes.

In previous contracts the degree of compaction of the cohesive material was judged by its air voids in the compacted layers.

The cohesive fill is deemed to be satisfactorily compacted when the air voids are less than 5% and granular material is less than 10%.

The DTp Specification (1976) and MOT (1969) specifies the maximum depth of compacted layers and minimum number of passes according to the soil group and mass and type of compacting equipment. Some Soils Engineers still prefer the end product specification (air voids) because, if the material is reasonably wet, it needs less compaction than a relatively dry one. Excessive compaction will generate positive pore water pressures, which manifests itself by heave under the movement of the roller. The limitation of 5% air voids (used in previous contracts) perhaps reduces the risk of generating high pore water.

CBR and moisture content

It is important to repeat that the function of an embankment is to provide a stable sub-grade for the pavement. The strength of the sub-grade is usually measured by its shear strength and in the road construction industry the California Bearing Ratio (CBR) is generally used as a criterion in order to assess the required pavement thicknesses. The CBR values are plotted against the moisture content and the dry density, Figs.7,8 and 9; the dry density is also plotted against the moisture content.

It is pertinent, therefore, to note that the CBR and dry density decrease with increase in moisture content, and when the moisture content is more than 20% the CBR value drops below 5% with corresponding values for the dry density of over 1730kg/m^3 and at 12%, $\gamma_d \geqslant 1950$kg/m^3.

It appears that a linear relationship exists between γ_d and moisture content, between optimum moisture content and 20%.

CBR and dry density

The CBR is also a function of the dry density and Fig.8 suggests that the CBR increases with the increase in dry density. In order to control the compaction of the till, in addition to the method specification, the moisture content, dry density and air voids are usually monitored. Typical compaction test results are shown in Table 2.

Stress ratio factor (K) and construction pore pressure

It is accepted that excessive compaction of fill material with high moisture content will generate high positive pore pressures, which cause excessive heave under the movement of the roller. This is undesirable and may cause adverse results.

For instance, Pells (ref.10) found that the unusual behaviour of pore pressure in earth dams was caused by high horizontal stress in the clay fill placed at or below its optimum moisture content due to heavy compaction.

It is, perhaps, pertinent to reproduce Skempton's equation (ref.12) in terms of the total principal stress:

$$\Delta u = B(\Delta\sigma_3 + A(\Delta\sigma_1 - \Delta\sigma_3)). \qquad (1)$$

Pells (ref.10) expanded this equation as follows:

$$\Delta u = \Delta u_o = r_u . \gamma \Delta h. \qquad (2)$$

where r_u is the pore pressure ratio and is used here instead of B' as $\gamma . \Delta h$. may not be equal to "σ_1".

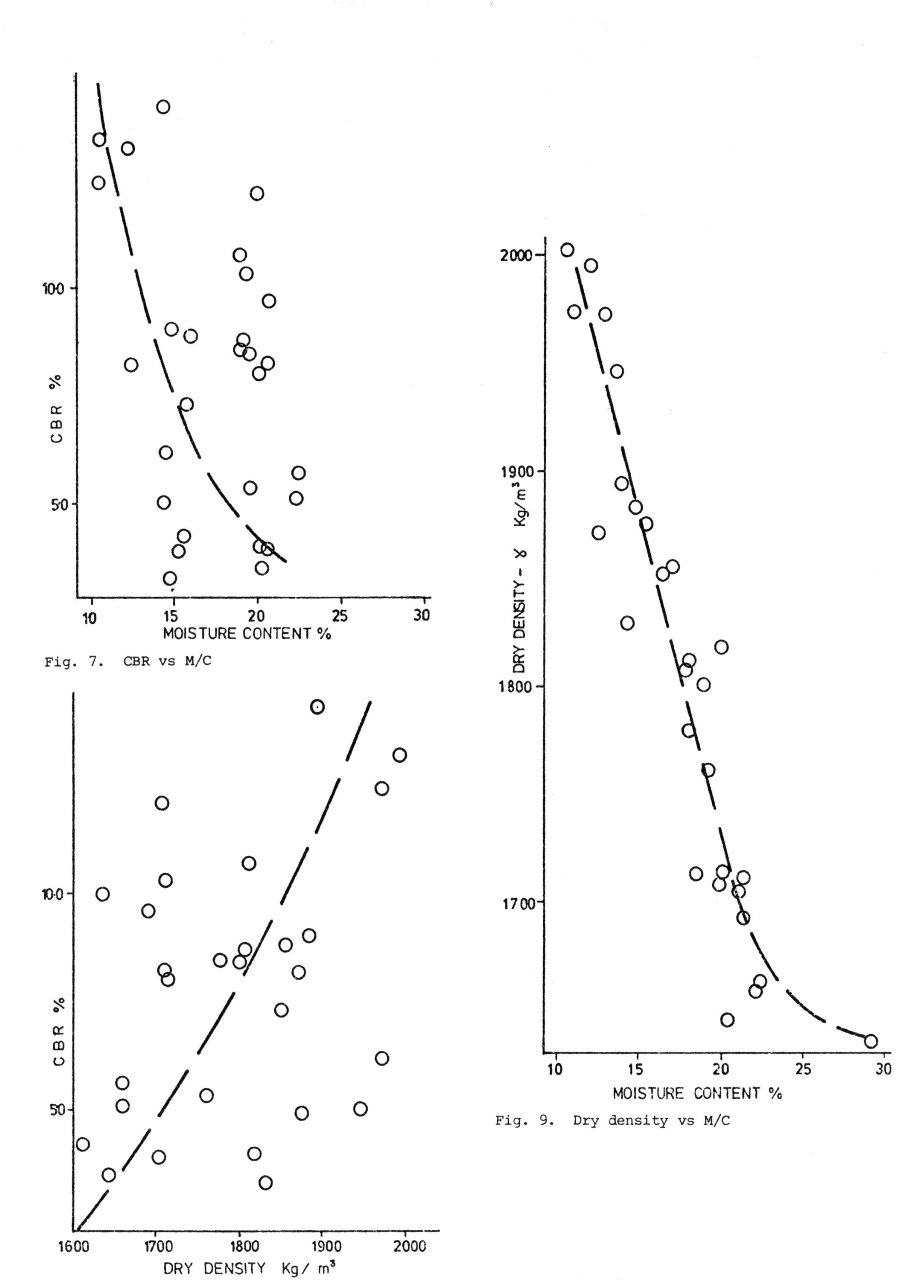

Fig. 7. CBR vs M/C

Fig. 8. CBR vs dry density

Fig. 9. Dry density vs M/C

To evaluate the stress ratio "K" in terms of the effective principal stress

$K = \sigma_3'/\sigma_1'$

Pells (ref.10) considered two cases:

1. When $\sigma_3' < \sigma_1'$ i.e. $K = \sigma_3'/\sigma_1' < 1.0$.

After compaction in a broad embankment, the centre of which will soon be under conditions of no lateral strain and in this case $K = K_o$. Assuming that the principal stresses in the fill will be σ_v and σ_h for the vertical and horizontal directions, then

$$\Delta\sigma_1 = \gamma.\Delta h. \tag{3}$$

$$\Delta\sigma_3 = K(\gamma.\Delta h - \Delta u) + \Delta u. \tag{4}$$

substituting 3 and 4 in equation 1

$$\Delta u = \frac{B.(K + A(1-K))}{1 - B(1-A)(1-K)}.\gamma.\Delta h. = r_u.\gamma.\Delta h. \tag{5}$$

2. When $\sigma_3' > \sigma_1$, $K > 1.00$.

In this case, an increase in the overburden pressure will result in stress changes, i.e.

$$\Delta\sigma_1 = \frac{\gamma.\Delta h - \Delta u}{K} + \Delta u \tag{6}$$

$$\Delta\sigma_3 = \gamma.\Delta h \tag{7}$$

$$\Delta u = \frac{B(K + A(1-K))}{K + BA.(1-K)}.\gamma.\Delta h. = r_u.\gamma.\Delta h. \tag{8}$$

Pells (ref.10) used equations 5 and 8 to evaluate A,B,K in any change of stress.

In the field, Blight (ref.3) determined the stress ratio σ_3'/σ_1' at Bridle Drift to be 7 at a depth of 2m below the crust and 3 at a depth of 10m below the crust.

The writer measured the insitu "K_o" for the normally to lightly over-consolidated stiff fissured Cheshire till and found it to fall between > 1.00 at 2.0m to 0.66m at 6.00 below ground level. Values for "K_o" outside these limits were also measured for different types of Cheshire tills markedly varying in shear strength and ground conditions.

Brooker and Ireland (ref.5) proved in the laboratory that when the OCR exceeded 20 the stress ratio σ_3'/σ_1' could exceed 2. Pells (ref.10) suggested that this could be actually achieved in the field on material compacted by a 10 ton tampered roller.

Pells (ref.10) was able to measure the residual total horizontal stress in embankments by placing earth pressure cells in a 20m layer of fill under compaction.

After 16 passes of a heavy grid roller and 5 passes of a 10 ton vibratory roller, the earth pressure cells (8cm below the surface) recorded a residual horizontal stress varying between 65 and 80kN/m^2.

Further, during the construction of the Bridle Drift (ref.3) it was observed that the marked decrease in the pore water dissipation was due to the high horizontal stress caused by heavy compaction of the fill material placed at or slightly below the optimum moisture content.

At Chester Southerly By Pass the lodgement till was usually placed at 5-6% above its optimum and received the prescribed number of passes by the specified roller (see Table 6/2, DTp specification). Typical test results are shown in Table 2.

These results show that the material is suitable in terms of the limit of 1.2PL and the average moisture content of the group is about 19% which, in this particular contract, is also suitable.

The dry density on average fell slightly below 90% of the AASHO and the average air voids of the group was about 2½%.

These typical compaction parameters are acceptable because they satisfy all conditions already discussed.

Dissipation of pore water pressure and the coefficient of passive earth pressure "K_p"

Field and laboratory work carried out on the lodgement till in Cheshire showed that pore water dissipated fairly rapidly and that the permeability of the disturbed material was only marginally lower than the insitu 10^{-8}mm/sec. However, high lateral stresses may reduce the rate of dissipation (ref.10).

It is, therefore, suggested that consideration should be given to the construction pore water pressure and the stress ratio coefficient "K" during the construction of an embankment, as well as the moisture content, plastic limit and shear stress of the compacted fill.

Construction pore water pressure under cyclic loading may be simulated in the laboratory by using a 150mm hydraulic oedometer in order to forecast its values and the performance of the till under field conditions (no lateral movement).

Excessive compaction will increase the stress ratio to the value of the coefficient of the passive earth pressure (K_p) which has been found in Cheshire tills to vary between 3.5 to 1.6 in the insitu material. K_p was measured from

$$K_p = \frac{2C'.\cos\phi'}{P} + \frac{1+\sin\phi'}{1-\sin\phi'} \tag{9}$$

Skempton (ref.13).

K_p in the compacted fill may be less than the insitu stress ratio due to the possible reduction in C'.

When K_o becomes equal to K_p, the material will undergo passive failure.

Another point to bear in mind is that there have been some occasions when the fill material, although acceptable in terms of moisture content and plastic limit, had very low shear strength. The compactibility of this type of material may cause more problems than can be resolved in terms of heave and settlement.

CONCLUSION

The formation conditions and past stress history of the lodgement till have given rise to its engineering properties in both aspects of foundations and earthworks.

The macro-structure of the lodgement till, fabric, and granular behaviour are principal factors with which the performance of its compactibility in embankment can easily be assessed and satisfactorily monitored.

Due to its low plasticity and expansive minerals it is considered, like the under-lying parent Triassic rock series, sensitive to moisture and vulnerable to storage and double handling.

It is suggested that consideration should be given to the construction pore water pressure and the stress ratio factor "K" in relation to the shear strength and moisture content of the fill during compaction.

Experience at Chester Southerly By Pass suggests that better results will be achieved in compaction and performance when the lodgement till is placed at a moisture content of less than 20% and the criteria of the compaction remains the 5% air voids level.

ACKNOWLEDGEMENT

The author wishes to thank Mr D F Dean, Director of NWRCU, Mr V A Knight, Chief Engineer of the RCU Cheshire Sub-Unit and Director of Highways & Transportation of Cheshire County Council, and his deputy, Mr B B Neilson, for permitting this paper to be presented.

Many thanks to both Dr A G Davis, Chef de division Sols-Recherches, CEBTP, Paris, France and Mr M J Lloyd, Chief Assistant Engineer for Construction, Cheshire Sub-Unit for their constructive criticism.

The views expressed do not necessarily conform to those of the Department of Transport.

REFERENCES

1. AL-SHAIKH-ALI M.M.H. Full scale pile testing to failure to determine the effect of fissuring in stiff boulder clay in the County of Cheshire. Symposium on the Engineering Behaviour of Glacial Materials, Birmingham University, 1975, 256-262.
2. BJERRUM L. The third Terzaghi Lecture. Proc. Am.Soc.C.Eng.(SM5), Soil Mechanics and Foundations Division, 1967, 1-49.
3. BLIGHT G.E. Construction pore pressure in two sloping core rock fill dams. I.C.O.L.D. Q.36, R.18, 1970.
4. BOULTON G.S. & PAUL M.A. The influence of genetic processes on some geotechnical properties of glacial tills.Q.J.Eng.Geol.Foundation on Quaternary Deposits,Vol.9,No.3,1976,159-194.
5. BROOKER E.W. & IRELAND H. Earth pressure at rest related to stress history. Canad.Geotech. J. Vol.11, No.1, 1964, 1.
6. CASAGRANDE A. The determination of the pre-consolidation load & its practical significance. First C.S.M.F.E. Vol.3,1936, Cambridge.
7. EVANS W.B., WILSON A.A. et al Geology of the Country around Macclesfield. 1968 HMSO.
8. JONES R.H. Contribution to the Symposium on the Engineering Behaviour of the Glacial Matls. Proc.Symp.Midlands S.M.F.E. 1975, 253-254.
9. KEELING P.S. The examination of clays by LL/MA. British Ceramic Society, Trans.Brit. Ceramic Abs., Vol.60, 1961, 217-244.
10. PELLS P.J.N. Stress ratio effects on construction pore pressure. Proc.8th Int.Conf.S.M. F.E.Moscow USSR, Session 1 Vol.1,1973,327-332.
11. POOLE E.G. & WHITEMAN A.J. Geology of the Country around Nantwich & Whitchurch. 1966 HMSO.
12. SKEMPTON A.W. The pore pressure coefficients A & B. Geotechnique 4, 1954, 143.
13. SKEMPTON A.W. Horizontal stresses in an over-consolidated pressure in London clay. Proc.Conf.Pore-Pressure & Suction 1961, 81-84.
14. TAYLOR B.J. PRICE R.H. et al Geol. of the Country around Stockport & Knutsford.1963 HMSO.
15. JANBU N. Soil compressibility as determined by Oedometer & Triaxial tests. Proc.3rd Eur.Reg. Conf.I.S.S.M.F.E. Settlement & Compressibility of Soils, Wiesbaden, Vol.1, 1963, 19-25.

E. J. ARROWSMITH, BSc, MICE, North Western
Road Construction Unit, Department of Transport

Roadwork fills—a materials engineer's viewpoint

The problems of sampling and testing boulder clay for its suitability as embankment fill are analysed and the development of suitability criteria described. Improved standards of compaction and the increasing weight of construction plant are shown to cause troublesome resilient strains in boulder clay compacted wet of optimum moisture content. The operation of the Method Specification is discussed. Finally, the difficulties in separating, depositing and compacting mixed sands and clays in embankments are considered.

INTRODUCTION

1. In the past 20 years over 300 miles of motorway and trunk roads have been constructed in the north west of England and the majority of the embankments consist of compacted clay fills. At the start little experience of major earthworks existed as the last major earthworks took place in this area in the railway era, over 100 years ago, and the builders had passed down remarkably little of their experience. Construction plant had changed so much, anyway, that the rules of the game were different.

2. Both designers and contractors have learnt much about the problems of clay fills and the Materials Engineer is in a unique position to appreciate the difficulties of both. It is his responsibility to test the fill produced by the Contractor, against the Specification written by the Designer. In this paper I propose to draw on the experience of many Materials Engineers employed by Consultants, local authorities, Road Construction Units and independent testing houses.

3. The drift cover of the north west consists mostly of a wide variety of boulder clays (the product of several ice ages) glacio-fluvial sands and gravels and alluvium. It is proposed to deal firstly with the boulder clays and then to introduce a section on mixed clay and sand fill which has particular problems of its own.

BOULDER CLAY

4. Boulder clay is currently not a popular term as the nomenclature has led to contractual difficulties because the size and quantity of boulders cannot be inferred from the description. The geologist prefers the term glacial till although this is no more explicit. Nevertheless, it is proposed to use the term boulder clay throughout this paper as it is frequently the interaction of the boulders and the clay that gives the material its particular characteristics.

Site Investigation

5. It is particularly important during the site investigation to carry out both boreholes and trial pits. From the trial pits an estimate of both size and the grading of the boulders can be made and the fabric of the clay inspected; the boreholes are advisable to recover undisturbed samples and record water entries.

6. It is frequently not possible to obtain undisturbed samples because of the boulders and in this case it is normal practice to recompact the soil into a mould before carrying out strength tests. It is generally agreed that these clays are insensitive and when recompacted give similar strength results as when tested in situ. Care must be taken, however, when recompacting at a moisture content greater than the optimum in the BS Compaction test (ref 1) as the build-up of pore water pressures can give low strength results.

Development of suitability criteria

7. On the Lancaster By-Pass, constructed between 1957 and 1960 the Ministry of Transport Specification required that 'soil approved for use as filling shall be compacted, if a cohesive soil at a moisture content within plus or minus 2% of its plastic limit'. The maximum size of the stone in the moisture content sample was not specified and the plastic limit test was, of course, carried out on soil passing the 425 micron sieve.

8. The method worked reasonably well for clays of low plasticity with a low stone content. Clays, classified as suitable for fill on this basis, could be handled and compacted by conventional plant which at that time consisted principally of towed scrapers. It was found, however, that for more plastic clays this Specification declared 'unsuitable' soil which was quite strong enough to form a stable embankment. On the other hand for stoney clays weak soil was declared 'suitable' as, although the moisture content of the bulk sample was not more than 2% above the plastic limit, the moisture

Clay fills. Institution of Civil Engineers, London, 1978, 25-36

content of the clay matrix was well above that figure.

9. The procedure developed on the Lancaster By-pass to find the moisture content of the clay matrix was inspired by a paper by H Smith on Concrete Mix Design at London Airport (ref 2). The stone sizes above 5mm were assumed to be spherical in shape, having a specific gravity of 2.55 and coated with a water film 0.23mm thick. The soil between 5mm and the 425 micron size was taken as equivalent to a sand aggregate with a natural moisture content of 9%. From a knowledge of the grading curve and the moisture content of the whole sample, it was possible to find, by proportion the moisture content of that part passing the 425 micron sieve. For each cohesive soil a factor was calculated with which to multiply the bulk sample moisture content to obtain the moisture content of the matrix, the material passing the 425 micron sieve. This corrected moisture content was then compared with the plastic limit to determine the suitability of the soil as fill.

10. On the M6 between Preston and Warrington constructed between 1961 and 1963 a similar specification, requiring compaction to be carried out at a moisture content not greater than 2% above the plastic limit, was used. From experience on this contract it was suggested that the upper limit should be +2% for clays of low plasticity, +4% for medium plasticity and +6% for high plasticity. This was refined by the Lancashire County Laboratory to a sliding scale from which the recommended upper limit of 1.2 multiplied by the plastic limit, quoted in the Notes for Guidance on the 1969 Ministry of Transport Specification for Roads and Bridgeworks, was derived. The principle of designing a suitability requirement for cohesive soils, related to the plasticity was established.

Factors affecting suitability

11. We now consider the factors which have to be taken into account in choosing suitability criteria. As settlement with boulder clay is fairly rapid, the suitability criterion is determined from consideration of

a. Embankment Stability

b. Trafficability by earthmoving plant

Embankment stability

12. An adequate shear strength for embankment stability is determined from a total shear strength analysis using Taylor's curves. Fig 1 relates the required shear strength to the embankment height for different side slopes. This is a simplified analysis which assumes that the potential slip occurs in the embankment fill. If the embankment is to be built on weak material a more detailed analysis will be required. Having decided on a minimum shear strength for stability, which will vary with the circumstances of each scheme and depends not only on the embankment height but also on such factors as the balance of cut and fill, the shear strength has to be related to a maximum moisture content to include in the specification. Fig 2 is a typical 'measles' plot from the Hyde By-pass from which a moisture content/plastic limit ratio (MC/PL) has to be deduced. This ratio multiplied by the plastic limit of any soil will give the maximum moisture content of that soil at which it will be suitable for forming an embankment. In this particular case the minimum shear strength selected was 70 Kn/m^2 and an MC/PL ratio of 1.2 was chosen for the following reasons:-

(1) It will be seen from fig 2 that there are a minimum number of samples in the top right quarter which have shear strengths over 70Kn/m^2 and MC/PL ratios over 1.2 ie a minimum quantity of suitable material is declared 'unsuitable'.

(2) There are a minimum number of samples in the bottom left quarter which have shear strengths below 70Kn/m^2 and MC/PL ratios below 1.2 ie a minimum quantity of unsuitable material is declared 'suitable'.

13. The wide spread of results in the shear strength - MC/PL ratio graph suggests that there is no unique ratio corresponding to a particular strength. In fact for a shear strength of 70Kn/m^2 it can be seen from fig 2 that the ratio varies with each cutting, as follows:-

Woodside	1.42
Godley	1.38
Miniature Castle	1.15
Hyde	1.29
Denton	1.25

In these variable boulder clays the method of specifying suitability by requiring that the moisture content should not exceed some factor multiplied by the plastic limit is, to say the least, a blunt instrument and many test results are required to enable a reasonable choice of limiting MC/PL ratio to be made.

14. The plastic limit test itself takes some time to carry out if the requirements of BS1377 (ref 1) are followed precisely and it is not very reproducible. Sherwood(ref 5)has shown that one third of the results will be more than 3 units from the actual value. Referring to the mean line on fig 2, a fall in the plastic limit from say 20 to 17 at an MC/PL ratio of 1.2 would increase the ratio to 1.42 making a difference of 28Kn/m^2 and causing a material to appear suitable when it was really unsuitable.

15. It is more logical to use the shear strength criterion for suitability directly rather than through the MC/PL ratio. A contract is being let in the north west on the Haslingden By-Pass on this basis. The unconfined compression test on laboratory recompacted samples will be used as the vane test is not suitable for many boulder clays. The procedure will consist firstly of taking a bulk sample and carrying out a rapid moisture content by use of infra-red driers.

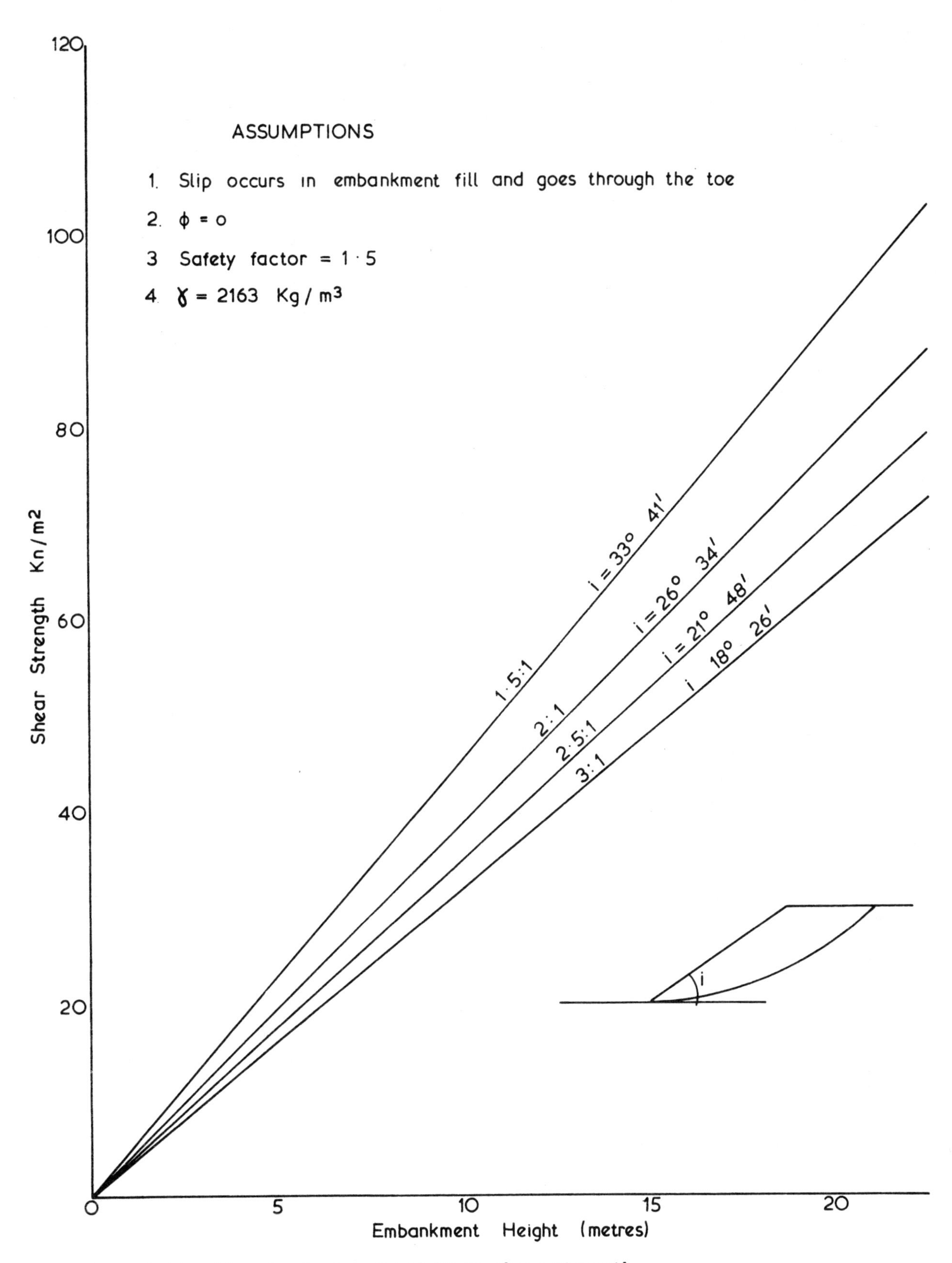

Fig. 1. Relationship of embankment height to shear strength

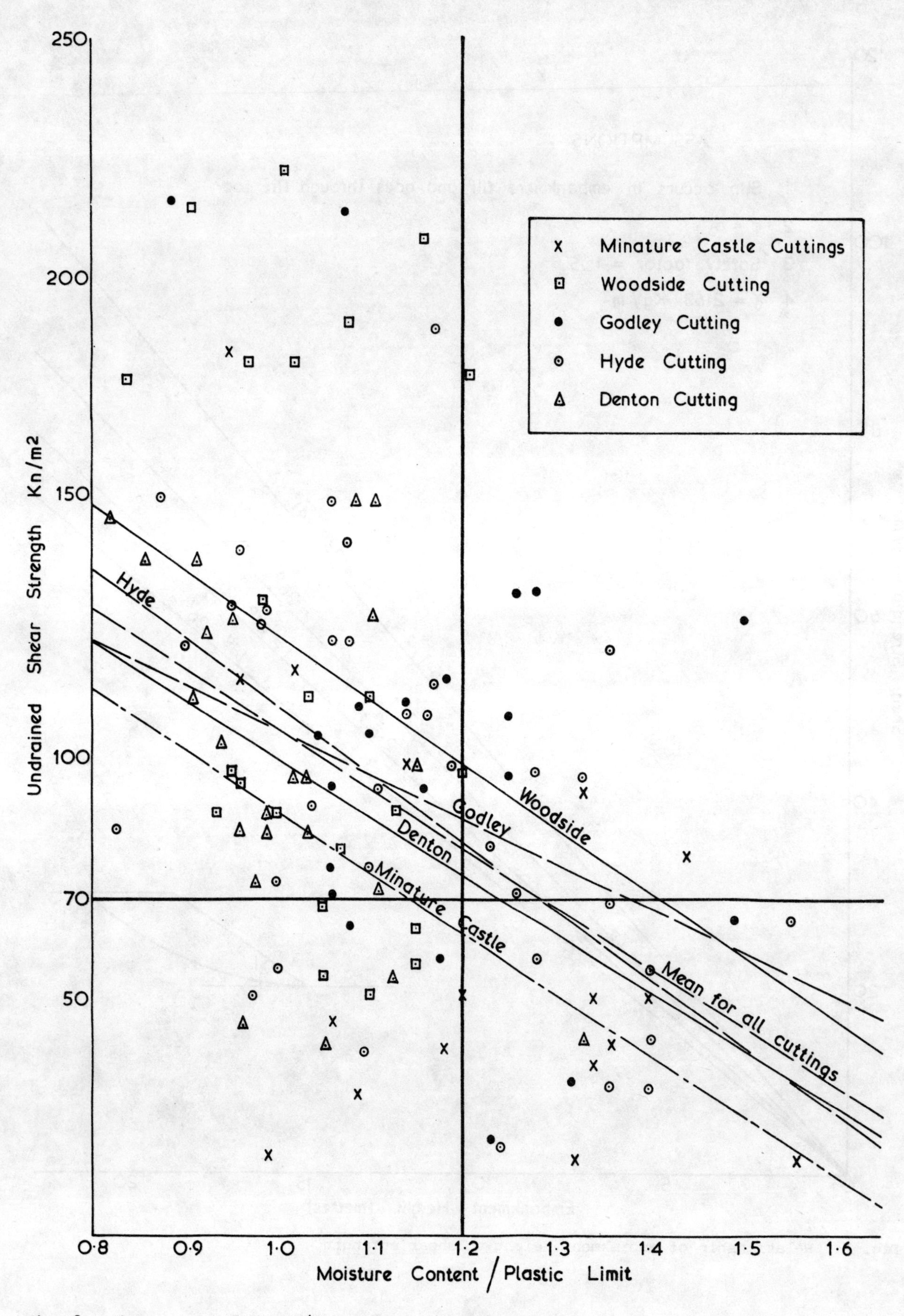

Fig. 2. Shear strength vs MC/PL ratio on the Hyde By-pass

The soil will then be statically compacted to an air voids content of 5% in a 150mm dia mould. Test 16 of BS1377 (ref 1) requires that where the air voids content is less than 5%, the specimen shall be allowed to stand for 24 hours before carrying out a California Bearing Ratio (CBR) test so that pore pressures built up by the compaction process can dissipate. A small pilot test on the Bury By-Pass on similar boulder clay as that expected at Haslingden has indicated that the shear strength is essentially unaltered if the test is carried out without the 24 hour delay so 38mm dia samples are taken from the 150mm mould and tested immediately in unconfined compression. The shear strength is taken as half the compressive strength and in this case a minimum compressive strength of 100 Kn/m^2 is required by the Specification for material suitable for forming embankments.

16. In sandy silty clays it is possible to use the vane test to determine the shear strength and on the Hyde By-Pass five different materials were sampled and remoulded at various moisture contents in the CBR mould. 38mm sampling tubes were driven into the material in mould and shear vane measurements were taken around the outside of the sampling tubes. The sampling tubes were then removed and the samples extruded for unconfined compressive strength determinations.

17. In fig 4 vane shear strength is plotted against the shear strength determined from the unconfined compressive strength. A linear relationship is shown to exist for the clays such that vane shear strength = $1\frac{1}{2}$ x shear strength as determined from unconfined compressive strength tests. For the silt (sample No 5) the relationship is more complex.

18. A variation of the suitability criterion was adopted on the Keswick By-Pass. A maximum moisture content for suitability of 1.1 multiplied by the BS light Optimum moisture content (Test 12 of BS1377 ref 1) was chosen. This figure was decided on from consideration of a large number of bulk samples on which CBR tests had been carried out at different moisture contents. A graph of CBR against moisture content/BS optimum moisture content was plotted for each sample and it was found that at a moisture content/BS optimum moisture content ratio of 1.1 the CBR value of nearly all the samples was above 1 per cent. The pavement was designed for a CBR of 1% and the boulder clay was strengthened by a selected fill layer of gravel 550mm thick to bring up the CBR to a minimum of 6% below the sub-base. A further advantage of relating the moisture content for suitability of boulder clay to BS optimum moisture content is that the BS compaction test is carried out on material passing the 20mm sieve. The moisture content of a bulk 20mm down sample in the field can be related directly to that on which the suitability test was carried out. Moisture content criteria based on plastic limits cannot.

19. The boulder clay wetter than 1.1 x BS optimum was not discarded but fill areas were designated where the ground was reasonably level and depth of fill was greater than 3m. The design incorporated horizontal drainage layers at 3m intervals. This device to maximise the use of boulder clay as fill had been used successfully on the M6 in Cumbria. In the event the summer was exceptionally dry and the wet boulder clay became dry and complied with the specification of the previous paragraph so the drainage layers were not required.

Trafficability by earthmoving plant

20. The trafficability of earthmoving plant was not originally a problem as caterpillar tractors and scrapers were able to excavate and place any material that could be compacted. However, as self-propelled rubber-tyred scrapers were introduced and became progressively larger, the shear strength requirements of the soil to support them became larger also. From observations on the M6 between Preston and Lancaster and from a limited field trial on the M61, the following minimum shear strength requirements emerged:-

Caterpillar tractors and scrapers - 35Kn/m^2

Large rubber-tyred scrapers - 50Kn/m^2

FARRAR and DARLEY (ref 3) have suggested rather higher figures and these are supported by later work carried out on the Hyde By-Pass on large scrapers and shown in fig 3. It is likely that the minimum requirement to support large plant rises steeply as the plant weight increases. This is analagous to the logarithmically increased damage caused to pavements by heavy axle loads.

21. Specifications tended to be written to permit the large rubber-tyred scrapers to be used and much otherwise suitable material was disposed of to tip. The Department of Transport now instructs that this practice should be stopped and contractors should tailor their plant to meet the other limitations of stability and settlement. Department of Transport Specification 1976 Clause 601.3 (ref 4).

22. On the M58 Aintree - Skelmersdale contract currently under construction, the suitability requirement has been based on a minimum shear strength of 35 Kn/m^2. It is realised that this will restrict the weight of plant able to operate and will exclude the heavier rubber-tyred scrapers. However, the alternative is to import large quantities of expensive imported fill much of it inevitably through built-up areas.

Compaction

23. On the early motorways an end product Specification for Compaction was used requiring 10% air voids to the main mass of fill to embankments and 5% air voids in the top 1m of fill.

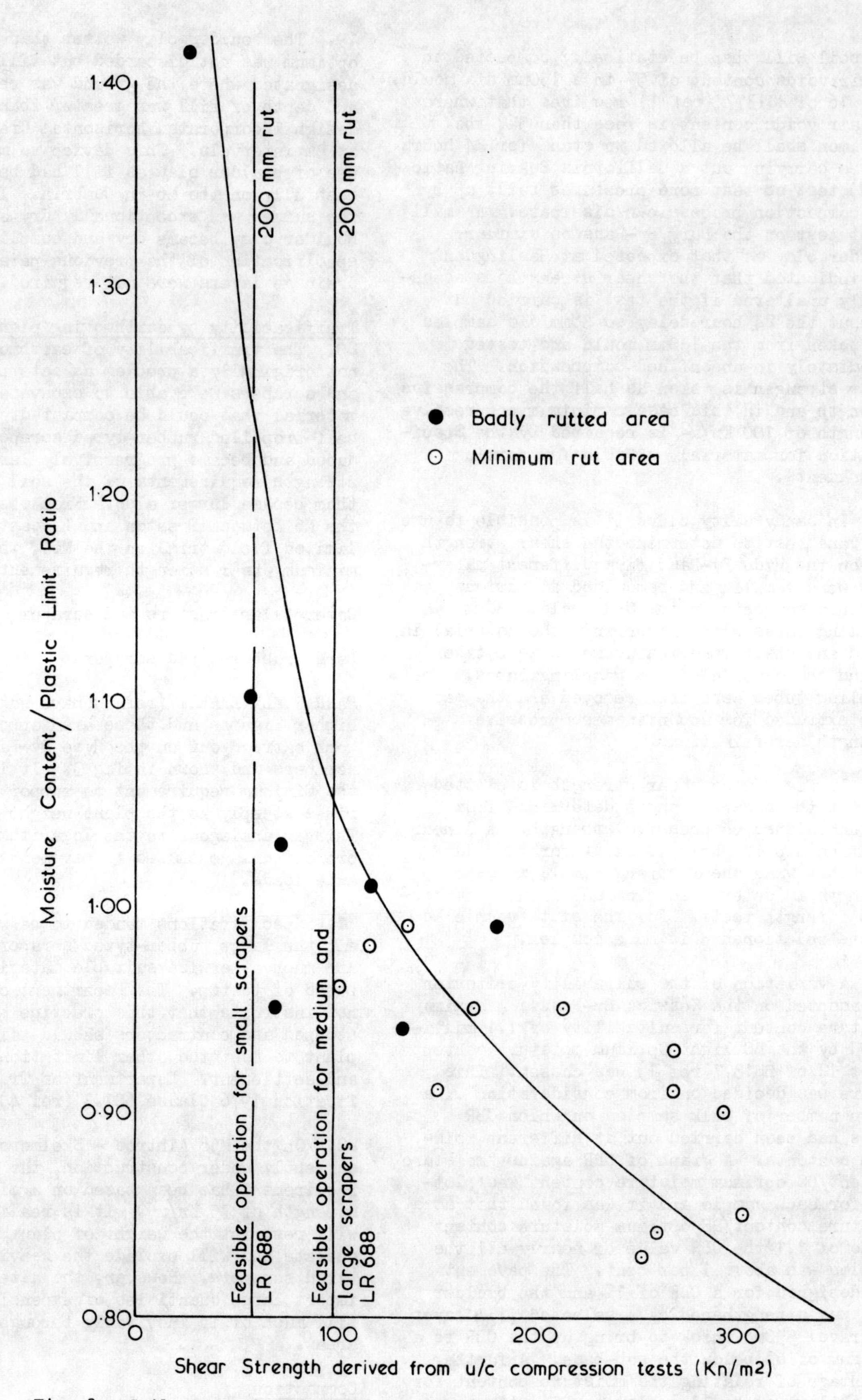

Fig. 3. Soil strength and trafficability for large scrapers

Sample No	Key	Description	Casagrande Classification	LL	PL	PI
1	x	Thin dark brown sandy silty clay	C1	38	19	19
2	⊙	Stiff dark brown sandy silty clay	C1	36	17	19
3	⊡	Firm sandy silty clay	CL	32	16	16
4	△	Stiff silt/clay	CI	44	22	22
5	■	Light brown laminated silt	ML	33	26	7

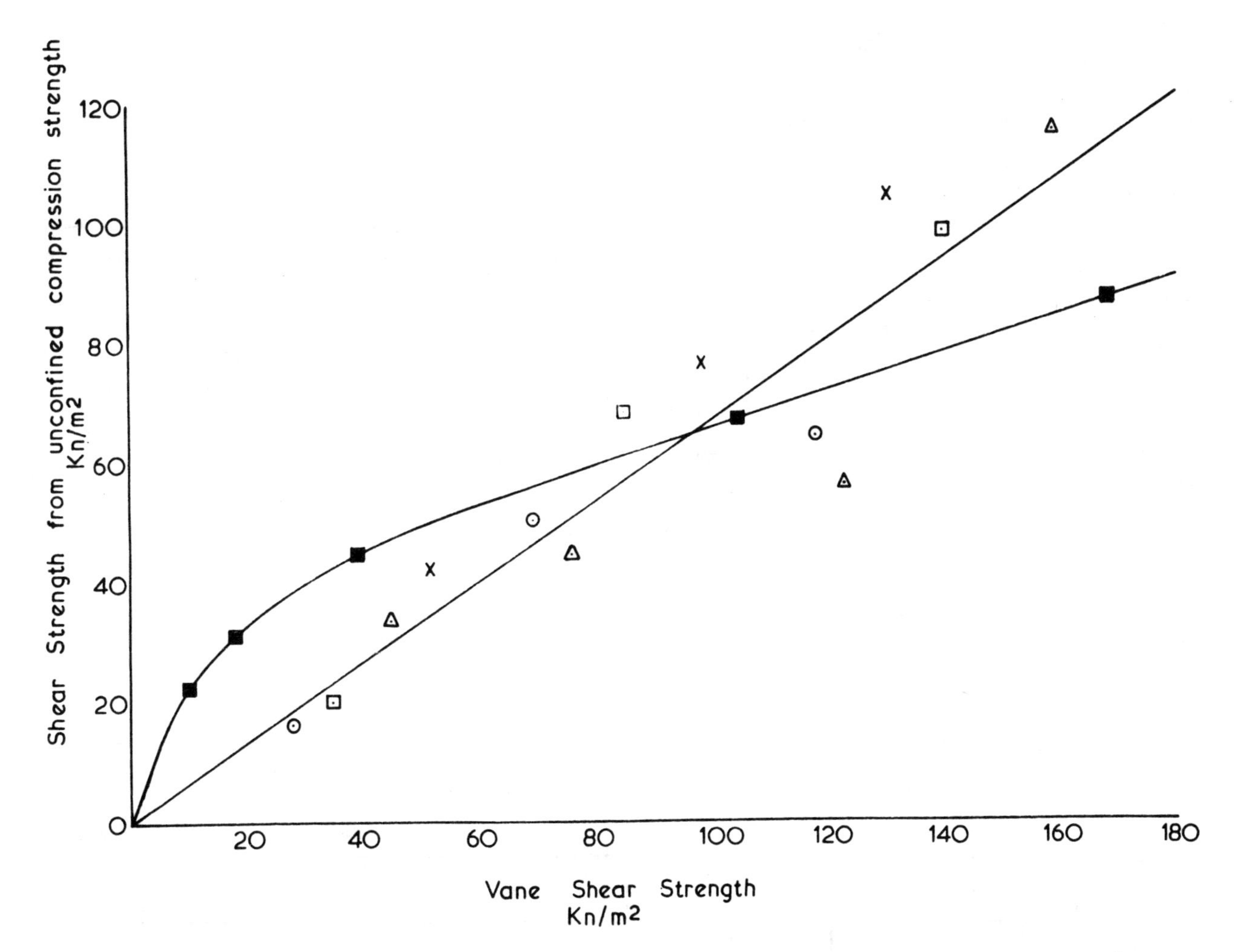

Fig. 4. Relationship of shear strength from the unconfined compression test to vane shear strength

The procedure consisted of setting up compaction trials at the beginning of each major cutting by spreading a 225mm thick loose layer of soil, compacting it with 2, 4, 6 and 8 passes by different rollers and carrying out density tests to compare the results. It was normally possible to get the air voids of clay fills below 10% by 4 passes of a 10 ton rubber-tyred roller or a 10 ton steel-tyred roller providing that the moisture content was within the specified limits of ± 2%. In Table 1 the average results of all tests on clay fill on the Lancaster By-Pass and the M6 in Westmorland are given. The CI Clay results represent a relatively small number of tests and the high air void content can be accounted for by the low moisture content, compared with the plastic limit. For the remainder of the boulder clays at Lancaster 10% air voids at a moisture content close to the plastic limit corresponded closely to 90% of the BS Light Compaction, Test 12 of BS 1377 (ref 1).

24. Some years later (1968 - 1970) the M6 was extended northwards from the Lancaster By-Pass through the same boulder clays but with the passage of time the weight of the construction plant had increased considerably and better densities were demanded. The excavating plant was generally rubber-tyred scrapers (Caterpillar 657's or Terex TS 14's) and compaction was carried out generally with 4 passes of a 4 ton vibrating roller. The minimum dry density required was 95% of the maximum found in the BS Light Compaction test. In Table 1 the M6 test results for the section nearest to the Lancaster By-Pass are shown in the last line. It will be noticed that the boulder clay is virtually identical to the SC material at Lancaster but that the average relative dry density in the field is increased from 91.7% to 97%. This increase in density had very significant effects.

25. Some 'heave' or more properly, resilient strain, had been experienced at Lancaster but it was not a major problem. On the M6, however, this visco-elastic strain which was recoverable but time - dependent often reached 50-75mm under the roller at road formation level. It was feared that if permanent construction was proceeded with repetitive traffic loads would cause fatigue failure of the road base or the unstable foundation would inhibit the compaction of the sub base and base. It was necessary to stabilise the foundation with considerable thicknesses of rockfill and/or replace the granular sub base with a stiffer alternative cement or bitumen-bound material.

26. It is suggested that the heavier compaction on the M6 at a moisture content wet of optimum, compared with the Lancaster compaction standard, was responsible for this troublesome resilient strain. Seed et al (ref 7) have shown that at the same moisture content, providing that the compaction is carried out wet of optimum, the resilient strain is greater, the higher the standard of compaction. It is possible therefore, to over-compact. Furthermore, they show that if the soil is initially at a low density and is wetted up the resilient strain is smaller than if the initial density was greater. A sub-grade is often exposed to weather conditions which would cause wetting up and subsequent trafficking could cause resilient strains to develop.

27. An explanation of this 'heave' phenomenon in boulder clays of low plasticity, compacted wet of optimum is due to HIGHT (ref 8). He postulates that at the top of the embankment the clay is partially saturated and contains both air and water voids. The smaller pores are filled with water and the larger voids contain occluded air which cannot escape because of the low permeability. On being transiently loaded the pore water pressure increases and the air is compressed like a spring. When the load has passed rebound occurs, and the air expands to its original volume. The pore water pressure does not entirely dissipate and the soil is weakened by the next loading causing a greater deformation. This phenomenon of increasing deformation on embankment up to a maximum was noticed in the field. In time the pore pressures in the embankment dissipated and the soil regained strength.

Table 1

Location	Soil Classification	LL	PL	PI	Density/Moisture Content: Kg/m³/% BS Light Compaction	Density/Moisture Content: Kg/m³/% In the field	Relative Density %	Air Voids %
Lancaster By-Pass	SC	25.5	18.9	6.6	1989/11.1	1874/12.0	91.7	8.8
	CL	30.1	17.9	12.2	1896/12.6	1704/15.0	88.5	10.5
	CI	44.0	21.8	23.2	1779/14.4	1608/16.5	90.4	13.7
M6 Westmorland	SC	25	18	7	2002/11.2	1942	97	

When the whole pavement had been placed the pressures at sub-grade level would be minimal so the problem was a construction one. Just as the speed of modern construction plant creates stability problems with rising pore water pressure, the increasing weight of plant has also brought problems.

Method Specification

28. Since the introduction of the Ministry of Transport 1969 Specification (ref 9) a Method Specification has operated to control compaction. The Contractor is presented with a table from which he can match the type of compaction plant with the type of material and obtain the maximum layer thickness and the number of passes required. It is implied that the controlling staff will actually count the passes but this does not happen in practice except perhaps at the commencement of a new embankment. The tendency is to supply control staff also with a table which match compacting plant to the plant depositing the fill. See Table 2 from which it is possible to check the plant balance.

If the speed of rolling of the actual roller is different to that assumed in the table, then corrected capacity can be calculated from the formula below:-

$$\text{Capacity (cu m/hour)} = \text{capacity in Table 2} \times \frac{\text{Actual speed}}{\text{Assumed speed}}$$

It should be noted that the capacity thus measured is the approximate amount that can be compacted per hour of rolling and not necessarily per hour. (Rollers may only work 50 mins in the hour).

The Clerk of Work checks that:

1. layer thickness is not exceeded
2. rollers cover the area evenly

and records:

1. type of delivery vehicle
2. rate of delivery in loads/hour
3. the number of rollers involved
4. the total time actually rolling
5. the estimated average rolling speed

The earthworks controller assesses the capacity of the delivery vehicles and can then use Table 2 (modified if necessary as above) to check that the compaction plant is adequate.

29. There is no evidence that this method specification has resulted in less control staff as the original practice of carrying out a compaction trial to determine the amount of compaction required had exactly the same effect of creating a method of working. The only difference is that in the Method Specification the contractor can price exactly for the number of passes stated, whereas in the End Product Specification he has to rely on experience. There is no evidence in the field that standards are any lower. In fact the problem of over-compaction at too high a moisture content is probably more serious than under-compaction in road embankments. In earth dam construction the priorities are different and this is unlikely to be true.

MIXED SANDS AND CLAYS

Excavation

30. Mixed sands and clays occur frequently in glacial deposits. Sometimes they occur in distinct layers but more frequently as pockets or lenses. Separation of the two materials on excavation may be very difficult but should be attempted by the correct choice of plant. Scrapers are, for instance, more suitable for excavating distinct layers than face shovels. It will, however, rarely be possible to separate material closer than 0.5m to a sand-over-clay interface. The top layers of clay are likely to be soft and unable to give adequate support to the wheels of the scraper.

Deposition

31. The Department of Transport Specification for Roads and Bridgeworks Clause 608.7 (ref 4) gives instructions for placing sand and clay separately but only so that the prescribed compaction by the method specification can be carried out separately on each material. It is not concerned with the physical juxtaposition of these materials which can cause problems, some of which are illustrated in fig 5 and described below:-

a. Sand fill overlying clay - this is a frequently met condition which in the long term is quite stable. During construction however, before the impermeable pavement is laid, fine sand can suffer severe erosion. This would not occur if the clay was placed above the sand.

b. Clay containing a sand core - this is also a constructional problem but slips have occurred in this situation due to infiltration of water behind the clay blanket.

c. Slip on inclined interface between sand and clay.

d. Slope comprising a thin sand layer - slips have been caused by water movement sometimes fed by leaking drains.

e. Irregular deposits of sand in a clay embankment - water movement from french drain via inter-connecting permeable sand causing slip.

None of these problems are imaginary as they have all occurred on motorway construction sites. They are difficult to control with a

Table 2

Type of Roller	Roll Width (m)	Assumed App Roll Speed m/sec	Cohesive Soil			Well-Graded Granular			Uniformly Graded		
			Max Compact layer depth (mm)	Min No of passes	App Max Output m^3/hr	Max Compact layer depth (mm)	Min No of passes	App Max Output m^3/hr	Max Compact layer depth (mm)	Min No of passes	App Max Output m^3/hr
SMOOTH-WHEELED											
Three-point 'footpath'	1.07	1.22	125	8	73	125	10	60	–	–	–
Unballasted towed	3.05	1.53	125	8	265	125	10	215	125	10	215
Unballasted three-point	1.83	1.22	125	6	160	125	8	125	–	–	–
Ballasted towed	3.05	1.53	125	6	350	125	8	270	125	8	270
Tandem - Marshall TLJ *	1.32	1.22	125	6	120	125	8	90	–	–	–
Ballasted three-point 10-12 ton *	1.83	1.22	150	4	305	150	8	155	–	–	–
GRID (Hyster Model D)											
Unballasted	1.63	1.53	150	10	135	–	–	–	150	10	135
Partially ballasted	1.63	1.53	150	8	170	125	12	95	–	–	–
Fully ballasted	1.63	1.53	150	4	340	150	12	115	–	–	–
VIBRATORY											
Pedestrian Operated											
BOMAG double drum BW65	0.66	0.46	–	–	–	75	8	11	150	8	21
BW35	0.38	0.46	–	–	–	75	6	8	150	6	15
BW605	0.61	0.46	–	–	–	75	6	13	150	6	25
BW75 & 75S	0.76	0.46	–	–	–	75	6	16	150	6	32
BW90L & 90SL	0.91	0.46	–	–	–	75	6	19	150	6	38
Stothert & Pitt 28W (single drum)	0.71	0.18	–	–	–	75	12	3	150	12	6
Towed											
Single 54" - 54T2(S & P)	1.37	0.67	100	12	28	125	12	35	150	6	85
.. 72" - 72T "	1.83	0.67	150	4	170	150	4	170	225	12	85
.. 72" - T182 "	1.83	0.67	175	4	195	175	4	195	250	10	110
.. 75" - CH44,CH45 (Winget Dynapac)	1.91	0.67	175	4	200	175	4	200	250	10	115
.. 72" S & P,T182SF	1.83	0.67	200	4	225	200	4	225	275	8	155
.. 82" S & P,Vibroll T208	2.08	0.67	300	4	380	300	4	380	300	4	380
Self-propelled											
Tandem (S & P) 32RD	0.81	0.40	100	12	10	125	12	12	150	6	29
Double Roll Bomag BW200	2.00	0.27	125	4	64	150	4	75	–	–	–
Tandem 2 ton	1.02	0.40	125	8	23	150	8	27	–	–	–
Pacmaster SP54	2.16	0.67	250	4	330	250	4	330	–	–	–

* with or without ballast

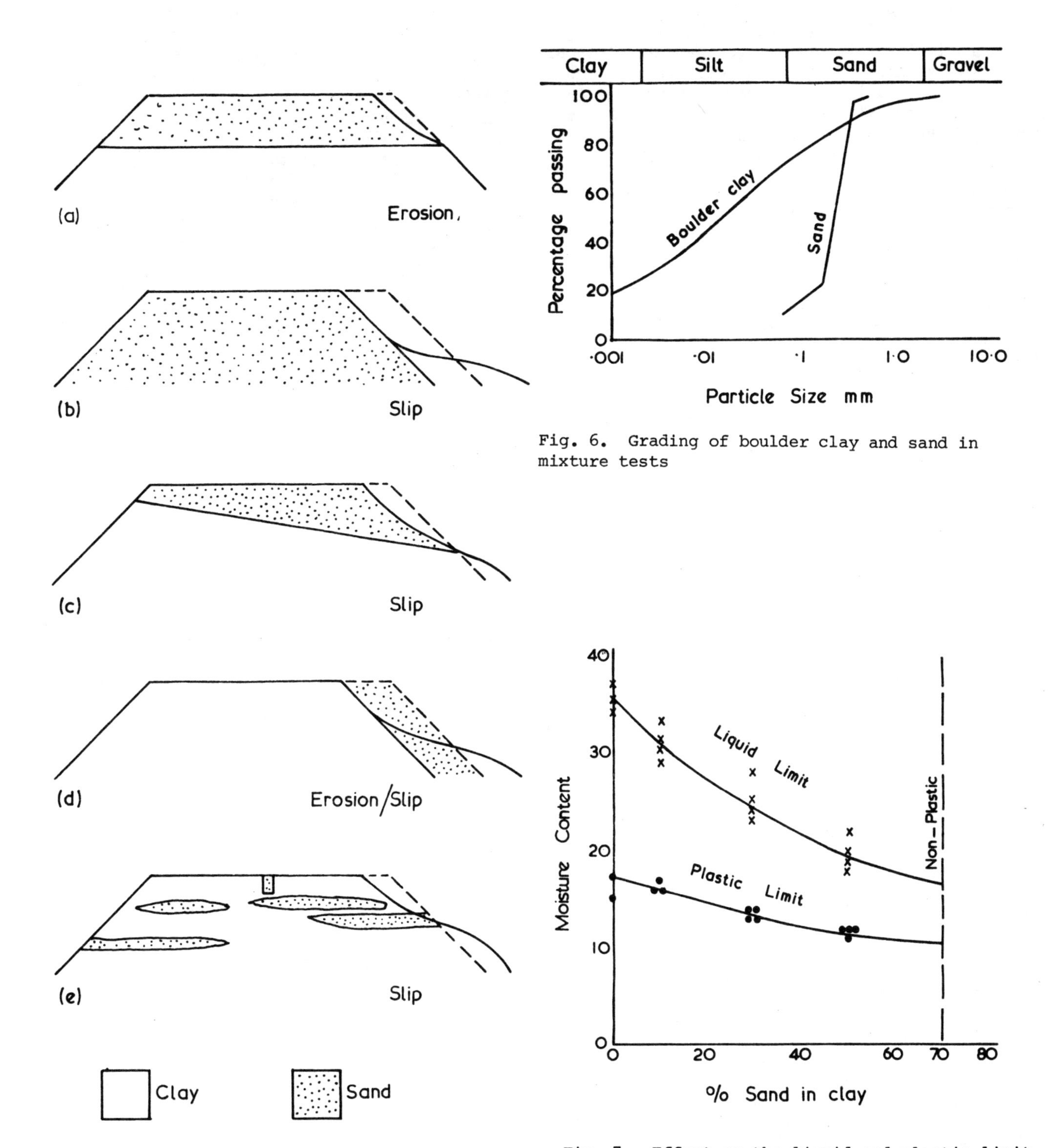

Fig. 5. Unstable sand/clay fills

Fig. 6. Grading of boulder clay and sand in mixture tests

Fig. 7. Effect on the liquid and plastic limit of clay of adding a percentage of sand

specification clause but it is sometimes possible from a knowledge of the soil survey and a close scrutiny of the Contractor's programme to anticipate the problems and avoid them. It is suggested that sand or clay should be placed in reasonably level layers across the full width of the embankment. If it is unavoidable that the top layers of an embankment are constructed in erodable material then steps should be taken to protect the side slopes and upper layer from the ingress of water as soon as possible.

Compaction

32. The compaction factor can be defined as a constant which when multiplied by the volume of soil in cut equals the volume of soil in compacted fill. From time to time this factor has been obtained and the results from the Lancaster By-Pass and the Rochdale Oldham Route A627M, are given in Table 3.

Table 3

Site	Type	Description	Compaction factor
Lancaster By-Pass	CL	Shale & blue clay	1.08
	GF	Brown gravelly soil	1.08
	SP	Coarse sand	0.83
	CL	Brown boulder clay	1.02
	SC	Brown gravelly soil with clay	0.92
Rochdale - Oldham	SF	Sand	0.80-0.87
Route	CI	Clay	1.01

33. It will be noticed that the sand, deposited originally in a loose natural state is subsequently recompacted by heavy plant to a density 13 - 20% greater. The clay on the other hand, has very similar densities when recompacted. The effect of mixing sand and clay, has a surprising effect on the compaction factor. On the Rochdale Oldham Route it was found that the compaction factor varied from 0.85 for 75% sand to 0.89 for 50% sand. It was a uniform fine sand and it is assumed that the clay occupied the voids in the sand. On a normal site there is a considerable waste of fill material due to overfilling, creating haul roads and removal from site after contamination by rain. This waste can be as great as 10 - 15% and should not be confused with the compaction factor which is related only to the in-situ and the recompacted densities.

Suitability as fill

34. Suitability criteria in the current Department of Transport's Specification (ref 4) are related to moisture content and as has been explained above for cohesive materials an upper moisture content limit of some factor multiplied by the plastic limit is normally specified. For non-cohesive materials the upper moisture content is usually related to the optimum moisture as found in the standard compaction test (ref 1). In Fig 7 the effect on the liquid and plastic limits of adding a proportion of uniform fine sand to three different samples of clay is shown. The effect of adding 50% sand is to reduce the plastic limit from 18 to 12%. If the mixed material is classed as cohesive because it has a plastic limit then the upper limit of suitability will be depressed and the suitability criteria rendered incorrect. The gradings of the boulder clay and sand are shown in fig 6.

ACKNOWLEDGEMENTS

The Author wishes to thank Mr D F Dean BSc, FICE, Director of the North Western Road Construction Unit for permission to publish this paper and the following Materials Engineers who have provided material without which the paper could not have been written.

Atkinson P A	North Western RCU
Cotton R D	Messrs Sandberg
Day J B A D	Lancashire County Council
Garrett C	Kent County Council formerly Sir Wm Halcrow & Ptns
Morris A G	North Western RCU
Smith D P	Merseyside CC, formerly NWRCU
Sutcliffe M A	Cheshire CC, formerly Lancs CC
Ward J	North Western RCU

The views expressed are personal and are not necessarily those of the Dept of Transport.

REFERENCES

1. BRITISH STANDARDS INSTITUTION, Methods of test for soils for Civil Engineering purposes BS 1377, 1975.

2. SMITH H, Gravel Compaction and Testing and Concrete Mix Design at London Airport. Proceedings of the Institution of Civil Engineers, 1952, February, 1 - 54.

3. FARRAR and DARLEY. The operation of earthmoving plant on wet fill, Transport and Road Research Laboratory, LR 688, 1975.

4. DEPARTMENT OF TRANSPORT, Specification for Roads and Bridgeworks, HMSO 1976.

5. SHERWOOD P T, The Reproducibility of the Results of Soil Classification and Compaction Tests, Transport and Road Research Laboratory LR 339.

6. COCKSEDGE J E & HIGHT D W, Some Geotechnical aspects of Road design and construction in tills, Symposium on the Engineering Behaviour of glacial materials, Birmingham 1975.

7. SEED H B, CHAN H K & LEE C E, Resilience characteristics of subgrade soils and their relation to fatigue failures in Asphalt pavements, Proc 1st Int Conf on the structural design of Asphalt pavements, 1962.

8. HIGHT D W, The resilient behaviour of soil under traffic loading, MSc Dissertation, University of London, 1971.

9. MINISTRY OF TRANSPORT, Specification for Roads and Bridgeworks, HMSO, 1969.

W. BLACK and N. W. LISTER, Pavement Design Division, Highways Department, Transport and Road Research Laboratory

The strength of clay fill subgrades: its prediction and relation to road performance

1. INTRODUCTION

The stiffness and strength of soil subgrades play an important part in determining the performance of flexible roads. Deformation in the subgrade is controlled by the strength of the subgrade in relation to the traffic stresses transmitted to it and these are greatly influenced by the stiffness of the subgrade itself. The stiffness of the subgrade also influences the stresses generated by traffic within the layers of the pavement.

In the design of new roads in the United Kingdom the design parameter used to describe the soil is a measure of both its stiffness and strength. The California Bearing Ratio or CBR strength is determined with the soil at the equilibrium value of moisture content that is considered to persist under the road pavement throughout the life of the road. In actual practice the moisture conditions during construction are often much more adverse and seasonal variation also occurs under the in-service road.

Because the relation between soil suction and the wetting and drying of soils exhibits hysteresis, a subgrade which becomes wet during construction can be shown to attain a higher equilibrium moisture content than one which has been protected from the weather. The consequent reduction in subgrade strength (and stiffness) shortens the life of the road. Low subgrade strength during construction can result in poorly compacted granular sub-bases and road-bases which deform relatively easily under traffic. If a poorly compacted sub-base placed on a weak-subgrade does not provide an adequate platform for the effective compaction of cement-bound and bituminous bound roadbases road performance will again be adversely affected.

Because of the importance of subgrade strength and stiffness on both the structural quality and the in-service performance of road pavements this paper will attempt to quantify the likely changes in moisture content and strength of the subgrade under a range of site conditions. The changes in moisture content are relatively independent of whether the soil is undisturbed in well compacted fill and the Suction Index method is used to assess the associated changes in soil strength. The effect of these changes on the structural life of pavements is assessed using relationships established between the structural performance of road pavements and their deflection under a rolling wheel measured in a standard manner.

2. THE SUCTION INDEX METHOD

The Suction Index method estimates the strength and California Bearing Ratio, CBR, of remoulded cohesive soils from plasticity data[1]. It is based on experimentally established correlations between the plasticity indices, remoulded soil suctions and effective friction angles of soils. It was shown in Reference 1 that the CBR correlated well with the product of the soil moisture suction (in lb/in^2) and one tenth of the bearing capacity factor when these two factors were estimated from plasticity data. Comparison of estimated and measured CBR values for various British and African soils showed that the method was capable of predicting changes in strength due to changes in soil moisture content and soil type with considerable accuracy and gave a reasonable idea of the absolute CBR to be expected. If absolute values rather than relative values of CBR were required some calibration was considered to be desirable.

In order to use the same curves relating CBR and plasticity data for estimating shear strengths the relation between CBR and remoulded soil and the undrained shear strength, Cu was required.

In metric units the CBR of a soil is defined as the pressure on the CBR plunger in kPa, at 2.5 mm penetration divided by 69. (1)

From model footing studies[2] the pressure on a footing the size of the CBR plunger at a penetration of 2.5 mm is equal to about half of the pressure, qu, at ultimate bearing capacity of the soil.

$$\text{Hence CBR} = \frac{\frac{1}{2}\,qu}{69} \quad (2)$$

As the undrained shear strength is related to the ultimate bearing capacity of cohesive soils by the equation $qu = 6Cu$ equation 2 becomes

$$\text{CBR} = \frac{Cu}{23} \quad (3)$$

This applies when the CBR is measured on remoulded clays. When the CBR is inferred by the Suction Index method as in Reference 1

Fig. 1. Relation between shear strength CBR and plasticity data

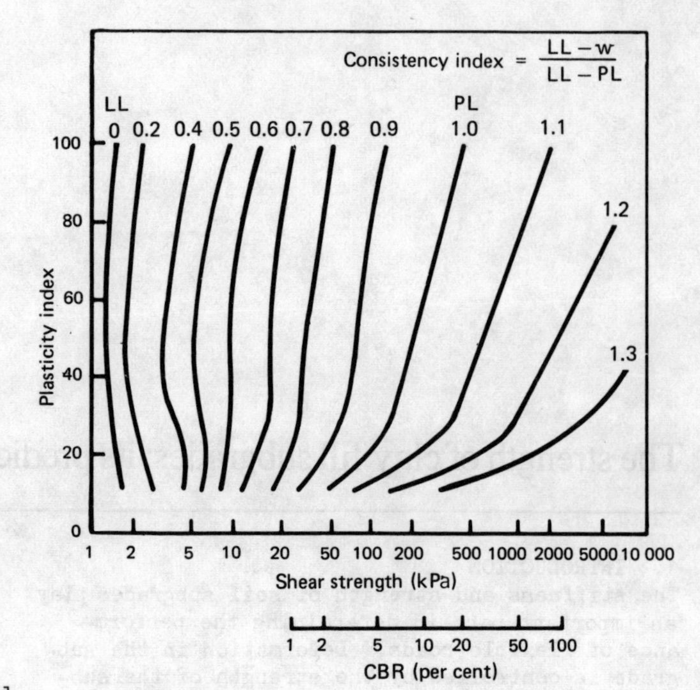

Fig. 2. Comparison of measured shear strengths with values estimated by the suction index method

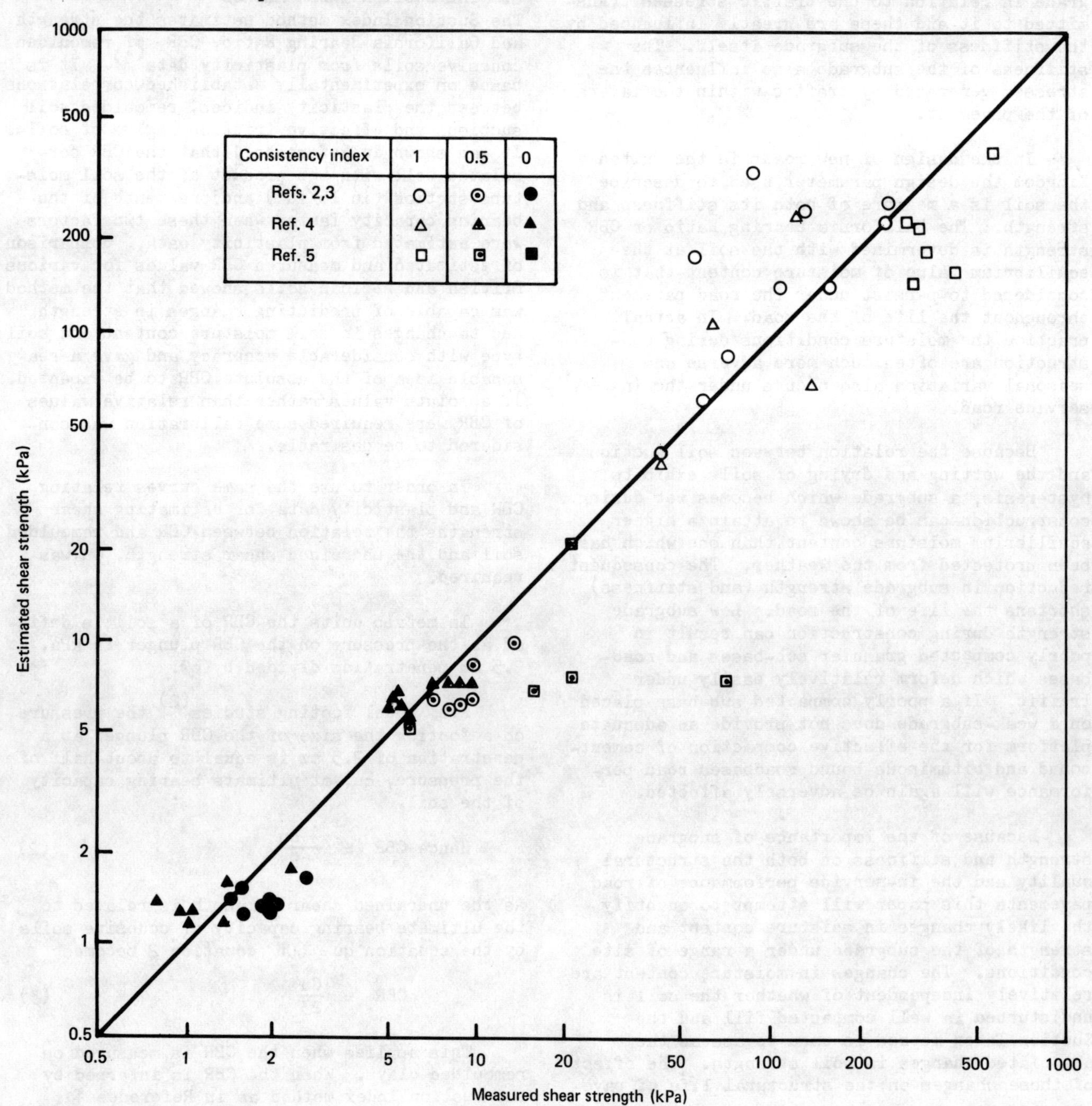

better correlation was found between estimated and measured shear strength if the equation was written

$$Cu = (23 \text{ CBR} + 1) \text{ kPa} \qquad (4)$$

This is a negligible correction at strengths typical of road subgrade soils but is necessary for correlation with strength at the liquid limit. It may reflect the fact that at zero suction, such as when a soil is immersed in water, moist soils still cohere; this implies that some force other than suction is operating. The relations obtained between Cu, CBR and plasticity data are given in Fig 1.

The prediction ability of the method has since been further evaluated on a wide range of soils whose index and strength properties are described in References 3, 4, 5 and 6. Reference 3 comprised a comprehensive series of vane shear strength measurements on artificial soils created by mixing different proportions of sand with Kaolinite and also Montmorillonite minerals. The plasticity indices of these soils ranged from 9 to 91 per cent; Reference 4 and 5 compared vane shear tests on remoulded British soils, London Clay, Brickearth, Keuper Marl etc; Reference 6 includes similar tests including tests on some continental soils. The comparison between measured strengths and strengths estimated by the Suction Index method is given in Fig 2. Strength comparisons are made at the liquid limit and the plastic limit (some by extrapolation) and also at moisture contents mid-way between these limits. The 45° line in Fig 2 which is the line of equality between measured and estimated strengths is quite a good fit through the points. It confirms that the estimation method, on average, correctly estimates changes in shear strength resulting from moisture content changes, although, if a close estimate of strength is required it may require calibration for any particular soil.

A method of estimating of the moisture changes which bring about changes in soil strength during and after construction is also required if the Suction Index method is to be used to predict these strength changes. This is considered below, on a comparative basis.

3. ESTIMATION OF MOISTURE CONTENT CHANGES

The relationship between moisture content, overburden pressures, and water table positions has been fully dealt with elsewhere[7]. It has been shown that moisture content changes with the suction in the pore water of the soil and suction can be calculated using the equation

$$\alpha P + S = U \qquad (5)$$

U is the pore water pressure in the soil which is controlled by the position of the water table when equilibrium has been attained.

S is the soil moisture suction, or pore water pressure at zero overburden pressure

P is the overburden pressure

α is the proportion of the overburden pressure which is effective in changing the pore water pressure (it is 1 for a clay and 0 for sand).

Except in the case of sands soil suction at equilibrium is therefore controlled both by depth of the water table and by the surcharge imposed by the pavement.

Moisture content changes in the subgrade can then be established from the results of extensive laboratory studies made of the relations between soil suction and moisture content of a range of soils, described in Reference 8. This information has been used to develop a series of suction/moisture content relations for soils with plastic indices of between 5 and 90 per cent. Different curves are obtained depending on whether the soil is drying-out or wetting-up and the characteristic hysteresis shape accounts for the important fact that the moisture content of a soil in equilibrium with a water table at a given depth under a given pavement can be at one of two values as the result of one cycle of suction change. In practice an infinite number of hysteresis loops are possible, depending on the range and absolute values of suction at the ends of the wetting and drying phases.

It should be noted that if a soil is at any given suction and the soil is wetted and dried back to the same suction it will have a moisture content at the end which is greater than that at the beginning. The amount of the hysteresis usually lies between 2 per cent and 3 per cent moisture content for all soils. In the case of low plasticity soil this amount of hysteresis is a large proportion of the moisture necessary to change a soil from its plastic limit to its liquid limit and therefore has a dominant effect on the ultimate equilibrium strength of low plasticity soils subjected to different amounts of wetting during road construction. An example of two drying curves which start from two different suctions for a soil of plasticity index of 10 per cent are shown dotted in Fig 3.

The suction/moisture content curves in Fig 3 are typical average values for soils which satisfy a line on the Casagrande Plasticity chart typical of cohesive soils with an equation

$$PI = 0.838 \text{ LL} - 14.2 \qquad (6)$$

when soils satisfy this equation, the moisture content of the soil is an "effective moisture content"[1] and suction can be read directly from Fig 3 from measured moisture contents. If soils do not satisfy equation 6 then an effective moisture content must be calculated, which differs from its actual moisture content. It can be calculated from the equation:-

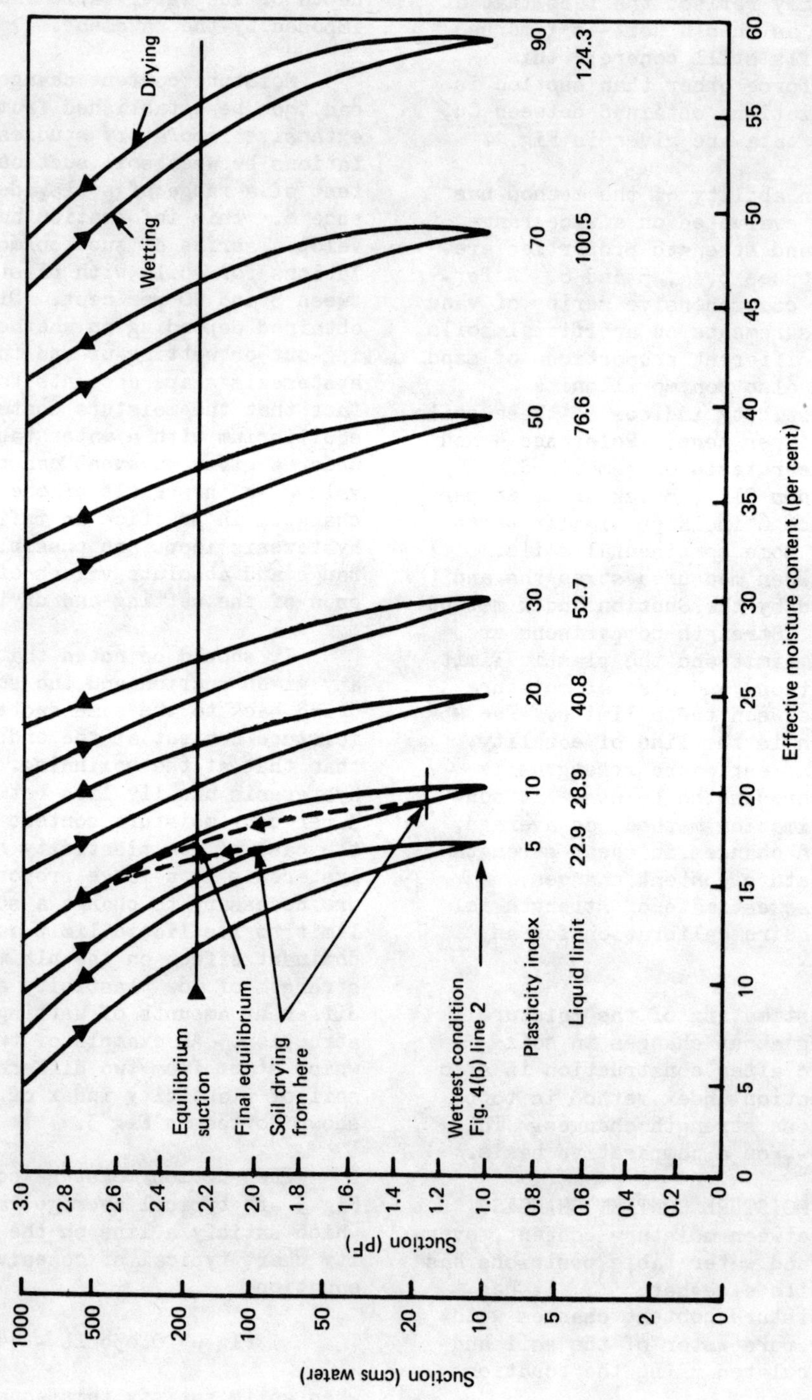

Fig. 3. Relation between suction and moisture content for various soils

Effective moisture content = actual moisture content + $\left(\frac{P1 + 14.2}{0.838}\right)$ - LL (7)

4. PREDICTION OF EQUILIBRIUM STRENGTHS OF ROAD SUBGRADES

4.1 The Suction Index method can be used in conjunction with the equation $\alpha P + S = U$ and the curves of Fig 3 to estimate strength of various soils which are subjected to different degrees of wetting or drying during and after construction. It is assumed in each analysis that the final equilibrium suction under the pavement in winter under all road pavements and in all soil types is 180 cm of water. The justification for UK conditions is discussed in Reference 1. This suction corresponds to the highest winter water table level normally permitted in good construction practice ie the water table should not be closer to the top of the subgrade than about 600 mm.

The first case considered is that of subgrades which are drier than their equilibrium suction at the earthwork stage and remain drier throughout construction of the pavement. Ultimately the soil wets up and comes into equilibrium with the water table. The predicted equilibrium strength of soils of various plasticity index is shown in Fig 4a. This curve represents the strength curve, combined with good weather conditions which can be obtained with the best construction practice corresponding to the highest position of water table in winter. This is however unlikely to be often achieved in practice because of the vagaries of our weather. Equilibrium strengths likely to be achieved as the result of construction conditions more typical of the United Kingdom are therefore calculated and compared with the ideal conditions represented by line 4a. Two cases are considered, (1) the effect of very wet weather during construction and (2), the condition on a well drained site.

4.2 The effects of very wet weather

In normal practice a road sub-base can not be laid unless the subgrade has adequate strength to carry the construction machinery without gross distortion of the soil surface. Hence the effects of wet weather on the soil need not be considered at this stage as any wetted soil will normally be removed and replaced with selected fill before construction begins. However, heavy rain after the sub-base has been laid must be considered.

It will be assumed in the analysis that the sub-base thickness has been laid according to Road Note 29 ie there will be a minimum sub-base thickness of 150 mm on low plasticity soils which we assume to be of high strength, increasing to a thickness of between 500 and 600 mm on the heaviest soils, the exact thickness of course depending on the expected amount of traffic on the completed road. It is also assumed that sufficient rain had fallen to create a water table in the sub-base 150 mm above formation level; this would require about 50 mm of rain. Even on an embankment this is a possibility because the water table has to be above formation level before there is sufficient head for it to run off over the sides of the embankment. The lightest soils are known to be sufficiently permeable to reach moisture equilibrium to a depth of between 300 to 600 mm in a matter of days; in heavy soils the depth is much less. Equilibrium strengths at the surface of the subgrade are given as line 2 in Fig 4(b) the CBR strength of about 2 per cent obtained for most soil drops rapidly below that value as the plasticity of the soil decreases below 25 per cent. From experience it is known that at a CBR value of about 2 per cent the minimum sub-base thickness of 150 mm would be insufficient to carry construction machinery and under these conditions a minimum thickness of sub-base of about 300 mm is required. This implies that any soil of plasticity index of 25 per cent or less would be liable to produce difficult construction problems in wet conditions and this is, of course, borne out by site experiences.

As a consequence of hysteresis in the suction/moisture content relation, Fig 3, it can be seen that once the soil has become very wet it will follow a drying curve which ensures that the soil will always remain wetter at equilibrium than on the same soil which has never allowed to become so wet. Curve 3 of Fig 4b shows that the final equilibrium CBR of the finished pavement may be as little as one half of the ideal shown as curve 1 for low plasticity soils.

Very low strength soil beneath the sub-base may lead to soil being forced into the lower 50-75 mm of granular material during construction of the road sub-base and base. This will be particularly liable to happen with the higher grades of sub-base which are low in fines and are therefore easily penetrated by soil. Intrusion of clay into a sub-base virtually reduces its effective depth by the depth of the intrusion.

4.3 Well drained site

If both the effects of adverse weather and of penetration of the subgrade into the sub-base are to be avoided then unimpeded drainage would be required at the earthworks stage to maintain the water table at an adequate depth beneath the prepared subgrade; this is particularly so with lighter soils. In fill it is quite easy for a perched table to form and drainage of an embankment is as important as drainage in cut. In addition the soil surface could be laid to a good fall and waterproofed. It is not unlikely that a waterproofed and drained subgrade might reduce the urgency to lay the sub-base as soon as possible, which means that the soil would wet up from below with no counteracting effect on the suction, and hence strength, from the surcharging effect of the sub-base. The resulting CBR of the soil could then be as shown in line 4 in Fig 4C. At equilibrium under the finished road the resultant equilibrium CBR is likely to be as shown in line 5 of Fig 4C.

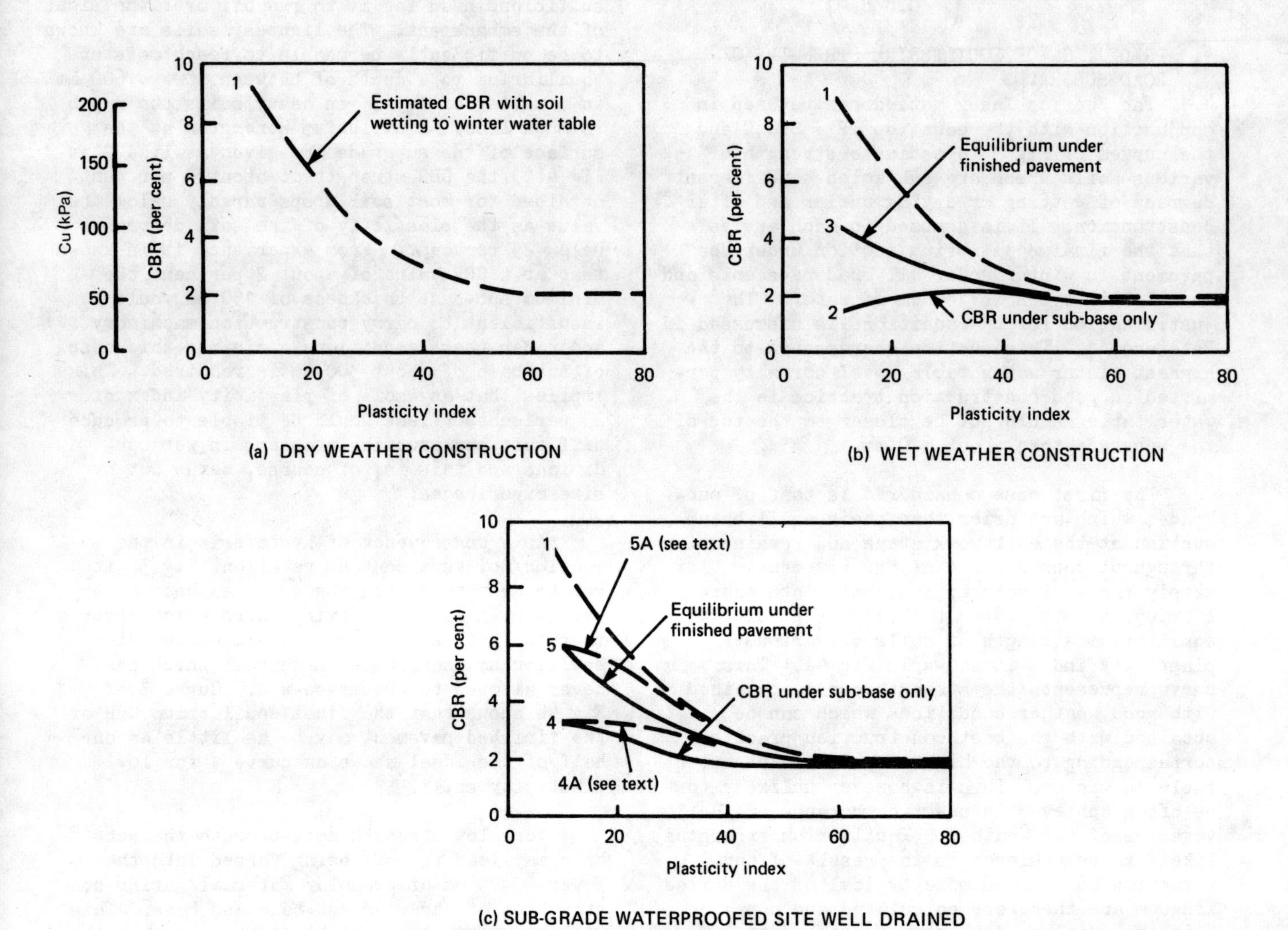

Fig. 4. Effect of weather and drainage on the subgrade strength during and after completion of pavement

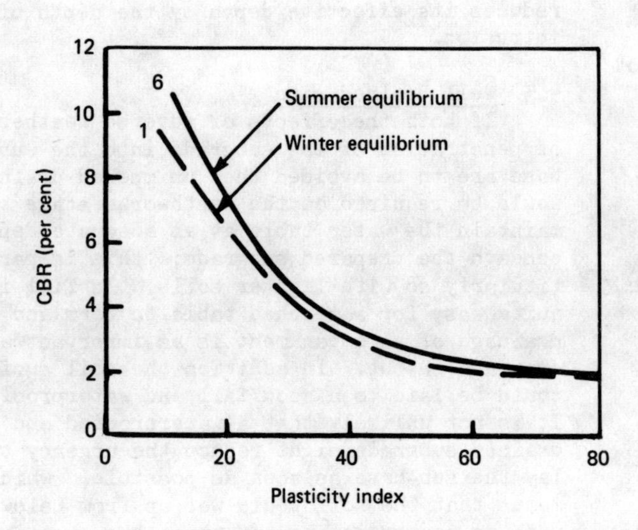

Fig. 5. Seasonal change in strength of soil beneath impermeable pavement

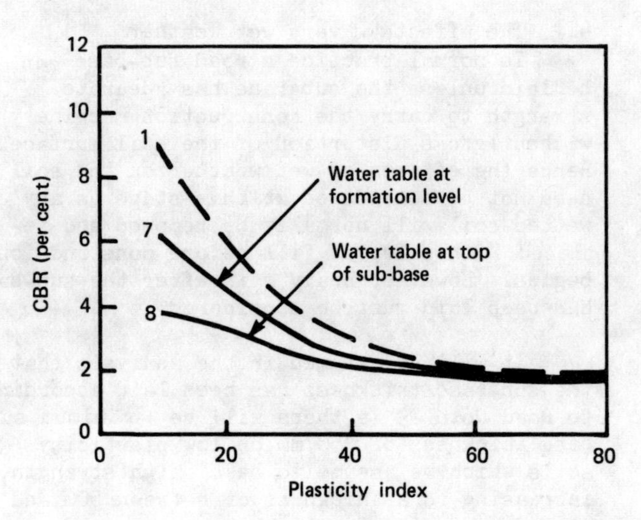

Fig. 6. Effect of water table above normal equilibrium level

Both of these curves are unlikely to occur in practice, the reason being the time necessary to attain equilibrium when the soil is fairly dry and dependent on drawing up water from the water table to achieve its equilibrium values. It has been assumed that only the lightest soil at a PI of 10 per cent will achieve these equilibrium values, because such soils are more permeable and will therefore more readily achieve equilibrium. Thereafter the curves shown dotted as Line 4A and line 5A have been sketched in as more probable relations.

Above a plasticity index of 50 per cent it can be seen that the various effects of wetting and drying, including those from the ideal and the very wet sites, have little effect on the final equilibrium strength because the moisture content difference between the wetting and drying curves is a small proportion of the amount of water necessary to change the soil from its liquid limit to its plastic limit.

If the weather during construction of a road is moderate, which is usually the case, the extreme conditions described in Fig 4b will only rarely occur. In our variable weather, alternating wet and dry weather would be unlikely to raise the water table much and some surface wetting would be quickly removed by natural drainage and by drying winds. It is suggested that under such conditions, the soil, not waterproofed, but surcharged with a sub-base related to the soil type would very likely attain equilibrium conditions during and after construction not dissimilar to the curves shown as 4A and 5A in Fig 4C.

An alternative to waterproofing the soil, or hoping for at least reasonable weather, would of course be to increase the depth of sub-base.

To achieve the same CBR values as a well drained waterproofed site would require a minimum depth of construction of 370 mm of sub-base instead of 150 mm. This alternative permits the relative economics of each solution to be investigated.

4.4 Seasonal strength changes

The equilibrium strength Line 1 in Fig 4 is appropriate to winter water table levels when the soil wets up to its equilibrium value. If the soil wets up to the equilibrium associated with the lower depth of water table which is possible in summer, the line 6 is appropriate. The latter line was based on the assumption that the summer water table under a pavement is up to 1 metre deeper than in winter and causes an increase in suction of the soil of one metre of water. This is typical of the change which can be expected in heavy clays[10] and in the absence of more definite information it has also been accepted for all clay soils. The difference between the winter and summer equilibrium strength only exceeds 10 per cent at plasticity indices of less than 20 per cent and this is the maximum change likely to occur. The change in strength due to seasonal rise and fall of the water will be less than this value because the slope of the solid wetting or drying curves of the suction/moisture content in Fig 3 used for the calculations will be considerably less than those associated with seasonal movements of water table (a drying curve is shown as a dashed line).

4.5 Effects of pavement leakage on soil strength

In the event of cracking of a pavement which results in water leaking into the subgrade it has been shown that[11] the water table in heavy clays can readily rise to formation level. In lighter soils the effect is not quite so great[10] but for the purpose of this analysis it was assumed that they would be equally affected. The reduction in strength can be seen in Fig 6 by comparing line 1, the ideal equilibrium strength line, and line 7 the possible equilibrium produced by leaks. In fact the effect of some leakage appears to create a soil strength similar to that of line 5A, Fig 4C which was considered to be typical of expected strengths at equilibrium under roads constructed in variable weather.

If drainage of the sub-base is impeded and leakage permits saturation of the sub-base the much lower soil strengths would be predicted as shown in line 8 of Fig 6. This approaches the expected equilibrium strength line experienced under pavement construction in adverse weather, line 3 Fig 4b. In lighter soils it can be seen that impeded drainage could reduce the strength of the soil by 60 per cent.

Sometimes rapid deterioration of roads in a cracked condition is attributed to this loss of strength of the subgrade but the subgrade is not always to blame. It is possible for the surface to be disrupted by high pore water pressures developing due to the action of traffic on some saturated layer of granular base or even of deteriorating cement bound bases without the lower layer of the road being involved.

4.6 Summary and conclusions on moisture conditions

Best estimates of the strength of soils at various plasticity indices during and after construction are summarised in Fig 7. The line numbers are consistent with those in Fig 4 to 6.

Lines 1 and 6. These are the likely strengths of soils wetting to winter and summer equilibrium respectively under a completed pavement when the subgrade has been prepared and the road built while the soil is drier than its final equilibrium. In our climate this is likely to be exceptional.

Line 5A. This is the estimated most likely relation between plasticity index and soil strength of roads constructed in moderate weather. The estimate is based on the highest water rate table being at least 600 mm below formation level. It also predicts the strength which the soil would more reliably attain if the site was well drained and the soil waterproofed.

Line 3. This is considered to be the

lowest equilibrium strength which is likely to occur under a pavement constructed in very adverse weather. It also estimates the likely strength in the soil if the sub-base becomes water logged through impeded drainage.

Line 2 and 4A. These lines represent strength in the subgrade soil during pavement construction. Line 2 represents the lowest likely strength which will occur if adverse weather follows the construction of the sub-base. For soils lighter than a plasticity index of 25 per cent no roadbase construction could take place without first permitting the site to dry out. Line 4A is the equilibrium strength of waterproofed soil with a controlled water table. It also probably represents the strength of the soil under a sub-base during variable weather when some rain alternates with drying conditions and could be considered typical for "normal" existing conditions.

5. COMPARISON OF ESTIMATED AND MEASURED CBR VALUES

The curves in Fig 7 which relate CBR and plasticity index can be compared with the equilibrium CBR values given in the official guide to the structural design of pavements for new roads, Road Note 29[9].

In Road Note 29 is given typical subgrade equilibrium CBR strengths for soils of different plasticities with the different water table conditions. The table was based on laboratory and field tests on soil at its natural moisture content[7], the moisture content which occurs in natural soil at a depth of 1 m or so, a depth at which there is normally little seasonal change of moisture content, or at the equilibrium condition measured under a pavement.

The Road Note values of CBR are reproduced as Table 1 below.

Table 1. Estimated laboratory CBR values for British soils compacted at their natural moisture content

Soil type	Plasticity index per cent	CBR per cent Depth of water table below formation level	
		Depth of water more than 600mm	600mm or less
1	2	3	4
Heavy clay	70	2	1
	60	2	1.8
	50	2.5	2
	40	3	2
Silty clay	30	5	3
Sandy clay	20	6	4
	10	7	5
Silt	-	2	1

The relations between CBR strength and plasticity index for the two water table conditions quoted in the Table are shown as dotted lines in Fig 7.

Line 5A, which has been labelled as the typical equilibrium which might on the average be expected in our climate, roughly follows, and is a lower bound to, the Road Note 29 strength for the deeper water table condition; Line 3, which is the lowest predicted equilibrium strength line for roads constructed in very adverse condition, approximates to the strength quoted in Road Note 29 for sites with shallow water tables, except for soils whose plasticity indices are greater than 50 per cent where Road Note 29 suggests very low strengths that are below that predicted.

It therefore appears that the equilibrium CBR values actually achieved in service are likely to be considerably lower than measured values used for design if adverse moisture conditions are encountered during construction.

The Suction Index method has not been properly calibrated below a plasticity index of 10 per cent and no comparison has therefore been made between estimates and measured CBR values in this range.

The very low strengths quoted for non-plastic silts in Road Note 29 reflect the fact that even in normal weather there can be sufficient rain during construction to so lower the strength of these materials that the road cannot be constructed to take advantage of the probable quite high equilibrium strengths. Sub-base thicknesses must therefore be designed for the construction phase to the very low strength which may occur during construction; allowance is made for this in Table 1 of Road Note 29.

6. EFFECT OF SUBGRADE STRENGTH ON THE STRUCTURAL PERFORMANCE OF FLEXIBLE PAVEMENTS

Any mismatch between the design strength of a subgrade and the equilibrium strength developed under the road will be reflected in the life of the road achieved under traffic. In Road Note 29 only the design thickness of sub-base is related to the subgrade strength and it is not possible to deduce the effect on pavement life of a mismatch between design and achieved strength.

A major in situ and Laboratory study is at present in progress to evaluate the equilibrium subgrade conditions under the major full scale road experiments built by the Laboratory. The aim is to characterize strengths in terms of conventional in situ CBR/shear strength as an essential input to revising design recommendations. Information is also being obtained relating to the fundamental behaviour of subgrades primarily by testing recovered samples in repeated triaxial loading designed to simulate as far as possible the stress conditions generated by moving traffic; the information is required as part of a wider programme aimed at developing a structural method of design based on a knowledge of the real behaviour of

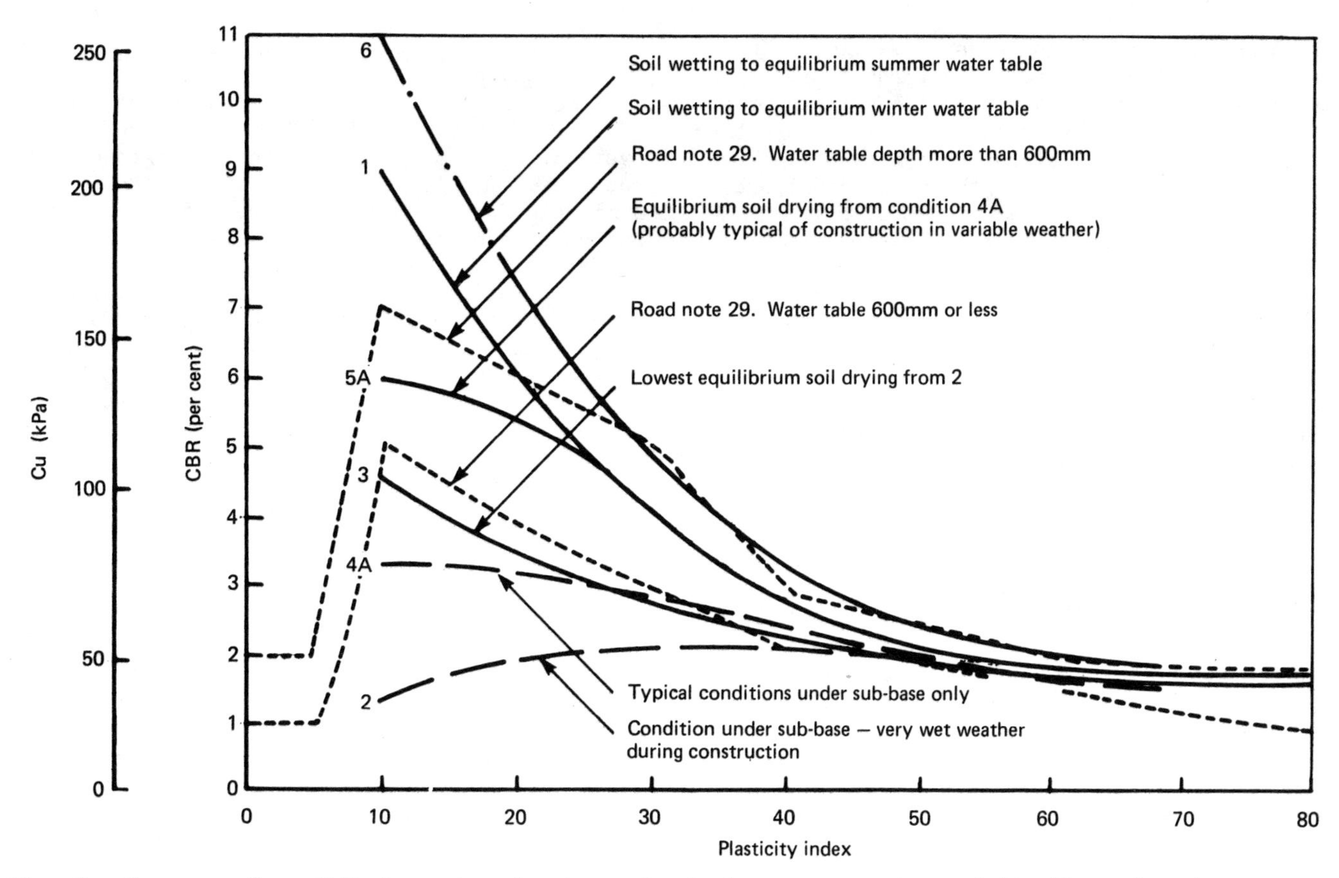

Fig. 7. Summary of equilibrium subgrade strengths during construction and finally under the finished load

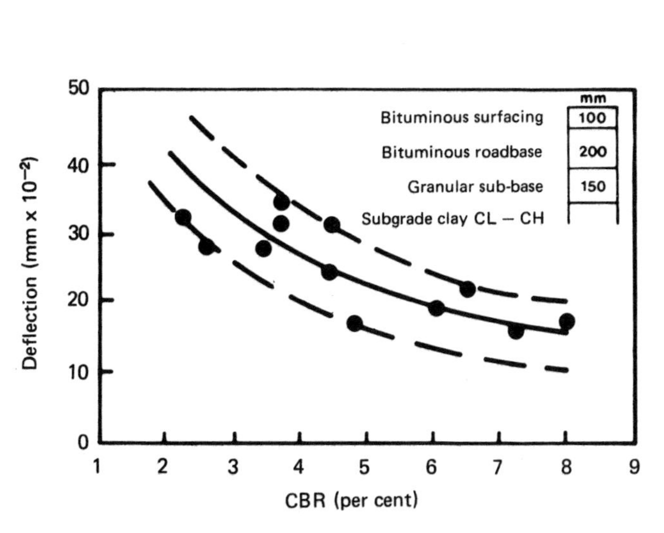

Fig. 8. Relationship between equilibrium deflection and equilibrium subgrade CBR for experimental roads with bituminous road bases

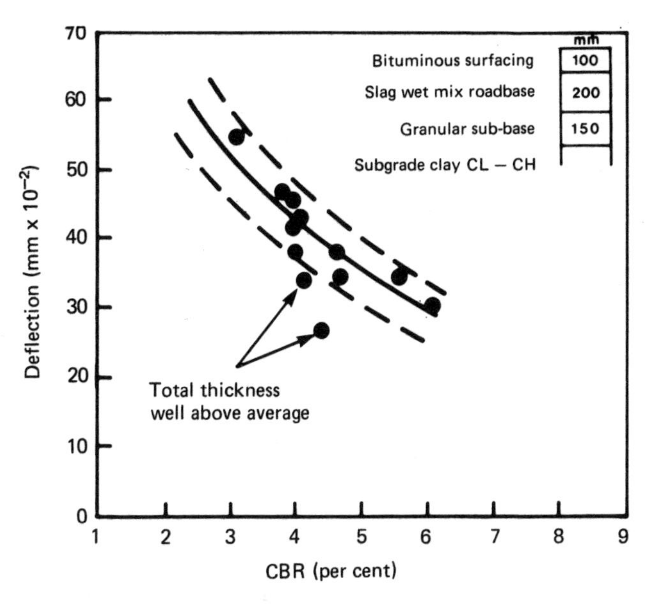

Fig. 9. Relationship between equilibrium deflection and equilibrium subgrade CBR for an experimental road with unbound road bases

road materials and subgrades in service. In relation to the subgrade the approach will take account of the fact that conditions in the road are very different from those obtained in either the CBR test or at failure in a conventional measurement of soil strength; the actual subgrade sees a range of magnitudes of soil stress repeatedly applied with the moisture conditions within it responding to the presence of a water table and to the applied stresses. Progress in this field should lead to better design but is beyond the scope of this paper.

The study in terms of in situ subgrade strength although in its early stages is yielding results which indicate the importance of subgrade strength in influencing the performance of flexible roads strength. The work is considered in the following sections of the paper.

6.1 Measurement of subgrade strength

Cores of 150 mm diameter have been taken from the wheelpath zones of a number of experimental sections constructed. Three or six cores in each section provided accurate measurements of construction thicknesses and also access to the subgrade whose strength was measured by means of a static penetrometer calibrated in terms of CBR values. Strengths were averaged over the 0.75 m of subgrade below formation level.

6.2 Assessment of associated pavement performance

The structural performance of a flexible pavement can be correlated with the deflection of its surface under a dual wheel load moving at creep speed[12]. From systematic measurements of deflection made with the Deflection Beam[13] on full-scale road experiments over a period of years deflection histories have been developed: these relate deflection to pavement deterioration under traffic expressed in terms of equivalent standard axles of 8,175 kg. Well defined relations between deflection and pavement life (in standard axles) up to the stage when the pavement has deteriorated to a critical condition requiring overlaying, ie when it retains much of its structural integrity and therefore does not require reconstruction, have been obtained; they are of the form

$$\text{Life} = \frac{A}{(\text{deflection})^n} \qquad (8)$$

where A is a function of the type of roadbase and n has a value close to 3, irrespective of the type of road base[13][14][15]: other pavement parameters such as the thickness of pavement layers and strength of the subgrade over a wide range have no significant effect on the deflection/performance relationship[13]. The deflection concerned is the value established in the early life of the pavement when moisture equilibrium in the subgrade has been established and after any traffic compaction and cementing action in the granular layers of the road has taken place.

6.3 Correlation between deflection and CBR strength

Equilibrium deflections measured on a number of experimental sections of similar type vary systematically according to the strength of the subgrade. The limited number of results so far available for pavements constructed with road bases of dense bituminous materials are shown in Fig 8. Not all the pavements have the same thickness of roadbase; the deflection values have, therefore, been adjusted to a standard roadbase thickness of 200 mm using the overlay correction curves in Reference 15. Fig 9 gives similar results for experimental pavements constructed with slag wet-mix roadbases and having the same nominal thickness.

A further check on the validity of the relations in Figs 8 and 9 has been obtained from sections of road experiments where the CBR strength of the subgrade at the time of construction and the associated deflection values were substantially different from the equilibrium values of these parameters. The actual change in deflection between construction and equilibrium will reflect the compaction and possibly cementing action in granular road layers as well as changes in subgrade strength and great accuracy cannot therefore be expected in using the relations in Figs 8 and 9 to predict the deflection at construction from knowledge of the changes in subgrade strength. However the results obtained on five experimental sections, given in Table 2 and plotted as the solid points in Fig 10, show good agreement.

The corresponding data for equilibrium conditions used to develop the curve in Fig 8 are also shown.

6.4 Implication for pavement life

Given the sensitivity of pavement life to the level of equilibrium pavement deflection indicated by equation (8), the relations in Figs 8, 9 point to significant changes in pavement life as the result of changes in CBR strength. The greatest likely change in CBR strength resulting from varying weather conditions during construction has been shown to be in soils of low to medium plasticity.

The example given in Table 3 indicates predicted equilibrium deflections and critical lives of a road having a road base of dense bituminous macadam of 150 mm thickness surfaced with 100 mm of rolled asphalt. It is constructed on a 180 mm thickness of granular subbase on a soil with a plastic index of 10 per cent. The estimated critical life for a road built in very wet weather is only about half of that of a road built under average conditions; in turn a road built in favourable conditions is capable of achieving a life which is double that of the 'average' road.

The thickness design curves in Road Note 29 relates to lives associated with total failure of the road base and surfacing; these would be expected to be about 30 per cent greater than the critical lives, which are related to the onset of deterioration in these pavement layers.

Table 2. Comparison of initial and equilibrium deflections

Section No.	Thickness of bituminous roadbase and surfacing	CBR		Deflections			
				Estimated		Measured	
		Initial	Final	Initial	Final	Initial	Final
1	180	9	2.6	28	72	35	57
2	250	3.8	2.2	36	52	36	43
3	350	12	8	11	13	13	13
4	300	12	7.2	14	17	16	16
5	250	8	6	20	25	22	25

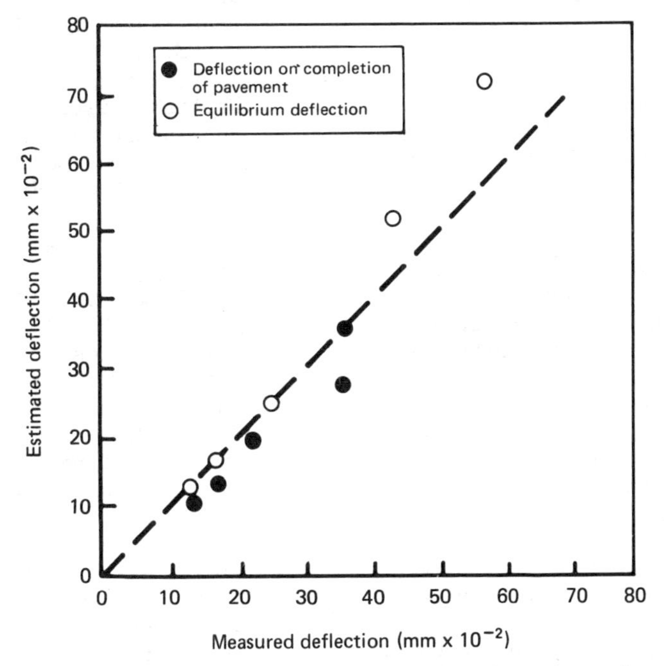

Fig. 10. Comparison of measured and estimated deflections

Table 3. Predicted deflections and critical lives for different construction conditions

Conditions during construction	Equilibrium deflection mm x 10^{-2}	Critical life standard axles x 10^6
Soil drier than equilibrium during road construction (Curve 1)	19	42
Average construction weather (Curve 5a)	25	22
Very wet weather during construction (Curve 4a)	31	13

The predicted critical lives are considerably greater than that indicated by Road Note 29; the latter, however, include a factor of safety whereas the critical lives quoted were derived from curves which estimate life in terms of a 0.50 probability of achievement.

7. SUMMARY AND CONCLUSIONS

1. Evidence is presented that the Suction Index Method, which relates the moisture content of a soil to its undrained strength, can be used in conjunction with relations between soil-suction and moisture content to analyse strength changes in subgrades under roads.

2. It has been used to quantify the effect of hysteresis observed in the suction-moisture content relations on soil strength resulting from various moisture conditions which may occur during construction and subsequently under the completed pavement.

3. For average weather conditions during construction, typical equilibrium CBR strengths are similar to the values recommended in Road Note 29 for deeper water table conditions.

4. Very wet weather during construction on a poorly drained site reduces markedly the ultimate equilibrium strength achieved in soils of low plasticity. Equilibrium strengths are less than half values associated with dry weather construction and design should always be for the most adverse drainage conditions indicated in Road Note 29. To guarantee that problems during construction in very wet weather are avoided the thickness of sub-base used on soils of low plasticity should be increased further; for soil of plastic index of 10 per cent thicknesses of about 370 mm are required.

Alternatively effective drainage and waterproofing can maintain strengths associated with, at least, average construction conditions.

5. Seasonal variations in the level of water table under the completed road change the strength of the soil by less than 10 per cent.

6. Very wet weather during construction can result in pavement lives which may be only about half those expected for roads constructed during average weather conditions.

8. ACKNOWLEDGEMENTS

The work described in this paper forms part of the programme of the Transport and Road Research Laboratory and the paper is published by permission of the Director.

9. REFERENCES

1. BLACK W. A method of estimating the California Bearing Ratios of cohesive soils from plasticity data. Geotechnique, Dec 1962. Institution of Civil Engineers, London 1962.
2. SKEMPTON A.W. The bearing capacity of clays. Building Research Congress, London 1951.
3. DUMBLETON M.J. and WEST G. The suction and strength of remoulded soils as affected by composition. Ministry of Transport RRL Report LR 306 Crowthorne 1970. (Road Research Laboratory).
4. LEWIS W.A. and ROSS N.F. An investigation of the relationship between the shear strength of remoulded cohesive soil and soil moisture suction. Department of Scientific and Industrial Research. Road Research Note No. RN 2389/WAL/NFR 1955 (unpublished).
5. SKEMPTON A.W. and NORTHEY R.D. The sensitivity of clays. Geotechnique 3 30-53 1952. Institution of Civil Engineers, London.
6. SCHOFIELD A. and WROTH P. Critical state soil mechanics McGraw Hill, London 1968.
7. CRONEY D. The design and performance of road pavements HMSO, London 1977.
8. CRONEY D., COLEMAN J.D. and RUSSAM K. The suction and swelling properties of some British soils. Department of Scientific and industrial research. Road Research Note RN 1964/DC JDC.Kl. 1953 (unpublished).
9. DEPARTMENT OF THE ENVIRONMENT ROAD RESEARCH LABORATORY. A guide to the structural design of pavements for new roads. Road Note 29 Third Edition London 1970. (HM STATIONERY OFFICE).
10. BLACK W.P.M., CRONEY D. AND JACOBS J.C. Field studies of the movement of soil moisture DSIR Rd. Res. Tech. Pap. No. 41 HMSO London 1958.
11. FARRAR D.M. The effectiveness of subsoil drainage of paved areas on two heavy clay sites. Ministry of Transport RRL Report No. LR 186, Crowthorne 1968. (Road Research Laboratory).
12. LISTER N.W. Deflection criteria for flexible pavements. Department of the Evironment. TRRL Report LR 375. Crowthorne 1972. (Road Research Laboratory).
13. KENNEDY C.K., FEVRE P. and CLARKE C.S. Pavement deflection: equipment for measurement in the United Kingdom. Department of the Environment Department of Transport. TRRL Report LR 834 Crowthorne, 1978. (Transport and Road Research Laboratory).
14. LISTER N.W. and KENNEDY C.K. A system for the prediction of pavement life and design of pavement strengthening. Proc. 4th Int. Conf. Struct. Des. Asphalt Pav. Vol 1, 1977 (University of Michigan Ann Arbor) pp 629-648.
15. KENNEDY C.K. and LISTER N.W. Prediction of pavement performance and the design of overlays. Department of the Environment, Department of Transport TRRL Report LR 833 Crowthorne 1978.(Transport and Road Research Laboratory).

P. BOMAN, Civil Engineer, and B. B. BROMS, Professor,
Department of Soil and Rock Mechanics, Royal Institute
of Technology, Stockholm

Behaviour of a road embankment constructed of soft clay and provided with drain strips

SYNOPSIS

The disposal of soft clay is both difficult and expensive in Sweden due to environmental considerations. If soft clay can be used in embankments instead of sand and gravel close to the place where it is excavated, then the costs can be reduced by 50 to 100 Sw.Cr/m^3;5 to 10 f/m^3. To examine this possibility a 42 m long and 12 m wide road embankment was constructed of soft clay about 10 km north of Stockholm. The thickness of the clay fill was 1.8 m. The embankment was opened for traffic in May 1973, eleven mont after the placement of the clay. The road embankment have behaved satisfactory up to the present (1978).

The clay was so soft when it was placed in the embankment that it behaved almost as a liquid. The clay had initially a shear strength of about 10 kPa but in the embankment the shear strength was less than 3 kPa due to the remoulding. Plastic-paper drain strips (Geodrains) were placed in the fill two months after the construction of the fill. The 10 cm wide drains were placed either horizontally or vertically in two test sections. The third test section was used as reference. The clay embankment was loaded by a 1.0 m high gravel fill which was placed on the clay. After nine months 0.4 m of the fill was removed and the area was paved.

The settlements of the 1.8 m thick clay fill have been measured as well as compression of different laye Also the distribution of the pore water pressure and of the lateral earth pressure in the different sections during the consolidation was determined as well as the increase of the shear strength and of the bearing capacity of the clay with time.

INTRODUCTION

The soft clays in the central and western parts of Sweden have as a rule a very low shear strength, a high compressibility and a high sensitivity ratio. These clays are normally or slightly overconsolidated. The water content is close to or exceeds the liquid limit. It is usually not less than 50 percent and can sometimes be 100 percent or more. The sensitivity ratio is often 10 to 15 but can be as high as 100 to 200 or more. Such sensitive clays behave essentially as a fluid when they are excavated and remoulded. They cannot be compacted due to the high water content and low shear strength. It can take several months or even years before it is possible to walk on a fill with soft clay. It is thus both difficult and expensive to dispose of soft clay or to use soft clay as construction material.

The need of sand and gravel for fills is increasing in most urban areas in Sweden. Existing sand and gravel pits are rapidly being depleted and there are severe restrictions on the operation of new pits due to environmental considerations.

The costs for the disposal of soft clay can be considerable reduced if the clay can be used e.g. in road embankments instead of sand and gravel. An experimental 42 m long road embankment was therefore constructed of soft clay close to Stockholm. Drains were placed either vertically or horizontally to increase the consolidation rate of the clay fill. The results are described in this paper.

TEST EMBANKMENT

The 42 m long and 12 m wide test embankment of very soft clay was constructed about 10 km north of Stockholm at Akalla. The embankment was constructed by tipping the excavated soft clay on a 0.2 to 0.5 m thick layer of till. The clay was contained on one side by a 2 m high timber wall and by gravel dikes as shown in Fig. 1. Earth pressure cells were attached to the wall so that the lateral earth pressures could be measured during the placement of the clay and during the consolidation.

The clay was excavated about 200 m to the north of the test embankment. The excavation can be seen in the background of Fig. 1. A soil profile at the excavation is shown in Fig. 2. The soil consists from the ground surface of grey-brown organic clay down to a depth at 1 m and of a brown-grey varved clay down to a depth of 5 m. The ground water level was located 0.2 m below the ground surface. The undrained shear strength as determined by field vane tests, fall cone tests and unconfined compression tests increased with increasing depth. The average shear strength was slightly less than 10 kPa. The sensitivity ratio was 7 to 25. The consolidation characteristics as determined from oedometer test on 50 mm diameter samples is shown in

Fig. 1. Test embankment before placement of clay

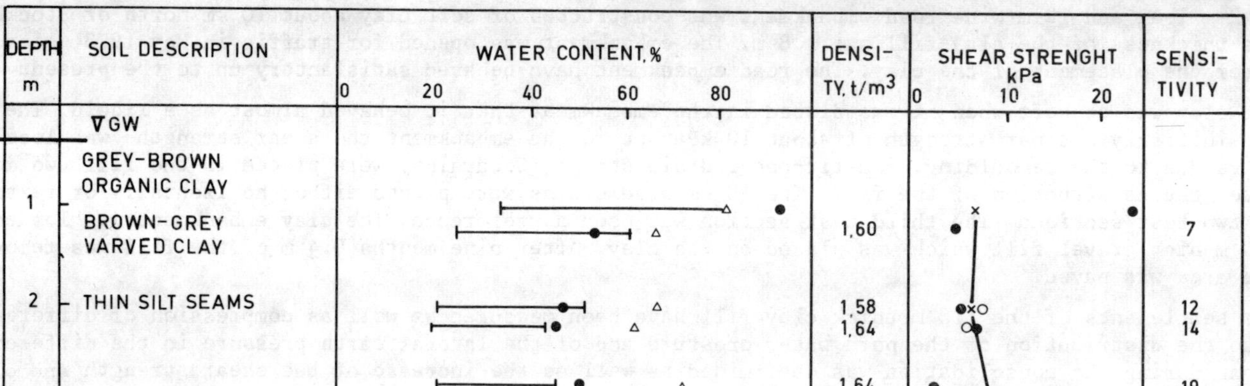

DEPTH m | SOIL DESCRIPTION | WATER CONTENT, % | DENSITY, t/m³ | SHEAR STRENGHT kPa | SENSITIVITY

∇GW

GREY-BROWN ORGANIC CLAY

BROWN-GREY VARVED CLAY

THIN SILT SEAMS

SILT SEAM

ROCK OR TILL

Density, t/m³	Sensitivity
1,60	7
1,58	12
1,64	14
1,64	19
1,69	12
1,68	25
1,71	10
1,68	18

PLASTICITY LIMIT W_p — WATER CONTENT W_n — W_L LIQUID LIMIT — W_L (CONE) LIQUID LIMIT

× VANE TEST
● CONE TEST
○ UNCONFINED COMPRESSION TEST

Fig. 2. Soil conditions

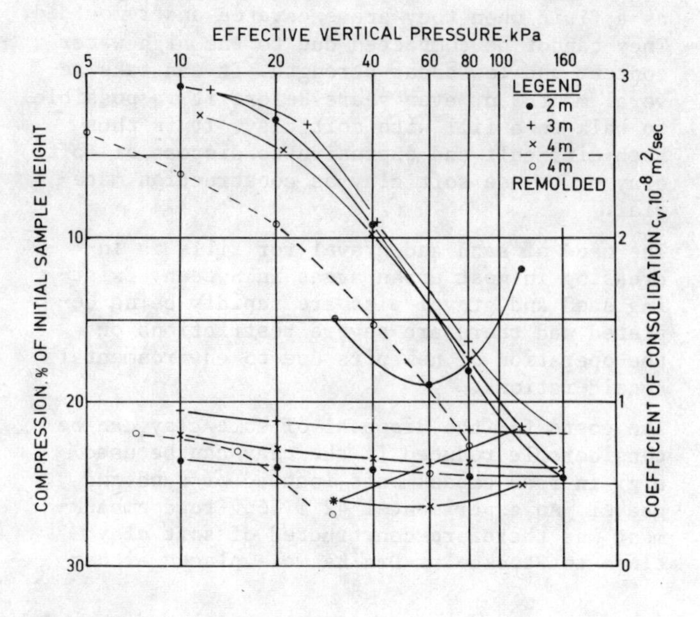

Fig. 3. Consolidation characteristics of undisturbed and remoulded clay

Fig. 4. Excavation of clay. It was not possible to construct a road on the original clay. The piles in the background were driven for a future road embankment

Fig. 5. Placement of clay

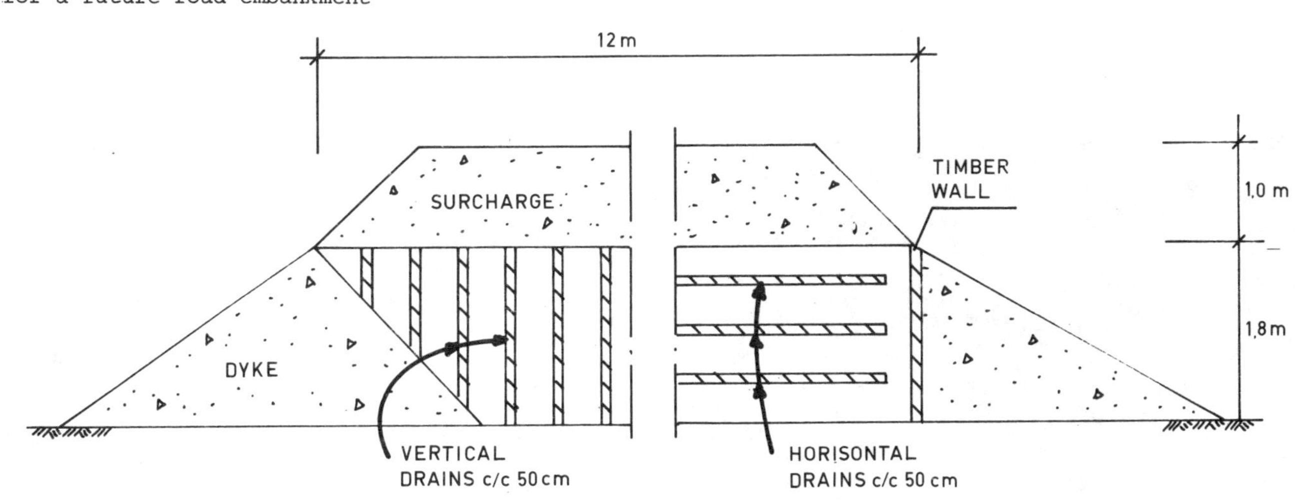

Fig. 6. Cross section of test embankment

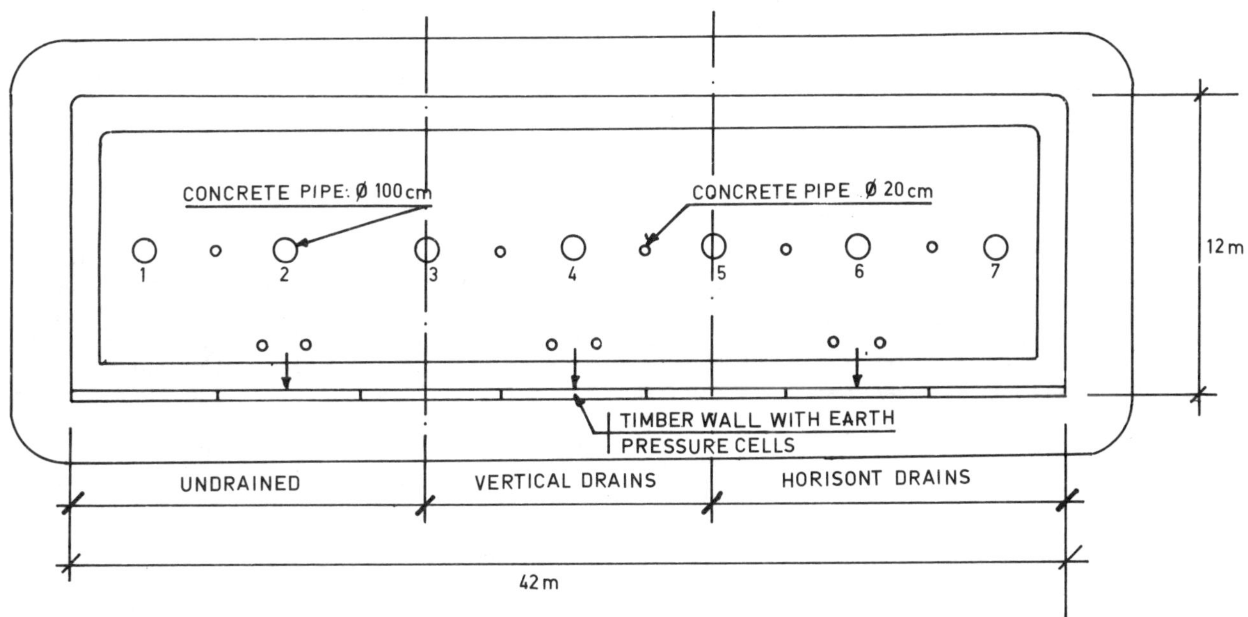

Fig. 7. Plan of test embankment

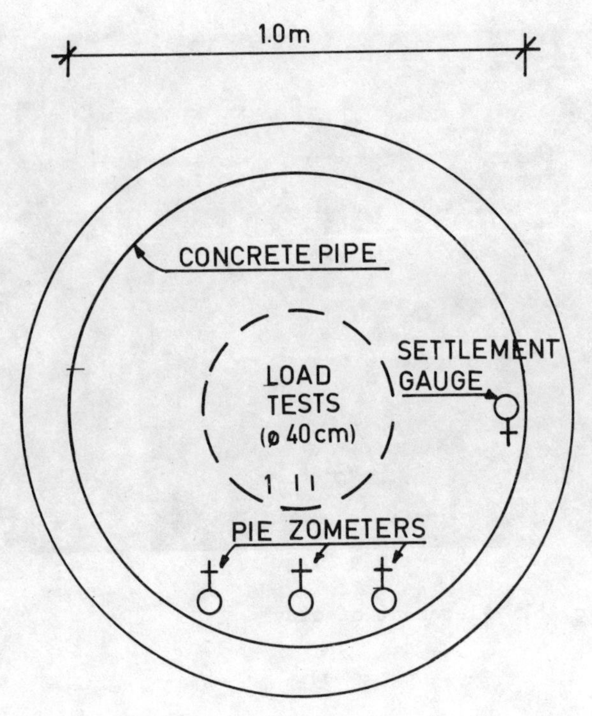

Fig. 8. Location of settlement gages and piezometers in the concrete pipes

Fig. 9. Location of earth pressure cells

Table 1. Geotechnical conditions of clay fill

	Horizontally drained section	Vertically drained section	Undrained section
Water content,%	60	65	70
Shear strength, kPa	4,3	1,9	2,4
Compression of doubled load %	4,3	4,9	5,5
Coefficient of consolidation $.10^{-8}$ m^2/s	0,4	0,5	0,7

The clay was excavated below water using a clam shell bucket as shown in Fig. 4, and loaded on lorries. The clay was dumped by end tipping between the dikes and the timber wall. The remoulded clay flowed like a liquid. The slope of the surface of the clay fill was 1:20 (Fig. 5). The final height of the fill was 1.8 m. The test embankment was divided into three 12 to 15 m long test sections. Plastic-paper drains (Geodrains,Ref.1)were installed in August 1972,two months after the placing of the clay fill,in two sections either horizontally or vertically. The spacing of the drains was 0.5 m as shown in Fig. 6. The horizontal drains were installed with a specially constructed plow which was dragged across the fill. The vertical drains were pushed down by hand into the clay using a short steel rod. The third test section of the embankment was undrained and served as a reference area.
The clay embankment was loaded by a 1 m thick gravel fill which was carefully placed in layers on the clay. After nine months 0.4 m of the gravel fill was removed and the embankment was paved.

INSTRUMENTATION AND MEASUREMENTS

Seven 1.0 meter long concrete culvert elements (concrete rings), segments (1 through 7) with 1.0 m diameter were placed in the gravel fill so that the bearing capacity of the clay could be determined as a function of time during the consolidation. The culvert elements extended down to the clay fill. The location is shown in Fig.7. The elevation of the elements was measured. Piezometers (vibrating wire type NGI) and settlement gages (Ref.2) were installed as illustrated in Fig. 8 in three element (2, 4 and 6). The compression of the clay fill was measured every 30 cm after the embankment had been completed and the gravel fill had been placed.

Plate load tests were carried out on the clay in three of the culvert elements. The diameter of the plates was 40 cm. The plates were placed on a 20 cm thick sand layer at the bottom of the culvert elements. The concrete rings were filled with sand between each load test. The weight of the sand corresponded to half of the weight of the gravel that was replaced by the concrete elements. The concrete elements were also used for sampling.

TEST RESULTS

Results from oedometer tests on 50 mm diameter samples from the clay embankment which were taken just after the gravel fill had been placed are shown in Fig. 10. The results are similar as those for the remoulded clay shown in Fig. 3. The relative compression of the clay was about 5% when the effective stress was doubled. The geotechnical data are presented in Tab. 1.
The lateral earth pressure against the timber wall (Fig. 7) were measured at each test section with six earth pressure cells type Glötzl mounted on the wall (Fig. 9).
A slide occurred during the excavation of the soft clay, when the clay embankment was half completed (See Fig. 5). Gravel and small stones were then mixed with the clay which affected the

field vane tests and the scatter of the results was large. Soil samples were taken of the clay fill at the bottom of culverts 2, 4 and 6.

The settlements of the test embankment and the compression as a function of depth are shown in Figs 11 and 12. It can be seen that the settlements of the two drained sections were larger than for the undrained section as expected and that the settlements of the section with vertical drains were larger than for the section with horizontal drains. This difference can partly be explaiend by the difference in drainage conditions and partly by the difference in the compressibility between the two sections. The measured total settlement of 13 cm for the two drained sections corresponds to a relative compression of 7.2%.

The compression of the clay fill with depth (Fig. 12) indicates that the fill has partly been dried out and drained at the surface and the bottom during the two month period after the placement of the clay fill and before the placement of the 1 m thick gravel layer.

The settlements of the different layers increased rapidly with time. After eight months the settlements at the center of the clay fill was somewhat larger for the reference section (about 13%) than for the two drained sections (about 10%).

The consolidation of the clay fill with time at 2, 4 and 6 is shown in Fig. 14. The consolidation of the two drained sections has been faster than

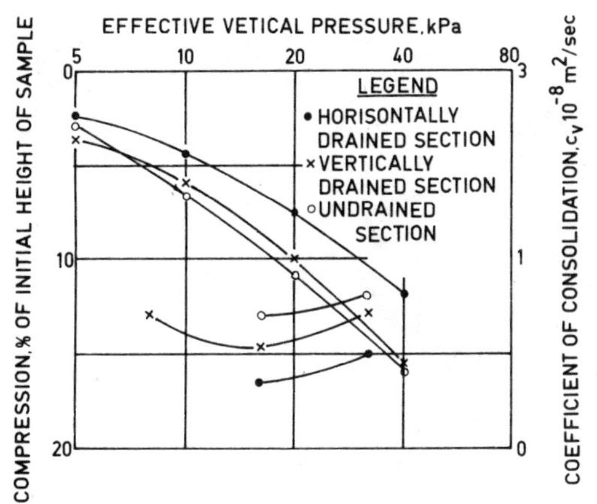

Fig. 10. Consolidation characteristics of the clay in the fill

for the undrained section as expected. The consolidation of the section with vertical drains was completed after about 100 days while about 230 days were required for the section with horizontal drains. The consolidation of the undrained area was not completed even after 8 months. These results indicate that the vertically placed drains have been more effective than the horizontal drains, possibly due to the difference in drainage conditions.

The pore pressure distribution in the three test sections is indicated in Fig. 14.

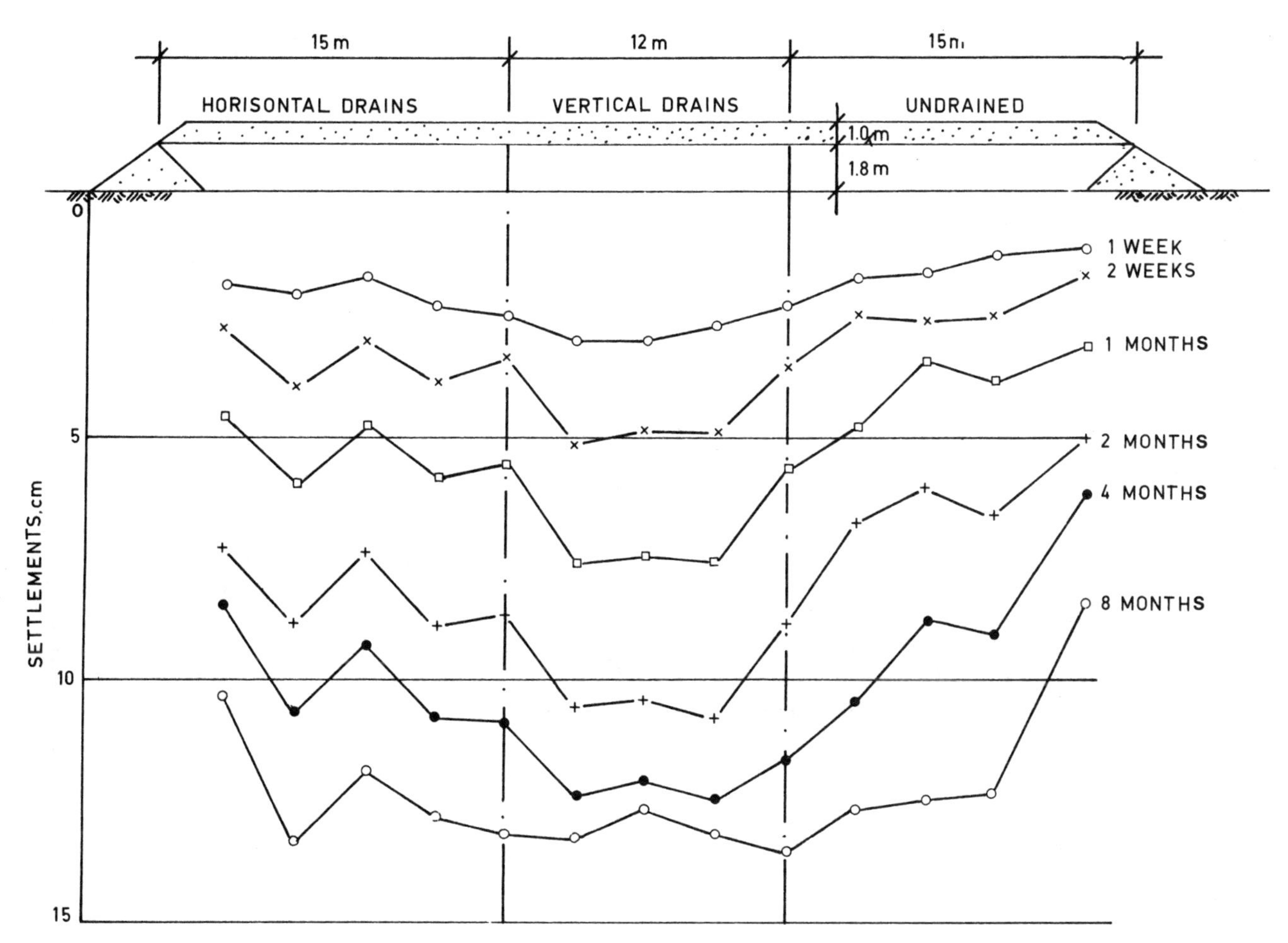

Fig. 11. Settlements of the clay embankment

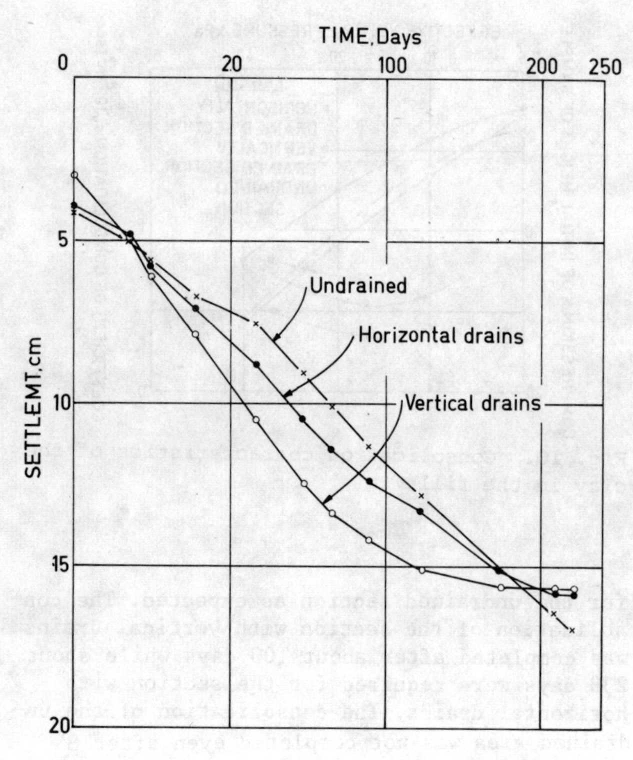

Fig. 12. Variation of settlements with depth

The measured pore pressures corresponded initially to the total overburden pressure as expected. The dissipation of the excess pore water pressures occurred more rapidly for the section with the vertical drains compared with the section with horizontal drains. After eight months the pore water pressure distribution corresponded to the hydrostatic pressure with respect to the surface of the fill. For the undrained area some excess pressures still remained after eight months.

The lateral earth pressures on the timber wall and the modulus of elasticity determined from the plate load tests are shown in Figs 15 and 16, respectively. The total lateral earth pressure decreased with time. It was the smallest for the section with the vertical drains. The measured lateral earth pressures have probably been affected by the friction along the timber wall which has reduced also the vertical overburden pressure in the clay fill close to the wall.

The modulus of elasticity of the clay fill has been calculated from the plate load tests using the following relationship (Ref. 3)

$$\delta = \frac{3}{4}\frac{qD}{E}$$

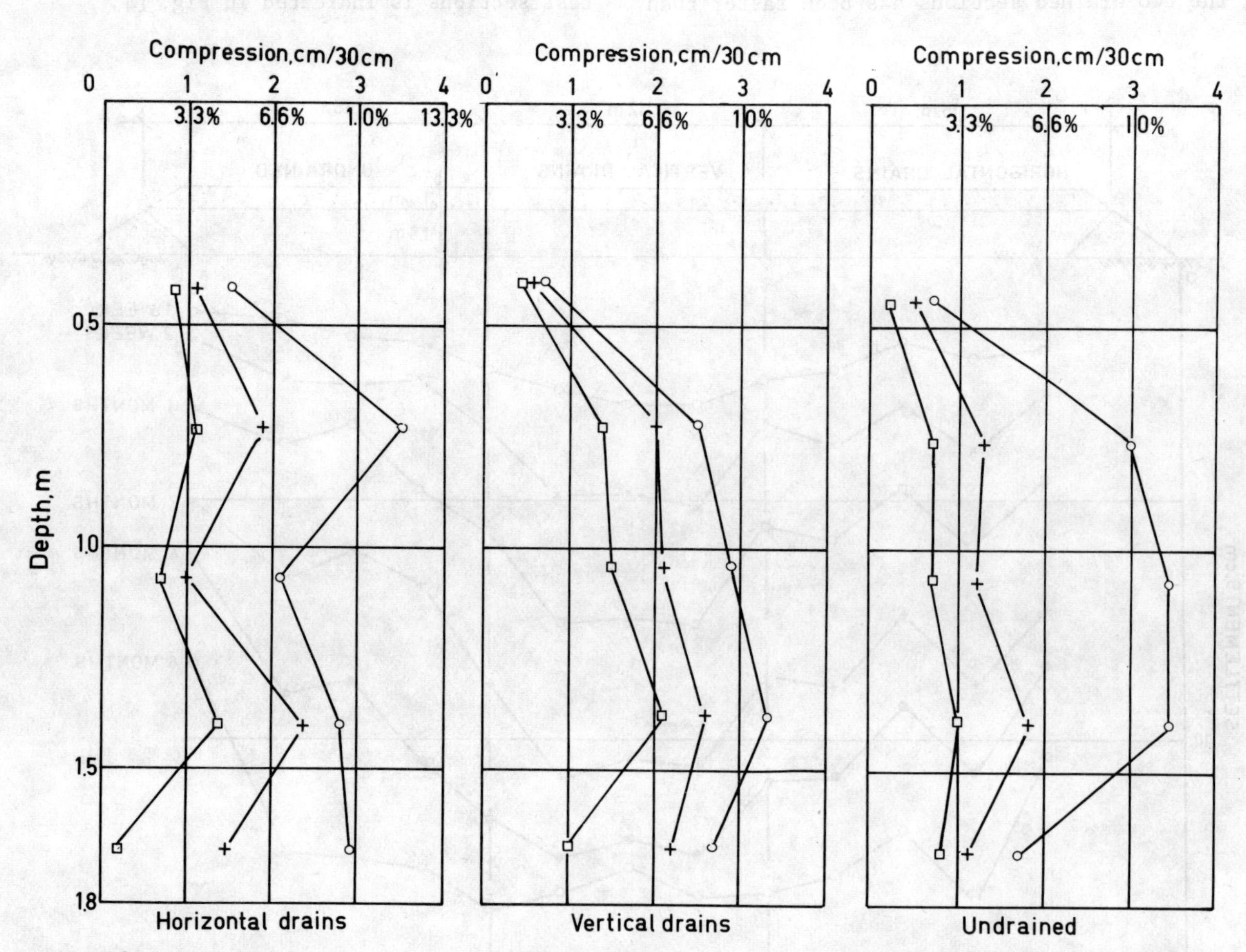

Fig. 13. Consolidation with time at the center of the three sections

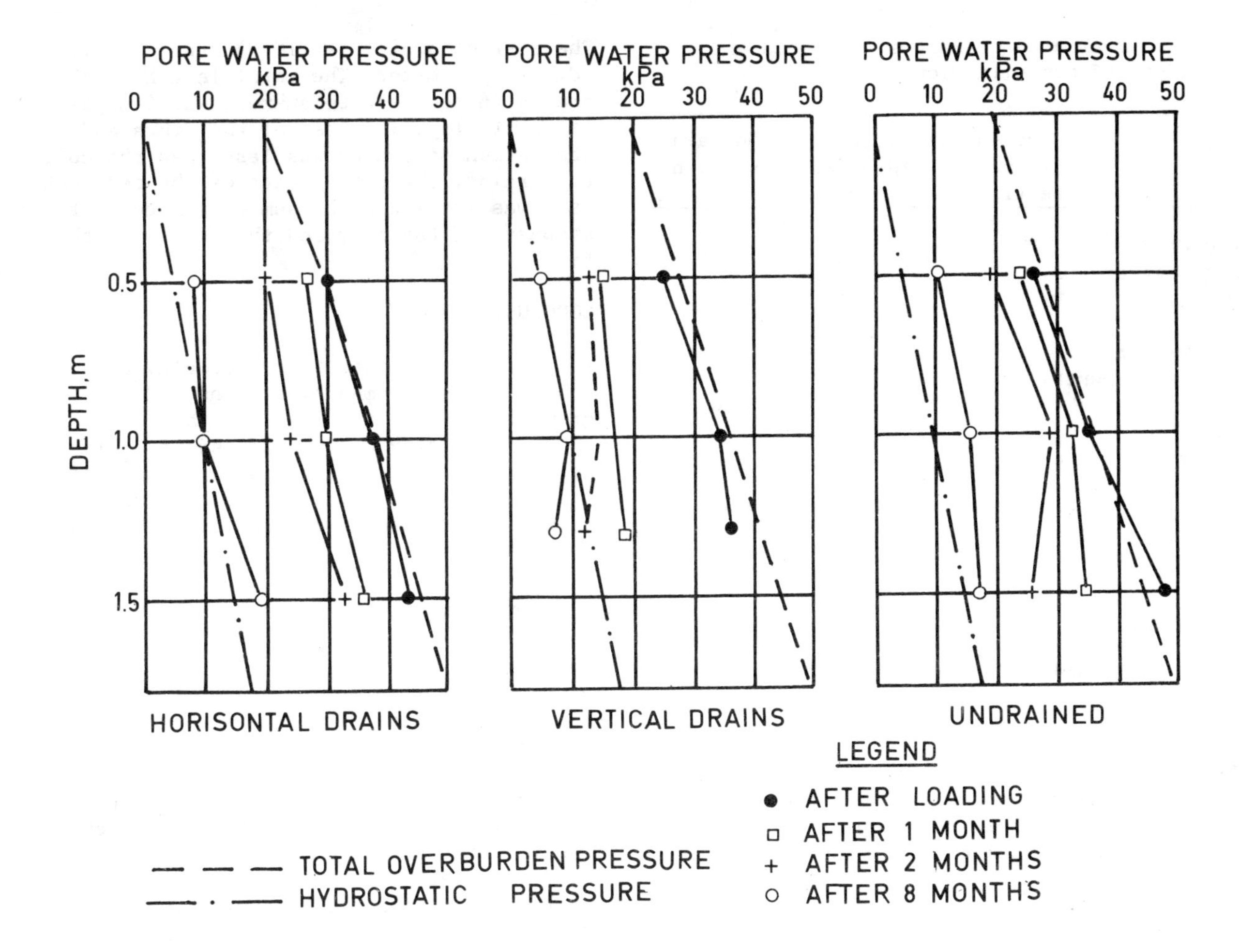

Fig. 14. Pore pressure distribution

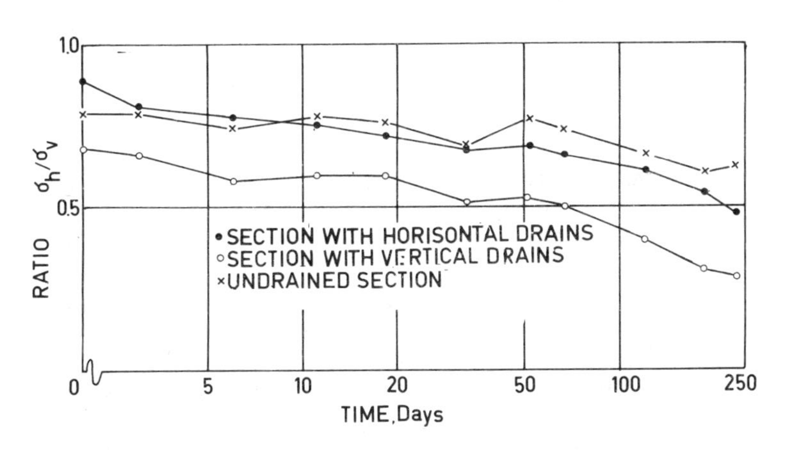

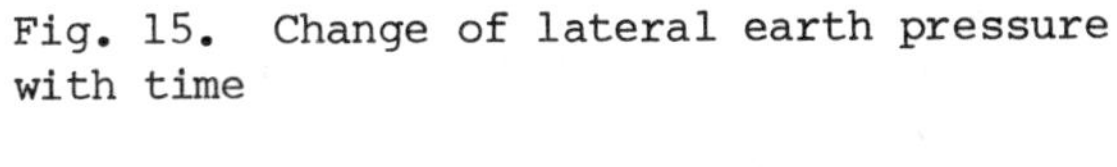
Fig. 15. Change of lateral earth pressure with time

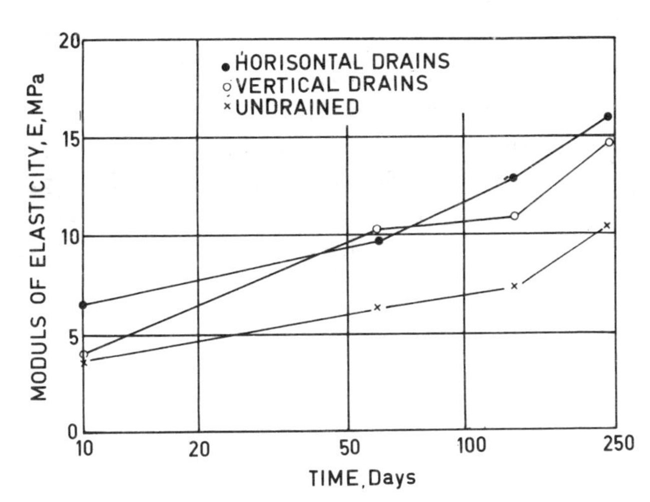

Fig. 16. Results from plate load tests

Table 2. Evaluated primary settlements and coefficients of consolidation

	Horizontally drained section	Vertically drained section	Undrained section
Predicted primary settlements, cm	13	15	18
Observed primary settlements, cm	13	13	16
Observed elastic settlement, cm	4	3	4
Coefficient of consolidation $\times 10^{-8}$ m^2/s	1.1	1.5	0.8

Bottom layer assumed to be impervious.

where δ is the settlement of the plate, q is the applied unit load, D is the diameter of the plate (0.4 m) and E is the modulus of elasticity. It has been assumed that Poisson's ratio is 0.5.

DISCUSSION OF TEST RESULTS

The properties of the remoulded clay in the three test sections of the embankment were not the same. The compressibility of the clay in the section with horizontal drains was slightly less than for the other two sections possibly due to the gravel and the stones which were mixed with the clay during the excavation (see above). The clay in the undrained section and in the section with vertical drains was about the same.

The effectiveness of the drains is shown in Tab.2, where the compression datas have been fitted to the consolidation equations derived by Terzaghi (Ref.4) and by Barron (Ref.5) for one dimensional consolidation and radial drainage, respectively.

The modulus of elasticity for the drained section of the embankment increased about 300% in eight months and was about 50% higher than that of the undrained section.

About 40 cm of the gravel was removed eight months after the gravel fill had been placed on the clay embankment. The remaining gravel fill was compacted and paved. The road has been open for traffic since June 1973 and has behaved satisfactory.

The cost of drains is (1978) about 5 Sw.Cr. (0.5 f) per meter. The total length of the drains in the test embankment was 4 m per cubic meter of clay. The cost of the dikes and installation of drains was less than the costs of the drains. The total costs of the test embankment was less than the sum of the costs for the disposal of the clay and the costs for the gravel fill.

CONCLUSIONS

The results indicate that soft clay can be used economically in embankments instead of sand and gravel. The time required for the reconsolidation of the clay can be reduced if drains are used. Vertical drains seems to be more effective than horizontal drains. The consolidation rate is underestimated if the results from oedometer tests are used in the analysis.

ACKNOWLEDGEMENT

The authors acknowledge with many thanks the financial support provided by the National Swedish Board of Building Research. The research was performed by the Swedish Geotechnical Institute, the test embankment was constructed by the Stockholm Public Works Office and the drains were provided by the company Terrafigo AB, Stockholm. The load tests were made by the National Swedish Road and Traffic Research Institute with financial support from the National Swedish Road Administration.

REFERENCES

HANSBO, S. and TORSTENSSON, B.A. Geodrain and other vertical drain behaviour. Proc. of the IX. Int. Conf. on SMFE, Tokyo 1977. Vol. 1.2, 533-540.

WAGER, O. Bellow settlement gage for measurements of vertical movements in soil. (In Swedish with English summary). Proc. of NGM 72, Norwegian Geotechnical Institute, Oslo 1972, 95-98.

ODEMARK, N. Investigation as to the elastic properties of soils and design of pavements according to the theory of elasticity. (In Swedish with English summary) National Road and Traffic Research Institute, Meddelande 77, Stockholm 1948, 100 pp.

TERZAGHI, K. Theoretical soil mechanics. John Wiley & Sons, New York 1943, 510 pp.

BARRON, R.A. Consolidation of fine-grained soils by drain wells. Transactions ASCE, Vol. 113, New York 1948, 718-754.

G. CEDRUN and D. SIMIC, Civil Engineers, Intecsa, Madrid

Compaction and behaviour of embankments built with Eocene marl rock at Navarra

ABSTRACT

It is described the geotechnical characteristics and compaction method used for construction of Navarra highway embankments in Spain, and also its behaviour after placement. The material was obtained from the cuts excavated in eocene blue marls. This rock, of aeolian origin, has approximately 45% of CO_3Ca and the compression strength of the sane rock varies between 100 and 200 kg/cm^2. Blasting was necessary for its excavation. The Atterberg limits of the marl are 16 and 35. Mineralogical analysis of the sane marl gave calcite, quartz, mica of illitic-muscovitic type, chlorite and small quantities of kaolin. By the rock weathering chlorite transforms by hidratation into vermiculite yielding to the rock fracturation by a stress increment due to volume increase. During the embankments construction, the compaction was performed with impact-compactor CAT-825, employing 10 passes at a speed of 7 km per hour. By these means, it was reached the 90% of the maximum density (of Modified Proctor Test), with water contents between the optimum and two points below. The deformation module obtained from load plate tests varied between 300 and 550 kg/cm^2. The behaviour of these embankments has been satisfactory after two years of road traffic.

INTRODUCTION

The Navarra highway will link the villages Irurzun, Pamplona and Tudela with a total length of 105 Km. It lies across two main lithological groups of materials: continental deposits ranging for Lower Tertiary (oligocene) to Quaternary and marine deposits showing flysch-type alternation ranging from Upper Cretacic to Quaternary.

The continental materials belong to the great geological province of Ebro's basin. The deposits filled this area during the Tertiary in a continental endorreical way, giving large amounts of evaporitic rocks (gypsum, anhidrite, salt, etc.). The marine deposits consist of deep sediments: flysch marl-limestone alternations and occasional levels of sandstones. This deposits belong to the edges of the Pyreneal geosynclines.

For an economical design of the highway, the volumes of excavation and embankments are to be balanced. This requires heights of embankment generally less than 10 m., reaching a maximum of about 20 m. For the project section near Pamplona a question arose about the possibility of using the tufa in highway embankments and, if so, what methods of construction are desirable.

At the Pamplona neighbourhood between the village of Tiebas and Erice, the highway traverses 25 km. on Eocene blue marls of marine origin, locally called "tufa" These are very regular, containing about 50% of calcium carbonate and showing little stratification. They are easily weathered yielding to a very plastic clayey soil due to decalcification.

The upper portion (weathered tufa) can be excavated with tractors and rippers but the lower portion (unweathered tufa) must be blasted. After excavation both the weathered and sound tufa disintegrates rapidly in the atmosphere.

CHARACTERISTIC OF THE MATERIAL

The tufa is a soft clayey sedimentary rock, that has been transformed from clay into soft rock by high pressures but has not been sufficiently cemented by the geologic processes to have formed a hard shale. It can be considered as a highly consolidated clay.

The compression strength of the sane rock varies between 100 and 200 kg/cm^2. The RQD is generally over 80%, falling seldom to 40%. The Atterberg limits of the tufa are 16% (plastic limit) and 35% (liquid limit). Mineralogical analysis of the sane tufa gave calcite, quartz, mica of illitic-muscovitic type, chlorite and small quantities of kaolin. By the rock weathering, chlorite transforms by hydratation into vermiculite yielding to the rock fracturation by a stress increment due to volume increase.

Table 1. Partially weathered tufa

	DENSITY (T/m^3)	MOISTURE (%)
Field value	1,96	9,5
Optimum Normal Proctor	1,96	12
Optimum Modified Proctor	2,88	11

	SAMPLE NUMBER	DENSITY (T/m^3)	PERCENTAGE ON N.P.	MOISTURE %	DEFORMATION MODULI LOAD PLATE TEST (Kg/cm^2)
Section 20 cm. thick	1	1,96	100	5,5	193
	2	1,84	94	9	
	3	1,93	99	7,3	
	4	1,89	97	10	
	Average	1,91	98		
Section 30 cm. thick	5	1,86	96	8,8	
	7	1,71	88	10,9	
Section 40 cm. thick	6	1,86	96	8,4	319
	8	-	--		

Table 2. Grading after compaction on partially weathered tufa

SIZE	1 1/2"	1"	3/4"	1/2"	3/8"	ASTM No.4	ASTM No.10	ASTM No.40	ASTM No.100	ASTM No.200
% of weight passing.	96,3	88	78,2	63,4	52,6	32,7	18	3	1,6	1,2

Table 3. Sane tufa

	DENSITY (T/m^3)	MOISTURE (%)
Field value	2,35	5,1
Optimum Normal Proctor	1,99	11,9
Optimum Modified Proctor	2,08	9,3

	SAMPLE NUMBER	DENSITY (T/m^3)	PORCENTAGE ON N. P.	MOISTURE %	DEFORMATION MODULI LOAD PLATE TEST (kg/cm^2)
Section 25 cm. thick	1	1,91	96	6,9	565
	2	1,87	94	8,6	
	3	1,89	95	11,0	
	4	1,80	90	8,1	319
	5	1,94	97	7,0	
	6	1,90	95	6,3	
	Average	1,89	95	8	

Table 4. Grading after compaction of sane tufa

SIZE	1 1/2"	1"	3/4"	1/2"	3/8"	ASTM No,4	ASTM No.10	ASTM No.40	ASTM No.100	ASTM No.200
% of weight passing.	77,9	62	51,4	36,1	27,3	12,0	5,2	1,5	1,0	0,4

The loss by use of SO_yMg, in five cycles of inmersion is 100%. Tests of satura - tion-dessication lead to sample disintegration.

The sane tufa has a natural density of 2,35 T/m^3 and a moisture content of 5,1%. After compaction with Proctor energy, the maximum densities were 1,99 T/m^3 (for standard Proctor) and 2,08 T/m^3 (for modified Proctor) and the optimun water content was respectively 11,9% and 9,3%.

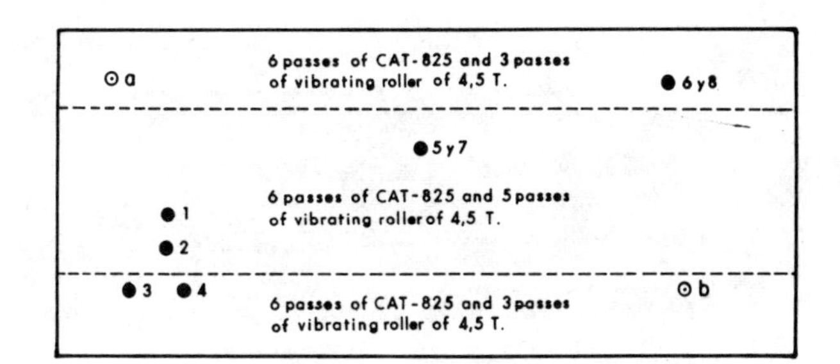

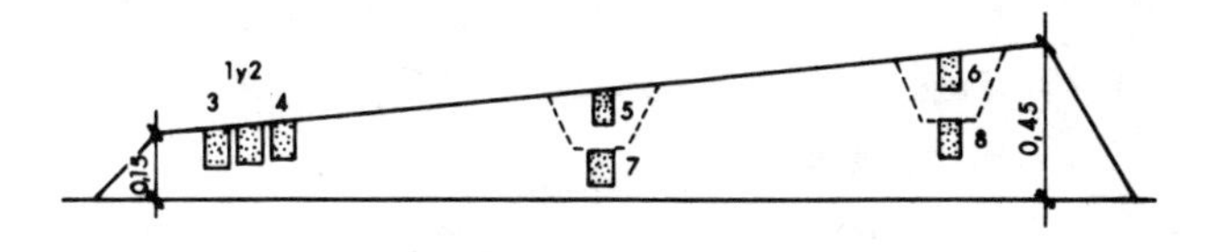

Fig. 1

THE USE OF TUFA IN EMBANKMENTS

Due to the geotechnical characteristics of the tufa mentioned above, that is to say: (1) soft rock behaviour and (2) rapid weathering with atmospheric exposure, the construction of rock-fill embankments using this material will probably lead to progressive deterioration and settlement with time.

In order to use such rocks for embankments which do no deteriorate and settle with time, methods of construction are used which break down the material into the form of gravel-clay soil so that the embankment formed is relatively impervious and all the voids between remaining rock fragments are filled with compacted clayey soil.

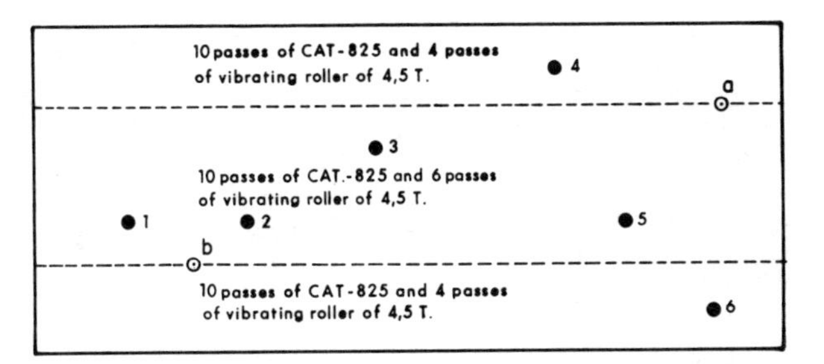

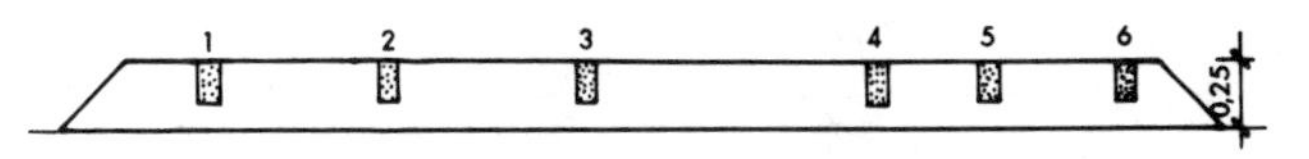

Fig. 2

The excavation and construction methods of the embankments ought to meet the following basic requirements, according to a Sherard report: (1) No rock fragment remaining should be larger than 15 centimeters, and (2) the density of the portion of the embankment finer than 3/4 inches should be at least 95% of the standard Proctor maximum density. It would be also desirable that this portion would represent at least 65% or 70% of all material weight.

In order to accomplish an adequate break down of the tufa pieces during the embankment compaction, impact roller and/or vibrating sheep foot roller should be used. Rollers 'speed and layers' thickness were to be determined from compaction tests strips.

In order to ascertain the special characteristics of such embankments, three compaction tests strips were performed.

The first one was executed with partially weathered tufa obtained by ripping. Schema in figure No. 1 shows a plant and a profile of the test strip. There are also shown the type of compaction employed on each zone and the situation of the samples taken.

Table No. 1 shows the values of field density attained. There are also shown the values of the moisture content and the results of the load plate tests performed.

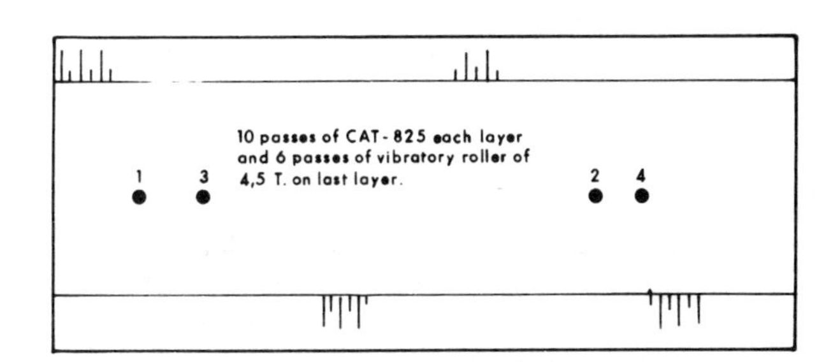

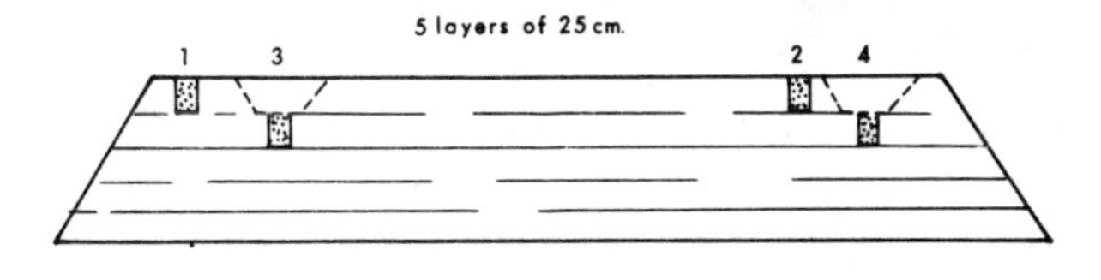

Fig. 3

Fig. 4. Core samples obtained from a borehole on Navarra blue marl. The little fracturation of the rock in its natural state can be seen

Fig. 5. Aspect of a mass of the marl, showing the weathering due to atmospheric agents

Fig. 6. Impact roller employed for the embankment construction

The granulometric analysis of a sample taken after the compaction gave the distribution shown in Table 2.

The second test strip was executed with sane tufa obtained by means of blasting, extension (18 minutes) and crusching (8 minutes) by a CAT D-96.

Schema in figure 2 shows the geometry of the test strip, the situation of samples taken and the type of compaction performed.

The roller employed was an impact type CAT-825 at speed of 7,1 km/h and a pulled vibrating roller of 4,5 T/m at a speed of 1,5 km/h.

The results of the test are shown in Table 3 and Table 4.

Finally, the third test strip was executed with the clay resulting from tufa weathering.

Schema in figure 3 shows the geometry of the test. The roller employed was a CAT-825.

Although it was intended to perform 10 passes on each layer, this was not fully accomplished throughout all the strip due to problems with the roller. Table 5 shows the results of the test.

The compaction strip tests perfomed show

(1) Both sane and weathered tufa obtained by blasting and crushing or by ripping give a well graded material after the embankment compaction, consisting of pieces of rock and soil with a maximum size of the order of 10-15 centimeters, with an appreciable percentage of sizes between 4 cm. and 1 cm. and a very small percentage passing of #200 sieve.

(2) Compacting layers of 25 cm. thick, a percentage of the 95% of the standard Proctor maximum density is guaranteed. During the construction of the highway embankments, the compaction was performed on layers 30 cm. thick by means of an impact type roller CAT-825, employing 10 passes at a speed of 7 km/h. It was reached the 90% of the Modified Proctor maximum density with water contents between the optimum and two percentage points below. The deformation moduli obtained from load plate tests varied between 300 and 550 kg/cm^2.

Table 5. Clay resulting from tufa weathering

SAMPLE NUMBER	DENSITY (T/m^3)	PERCENTAGE ON N.P.	MOISTURE % *
1	1,82	91	15
2	1,82	91	14
3	1,64	82	16
4	1,70	85	16

* Data not reliable, due to a rainfall during the test execution.

CONCLUSIONS

The soft clayey sedimentary rocks as the described tufa can be employed for embankments. In order to prevent the embankment deterioration methods of construction are used which break down the material into the form of a gravel-clay soil so that the embankment formed is relatively impervious and all the voids between the remaining rock fragments are filled with compacted clayey soil. Compaction on the dry side shows a good improvement in load capacity.

The behaviour of these embankments has been satisfactory after two years of road traffic.

AKNOWLEDGEMENTS

The authors wish to thank the Enterprises AUDENASA, Owner of the highway; DRAGADOS Y CONSTRUCCIONES, Contractor; and INTECSA, Office of Projects for their skill during the work. They would also like to thank Professor Sherard for his advice throughout the project.

J. A. CHARLES, PhD, MSc(Eng), MICE, Building Research Establishment, E. W. EARLE, MICE, MIMunE, Corby Development Corporation, and D. BURFORD, DIC, Building Research Establishment

Treatment and subsequent performance of cohesive fill left by opencast ironstone mining at Snatchill experimental housing site, Corby

A carefully monitored experiment is being carried out involving the treatment of backfilled land left by opencast ironstone mining and the subsequent erection of groups of houses on the treated ground. The overburden at the site had been excavated by dragline during the mining operation and the resulting backfill was loose, unsaturated and about 24 m deep, the upper part being predominantly cohesive. Each of three areas of the site was subjected to a different form of ground treatment and subsequently groups of two-storey dwellings of standard design with trench fill foundations have been built on each of these areas and on a fourth control area of untreated ground. The three types of ground treatment used were pre-loading with a surcharge of fill, 'dynamic consolidation' and inundation. Borehole settlement gauges of the magnet extensometer type were installed prior to ground treatment to monitor the compression produced at different depths within the soil.

Detailed results from the field experimental work are presented and a brief description is given of a programme of laboratory tests in which the susceptibility of the cohesive fill to further settlement on saturation was investigated. The effectiveness of the different ground treatment methods in compressing the fill is discussed and a preliminary assessment is made of the performance of the houses built on the treated and untreated areas. The implications of the experimental results for evaluating the load carrying properties of similar fill materials are discussed.

INTRODUCTION

Ironstone has been extracted in the Corby area for about a hundred years (ref 1) and currently 3 million tonnes of iron ore are excavated annually using large walking dragline excavators. In the vicinity of Corby over 1000 ha of land affected by opencast mining have been restored to agricultural use. The expansion of the new town involves housing and industrial developments on land that has been previously worked for ironstone by opencast methods. Over a number of years the Building Research Station has cooperated with the Development Corporation in the investigation of the problems involved in building on this restored ground. In 1963, 24 experimental houses were built at White Post Court. Four different types of foundation design were used and the performance of the houses was monitored (ref 2). Large differential settlements were measured although the damage was quite limited. During 1973 and 1974 factories were built on restored ground at the Earlstrees Estate. Part of the area had been treated with dynamic consolidation and part by vibroflotation. Structures on both types of treated ground suffered some damaging settlements in lightly loaded office blocks, but not in the more heavily loaded production areas. Following these developments it was felt that further investigations were necessary. The purpose of the current work at the Snatchill site is to investigate the usefulness of a number of different forms of ground treatment in improving the load carrying characteristics of loose fill left by opencast ironstone mining. It was hoped that a form of ground treatment could be found which would enable houses without special foundations to be successfully built on restored ground.

At the Snatchill site the loose unsaturated fill left by opencast mining is 24 m deep and the upper part is predominantly cohesive. Each of three 50 m x 50 m square areas was subjected to a different form of ground treatment and subsequently two-storey dwellings of standard design, with cavity walls on trench fill foundations, were built on the three treated areas and on a fourth of untreated ground. The ground treatment methods used were dynamic consolidation carried out by a specialist contractor, inundation via trenches dug across the site, and preloading with a static surcharge of 9 m of fill. Borehole settlement gauges were installed in each of the areas prior to ground treatment so that the compressions produced at different depths within the fill could be measured. Movements in the fill have been monitored before, during and subsequent to ground treatment. Settlement of the houses has been monitored by precise levelling.

The effectiveness of the different forms of ground treatment can be assessed in two ways. Firstly from the compression of the fill produced during treatment and secondly from the performance of the houses built on the treated ground. The behaviour of the ground during treatment gave valuable information about the condition of the fill as left by the opencast mining operations and a supplementary programme

Aerofilms

Fig. 1. Aerial view of Great Oakley quarry, restored ground and Snatchill experimental site

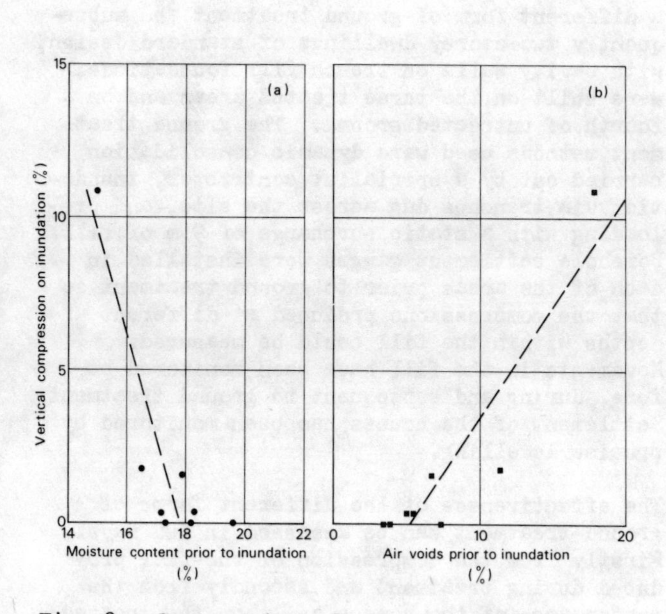

Fig. 2. Compression of cohesive fill on inundation in one metre diameter oedometer

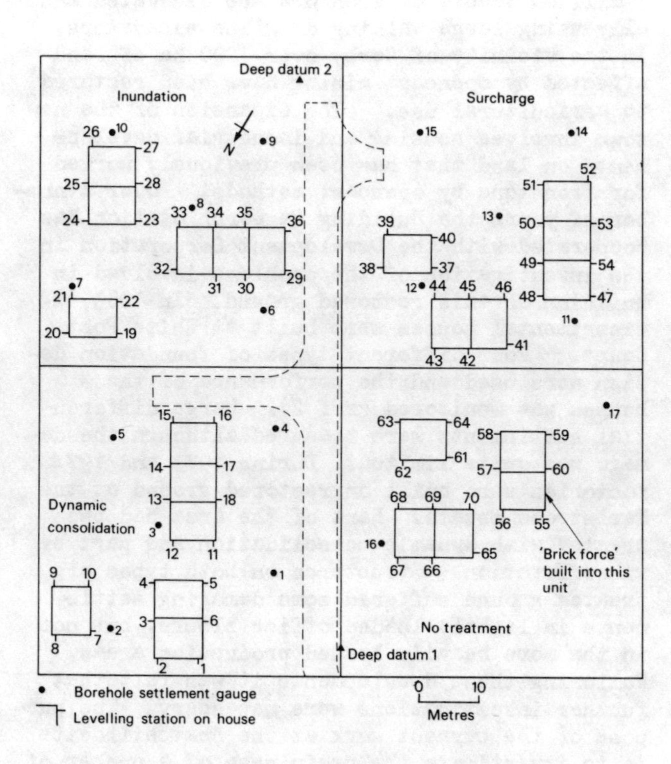

Fig. 3. Plan of Snatchill site

of laboratory tests was carried out.

OPENCAST MINING AT THE SITE

The Great Oakley quarry is on the south side of Corby and has an area of over 200 ha. Excavation began in 1963 and the quarry has been worked in 20 m wide strips (Fig 1). The succession of strata from the surface downwards is as follows: boulder clay, Oolitic limestone, Lower Estuarine sands and clays, Northampton Sand ironstone bed. The ironstone bed is up to 6 m deep but because of its chemical composition only the top 3 m are worked. The total overburden thickness is up to almost 30 m.

The British Steel Corporation uses a walking dragline excavator to strip the overburden and the ironstone is then dug out by a face shovel. The dragline is equipped with a 20 m^3 bucket and stands on the limestone layer some 11 m above the ironstone bed. Overburden stripping is carried out in two stages by the dragline. The boulder clay above the dragline is excavated and then the limestone, which requires blasting, and the other strata below the machine are removed. Excavated material is dumped directly into that part of the opencast mine where the ironstone has already been removed. The limestone is placed at the bottom of the tip and the upper part of the restored ground is predominantly cohesive. The fill is left in a characteristic hill and dale formation (Fig 1) and at this stage is surveyed by aerial photography.

The site is levelled by scrapers at a later date and the land is usually restored for agricultural use. The 100 m x 100 m experimental site at Snatchill was restored in 1970. The restored ground surface is fairly level at approximately 117 m AOD. Bedrock is at about 93 m AOD. During the opencast excavation drainage was installed to de-water the site and in this part of the quarry the drainage is by gravity. The drainage was left in place during backfilling and consequently the water table has remained below the level of the fill and is about 1 m below rockhead.

LABORATORY INVESTIGATION OF THE BACKFILL MATERIAL

The installation of instrumentation at the Snatchill site involved drilling a number of boreholes through the 24 m depth of backfill into bedrock. Borehole logs confirmed that the upper part of the backfill was predominantly cohesive. In some of the boreholes a distinct change in fill type was observed at a depth of about 15 m with a more granular fill below this depth. However in many of the boreholes no marked change in fill composition at a particular depth was found.

Some 100 mm diameter open drive samples were obtained from the boreholes and these indicated a wide range of moisture content and density in the fill. Moisture content varied between 7 per cent and 28 per cent with a mean value of 18 per cent. The corresponding range of dry density was from 1.5 Mg/m^3 to 1.8 Mg/m^3 with a mean value of 1.70 Mg/m^3. Standard Proctor compaction tests gave a maximum dry density of 1.92 Mg/m^3 at an optimum moisture content of 14 per cent. A particle size analysis indicated that 54 per cent of the sample was finer than 0.075 mm and that the clay fraction was 19 per cent. Typical values for plastic and liquid limit were 17 per cent and 28 per cent respectively. Undrained shear strengths were of the order of 100 kN/m^2. It is clear that most of the cohesive fill was significantly wetter than standard optimum moisture content. The calculated air voids of the samples ranged from 1 per cent to 20 per cent with a mean value of 5 per cent. It can be expected that this calculated average value of air voids from 100 mm diameter samples underestimates the actual average air voids in the fill as clearly no samples would be taken where there were large voids.

An uncompacted cohesive backfill containing a fairly high percentage of air voids can be expected to compress immediately when loaded and long-term consolidation due to slow dissipation of excess pore water pressures is unlikely to be a problem. However in these circumstances inundation may lead to sudden 'collapse' settlement of the fill under constant loading conditions. To investigate the likelihood of this phenomenon occurring in the field, a series of laboratory tests was carried out on samples of the cohesive fill placed at different initial moisture contents and with different initial air voids. The tests were carried out in a large oedometer which takes samples 1 m in diameter and 0.5 m high (ref 3). The samples were loaded under a vertical stress of about 100 kN/m^2 and, when compression under this load had virtually stopped, the samples were flooded and the amount of extra compression was measured. Figure 2(a) shows the amount of collapse settlement that occurred on inundation plotted as a function of moisture content immediately prior to inundation. The results show some scatter but do suggest that collapse settlement on inundation is unlikely, at a vertical stress of 100 kN/m^2, where the moisture content of the fill is greater than 19 per cent. Figure 2(b) shows collapse settlement plotted as a function of the air voids of the sample immediately prior to inundation. With air voids smaller than 5 per cent no sample compressed further on inundation.

THE FIELD INVESTIGATION AT SNATCHILL

The experimental site at Snatchill was divided into four 50 m x 50 m square areas (Fig 3). Each of three of these areas was treated by a different form of ground treatment and the fourth area was left untreated to act as a control. Five borehole settlement gauges were installed in each of three areas prior to ground treatment and two gauges were installed in the untreated area. The borehole settlement gauges were of the magnet extensometer type (ref 4) adapted for use in opencast backfills (ref 5). The extensometers were installed in 0.2 m diameter boreholes drilled through the 24 m of fill into bedrock. Ring magnets act as markers and are anchored to the sides of the borehole by strong springs. The magnets are detected and settlements measured by a reed switch sensor lowered with a steel tape attached down a central rigid access tube. This rigid tube is

(a) (b)

(c)

Fig. 4. Ground treatment at Snatchill; (a) dynamic consolidation, (b) inundation, (c) placing 9 m high surcharge

isolated from the ground by surrounding it with a flexible outer tube. The extensometer is installed by lowering it down the borehole with a rope attached to a heavy base weight and with the springs held back by nylon cords which can be cut by a small explosive device when the spring is in the correct position. The base weight contains a coarse porous stone and consequently the central access tube can be used to monitor water table level as well as for settlement measurements. In the lower part of the fill magnets were installed at 6 m vertical intervals but with a reduced vertical interval nearer ground level. The bottom magnet of each gauge was installed in bedrock and acts as a datum for the gauge.

The settlement gauges were installed during April and May 1974 several months before any ground treatment began so that there was an opportunity to observe any settlements occurring in the ground prior to treatment. Considerable variations in measured settlements were found across the experimental site. The average rate of surface settlement was about 1 mm/month and seemed to be mainly due to compression occurring in the top 5 m of fill. At the time of these measurements the ground had been restored for four years.

DYNAMIC CONSOLIDATION

Dynamic consolidation is a ground treatment technique in which deep compaction of soils is attempted by repeated impacts of a heavy weight onto the ground surface (ref 6). At the Snatchill site a 15 tonne weight with a base area of 4 m^2 was dropped from heights of up to 20 m by a heavy crawler crane (fig 4(a)). The work was carried out during September and October 1974 by Cementation Ground Engineering in a joint venture with Menard Techniques. The contractor used his own method specification to produce an acceptable end result in terms of the subsequent performance of houses built on the treated ground. Some preliminary information about the effectiveness of this method at Snatchill has been presented previously (refs 7 and 8).

At each point on a 10 m grid the 15 tonne weight was dropped repeatedly from a height of 20 m producing a hole some 2 m deep. Over 10 blows were usually required at each grid point. Each hole was backfilled using the surrounding material. The second stage of the treatment was to repeat the process using the same compaction at the same grid points. In further stages of the process, the two initial stages were repeated on a grid off-set by 5 m from the original grid. There was a final general tamping of the whole area with a reduced fall of the weight.

It is clear that the method of compaction results in the surface layers of the fill being completely rearranged to a depth of over 2 m. In general the settlement gauges did not have the weight dropped directly on them and as a consequence show heave of up to about 0.3 m in the top 2 m of fill. Settlements measured below 3 m depth are plotted in Fig 5. Large variations in settlement were measured at 4 m depth; the maximum settlement of 157 mm was measured at a gauge only 1 m away from a primary tamping point and 106 mm of this settlement occurred during the first stage of tamping. Below 10 m depth all settlements were smaller than 5 mm. The contractor took levels on the ground surface before and after treatment which, when corrected for the volume of sand imported to form a working platform for the crane, indicated a surface settlement of 240 mm (Fig 5). In the 6 months immediately following dynamic consolidation further movements of about 5 mm were measured.

The contractor monitored the work using a 63 mm diameter Menard pressuremeter. Tests were carried out in pre-drilled boreholes before and after treatment and average results showed improvement in pressuremeter modulus and limit pressures to a depth of 5 to 6 m. This agrees well with the settlement measurements, which also suggest that the treatment process had little effect below about 6 m depth.

INUNDATION

Five trenches about 1 m deep were dug across a 50 m x 50 m square area at 10 m centres during February 1975 (Fig 4(b)). They were first filled with water on the 21 February 1975 and were finally backfilled in June 1975 having been kept open and filled with water for four months. It is estimated that during the first ten days of the experiment some 90 cu.m of water were absorbed by the fill. Comparatively little water was absorbed subsequently. If it is assumed that the water penetrated uniformly over the area to a depth of about 5 m, then it would represent an increase in moisture content of the fill of about 0.5 per cent. In practice the penetration of water into the fill was very non-uniform over the area. The volume of water which was lost from a trench about 15 m from the boundary with the dynamic consolidation area was equal to the volume lost from all the other four trenches added together. Three weeks after the trenches were first filled with water, a number of 2 m deep boreholes were drilled through the bottom of the trenches in an attempt to encourage the water to penetrate deeper into the fill. This did not appear to have any significant effect on settlement.

Settlement of the fill occurred rapidly during the first weeks of the inundation and then more slowly (Fig 6). By the end of August 1975, movements that could be clearly attributed to the inundation had ceased and settlement rates were similar to those before the experiment began. The settlements measured at all five gauges in the inundation area during the period from first filling the trenches up to the end of August 1975 are shown plotted against depth in Fig 7. Four of the gauges showed surface settlements of between 77 mm and 96 mm while the other gauge, gauge 7, underwent a considerably larger surface settlement of 167 mm. Gauge 7 was within 2 m of the trench through which the largest volume of water penetrated into the soil. The settlement caused by inundation seemed to be confined to the upper 5 m of the fill.

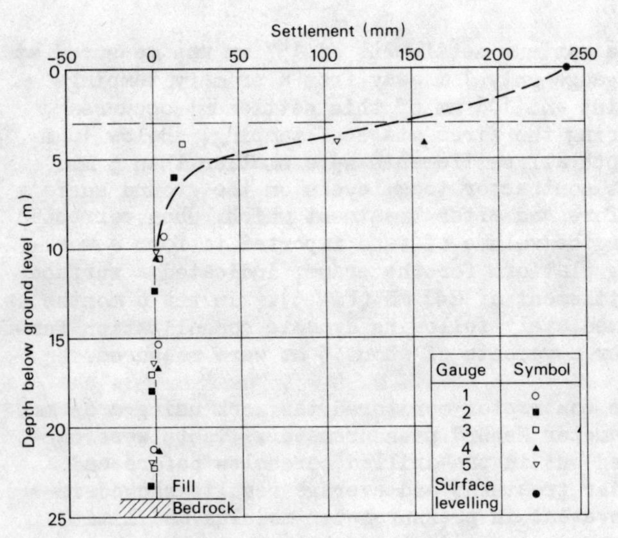

Fig. 5. Settlement produced by dynamic consolidation (Sept-Oct 1974)

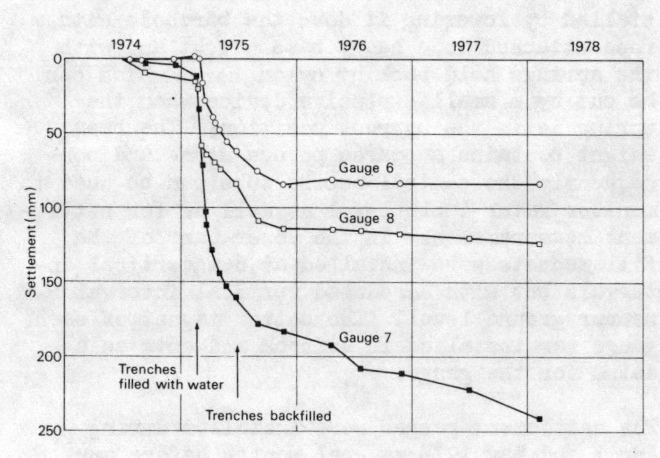

Fig. 6. Inundation; surface settlement v time

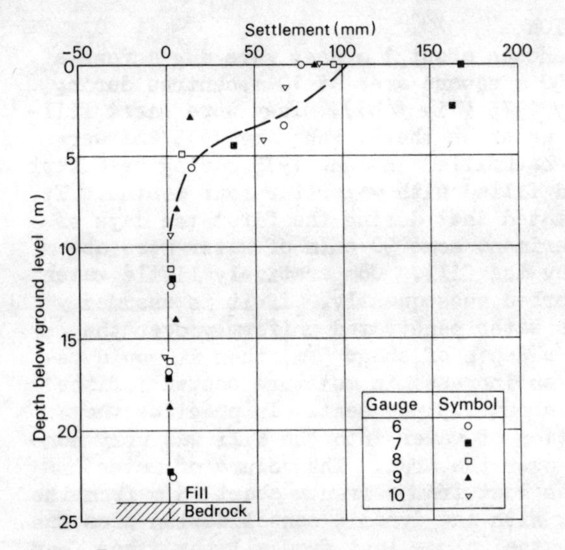

Fig. 7. Inundation; settlement produced Feb-Aug 1975 v depth

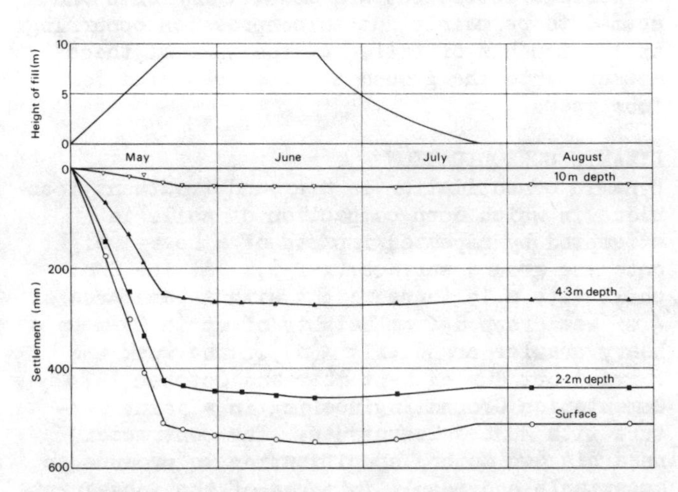

Fig. 8. Surcharge; settlement produced at gauge 14 v time

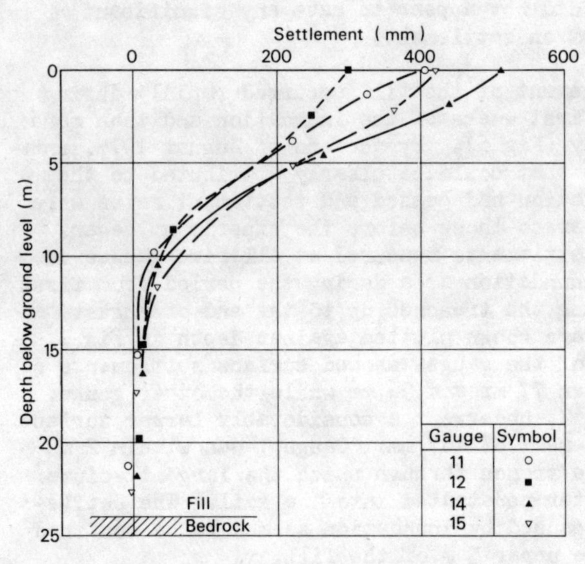

Fig. 9. Surcharge; settlement produced May-Aug 1975 v depth

Fig. 10. Experimental houses at Snatchill

PRE-LOADING WITH A SURCHARGE OF FILL

A 9 m high embankment of cohesive fill was placed over a 50 m x 50 m square area using tractor-drawn scrapers during a three week period in May 1975 (Fig 4(c)). The fill was brought from the adjacent existing opencast ironstone workings. The borehole settlement gauges were extended upwards through the surcharge fill so that settlement could be monitored as the fill was placed. The surcharge fill was left in position for a month and removed during July 1975.

Figure 8 shows the settlement that occurred at different depths at gauge 14 during the various stages of the surcharge experiment. It is clear that most of the settlement occurred as the fill was being placed and that the movements that occurred subsequently while the fill was left in place were much smaller. On removing the surcharge a slight heave took place.

The net settlements caused by surcharging are plotted against depth in Fig 9 for four settlement gauges. Unfortunately the remaining gauge, gauge 13, became blocked and no measurements could be made. Figure 9 shows that the surcharge produced an average vertical compression of about 4 per cent in the top ten metres of the fill. Below 10 m depth settlements were very small. The differences in surface settlement indicate the variability of the fill; gauge 14 had a surface settlement of 512 mm while gauge 12 had a surface settlement of only 299 mm.

EXPERIMENTAL HOUSES

The houses were erected by Wilcon Construction of Northampton in a package deal, and were their standard private market units. The foundations were not modified; trench fill was used, 375 mm x 900 mm deep with the top of the concrete 75 mm below ground level.

The three areas of treated ground each had a detached house, a semi-detached pair and a terrace of three houses built on them (Fig 3). The untreated area had one detached and two semi-detached pairs of houses built on it. One of the semi-detached units had 'brickforce' reinforcement built in at dpc and plate level. Levelling stations (ref 9) were installed when house construction reached dpc level and precise levelling has been carried out using a Wild NA2 automatic level with parallel plate micrometer adjustment and an invar staff. Deep datums as described in ref 9, but using a more substantial datum rod, were founded in bedrock underlying the fill. House building began in autumn 1975 and all houses had their roofs on by the end of March 1976 (Fig 10).

The table below summarises movements that have been measured during and subsequent to construction. Settlements have been smallest on the area previously surcharged where the average total settlement to date of all the levelling stations is only 4 mm. The untreated area has shown surprisingly little settlement and average settlements on the dynamic consolidation and inundation areas are considerably greater than on the untreated area. Damage to houses is caused by differential rather than absolute settlement and some information on this has been included in Table 1. Maximum and minimum settlements of levelling stations on each area are listed as is the maximum differential settlement between adjacent levelling stations which are from 6 m to 8 m apart. Also quoted is a standard deviation for the settlements measured in each area which gives a measure of the variability of the results. Differential movements have been greatest in the area previously treated by inundation and least in the pre-loaded area. A recent inspection of all the houses showed no indication of damage attributable to settlement.

DISCUSSION

Field measurements of settlement within the fill produced by the three types of ground treatment have given insights into the nature and behaviour of the cohesive fill at the Snatchill site. The backfill, which had been replaced by dragline during opencast mining, was generally significantly wet of standard Proctor optimum moisture content and had an undrained shear strength of the order of 100 kN/m^2, yet due to the lack of compaction during placement it has behaved as a loose unsaturated fill. All three treatment methods had some success in compacting this cohesive fill. Details of the settlements produced by the ground treatment methods

Table 1. Settlement of experimental houses (in mm)

	Settlement during construction					Total settlement during and subseq. to construction at 10.5.78				
	Mean	Stand. dev.	Max.	Min.	Max. diff.	Mean	Stand. dev.	Max.	Min.	Max. diff.
Dynamic consolidation	7.0	1.5	9.2	3.2	5.9	24.1	4.9	30.7	10.9	15.8
Inundation	6.1	2.9	14.3	2.8	6.8	22.0	16.3	73.2	7.1	44.7
Pre-loading with surcharge	1.4	0.9	3.0	-0.4	2.3	4.5	2.5	11.0	1.2	4.3
No treatment	2.7	1.3	6.8	1.4	2.8	9.8	4.3	21.0	4.4	8.5

have been presented in Figs 5 to 9, but for ease of comparison average settlements produced at different depths within the fill are summarised in Table 2. An approximate figure for the cost of each treatment at that time is also included in the Table.

Table 2. Average settlements produced by ground treatment and cost of treatment

	Dynamic consol- idation	Inundation	9 m high surcharge
At surface	240 mm	100 mm	410 mm
At 4 m depth	90 mm	40 mm	230 mm
At 10 m depth	<10 mm	<10 mm	40 mm
Cost*	£15,000**	£2,500	£12,000+

NOTES:

*Cost of treatment of 50 m x 50 m square area at late 1974 prices.

**Dynamic consolidation could have been considerably cheaper if a larger area had been treated.

+Average haul distance was 100 m for placement of surcharge.

The dropping of a heavy weight on the ground surface resulted in significant compaction of the upper layers of the fill. The average energy input of 2800 kN.m/m^2 produced an average vertical compression of 4 per cent in the top 6 m of the fill. Both the settlement measurements and pressuremeter results indicated that the compaction had negligible effect on the fill below 6 m depth. To assess the likely depth of effectiveness of dynamic consolidation, Menard and Broise (1975) suggested the relationship:

$$WH > Z^2$$

where W is the weight of the pounder in tonnes,
H is the height of fall in metres
and Z is the thickness of the compacted layer in metres.

This relationship can be re-written as:

$$Z < \sqrt{WH}$$

The Snatchill results correspond to,
$Z = 0.35\sqrt{WH}$.

The inundation experiment appears to be the least successful of the three ground treatment methods used at Snatchill. Inundation produced least compression of the fill during treatment and subsequent differential movements undergone by the houses were largest on this area (Tables 1 and 2). The ineffectiveness of inundation as a method of ground treatment is probably due to the difficulty of saturating the fill by the addition of water from the ground surface. The water tends to run away down the largest voids and fissures and fails to provide a uniform treatment over the area. The inundation experiment might have produced more settlement if the trenches had been closer spaced and deeper, but it may be that satisfactory inundation would only be achieved by a ground water table rising up through the fill. Although ineffective as a ground treatment, the inundation experiment has illustrated how easily substantial movements can be caused by water penetrating into the body of the fill via trenches cut through the impermeable surface crust. Clearly this can present a major hazard in the development of such a site, as trenches for services and drainage are likely to be dug close to buildings both during and subsequent to house building. The inundation experiment has also given an indication of the likely magnitude of such movements caused by collapse settlement on inundation. Laboratory tests indicated that susceptibility to collapse settlement is linked to the percentage of air voids in the cohesive fill, and as the average compression of the fill produced by both dynamic consolidation and preloading with surcharge is greater than the maximum compression measured at any of the settlement gauges in the inundation area, it might suggest that both these methods have considerably reduced the potential for collapse settlement in the upper layers of the fill.

Of the three treatment methods used at Snatchill only surcharge loading has led to a demonstrably better performance of houses built on the treated ground when compared with the houses built on untreated ground (Table 1). The superiority of this form of ground treatment was shown also by the greater enforced settlements produced during the treatment (Table 2). The practicality of pre-loading with a surcharge is heavily dependent on the availability of fill material, as the cost of the operation will be largely a function of haul distances. As compression of the fill mainly occurs as the surcharge is being placed, there is no need to leave it in position for a long period and, with a large area of land to be treated, the same surcharge of fill could be moved around the area in a continuous earthmoving operation. As pre-loading with a surcharge can reduce subsequent settlement in several distinct ways, there is no simple calculation for the required height of surcharge. The surcharge can pre-load the ground at stresses greater than any that will be subsequently applied by the foundation loads, thus largely eliminating movements due to foundation loading. A second effect of the surcharging can be to greatly reduce creep movements due to self-weight of the fill. Finally, the surcharge can reduce or eliminate the potential for collapse settlement on inundation. The maximum vertical stress applied to the fill by the Snatchill experimental houses is about 100 kN/m^2, compared with the surcharge loading of 200 kN/m^2. Consequently settlement of the houses during construction on this area was negligible (Table 1). The rate of settlement of the houses subsequently has been much less than the general creep rates measured over the site before ground treatment began. The compression measured in the fill during treatment suggested that the surcharge effectively compressed the fill down to a depth of about 10 m, ie a depth approximately equal to the height of the surcharge. At that depth the surcharge would double the existing vertical stress in the ground. It seems likely that pre-loading will prevent collapse settlement occurring in this top 10 m of the fill. This means that percolation of surface water into the fill through trenches should not be a hazard. However, a rising ground water table could cause substantial settlements by collapse compression of the lower part of the fill. In that eventuality the top 10 m of treated ground should serve to reduce differential movements by arching action.

The measured settlements of the houses built on untreated ground have been surprisingly small (Table 1), the largest settlement during the first two years was 17 mm compared with about 150 mm maximum settlement measured during a similar period in the earlier experimental work at the White Post Court (ref 2). The movements of the houses on the untreated area at Snatchill are so small that they fail to establish the need for ground treatment. However, in view of the fact that settlement and damage had occurred on similar sites in the Corby area, special measures to give greater certainty of successful building development are clearly desirable. As the behaviour of the fill is erratic, ground treatment to rectify the situation appears preferable to designing structures to survive estimated extreme movements.

The site investigation prior to building development on an area like Snatchill should be related to the particular character and behaviour of loose unsaturated fill typically left by opencast mining. Initially as much information as possible should be obtained about the details of the opencast mining. The small foundation loads applied by two-storey housing to the ground mean that the undrained shear strength and compressibility of the cohesive fill are unlikely to be of primary importance. The susceptibility of the fill to collapse settlement due to increase in moisture content and long-term creep compression in the fill due to self-weight are two considerations that may be of major significance. Precise levelling of surface settlement stations over a reasonable period of time can give some idea of the rate of creep settlement and differential movements occurring at the surface of the fill. The installation of borehole settlement gauges to measure settlement at different depths within the fill can give useful supplementary data. Piezometers will enable the position of any water table to be established. If the water table is close to ground level, clearly collapse settlement on inundation will not occur as the fill will be already saturated. As a major problem with such sites is variability of the fill, trial pits should be excavated so that visual inspection can be made of relatively large areas of the fill. Such pits can also be utilised for field inundation tests to determine whether the addition of water into the fill via a trench causes settlement of the surrounding fill. Moisture content, dry density and specific gravity of small samples can be measured in the laboratory and air voids estimated. Index properties can also be usefully determined and oedometer tests to measure collapse settlement on inundation can be carried out.

CONCLUDING REMARKS

1. The cohesive fill left by opencast mining at the Snatchill site contains a comparatively high percentage of air voids, although the fill is significantly wet of standard Proctor optimum moisture content. Consequently its behaviour is different from that of a similar fill in a well compacted state. Under a static surcharge loading the cohesive fill compresses immediately; the fill can be compacted by dropping a heavy weight on its surface; when water penetrates into the fill, collapse settlement occurs.

2. As could be expected from the uncontrolled method of placement, the composition and behaviour of the untreated fill are very variable. The most significant characteristics from the standpoint of building development on the site are firstly the susceptibility to collapse settlement on inundation and secondly the creep compression in the fill due to self-weight under conditions of constant stress and moisture content.

3. Where an opencast backfill has been replaced in thin layers with adequate compaction, building on the restored ground should present relatively few problems. Where building development has to take place on a site where opencast backfill has been tipped without compaction and where piled foundations are not an economic solution, two other approaches to the problem are possible. Firstly, building can be designed to withstand estimated differential movements, and secondly, some form of pretreatment process can be applied to improve the load carrying characteristics of the fill. Where, due to the variability of the fill, there is the possibility of large differential movements occurring, the use of ground treatment to reduce such damaging movements has obvious advantages.

4. On the cohesive backfill at the Snatchill site the settlement of houses built on untreated ground has been surprisingly small. Much larger movements have been measured on similar backfills in the Corby area and clearly it cannot be assumed that the settlement of untreated ground will always be this small. The only type of ground treatment that appears to have led to a significantly improved foundation behaviour is pre-loading with the 9 m high surcharge of fill. Clearly the economics of this form of treatment depend on the nearness of a supply of fill material. It must be emphasised however, that the experiment is continuing and the results presented in the paper are of a preliminary nature.

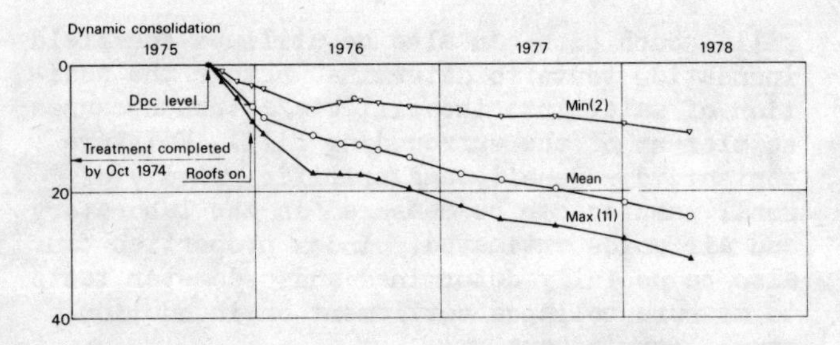

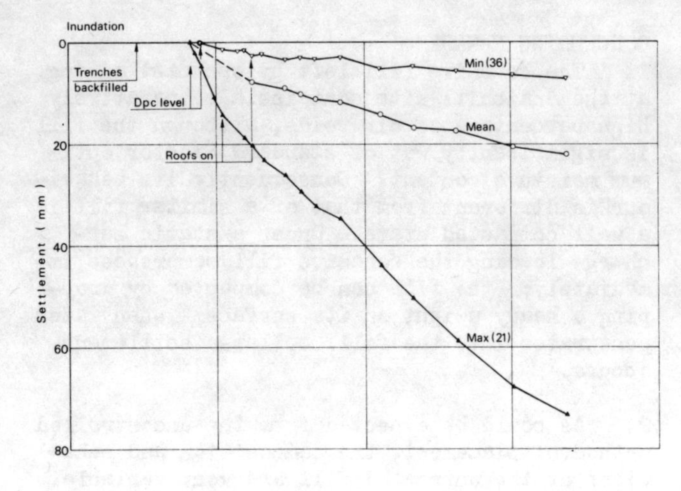

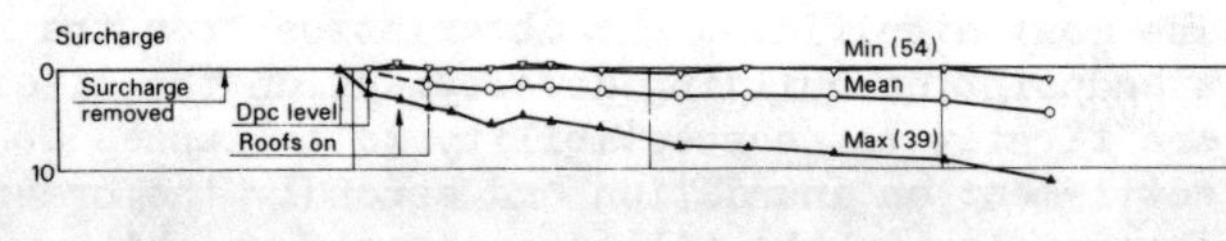

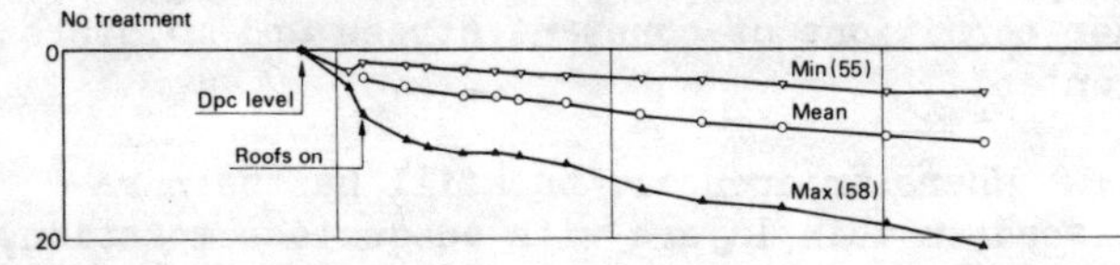

Fig. 11. Settlement of experimental houses

ACKNOWLEDGEMENT

The work described has been carried out as part of the research programme of the Building Research Establishment of the Department of the Environment. This paper is published by permission of the Director of the Building Research Establishment and the General Manager of the Corby Development Corporation. Encouragement given during the course of the work by Mr E W Godwin of Corby Development Corporation and Dr A D M Penman of the Building Research Establishment is gratefully acknowledged. The cooperation of the British Steel Corporation is acknowledged and in particular the assistance provided by Messrs R Cowan and G Naylor.

REFERENCES

1. POCOCK D C D. Iron and Steel at Corby. East Midland Geographer, 1961, vol 2, pt 15, pp 3-10.
2. PENMAN A D M and GODWIN E W. Settlement of experimental houses on land left by opencast mining at Corby. Brit Geotechnical Soc Conf on Settlement of Structures, Cambridge, 1974, pp 53-61.
3. PENMAN A D M and CHARLES J A. The quality and suitability of rockfill used in dam construction. 12th Int Conf on Large Dams, Mexico, 1976, vol 1, pp 533-566.
4. MARSLAND A and QUARTERMAN R. Further development of multi-point magnetic extensometers for use in highly compressible ground. Geotechnique, 1974, vol 24, no 3, pp 429-433.
5. CHARLES J A, NAISMITH W A and BURFORD D. Settlement of backfill at Horsley restored opencast coal mining site. Conf on Large Ground Movements and Structures, UWIST Cardiff, 1977. (Building Research Establishment Current Paper 46/77).
6. MENARD L and BROISE Y. Theoretical and practical aspects of dynamic consolidation. Geotechnique, 1975, vol 25, no 1, pp 3-18.
7. PEARCE, R. Contribution to discussion. Ground Treatment by Deep Compaction. 1st Geotechnique Symp in Print, Instn of Civil Engrs, London, 1975, pp 98-101.
8. CHARLES J A. Contribution to discussion. Ground Treatment by Deep Compaction. 1st Geotechnique Symp in Print, Instn of Civil Engrs, London, 1975, pp 96-98.
9. CHENEY J E. Techniques and equipment using the surveyors level for accurate measurement of building movement. Brit Geotechnical Soc Symp on Field Instrumentation in Geotechnical Engineering, Butterworths, London, 1973, pp 85-99.

D. G. COUMOULOS, PhD(Cantab), and
T. P. KORYALOS, MSc, Kotzias-Stamatopoulos Co. Ltd,
Geotechnical Engineering Consultants, Athens

Performance of the clay core of a large embankment dam during construction

The Paper describes the performance during construction of the clay core of a large embankment dam in South Europe. Placement quality control of the core is described and the characteristics of the core material are presented. Typical results from electrical earth pressure cells, electrical piezometers and inclinometers are presented and discussed briefly. Vertical settlement data are analyzed and compared with results from one-dimensional consolidation tests on the core material "as built". Constrained modulus values obtained from oedometer tests are compared with Young's modulus values determined during the consolidation of the core.

INTRODUCTION

1. The compacted clay core of the dam is symmetrical about the centerline. It has a maximum height of about 90 m, a crest length of approximately 450 m and a total volume of approximately 900,000 m^3. Placement of the core was completed within 14 months. The shells upstream and downstream consist of compacted river sand and gravel material which yielded excellent permeability properties. A simplified cross section of the dam at its maximum height is shown in Fig. 1.

DESCRIPTION OF THE CLAY CORE

2. The core of the dam consists of compacted brown clay of the group (CL) of the Unified Soil Classification System. The range of gradation and the index properties of this clay are shown in Fig. 2 and 3.

3. Core placement and compaction were controlled throughout the period of construction by a field quality control laboratory. Determination of percentage of compaction and checking of placement water content were done either by direct comparison to standard Proctor compaction or with the aid of the "rapid compaction control" method as described by designation E-25 of the USBR (ref. 1).

4. The specification for the compaction of the core called for (a) dry unit weight to be at least 98 percent of standard Proctor (ASTM D-698) and (b) placement water content to be within -3 and +3 percent of optimum. Quality control during construction was aiming at spotting and correcting defficient or doubtfull material. When there was no doubtful areas, sampling aimed at verifying acceptability. No defects were accepted. This sampling is called "purposive" (ref.2) and is fundamentally different from random sampling methodology which is used in industrial quality control (ref. 3). The merits of purposive sampling and the associated statistical techniques are discussed in some detail by Kotzias and Stamatopoulos (ref. 4).

5. Based on the results of the field quality control, the characteristics of the core are as follows:

(a) Wet unit weight as placed: 2.06 t/m^3
(b) Placement water content: 0.1 percent dry of optimum
(c) Dry unit weight: 1.77 t/m^3
(d) Percent compaction attained: 99.7 of standard Proctor

6. A number of block samples were taken from the dam core at various elevations. Specimens were cut horizontally from these block samples and tested for compressibility in 63.5 mm oedometer rings. Results of the consolidation tests are summarized in Fig. 4 plotted in terms of constrained modulus against vertical stress. The constrained modulus D can be computed from the void ratio versus vertical stress plots with the aid of the following equation (ref. 5):

$$D = \frac{\Delta p}{\Delta e}(1+e_o) \qquad (1)$$

where, Δp, denotes the increase in vertical stress,

Δe, denotes the decrease in void ratio for Δp

e_o, denotes the initial void ratio of the tested specimen

The plots of Fig. 4 are based on oedometer strains measured 24 h after applying a load increment.

7. In addition to these tests on the core material "as built", a number of permeability tests were carried out in the laboratory on specimens cut horizontally and vertically from the block samples. Permeability values for flow in the vertical and horizontal directions were found to be of the order of 10^{-9} m/s and $3x10^{-9}$ m/s respectively.

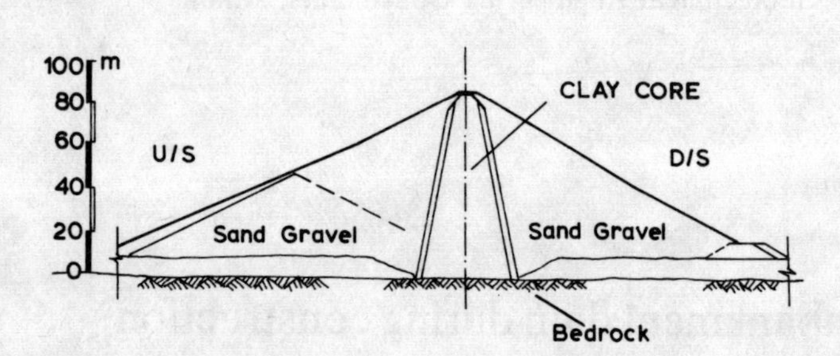

Fig. 1. Maximum cross section of the dam

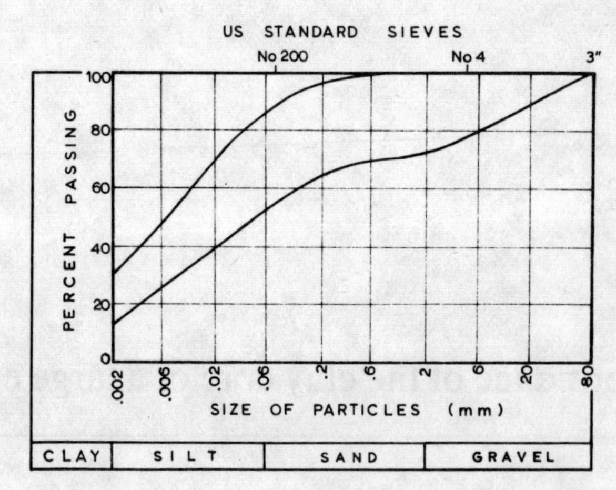

Fig. 2. Range of gradation of the clay core material

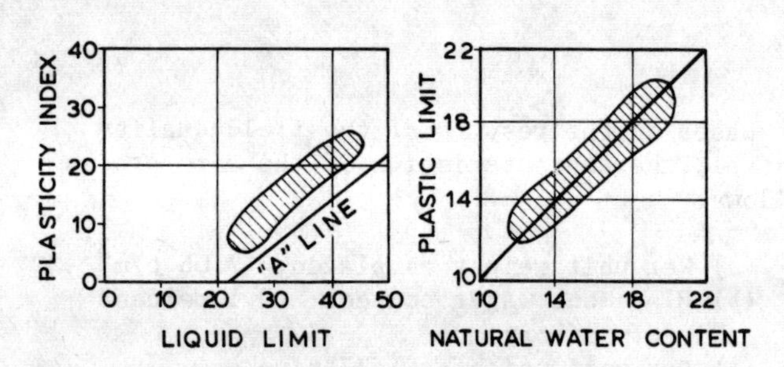

Fig. 3. Index properties of the clay core material

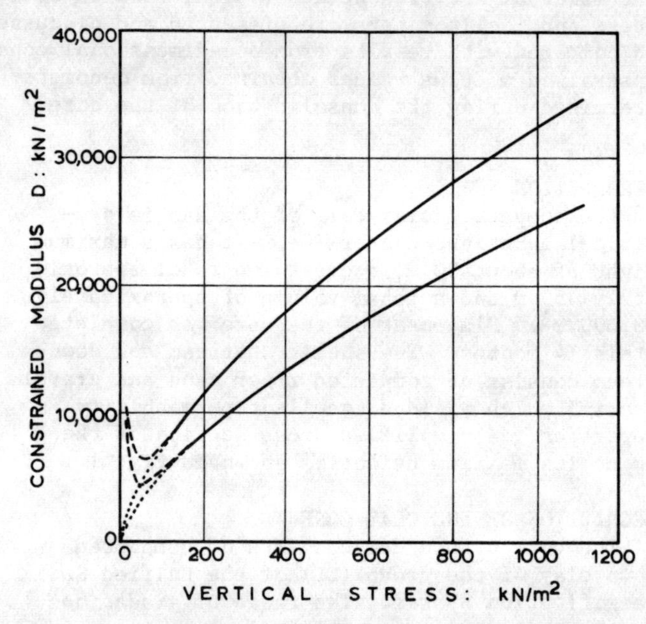

Fig. 4. Range of constrained modulus results vs. vertical stress from oedometer tests on the clay core material "as built"

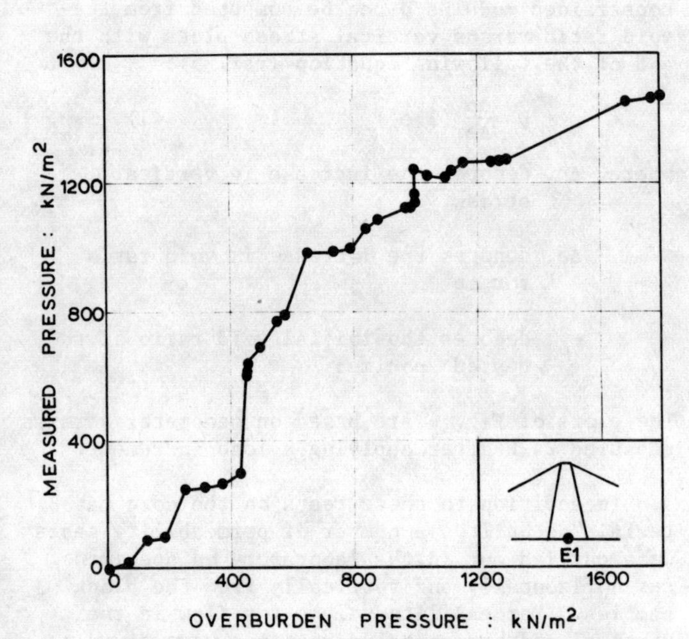

Fig. 5. Measured earth pressures during construction by an earth pressure cell placed on the bedrock

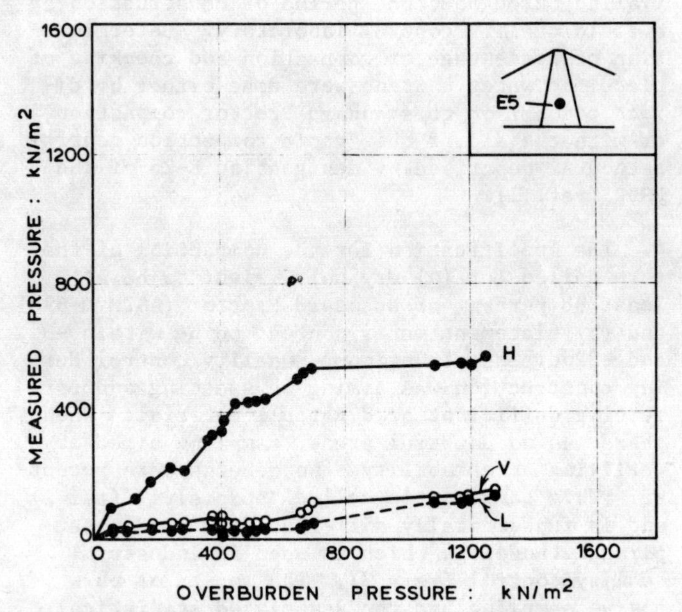

Fig. 6. Measured earth pressures during construction by a cluster of three earth pressure cells in the clay core

RESULTS FROM INSTRUMENTS IN THE DAM CORE

8. In the following paragraphs typical results from electrical earth pressure cells, electrical piezometers of the vibrating wire type, inclinometers and vertical movement devices, are presented and discussed briefly. The settlement data are further analyzed and compared with results from one-dimensional consolidation tests carried out in the laboratory on specimens from the compacted clay core.

Earth pressure cells

9. The electrical earth pressure cells installed in the core were of the earlier type of these cells with a ratio of diameter to thickness of about 2 to 1, and a modulus of elasticity of the order of 5,000 MN/m^2.

10. Typical results of measured earth pressures are shown plotted against calculated overburden pressures in Fig. 5 and 6. Fig. 5 shows the results from an earth pressure cell placed against the bedrock. Fig. 6 shows the results from a cluster of three earth pressure cells embedded in the compacted clay core. In this figure, H,V and L refer to earth pressure cells subjected to vertical stresses and horizontal stresses perpendicular and parallel to the dam axis respectively.

11. Readings from earth pressure cells are a measure of the response of each instrument to the stresses acting on it as affected by the relative stiffness of the soil and the cell, the interface conditions, the effect of embedment depth and the installation procedure (ref. 6,7, 8). It is very difficult to place and compact the soil around the instruments so as to have the same properties as the embankment.

12. The uncertainties in controlling the above factors are reflected by the performance of these earth pressure cells. The scatter of the observations from earth pressure cell E1 (Fig. 5) at low pressures is possibly due to lack of sensitivity at the low range of operation of the instrument or to arching effects. Between about 400 kN/m^2 and 1200 kN/m^2 the instrument overestimates calculated overburden pressures and above 1200 kN/m^2, it underestimates overburden pressures. On the contrary, as can be seen from Fig. 6 the horizontally embedded earth pressure cell E5 (H) showed a better response at the low range of operation, up to about 400 kN/m^2. Above 400 kN/m^2 the cell underestimates considerably calculated overburden pressures. Finally, the two vertically embedded earth pressure cells E5 (V) and (L) showed similar performance.

Electrical Piezometers

13. Fig. 7 shows measured pore pressures during construction from two selected piezometers in the clay core plotted against calculated overburden pressures. These plots indicate that construction pore pressures were low. Ratios of pore pressures to overburden pressures varied during construction between 9 and 25 percent. Low construction pore pressures were observed in a number of other embankments (ref. 9,10) and may be explained by the fact that placement water content of the core material was slightly drier than optimum water content (par. 5).

Inclinometers

14. Analyzed data from one inclinometer installed in the clay core at the maximum cross section of the dam, are shown plotted in directions normal and parallel to the dam axis versus height of embankment in Fig. 8. The full line refers to the last reading at the end of construction and prior to filling of the reservoir. The block points represent the positions of the top of the casing at intermediate dates. These diagrams, which are typical for all inclinometers in the clay core, show that practically no horizontal displacements took place in the position of the casing during construction. Horizontal distances between top and bottom of the casing of the instrument were confirmed with the aid of triangulation survey.

Vertical displacements during construction of the embankment

15. The vertical movement devices placed in the clay core were crossarms of the USBR type (ref. 1). Fig. 9 shows typical results of vertical settlements from two instruments plotted against height of embankment for different dates. These plots provide a quick check for acceptability of measurements. Scatter in these plots usually indicates either that individual readings were not taken correctly or that there are errors in determining elevations of the crossarms.

FURTHER ANALYSIS OF SETTLEMENT DATA

16. Vertical movement device S1 was installed at maximum cross section, while S2 was installed half way between maximum cross section and left abutment. The diagrams of Fig. 9 indicate that maximum vertical displacements occur at approximately mid height of the embankment, only during the initial stages of construction. As construction proceeded the point of maximum vertical displacements moved progressively downwards and by the end of construction it came close to the one third point of the height of the embankment.

Comparison with laboratory test results

17. From the above settlement data it is possible to compute the compressibility characteristics of the clay core and compare them with the laboratory test results from one-dimensional consolidation tests on block samples obtained during construction (par. 6).

18. Fig. 10 and 11 show vertical strains from representative layers between crossarms from the settlement devices S1 and S2 plotted against overburden pressure. Vertical strains were determined for each layer as follows: distance between two successive crossarms was calculated by subtracting tape readings. This was done for different dates and each distance was compared with the corresponding distance obtained during the first measurement, i.e., the earliest available vertical distance between crossarms. Overburden pressures in Fig. 10 and 11 are expressed as the product of the wet unit weight of the core material as placed and the vertical height of the fill above mid point of each layer between crossarms.

19. The plots of Fig. 10 and 11 suggest that the

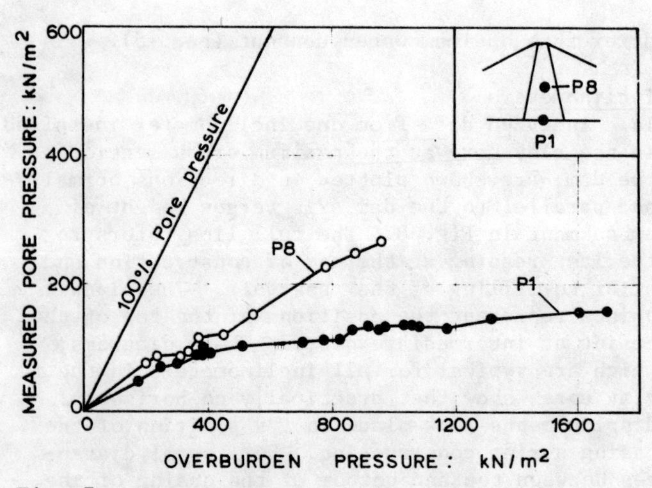

Fig. 7. Pore pressure observations in the clay core during construction

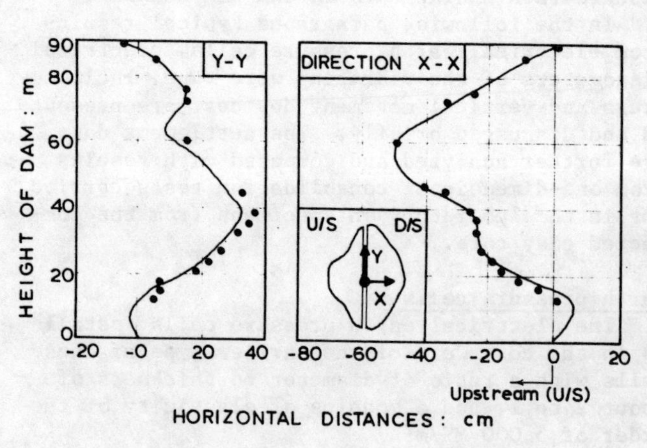

Fig. 8. Inclinometer data during construction

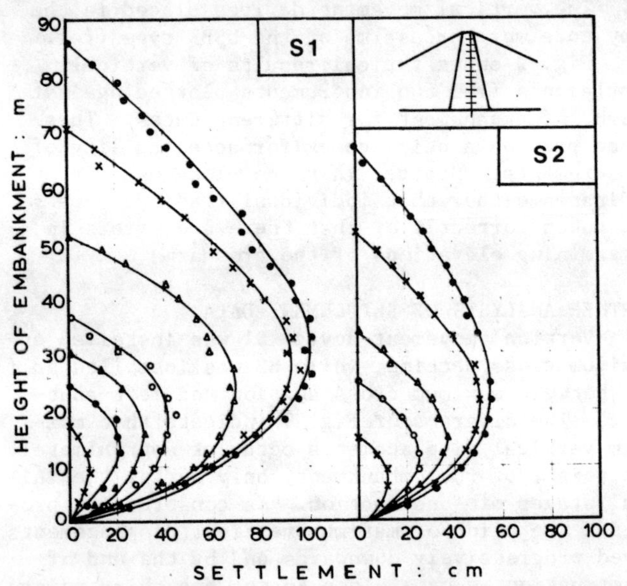

Fig. 9. Crossarm settlements during construction at two locations along the axis of the dam

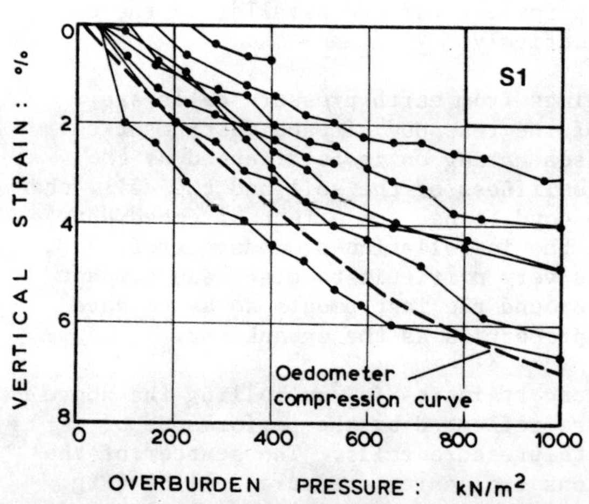

Fig. 10. Consolidation of the clay core during construction

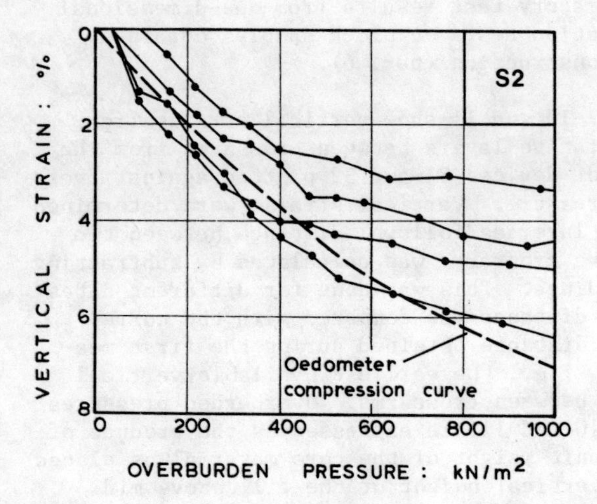

Fig. 11. Consolidation of the clay core during construction

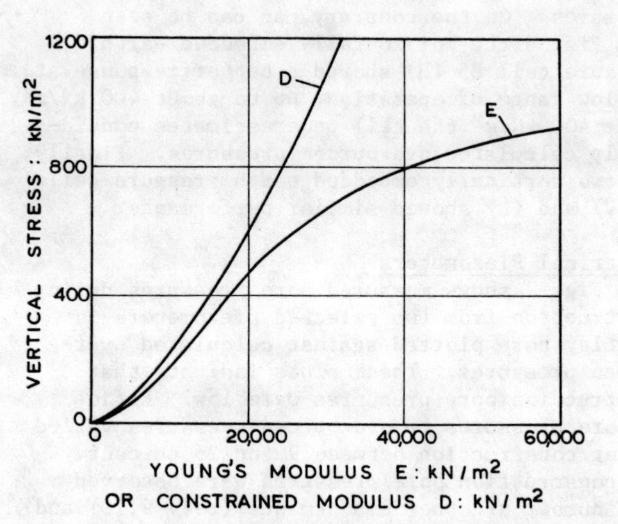

Fig. 12. Comparison of Young's moduli and constrained moduli plotted against vertical stress

condolidation behaviour of all layers between crossarms is in general similar. Scatter in these plots is basically due to erroneous readings. The reason that certain consolidation lines intersect the stress axis at values higher than the overburden pressure corresponding to the half layer, is because readings were not taken immediately after installation of the crossarms but at a later date, when elevation of the embankment was higher.

20. A comparison with the laboratory one-dimensional consolidation tests on the block samples from the core is also shown on the diagrams of the Fig. 10 and 11. The dashed line traced on these diagrams is the average compression curve of all oedometer tests carried out in the laboratory on core material "as built". The laboratory tests yielded at high vertical stresses a higher consolidation than that observed in the embankment. This is due to a faster dissipation of pore pressures in the oedometer than in the dam.

21. Mean values of Young's modulus E, derived from diagrams similar to those of Fig. 10 and 11 for all layers between successive crossarms from all settlement devices in the core, are shown plotted against vertical stress in Fig. 12. Constrained modulus values obtained from one-dimensional consolidation tests on core material "as built" are also shown in Fig. 12 by the line D which represents the average curve all results of Fig. 4. A comparison of E and D in Fig. 12 is attempted herebelow on the assumption that the central core of the embankment can be taken to be close to a condition of confined compression. This was confirmed by the inclinometer data which yielded no horizontal strains.

22. It can be seen from Fig. 12 that during the initial stages of construction of the core, constrained modulus values obtained from oedometer tests on core material "as built" are in close agreement with Young's modulus values derived from settlement data of the core. This close agreement between constrained moduli and Young's moduli, holds for this embankment up to a vertical stress of the order of 500 kN/m^2, which corresponds to approximately 25 m of fill.

23. On the basis of the above results it can be concluded that constrained modulus values from oedometer tests may be used to predict settlements of the core during the initial stages of construction of an embankment, provided construction pore pressures are low.

24. The application of the equivalent stiffness method which was developed at the Building Research Establishment by Penman Burland and Charles (ref. 11, 12) would yield, at the maximum embankment height for the clay core material, an equivalent Young's modulus E of the order of 27,500 kN/m^2 but this value of E would overestimate settlements. The equivalent stiffness method can be used successfully to predict settlements in the clay core during the initial stages of construction, during which maximum vertical displacements occur at approximately mid height of the embankment (see Fig. 9). Modifications of this method will be necessary as the embankment height increases and maximum vertical displacements do not occur any more at mid height of the embankment.

CONCLUSIONS

25. Constrained modulus values obtained from one-dimensional consolidation tests on the clay core material "as built"compare favourably with Young's modulus values obtained during the initial stages of construction of the central core of a large embankment dam.

26. Measured vertical displacements during construction of the core plotted against height of embankment indicate that maximum settlements occur at approximately mid height of the embankment, only during the initial stages of construction. As construction proceeded the point of maximum vertical displacement moved progressively towards the one third point of the embankment height.

27. Construction pore pressures in the clay core were low and varied between 9 and 25 percent of overburden pressures. Measured vertical earth pressures were not found to be in good agreement with calculated overburden pressures. Results from inclinometers indicated that no horizontal movements took place.

ACKNOWLEDGEMENT

28. The authors gratefully acknowledge the support received from Dr P.C. Kotzias and Mr A.C. Stamatopoulos while preparing the Paper.

REFERENCES

1. EARTH MANUAL, United States Department of Interior. Bureau of Reclamation (USBR), Denver, Colorado, 1963, 591-613 and 672-693.

2. YULE G.U. and KENDALL M.G. An introduction to the theory of statistics. Griffin, London, 1968, 14th ed., 370-384.

3. GRANT E.L. Statistical quality control. Mc Graw-Hill Book Co. Inc., New York, N.Y., 1964 3rd ed.

4. KOTZIAS P.C. and STAMATOPOULOS A.C. Statistical quality control at Kastraki earth dam. Proceedings of the American Society of Civil Engineers, 1975, 101, GT9, 837-853.

5. COUMOULOS D.G. and KORYALOS T.P. Correlation of constrained modulus with effective overburden pressure for settlement computations on soft clays. International Symposium on Soft Clay, Bankgkok, Thailand, 5-7 July, 1977, 223-230.

6. KRIZEK R.J., FARZIN M.H., WISSA A.E.Z. and MARTIN R.T. Evaluation of stress cell performance. Proceedings of the American Society of Civil Engineers, 1974, 100, GT12, 1275-1295.

7. PENMAN A.D.M., CHARLES J.A., NASH J.K.T.L. and HUMPHREYS J.D. Performance of culvert under Winscar dam. Géotechnique, 1975, 25, No 4, 713-730.

8. SKERMER N.A. Mica dam embankment stress analysis. Proceedings of the American Society of Civil Engineers, 1975, 101, GT3, 229-242.

9. SHERMAN W.C. and CLOUGH G.W. Embankment pore pressures during construction. Proceedings of the American Society of Civil Engineers, 1968, 94, SM2, 527-553.

10. LI C.Y. Construction pore pressures in three earth dams. Proceedings of the American Society of Civil Engineers, 1967, 93, SM2, 1-26.

11. PENMAN A.D.M., BURLAND J.B. and CHARLES J.A. Observed and predicted deformations in a large embankment dam during construction. Proceedings of the Institution of Civil Engineers, 1971, 49, May, 1-21.

12. PENMAN A. and CHARLES A. Constructional deformations in rockfill dam. Proceedings of the American Society of Civil Engineers, 1973, 99, SM2, 139-163.

D. W. COX, PhD, MSc, MICE, FGS, Polytechnic of Central London

Volume change of compacted clay fill

Clay fill is used frequently in earth dams, commonly in road embankments and sometimes as backfill below or within structures. The clay is taken from a borrow pit with a moisture content in equilibrium with its surrounding vertical and horizontal effective stresses. After placing and compaction new stresses will exist requiring a change in equilibrium moisture content. However clay volume also alters with variations in effective stress, the clay particles changing their spacing relative to each other. The research presented deals with the volume change resulting from an increase in the equilibrium moisture content where dry clay, shales or mudstones are used as fill and subsequently become wetter. The volume changes may be incompatible with a structure or with other fill materials resulting in a differential settlement or heave. While shear failure is the main cause for concern, the secondary movements due to volume change are a common cause of damage to earth dams and road and floor slabs.

THEORY

1. Clay fill is an aggregate of relatively undisturbed clay fragments of different particle sizes which are packed randomly together. A system of fine capillaries operates within each fragment, often fully saturated, with a negative pore pressure due to the relief of stress occurring during previous excavation from depth. Around the fragments is a second system of much larger capillaries and voids which are often full of air.

2. Volume change of fill occurs by two processes related to the quantity of water present and to the quantity of air entrained in the fill.

3. Initially a unit volume of saturated clay soil is excavated, broken up into fragments thus entraining air and then recompacted as a clay aggregate. External total pressures will then be increased as the height of overburden fill is raised. This causes compression of the fill and a consequent reduction in void size squeezing out air (or water, if the large voids have already become saturated). This is probably due to an adjustment of the contact area between the fragments, which depends on their shear strength.

4. On completion of the fill, moisture content changes will occur, either as a result of rapid inundation as in an earth dam, slow percolation of rain water as in a road embankment, or moisture transfer from the ground water as under a floor slab.

5. The volume changes are time dependant varying with the permeability of the soil, the drainage path, and the pore pressure gradients set up.

6. The dry clay initially has a high negative pore pressure. Supplying water at progressively lower negative pore pressures produces more and more volume change until an equilibrium is reached at zero pore pressure, equivalent to inundation.

7. The dry clay aggregate is highly stressed at the points of contact of the fragments but relatively unstressed around the larger air voids. The addition of water allows a softening at the points of contact and a rapid decrease in volume, known as collapse, occurs squeezing out air from the larger voids. This is a common problem in earth dams, Penman (Ref.1) and in the settlement of floor slabs on fill, N.H.B.C. (Ref.2).

8. As the wetting of the contacts proceeds individual fragments of clay forming the aggregate will swell. Since collapse is a surface effect it may preceed the swelling of the interior. The net volume change will be the sum of the two effects and may be positive or negative.

9. Swelling, shrinkage and consolidation are volume changes associated with minor changes in the spacing between individual clay particles. These are reversible to some degree. Collapse involves a major rearrangement of the fragments to a denser state of packing and is irreversible.

HISTORY

10. The problem of secondary movement in dams has long been recognised. The U.S. Committee on Large Dams 1966 Report suggested that the volume change characteristics of fill was the design field most urgently requiring further research. Recent failures in both the U.S. and U.K. suggest the problems are still with us.

Clay fills. Institution of Civil Engineers, London, 1978, 79-86

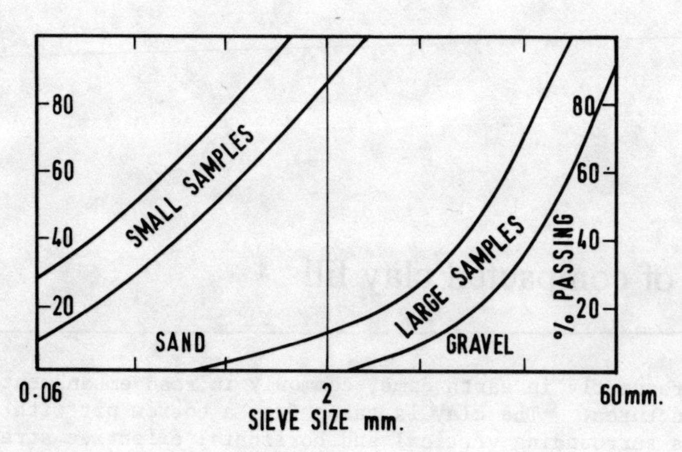

Fig. 1. Grading of shale fragments

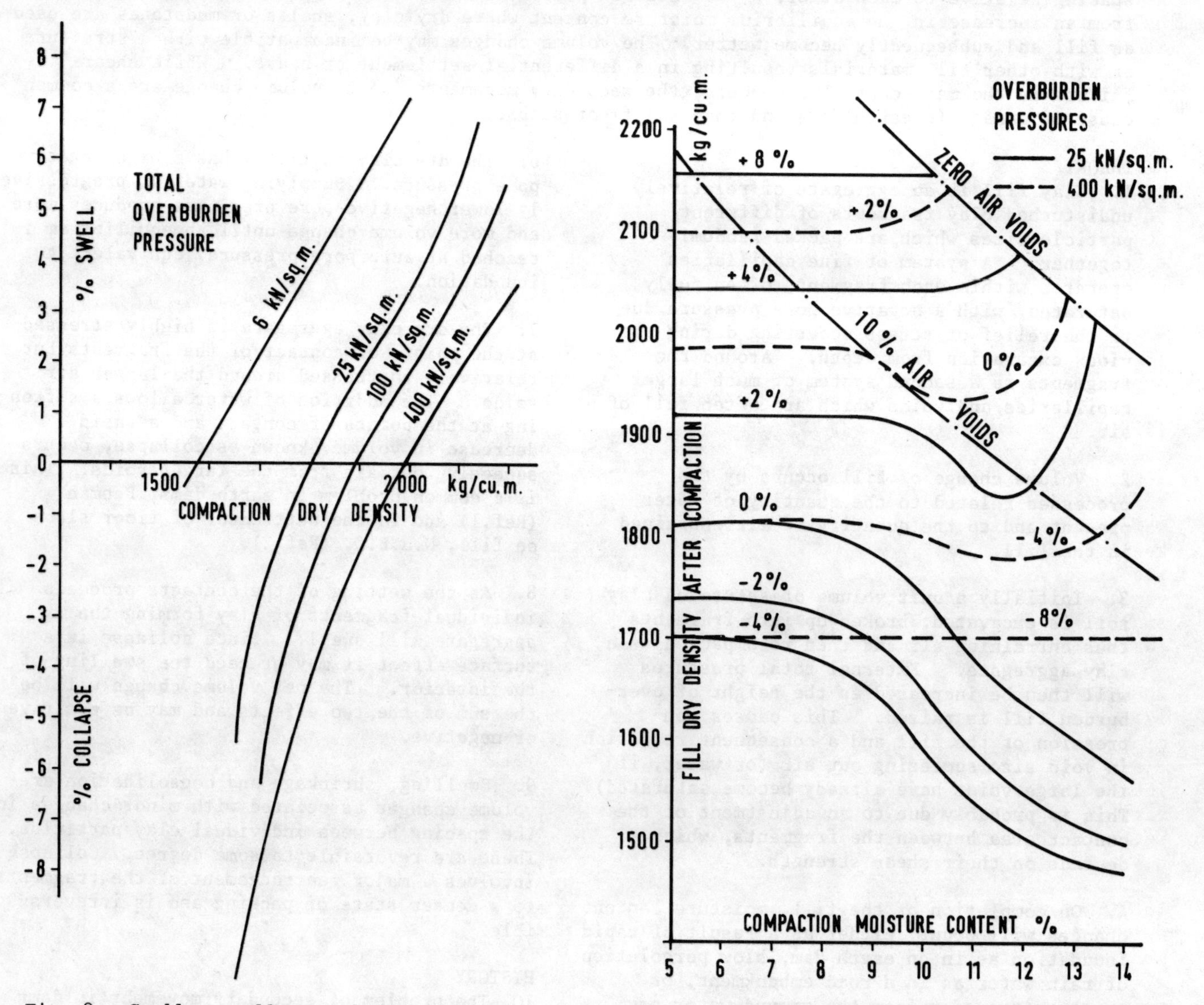

Fig. 2. Swell on collapse after inundation showing effect of initial dry density and overburden pressure (initial moisture content constant)

Fig. 3. Swell or collapse after inundation at varying initial moisture content and initial dry density (constant overburden pressure)

The Road Research Laboratory (Ref.3) suggested in 1948 that "as soon as new large scale road construction starts again - it is hoped to make detailed investigations in which the settlements of fill will be related to its relative compaction."

11. Volume change in partially saturated fill has been extensively researched. The swelling behaviour of clay has been related to its plasticity, clay content and type by several researchers. (Ref.4, 5, 6).

12. The relationship between initial dry density, moisture content and swelling has been given by Holtz and Gibbs (Ref.7) Barber (Ref.8) and Kassif and Zeitlin (Ref.9).

13. The effect of negative pore pressure on moisture content was extensively investigated by the Road Research Laboratory (Ref.10). The effect of negative pore pressure on swelling has been investigated by de Bruijn (Ref.11) and Blight (Ref.12).

14. Collapse has been examined by Jennings, Knight, Burland (Ref.13), Blight (Ref.12) Matyas and Radnakrishna (Ref.14), Barden Madedor and Sides (Ref.15). The work on collapse has often been concerned with stress path dependancy. Jennings showed that a soil could be inundated and then compressed wet or alternatively compressed dry and then inundated, both stress paths producing the same volume change.

15. The effect of overburden pressure has been investigated by McDowell (Ref.16).

16. The results presented now generally confirm the previous research listed above. The difficulty in making comparisons is that many of the papers related to small numbers of tests which examine one variable, sometimes without keeping the other variables constant or without recording their variation. The number of variables combined with the inconsistency of small soil samples requires a large number of tests to isolate a particular effect.

EXPERIMENTAL WORK

17. Laboratory tests were carried out to determine the components of volume change. Dry clay fill was compacted by a hydraulic press into rigid moulds, loaded vertically with various total (overburden) pressures and then wetted using water supplied at positive or negative pore pressures.

18. The total stress corresponds to the self weight of fill overburden. The water pressure corresponds to inundation in tests using positive water pressure as in a dam, and to moisture transfer in tests with a negative water pressure as in fill placed above the ground water level.

19. The first tests measured volume change on samples which were immediately inundated at different total (overburden) pressures. Additional tests with progressively smaller negative pore pressures indicated the way in which the volume change progressed.

20. Approximately 1500 tests were carried out, many using normal consolidation oedometer apparatus together with some large diameter (0.5m x 100mm) oedometers and oedometers constructed within suction plate apparatus in order to supply water at negative pressure. Various experimental techniques were developed. Cox (Ref.17).

21. The clay used for the main experiment was Keuper Marl, a Triassic mudstone taken from a motorway embankment near Bristol. The gradings of the dry mudstone fragments are shown in Fig.1. When wet graded by sedimentation the mudstone had a clay content of 27%. Mineralogical analysis showed a clay content of 67%. The difference is due to partial cementation.

22. The soil properties are:-
Plastic Limit = 16%) influenced by
Liquid Limit = 35%) cementation
Natural moisture content = 6% to 12%
Angle of friction = 27°
Specific gravity = 2.7
Composition:- Illite 47%, Chlorite 23%
Sepiolite 7%, Quartz 14%, Calcite 2%.
Dolomite 7%.
BS Heavy optimum moisture content = 14%
The properties are extensively detailed elsewhere, Davis (Ref.18), Chandler, Birch (Ref.19).

23. Samples were taken from borrow areas after ripping. They were stored at field moisture content, crushed and recompacted.

24. Further tests were conducted on Coal Measure Shales, Permian mudstones and Eocene London Clay (Ref.20).

25. To determine the rate of wetting soil was compacted into plastic tubes about 75mm. diameter and several metres long. Tubes laid horizontally were allowed access to water under a slight positive head. Tubes placed vertically were allowed to draw up water from the base. Samples were taken through the tube wall to determine the change in moisture content. This was repeated over a period of several months. See also Woolnough (Ref.21).

RESULTS

26. For a particular fill the internal soil properties which affected the amount of swell or collapse were the initial dry density and the initial moisture content. The external factors which affected the amount of swell or collapse were the total (overburden) stress applied and the pore pressure of the water supply.

27. The effect of the initial dry density is shown on Fig.2. After inundation dense samples swell while loose samples collapse. Volume change is also dependent on the total (overburden) pressure which restricts swelling and increases collapse.

28. Fig.3 shows the amount of swell or collapse after inundation for any combination of initial dry density and moisture content at the time of compaction. As the initial moisture content rises (for a constant initial dry density) the amount of swelling first increases then decreases. This is because swelling occurs first at lower moisture content than collapse. The total (overburden) pressure increases the amount of collapse as shown by the broken lines on the graph.

29. Fig.4 shows the swelling or collapse as a dry clay fill is progressively wetted. By using a negative pore pressure in the water supply the volume change at any stage of wetting prior to complete inundation may be determined. The contours indicate the behaviour of fills with different initial dry densities at compaction, and different negative water pressures as the moisture content increases. All the samples shown in this set of results were kept under the same total (overburden) pressure of 28 kN/m^2.

30. Fig.5 combining this information shows the change in dry density and moisture content of samples as they are progressively wetted under different total (overburden) pressures. A sample compacted to a high dry density swells by amounts depending on the total (overburden) pressure. A sample compacted to a low dry density collapses by an amount controlled by the total (overburden) pressure. Samples compacted to an intermediate density swell predominantly at low moisture contents until the amount of collapse exceeds the swelling at higher moisture contents.

31. Fig.6 illustrates the change in volume in relation to total (overburden) pressure and negative pore water pressure, as the moisture content increases. Fig.6(a) shows the behaviour of a dense soil and Fig.6(b) a loose soil.

32. Tests were also conducted using large diameter 500mm dia. x 120mm samples. The samples contained large mudstone blocks. The grading of the fragments is shown on Fig.1. These were inundated at various total (overburden) pressures. The swelling was similar to that for small samples. The amount of collapse was slightly less. However the number of large diameter tests (16) was not quite sufficient to be conclusive.

33. Tests on other clay soils showed similar behaviour to the Keuper Marl used in the main experiment. The amount of volume change appeared to increase with the clay content.

34. The results of the rate of wetting tests on soil columns are shown in Fig.7. In horizontal columns under a small positive head the moisture content increased as a relatively narrow wetting zone passed along the tube leaving a saturated zone behind. In vertical columns under a small negative head water diffused up the column leaving a wide transition zone of soil increasingly saturated with depth. The relative rates of progress of the wetting fronts are indicated on Fig.8. In both cases the rates are very low, 20 weeks being required for 1m. and an estimated 2m. in 80 weeks and 3m. in 180 weeks.

APPLICATIONS

35. Dry clays or mudstones are usually compacted at a uniform density and moisture content throughout an embankment. As the moisture content rises with inundation or moisture transfer the fill volume will alter depending on the total (overburden) pressure. The tests enable the type, rate and amount of volume change to be predicted.

36. The upper layers of an embankment will tend to swell while those at the base will tend to collapse. This will give a net increase or decrease in embankment height. The intermediate volume change may exceed the final one.

37. The sequence of movement will depend on the part of the embankment wetted first. An embankment wetted from the base will tend to collapse before swelling since this is the predominant behaviour at the base. The reverse is true for an embankment wetted from the top. The sequence may also depend on the rate of wetting. An earth dam rapidly inundated might collapse before swelling since the contacts between clay fragments would wet and shear first. An embankment slowly wetted should swell and then collapse because swelling occurs at a lower moisture content than collapse.

Earth Dams

38. Where collapse is accompanied by arching of the surrounding fill it is theoretically possible for larger voids to propagate through the fill at the wetted surface. This would substantially increase the local rate of flow. This aspect has not been investigated in detail, but is indicated by Fig.8 which shows substantial variations in horizontal permeability depending on density.

39. Zoned earth dams may have wet and dry clay fills adjacent to each other. This can lead to differential movement and the formation of voids at zone boundaries. On a large scale it is theoretically possible for dry mudstone shoulders to accept moisture vapour from a wet clay core. This would cause simultaneous shrinkage of the core and expansion of the shoulders. If the core dimensions allow arching between the shoulders then horizontal fissures will form through the core since the vertical strains in the two materials are opposed. See Vaughan (Ref.22). Where zones are inclined or tapered this risk is considerably reduced.

40. The permeability tests indicated that rapid flow could occur through the voids in a loose mudstone fill. Where mudstone is compacted in thick layers a density gradient occurs through the layer. The upper surface of the layer directly below the roller is

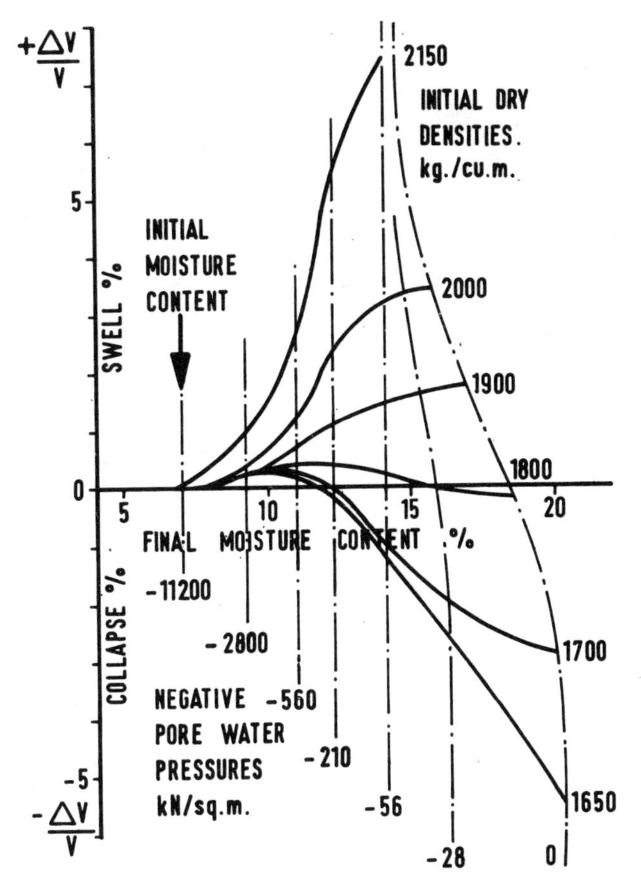

Fig. 4. Swell or collapse during progressive wetting (overburden pressure and initial moisture content constant)

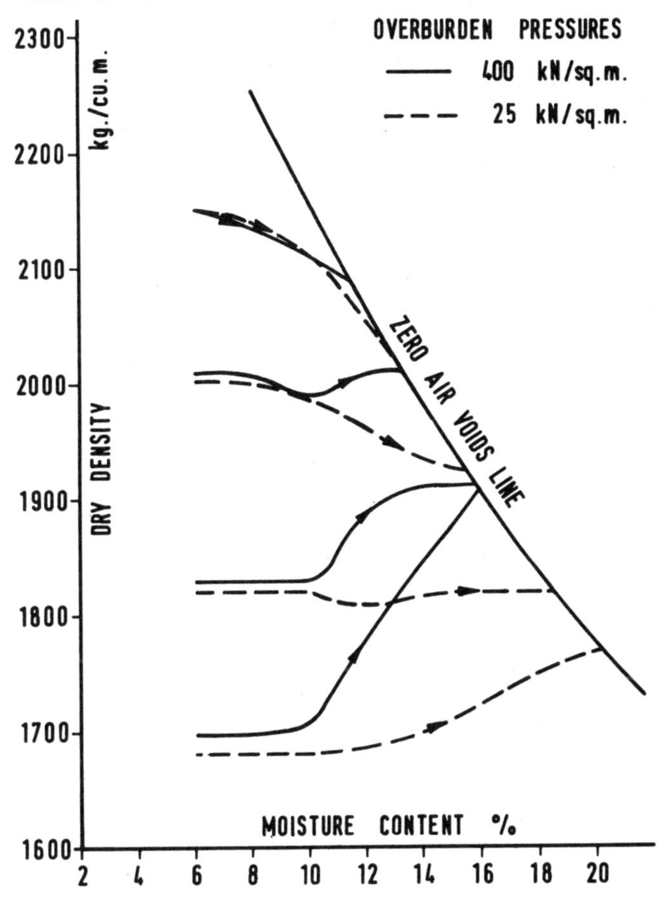

Fig. 5. Change in dry density and moisture content during progressive wetting (overburden pressure constant)

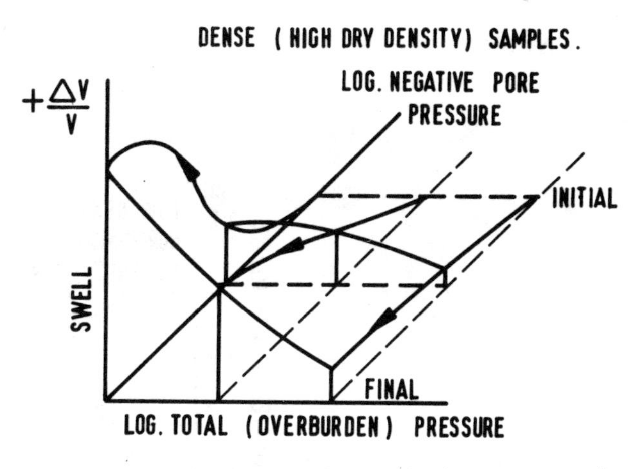

Fig. 6(a). Swell or collapse during progressive wetting (initial dry density and moisture content constant)

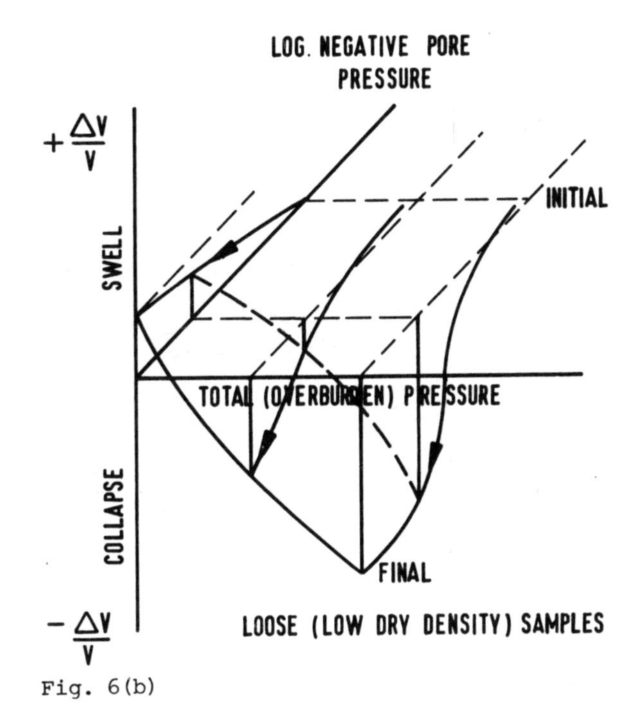

Fig. 6(b)

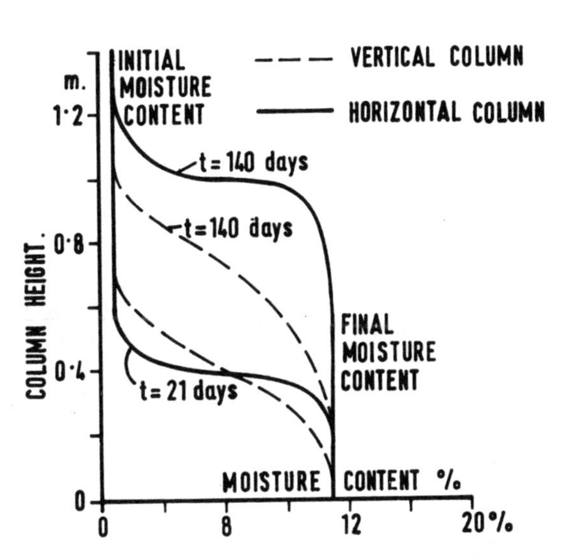

Fig. 7. Moisture increase in long soil samples wetted at base

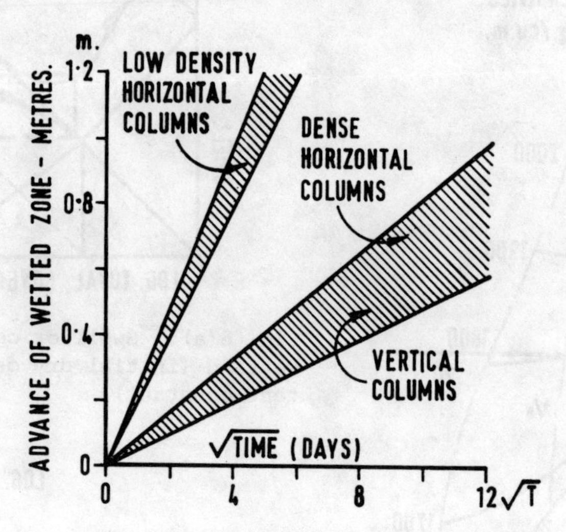

Fig. 8. Movement of wetted zone along soil column

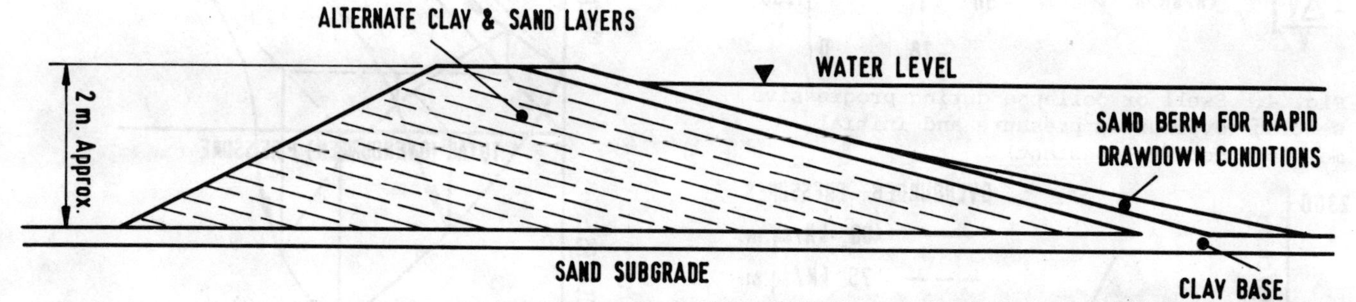

Fig. 9. Inclined compacted layer embankment for balancing pond, West Moors, Hampshire

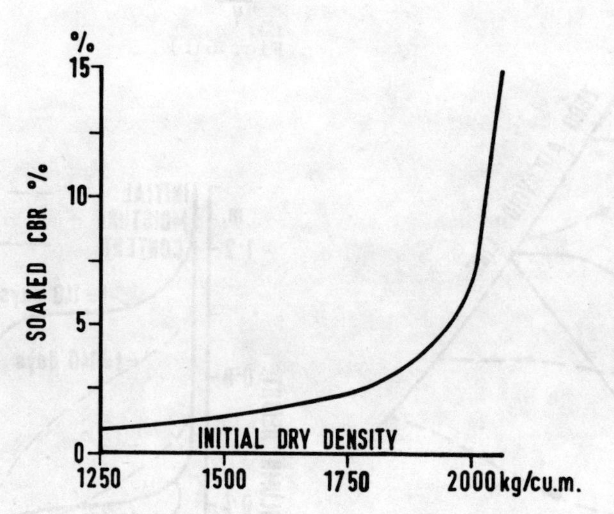

Fig. 10. CBR of samples compacted at various dry densities and soaked at various overburden pressures (initial moisture content constant)

dense and impermeable while the lower part of the layer may have larger voids, a lower density and is much more permeable. The more permeable base of each layer forms horizontal drainage paths through an embankment. A semi-vertical clay core is essential to arrest horizontal flow through such a dam.

41. An alternative method of construction is to tilt the compacted layers so that the impermeable compacted surface prevents horizontal flow. Fig.9 shows the cross section of a small embankment for a balancing pond recently constructed in this way at West Moors, Hants. To cope with the rapid drawdown stability condition and plant capability the upstream slope may have to be shallower than would otherwise be required. However the use of local material and the ease of construction may in some instances be cheaper than importing a smaller quantity of more suitable zoned fill from distance. Experiments with this type of structure are reported elsewhere. Skelton (Ref.23).

Road Embankments

42. Volume changes are not normally a problem within dry clay road embankments. However differential movements just below the subbase are a common cause of pavement failure. Usually surface water enters the fill via joints or the subbase. Wetting can then cause local heave and large differential deflections.

43. Another effect is the loss in the California Bearing Ratio (CBR) as the moisture content increases. This was examined by Birch (Ref.19). Figure 10 shows the CBR values after wetting for the soil used in the main experiment. The values were measured by CBR tests on large diameter oedometer samples after soaking at different overburden pressures. For comparison the CBR of a dry sample would be in excess of 30%.

44. Embankments may expand laterally and this has been recorded by Charles (Ref.24). This may cause minor longitudinal cracking of pavements.

Floor Slabs

45. Dry clay or shale fill under floor slabs will cause heave or settlement, though the final volume change may be less than at the intermediate stages. Floor slabs within structures are not generally suited to accommodating such movements unless special detailing is arranged. The alternatives are to fully suspend the floor to allow for the possibility of heave or settlement or to use a raft resting on the fill.

Piles

46. A related effect of volume change may occur with displacement piles driven into mudstone. At the pile tip mudstone fragments will be displaced by the pile, breaking up along joint planes. Since the displaced fragments will now be randomly arranged dilation must have occurred increasing the volume and leaving water or air filled voids between the mudstone blocks or fragments around the pile. With time the blocks will swell and collapse into the voids as would compacted fill. The increase in moisture content of the mudstone will result in some loss of bearing capacity. Such a loss with time under these conditions is reported by Tomlinson and Sherrell. (Ref.25).

ACKNOWLEDGEMENTS
Thanks are due to Dr. C.S. Dunn, Prof. J. Kolbuzewski, and the late Mr. J. Newey for their kind help and encouragement. The author would also like to thank the University of Birmingham, the Polytechnic of Central London, Messrs. Freeman Fox, Frank Graham and Howard Humphries, for providing facilities.

REFERENCES

1. PENMAN. The failure of Teton Dam 1977. Ground Engineering. Vol.10 No.6. Dec. 1977.
2. Guidance Notes for House construction. National House Builders Council 1977.
3. Road Research Laboratory. Technical Paper 11. 1948.
4. HOLTZ W.G. and GIBBS H.J. 1956 Engineering properties of expansive clays. Trans. A.S.C.E. Vol.121. P.641-663.
5. DE BRUIJN C.A., COLLINS L.E. and WILLIAMS A.A.B. 1956. The specific surface, water affinity and potential expansiveness of clay. Clay minerals bulletin Vol.3. No.17. 1956. P.120-128.
6. McDOWELL C. 1959. The relation of laboratory testing to design for pavements and structures on expansive soils. Quart.Journal of the Colorado School of Mines. Vol.54. No.4. P.122.
7. HOLTZ W.G. 1959. Expansive clays properties and problems. Soil Mechanics Conf. Golden, Colorado. Colorado School of Mines Quarterly, April 1959. (See Ref.4).
8. BARBER E.S. 1956. Discussion of paper by Holtz and Gibbs on "Engineering properties of expansive clays." Trans. American Society of Civil Engineers. Vol.121. P.669-673.
9. KASSIF and ZEITLIN. 1962. Lateral swelling pressure on conduits from expansive clay backfill. Highways Research Bull. 313. 1/12.
10. Road Research Laboratory.
RN 852, 944, 945, 1709, 1858, 1961, 1964, 1978, 2309, 3209,
LN 411, 634,
Tech. Papers 11, 24, 41.
11. DE BRUIJN C.M.A. 1963. Swelling characteristics of a decomposed soil profile at Onderstepoort, Transvaal. Proc. 3rd. Reg. Conf. for Africa on Soil Mechanics and Foundation Engineering.
DE BRUIJN C.M.A. 1965. Some observations on soil moisture conditions beneath and adjacent to tarred roads and other surface treatments in South Africa. Moisture equilibria and moisture changes in soils beneath covered areas. Butterworths 1965. Australia.
12. BLIGHT G.E. 1965. A study of effective stresses for volume change. Moisture equilibrium beneath covered areas. Butterworths.

London 1965.
13. JENNINGS J.E. 1960. A revised effective stress law for use in the prediction of the behaviour of unsaturated soils. Conf. on pore pressure and suction in soils. Butterworths, London.
JENNINGS J.E. and KNIGHT F. The prediction of total heave from the double oedometer test. Symposium on expansive clays. Trans. South African Institution of Engineers. Vol.7. No.9. Sept.1957. P.13-19.
JENNINGS J.E. and BURLAND J.B. 1962 Limitations to the use of effective stresses in partly saturated soils. Geotechnique Vol.12. P.125-144.
14. MATYAS and RADHAKRISHNA. Volume change characteristics of partially saturated soils. Geotechnique. Vol.18. No.4. P.432-448. Dec.1968.
15. BARDEN, MADEDOR, SIDES. Volume change characteristics of unsaturated clays. Journal Soil Mechanics and Foundation Engineering. Div. Proc. ASCE 95. 1969. No.SM1. P.33-51.
16. McDOWELL 1956. Interrelationship of load volume change and layer thicknesses of soils to the behaviour of engineering structures. Highway Research Board. Proc. 1956.35:754/770. See also Ref.6).
17. COX. Settlement Within Embankments. PhD Thesis Dept. of Transportation, University of Birmingham, 1970.
18. DAVIS. CIRIA Technical Reports 1972 on Keuper Marl.
19. CHANDLER, BIRCH. 1964-70. Internal research reports of Dept. of Transportation, University of B'ham. (listed fully in Ref.18).
20. Internal research reports by CHADWICK, CROWE, Polytechnic of Central London 1973, 74. (Also Ref.17).
21. J.R. WOOLNOUGH. 1973. Internal research report. Polytechnic of Central London.
22. VAUGHAN. 1970. Cracking of clay cores of dams. (Discussion at ICE). Proceedings I.C.E. Vol.46. P.115,116. May 1970.
23. SKELTON M, 1973. MSc. Report (uncompleted) Melbourne.
24. PENMAN and CHARLES. 1972. Deformations in rockfill dams. Building Research Station. CP.18/72 and CP.19/72. November 1972.
25. GEORGE, SHERRELL and TOMLINSON. 1976. The behaviour of steel H piles in slaty mudstone. Geotechnique. March 1976.

J. P. DENNEHY, BSc, MPhil, formerly Senior Soils Engineer, Ground Engineering Ltd

The remoulded undrained shear strength of cohesive soils and its influence on the suitability of embankment fill

This paper investigates some aspects of the use of cohesive materials as embankment fill. The restrictions imposed on the acceptable strength of fill by trafficability requirements of the plant employed during construction are examined. Field observations show that the depth of rut caused by trafficking plant can be related to the remoulded undrained shear strength of the soil. Equivalent strengths can therefore be obtained for the observed practical limits of plant operation. Tyre pressures and vehicle size are found to influence these limits.

From a study of the remoulded undrained shear strength behaviour of cohesive soils it is shown that there is a general relationship with moisture content for different soils. The principal influence on this behaviour is found to be that of liquid limit whilst the plastic limit is shown to be an inaccurate guide. A relationship is found to exist, at any remoulded undrained shear strength, between the liquid limit and the moisture content required to produce this strength. Universal limits in terms of a multiplicand of plastic limit are found to be inappropriate.

A direct undrained shear strength criterion is advocated as it relates to the functional limitations of plant and the immediate stability of embankments. It is thus easily adaptable to the conditions of usage, and has the benefit of being directly related to WL which can therefore be used as an indirect control parameter.

INTRODUCTION

1. In Great Britain, road embankment materials which are obtained from the upper strata of the ground probably contain more cohesive, rather than granular soils, the average natural moisture content of which frequently exceeds optimum. Additionally, exposure of cohesive materials to unfavourable climatic conditions can cause relatively rapid increases in soil moisture, particularly where the soil is being worked. As a consequence wet cohesive materials often constitute a high proportion of earthworks and have a considerable effect on the economics of earth moving. The most economic earthworks will however result when the maximum amount of such materials is utilised. It is therefore important to optimise their use by establishing design criteria which relate suitability to the actual conditions of usage. This paper is concerned with how cohesive soils behave when remoulded, and with establishing some practical limits for their use as embankment fill.

2. The work on this paper commenced in 1972 when the generally accepted arbitrary moisture content for road works in this country was 1.2 Wp. In 1976 the D.O.T (ref.1) suggested the principle that if materials can be excavated, transported and compacted they are safe for most earthworks, and quoted Farrar and Darley (ref.2a), who concluded that certain soils up to 1.3 Wp could be used. Previously Farrar (ref. 3) had indicated that a limit up to 1.4Wp might be feasible for some soils in embankments up to 5m high placed at 1:2 side slopes. Arrowsmith (ref.4) suggested that materials as wet as 1.3Wp have been used and could be suitable. However, Rodin (ref. 5) and Snedker (ref. 6) had indicated that an undrained strength criterion would be a more appropriate test of suitability, and that different restrictions might be appropriate for different soils. Of the few reported cases relating to field trials, McLaren (ref. 7) found that a boulder clay at the limit of vehicle traffickability formed a stable 6m high embankment at side slope of 1:2.

3. It is considered that there are three conditions which cohesive fill must meet in order to be suitable for embankment purposes. These are:

1) the ability to be excavated, transported, placed and constructed with normal equipment by normal methods.

2) the ability to form embankments with stable side slopes.

3) the capacity to have internal self-settlements which do not adversely effect the road profile.

4. Condition 1 is considered the most critical, Dennehy (ref.8), since mobility restrictions may prevent the effective implementation of an otherwise feasible embankment design. The first part of this paper is therefore concerned with an attempt to establish the traffickability limits of some typical earth moving plant. The second part of the paper investigates the undrained shear strength behaviour of remoulded cohesive soils. These

studies enable the current method of specifying fill suitability on the basis of a multiplicand of Plastic Limit to be examined in relation to trafficability limits.

PART 1: TRAFFICKABILITY

General

5. One of the major problems in using weak materials with high moisture contents is the difficulty experienced in driving heavy earthmoving machinery through them. The limit of trafficability for cohesive soils is often reached sooner when placing materials than when excavating them (Rodin, ref. 9) particularly for heavily over-consolidated clays (such as London Clay). When placing weak relatively loose and lubricated soil, the wheels of plant may sink in where the soil is over-stressed causing deep ruts as a result of excessive plastic deformation of the soil. Spinning and kneading by heavy tyred plant may at the same time cause slurrying of the soil. Considerable additional power will then be needed to move the vehicle forward against the soil drag.

6. The moisture content of the trafficked soil affects its undrained shear strength, which in turn affects the depth to which rutting of fill will occur. A study was therefore carried out to determine the affect of variations in remoulded undrained shear strength on the mobility of typical earthmoving plant as represented by rutting.

7. Observations were made at a number of sites of scraper-type vehicles placing wet cohesive fill in embankments. The rut depths caused by loaded plant trafficking levelled wet fill were noted together with tyre pressure details. In the ruts formed by each vehicle, several field hand vane tests were conducted with a Geonor type instrument. 76mm x 38mm diameter samples were also recovered from the rut bases for undrained triaxial compression and classification tests. These samples were tested at normal laboratory rates of strain. Since this rate is somewhat slower than the rate at which trafficking plant will impose load, the triaxial test results would be expected to produce under-estimates of the support available to trafficking plant.

8. The information collected was supplemented by some field vane data collected by Farrar and Darley, ref. 2b, for soils of a less wet nature. The combined data generally fell into two categories: that relating to medium to large scrapers having a struck capacity of $16m^3$ to $25m^3$ and tyre pressures between 340 and 380 kN/m^2 and that relating to medium to light and towed vehicles having a struck capacity of $15m^3$ or less, and having tyre pressures between 240 and 310 kN/m^2. From this data, possible tentative relationships between rutting and the undrained shear strength of fill for these two plant groups, are proposed in Fig. 1.

9. Fig. 1 explains why the common site expedient of reducing tyre pressures in soft soils is effective in increasing trafficability

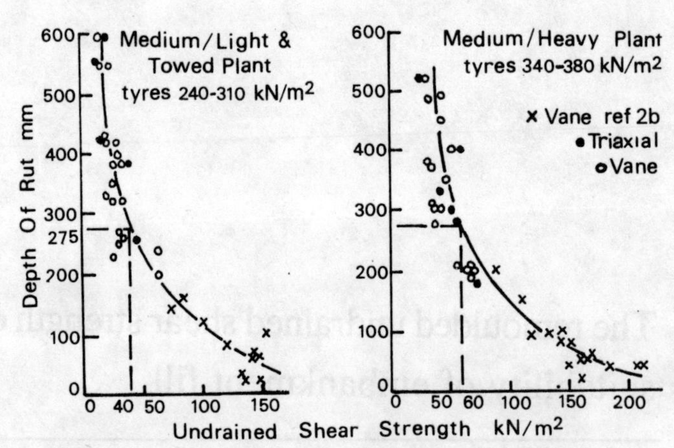

Fig. 1. Influence of shear strength of fill on wheel rutting of scrapers

In simple terms wheel contact pressure = wheel load ÷ wheel contact area, although some support is provided by the tyre carcase. Thus for the same wheel load, an increase in inflation pressure causes a decrease in the tyre contact area and increases the contact pressure. Higher stresses are therefore imposed on the near surface soils and a larger deformation results. If tyre pressure is kept constant and wheel load is increased, the contact area must also increase. Contours of stress distribution will therefore extend to a greater depth and more soil will be stressed to the same degree than for a wheel under lesser load, so again rutting would be expected to increase.

10. Small and medium sized vehicles generally carry less load per tyre than larger vehicles, (see Table 1), and operating tyre pressures are generally lower. Both these factors indicate that wheel sinkage will be greater for heavier machines with higher tyre pressure, and this has been observed in the field. However, it should be appreciated that large wheels and tyres are more efficient than small in soils that produce deep ruts. This is because the rolling resistance is less for larger wheels and there is greater facility for the bearing surface between tyre and soil to increase as the ruts are forming.

11. The spread of points in Fig. 1 is large reflecting to some degree the considerable

Table 1. Trafficability criteria of rubber-tyred vehicles (scraper type) on cohesive soils

Unsuit	Suitable	Tyre Pressure KN/m²	Typical maximum load per tyre Kg			
Shear Strength KN/m²						
<60	<80	340 to 380	9500	8500	7150	MH
			3150	2850	2400	ML
<40	<60	240 to 310	8000	7000	6000	MH
			2700	2400	2050	ML
200	275		16	48	54	
Rutting mm			Speed km/hr.			

Plant: MH Medium/Heavy; ML Medium/Light
Load data from manufacturers handbooks

number of variables involved:- soil type; vehicle load; non-uniformity of soil beneath the wheels etc. The suggested relationships are therefore only very general mean upper limits. Nevertheless such relationships, although requiring further justification, can be useful guides to the undrained shear strength of fill in the field.

Traffickability limit criteria

12. It is desirable that imposed restrictions to ensure traffickability are set no higher than those at which plant can work unassisted. Operating at lower criteria can result in severe damage to the formation as a result both of the deeper ruts caused and the subsequent attempts of scrapers to extricate themselves. It also becomes more difficult to reinstate the ruts. It is considered that the limit of traffickability occurs at the remoulded undrained shear strength at which fully loaded medium to heavy scrapers (capacity greater than $16m^3$) can just operate, unassisted, on fill. Site observations indicate this limit to be represented by ruts of about 275mm. (With assistance plant have been observed to work readily at up to 400mm). At depths below 200mm no traffickability difficulties were observed. Thus the area of marginal traffickability would be defined by ruts of about 200mm to 275mm. Even though they have been seen to work unassisted in deeper ruts, similarlimits are suggested for light to medium plant because, in general, such vehicles have smaller wheels, and the increased rolling resistance at greater rut depths restricts manouverability.

13. Using Fig. 1 these rutting limits can be related approximately to the required undrained shear strength of soils for the two tyre pressure ranges. It can be seen that for medium to light scrapers with low tyre pressures, soils with undrained shear strengths of $60kN/m^2$ or more would be suitable for trafficking whereas soils with strengths of $40kN/m^2$ or less would not. For medium to heavy scrapers at high tyre pressures, the respective figures for suitable and unsuitable would be $80kN/m^2$ and $60kN/m^2$. Table 1 summarises these limits.

14. The tyre pressure ranges can be expressed in terms of undrained shear strength Cu at these proposed limits of traffickability. Thus for vehicles operating at tyre pressures of 240 to 310 kN/m^2 the lower limit of traffickability, 40 kN/m^2, could be expressed as 6.0Cu to 7.75Cu. Similarly, the $60kN/m^2$ lower limit for tyre pressures of 340 to 380 kN/m^2 would be 5.7Cu to 6.3Cu. Although the actual pressure transferred to the ground will be less, because of the strength of the tyre carcase, it is apparent that the proposed limits correspond also to ultimate failure beneath the tyres.Thus the desirability of restricting ruts to less that 275mm is confirmed, since plastic failure is likely to occur above this limit.

Slope stability and Self Settlement

15. The adoption of any strength limit imposes certain conditions of moisture content and density on the fill, the influence of which on slope stability and self settlement should be checked. Experience indicates that for most cohesive soils, except perhaps those of low plasticity, neither self-settlement nor slope stability of embankments placed on competent ground are undue problems when fill is placed at undrained shear strengths of between $40kN/m^2$ and 60 kN/m^2. Although embankment slope stability is a long term consideration, short term analysis by a method such as Taylor's (ref.10), indicates that at these strengths embankments with side slopes of 1:1.5 or 1:2 placed on a hard foundation will be stable at heights between 9m and 14m. That is over a range of heights common in road works.

16. Self-settlement of an embankment may effect the riding quality of the road, alter the vertical curve or result in differential settlements at bridges or other structures. Within embankments differential settlements will be avoided if the fill is placed in layers of even thickness and ruts are progressively filled before the placement of the next layer. Additionally experience indicates that calculations of self settlement based on the consolidation resulting from the self weight of the fill tend to over-estimate self-settlements (Vaughan, ref.11), particularly for soils placed near the top of the fill which are not fully saturated. Within these limitations a certain amount of self-settlement can be tolerated. So although further research is required in this field, self-settlement is usually regarded as a non-critical factor, and at the strengths considered is usually within the 100mm - 150mm values quoted as being acceptable to the D.O.T. by Snedker (ref. 6.)

PART 2: THE REMOULDED UNDRAINED SHEAR STRENGTH BEHAVIOUR OF COHESIVE SOILS

17. Because soil placed as fill undergoes mechanical working, the remoulded undrained shear strength of a soil is considered to be of more relevance to the suitability of soils in embankment than the undisturbed strength. Traffickability and embankment stability depend on the support that can be provided by a soil and the remoulded undrained shear strength must therefore influence the suitability of cohesive soils as fill material.

18. A study was therefore carried out of the remoulded undrained shear strength properties of a wide range of soils (recovered by percussion boring) from various sites in England. A homogenous sample of about 100kg to 200kg was formed of each soil, at its as-recovered moisture content, using a soil mixer. From each composite sample several specimens were prepared by recompacting at pre-determined moisture contents, the material being either uniformly wetted-up or air dried. After compaction the specimens were thoroughly waxed and sealed, then left for a minimum of 24 hours before testing to allow equalisation of moisture content. Only soils wet of optimum were studied. In order to keep air voids to a minimum over the wide range

of moisture contents at which tests were required, and so facilitate comparison between materials, 'heavy' compactive effort was employed (BS 1377 test no. 13).

19. A full range of identification tests was performed on each soil. The mixing process produced consistent results and mean indices are given in Table 2. This table lists all the soils studied and includes other published information, although the method of preparation of the latter samples differs from the controlled tests and in some cases is unknown. Published data was obtained from Skempton and Northey (ref.13) and Dumbleton and West (ref.14). Information was also available from other sources. The soils listed in Table 2 cover almost the full range of clay soils encountered in Britain, and all plot above tbe Casagrande 'A' line.

Factors influencing behaviour

20. A series of undrained triaxial compression tests was performed at different moisture contents for each material investigated in order to relate undrained shear strength to moisture content. The results showed that there is a relationship of a similar form between undrained shear strength and moisture content for different clay materials when the air voids are 5% or less. Fig.2 gives a detailed typical result of testing, and indicates that the variation of cell pressure over the range used, 50 kN/m^2 to 200 kN/m^2, has only a small influence on the relationship because of the near-saturated nature of the specimens. Some test results, at high moisture contents, appear to diverge from the general trend. Such behaviour may be related to the grain and void sizes of a particular material. It could be that the relationship breaks down when there is little grain to grain stress transfer.

21. When the results are plotted on a log-log basis straight line relationships can be obtained between remoulded undrained shear strength and moisture content to good accuracy. The best straight line drawn through each set of data produces a series of approximately parallel lines. The results for all soils showed this form of relationship, even though some soils were tested unconfined to simulate field placement/trafficking conditions. Fig.3 which represents more than 150 individual sets of test results gives the relationship for the

Table 2. Summary of test results

No.	MATERIAL	W_L %	Wp%	W_I %	ACTIVITY WI/CLAY	SOIL AT 40kN/m^2			Cu(kN/m^2)	
						W%	$\frac{W}{W_p}$	L.I.	at Wp	at 0.5WL
1	SANDY CLAY	32	17	15	1.36	17.3	1.02	0.02	44	80
2	BRACKLESHAM BEDS	30	19	11	1.10	18.0	0.95	-0.09	30	91
3	BRICKEARTH, HARMONDSWORTH	31	18	13	-	18.5	1.03	0.04	52	92
4	HORTON CLAY	30	16	14	0.36	19.0	1.19	0.21	135	123
5	YELLOW BROWN BOULDER CLAY	35	18	17	-	20.0	1.11	0.12	70	77
6	BROWN SILTY CLAY	39	18	21	-	20.8	1.16	0.13	82	46
7	BROWN CLAY	44	19	25	0.71	22.0	1.16	0.12	83	42
8	UNWETHD LONDON CLAY WL<50	47	20	27	1.18	23.0	1.15	0.11	82	38
9	WEATHERED FOREST MARBLE	49	18	38	0.67	23.8	1.32	0.15	150	36
10	WEATHERED OXFORD CLAY	53	17	36	0.97	24.4	1.43	0.21	220	30
11	UNWEATHERED OXFORD CLAY	49	18	31	0.91	26.0	1.44	0.26	215	51
12	BAGSHOT BEDS	56	20	36	1.16	26.0	1.30	0.17	135	28
13	GREY CLAY +	49	20	29	-	26.8	1.34	0.23	163	51
14	REDDISH BROWN CLAY	59	21	38	0.76	27.0	1.28	0.16	125	30
15	WEATHERED LONDON CLAY	62	22	40	1.18	27.8	1.26	0.15	115	23
16	GAULT CLAY *	70	25	45	-	30.5	1.22	0.12	118	26
17	LONDON CLAY	73	25	48	0.96	32.5	1.30	0.16	135	23
18	UNWETHD LONDON CLAY W_L>70	78	26	52	1.13	33.5	1.29	0.14	125	19
19	GAULT CLAY	82	27	55	1.00	33.5	1.24	0.12	107	16
20	LONDON CLAY, HEATHROW	78	29	49	-	34.0	1.17	0.10	83	30
21	LONDON CLAY, FOREST HILL	87	27	60	1.15	35.4	1.31	0.14	140	15
22	SHELLHAVEN	97	32	65	1.27	40.0	1.25	0.12	115	16

1,9,10,11 Nuttall Geotechnical Services Ltd. (1973)
4,17,22 ref. 13. 3, 20 ref. 14. All others ref. 8.

* W_L by 4 point cone method
+ Mixture of two or three soils

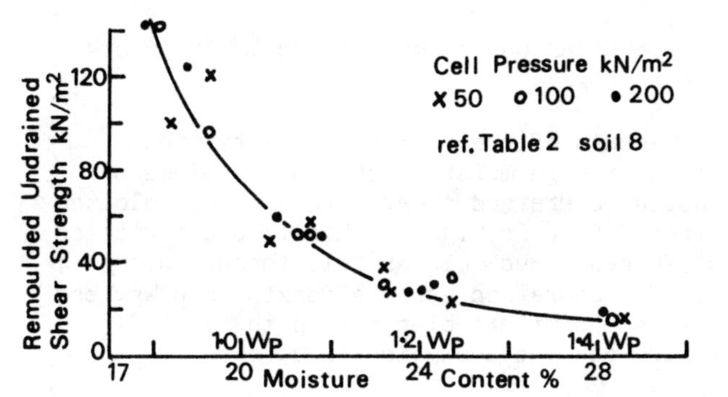

Fig. 2. Typical moisture content/strength curve

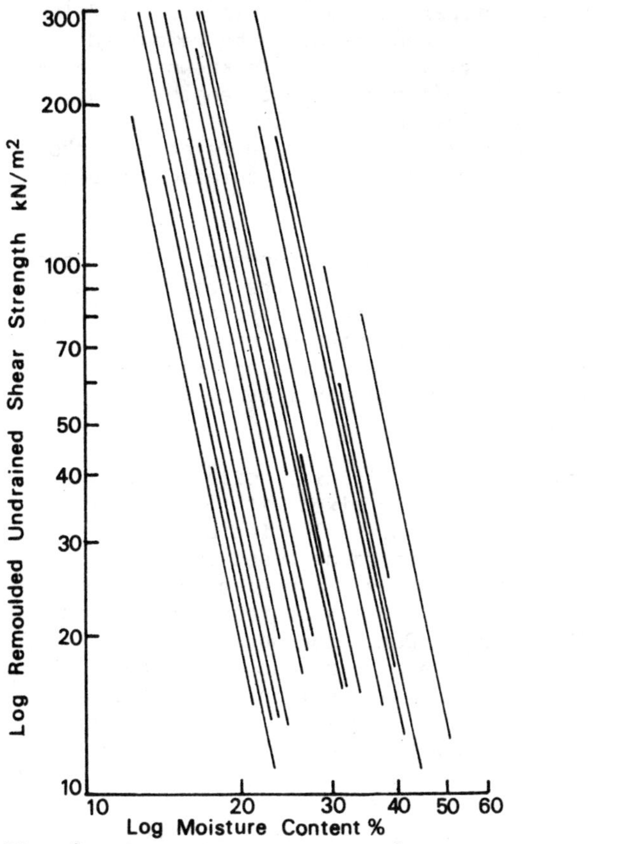

Fig. 3. Log moisture content/strength relationships

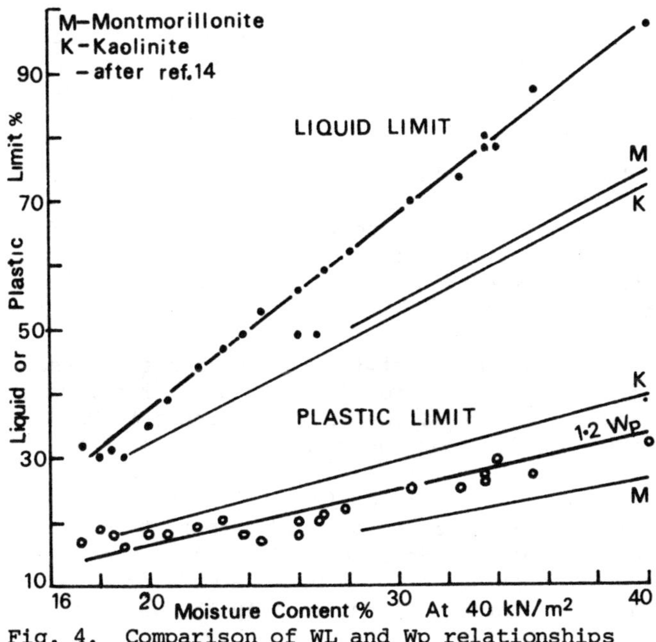

Fig. 4. Comparison of WL and Wp relationships with moisture content at a given strength

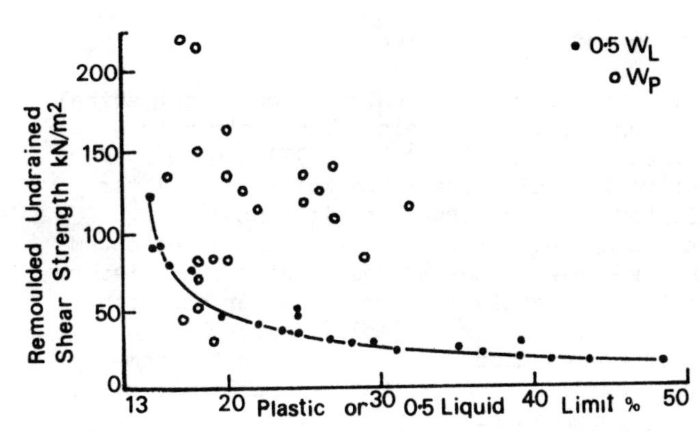

Fig. 5. Influence of Wp and WL on remoulded strength

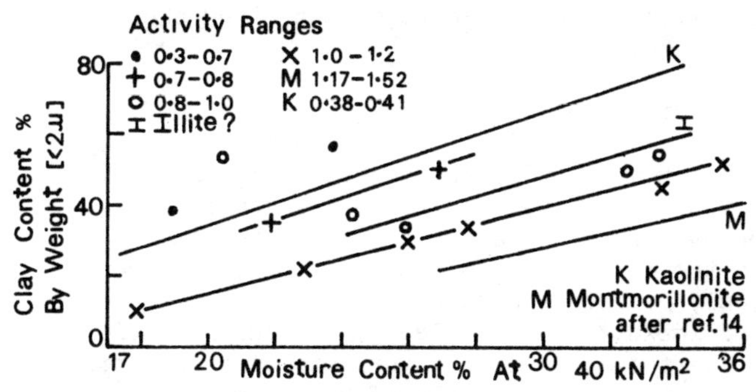

Fig. 6. Influence of activity on remoulded strength

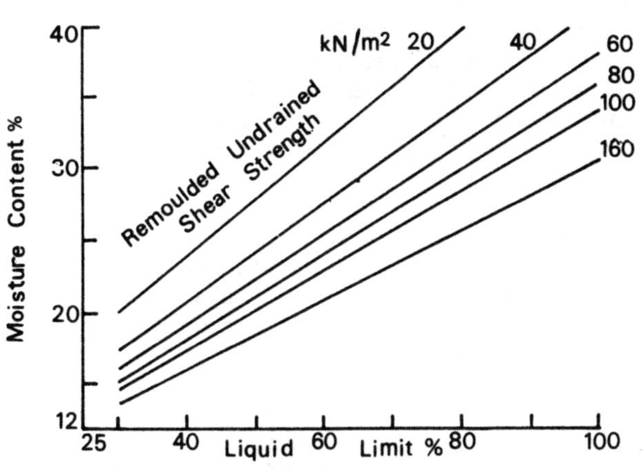

Fig. 7. Remoulded undrained shear strength contours

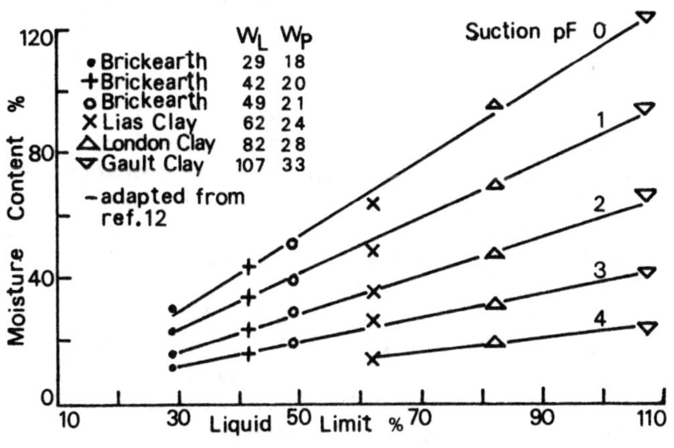

Fig. 8. Suction contours for remoulded soils

twenty two soils examined.

22. Rodin (ref. 5) calculated some hypothetical moisture content/undrained shear strength curves based on Black's (ref. 12) proposals for estimating CBR values. Snedker (ref. 6) later adopted Rodin's curves in his approach to assessing suitability. These curves indicated that the higher the Wp the greater the moisture content for a given undrained shear strength. However, reference to Table 2, where the materials examined are arranged in the same order as Fig. 3, shows that the materials do not align in definitive order of Wp. The W :Wp ratio, at, for example 40 kN/m^2, shows a fluctuating and wide variation between 0.93 and 1.44, although as Fig. 4 shows, there is an overall tendency for the moisture content to increase with the Wp for a given undrained shear strength.

23. Fig. 4 shows that for soils at 40 kN/m^2, a summary of behaviour could be approximated by the commonly utilised criterion of 1.2 x Wp. Whilst this line summarises the range of behaviour of different soils, it is clear that in the particular case this criterion cannot apply because the relationship is inaccurate and inexact. Particularly, for less plastic soils the 1.2 Wp line over estimates the strength, whereas for soils of higher plasticity a 1.3 Wp summary line would be more appropriate. Additionally, if a higher undrained shear strength of fill than 40 kN/m^2 was required even the 1.2 x Wp approximation would be an inappropriate generalisation - a smaller mutiplicand would be required. The lack of a satisfactory relationship between remoulded undrained shear strength and Wp becomes more obvious when Wp is plotted against strength at Wp, as in Fig. 5.

24. Other parameters, such as activity and liquidity index which have previously been thought to influence the remoulded strength behaviour of soils (e.g ref. 12), showed non-definitive tendencies when directly plotted against control parameters (see Table 2). Clay content and activity in combination may have some effect on this behaviour (refs. 14,8) as Fig. 6 shows. However, Fig. 6. would be an ineffective practical control, since each line represents a broad range of activities and natural soils are not simple one clay minerals. Their mineralogy is difficult to determine and so also is the clay percentage due to the nature of the sedimentation analysis.

25. Table 2 shows that of the simple classification parameters, it is the liquid limit which has the major influence on the sequence of curves in Fig. 3. This influence is very evident when WL is plotted against moisture content for any given remoulded undrained shear strength, as in Fig. 4 for 40 kN/m^2. The points plotted, which are taken directly from the linear relationships proposed in Fig. 3, produce a narrow spread which can be generalised by a straight line. General contours for soils of different remoulded undrained shear strength can be constructed in a similar manner to Fig. 4, with a similar degree of accuracy. These are shown in Fig. 7.

26. Because of the difficulty of testing samples at high moisture contents, values of remoulded undrained shear strength were plotted against 0.5 WL in Fig. 5. This figure indicates that WL has a much closer relationship with remoulded undrained shear strength than Wp, and reflects a similar relationship to that of moisture content shown in Fig. 2.

27. From data presented by Black (ref. 12) it is possible to prepare a plot similar to Fig. 7 for suctions rather than strengths of remoulded soils, as in Fig. 8. Since it is known that suction directly influences the strength of saturated soils Fig. 8 being analogous to Fig. 7 tends to confirm that WL can be related to the undrained shear strength of saturated remoulded soils.

28. The WL relationships proposed in Fig. 7 show a tendency to break down as the soils become coarser and the clay content decreases. This is illustrated by the scatter of points in the vicinity WL = 30% in Fig. 4. Conversely at higher WL's the reproducibility of results appears to be good.

29. Some work on artificial soils consisting of pure clay minerals mixed with various proportions of coarse materials, indicates that while the relationships proposed in Fig. 7 may be valid for the majority of British soils, where the clay minerals are mixed, they are not necessarily unique. Dumbleton and West (ref. 14) examined Kaolinite/quartz sand and Montmorillonite/quartz sand mixtures; interpretations of their results at a remoulded undrained shear strength of 40kN/m^2 are overdrawn on Fig. 4. Their results regarding WL fall below the line proposed for natural soils (on the safe side from a prediction point of view). However, since there is little deviation in the lines for Kaolinite and Montmorillonite, about 1% of moisture, differences with the proposed relationship for natural soils may be more a reflection of the distinction between natural and artificial soils than of mineral content. The mineral content, however, appears to have a considerable influence on Wp with the results for natural soils falling within the bounds of the results for artificial soils. The potential influence of mineral content and inaccuracy of the 1.2 Wp line at a remoulded undrained shear strength of 40kN/m^2 could be up to 10% of moisture.

30. Fig. 4 shows that for the Wp line the range of Wp's is small, from 16% to 32% and the spread produced is large. The WL line appears to be more accurate, and covers a wider range, from 30% to 97%. Inaccuracies in Wp therefore have a greater effect. Sherwood (ref. 15) found that one third of all Wp results could be more than ± 3 units in error. Reference to Fig. 4 indicates that this error corresponds to a wide range of moisture contents. Additionally the Wp is a more subjective test than the WL and appears more subject to mineral content influence. It is

clear therefore that the relationship of W_L to remoulded undrained shear strength is more accurate than that of W_p.

31. All but one of the determinations of WL in this paper were made using the Casagrande method. The introduction of the new preferred method of WL testing, recommendedin BS1377 (1975), using the fall cone penetrometer should enable even more accurate and reproduceable results to be obtained. (Sherwood and Ryley, ref.16; Sherwood ref. 15). This should lead to greater accuracy in deductions made in terms of WL. One soil, No. 16, where the cone method was employed produced values of remoulded undrained shear strength at different moisture contents which which fell exactly on the lines proposed in Fig. 7.

CONCLUSIONS

32. It has been shown that relationships of a similar form between moisture content and remoulded undrained shear strength exist for near-saturated cohesive soils. These relationships could not be related adequately to Wp although W at a given strength tended to increase as W_p increased possibly because of the large influence of mineral content. Consequently the method of specifying an upper moisture content limit as a rigid function of Wp, whatever multiplicand is used, appears to be erroneous and will not result in the optimum specification for individual materials. As Fig.4 shows for more than half the soils a 1.2 Wp stipulation would be uneconomic, whilst in other cases it would grossly over estimate suitability. The 1.2Wp criterion appears to be a crude guide to a remoulded undrained shear strength of 40kN/m^2, but if a greater strength than this was required in embankments, for example because large plant was to be employed, this criterion would be totally inappropriate.

33. A linear relationship was found to exist between the moisture content at a given remoulded undrained shear strength and the liquid limit of a soil, (Fig.4). This relationship appeared to be valid at any given strength, to be less influenced by mineral content, and was corroborated by comparison with work on remoulded soil suction.

34. The reason for the influence of WL on remoulded undrained shear strength and the inappropriateness of the Wp in this context may lie in the nature of these tests and in what they attempt to measure. In the liquid limit Casagrande type test the specimen is thoroughly remoulded, placed in the cup and then grooved. The liquid limit is then determined as the moisture content of the soil when a particular stress application (25 blows) results in 'slope failure'. Similarly, the fall cone test is a measure of resistance to an imposed stress. Intrinsically then, the liquid limit test itself would appear to be a measure of remoulded soil strength. On the other hand the plastic limit test does not appear to be stress related, and would appear to be at best a measure of malleability or bonding.

35. The suggestion is that WL rather than Wp or other classification parameters is the dominant behaviour identification factor, and that a knowledge of its value could be used to define the critical moisture content of a material which will ensure a given undrained shear strength.

36. It is suggested that the most appropriate way of specifying the suitability of clay materials for fill purposes is an undrained shear strength basis. Apart from the desirability of such a criterion to ensure immediate embankment stability, it has been shown that certain values of remoulded undrained shear strength are required in order to maintain plant trafficability. Problems of soil placement that have been experienced in the past must therefore be partly attributable to the irrational application of the 1.2 Wp criterion.

37. Employing blanket criteria such as 1.2 Wp or 1.3Wp ignores the design conditions under which soils are to be used, consequently material which might be suitable under different design conditions(e.g. in lower embankments) is being rejected. Additionally, there is the danger of unsatisfactory materials being used although they satisfy such criteria because at these moisture contents the undrained shear strength may not be sufficient to maintain embankment stability.

38. There appears to be a general relationship between rutting and undrained shear strength. Although this requires further substantiation, a useful guide to the undrained shear strength of fill can result by observing the depth to which levelled fill ruts, in conjunction with Fig. 1. The increasingly larger rutting that can result from small changes in undrained shear strength at values below about 80kN/m^2 can be explained in part by the influence of moisture at these strengths. Fig. 2. illustrates that below this strength small decreases in strength correspond to large increases in moisturecontent. This effect becomes more pronounced at the lower strengths. Consequently, in fills of low undrained shear strength, wet weather, whilst it may not significantly decrease strength, may result in excessive rutting requiring the suspension of works. But in fill of high undrained shear strength, the effect of wet weather will produce less severe increases in rutting, and work may continue. This is an important point, for it indicates the need for an undrained shear strength criterion to ensure that plant can work, in addition to the need for a minimum undrained shear strength to ensure immediate embankment slope stability.

39. The undrained shear strength criterion can be chosen in accordance with the engineering requirements of the soil. However, in order to prevent excessive damage to the formation it is desirable to set trafficability limits no lower than those at which plant can work unassisted. Suggested trafficability limits are given in Table 1. with the

important lower strength limits considered to be 40 kN/m^2 and 60 kN/m^2. Although few problems with embankment stability and self settlement are envisaged at these limiting strengths, these two factors should be checked before an earthworks strength criterion is adopted.

40. It is suggested that the relationship of WL with remoulded undrained shear strength provides a means whereby an assessment of soils suitability for earthworks can be more accurately made through correct and adequate testing at the site investigation stage. Nevertheless, it is considered essential that the validity of the WL relationships in Fig. 3 is checked at the site investigation stage since deviations may occur. Thus sufficient remoulded undrained shear strength tests at different moisture contents should be performed to establish the particular relationships for the principal soil types that will be extracted from cuttings for use as embankment fill. These relationships can then be used to relate the critical strength which ensures traffickability to easier control parameters. Sufficient classification tests should also be performed if a figure such as Fig 3 or 7 is to be used as a guide to soil strength and hence suitability. If relationships such as Fig.3 and 7 can be accepted they will provide a useful field control, with the advantage that rapid approximation of the critical moisture content can be obtained without the expense of resorting to triaxial testing except in dubious cases.

41. In discussing the adoption of a strength criterion for earthworks the concept of marginal materials has been retained to allow for natural variations in soils and the distribution of test results. According to Sherwood (ref 15), ond third of Casagrande WL results may be in error by more than 8%. Such a concept also caters for the contingencies involved in formulating the criteria, and allows the Engineer on site a certain degree of flexibility. It is suggested that in addition to classification tests, field controls could also include the use of hand vane tests. A design graph could be constructed for a particular site which could possibly be incorporated into contract specifications. Alternatively a direct undrained shear strength criterion could be specified.

42. The suitability of cohesive materials for embankments depends very much on their conditions of usage and the plant employed. The adoption of a strength criterion would enable the selection of suitable materials to be placed on a more scientific basis, but would not mean the inclusion of more 'marginal material' in embankments per se. This may mean that some unmanageable materials which are now being accepted should in the future be rejected. Similarly some materials now being rejected may in the future be used as long as they are manageable. However, more economical, consistent and better quality earthworks should result.

REFERENCES

1. ANON. Notes for guidance on the roads and bridge works specifications. Department of the Environment, 1969 and 1976.
2. FARRAR D.M.& DARLEY P. The operation of earthmoving plant on wet fill a) T.R.R.L. report 688, 1976. b) Draft report, 1972.
3. FARRAR D.M. A laboratory study of the use of wet fill in embankments. R.R.L. report LR406, 1971.
4. ARROWSMITH E.J. Earthworks on motorways-from the viewpoint of the design engineer. Highways and Traffic Engineering 1971, 39 (1736/7) 18-20
5. RODIN S. Ability of clay fill to support construction plant. Civil Engineering and Public Works Review. 1965, 60 (703/4) 197-202 and 343-345.
6. SNEDKER E.A. Choice of an upper limit of moisture content for highway earthworks. Highway Design and Construction, 1973, January.
7. McLAREN D. M6 Trial embankment at Killington. R.R.L. report LR238,1968.
8. DENNEHY J.P. A new method for assessing the suitability of cohesive soil for use as embankment fill. M. Phil thesis. Surrey University, 1976.
9. RODIN S. Earthworks- some practical aspects in the U.K. Part 2. Muckshifter and Bulkhandler, 1965, 22 (5) 343-345.
10. TAYLOR D.W. Fundamentals of soil mechanics J. Wiley, New York, 1948.
11. VAUGHAN P. Private Communication, 1976.
12. BLACK W.P.M. A method of estimating the C.B.R. of cohesive soil from plasticity data. Geotechnique 1962, 12 (4) 271-282
13. SKEMPTON A.W. & NORTHEY R.D. The sensitivity of clays. Geotechnique, 1952, 3(1) 30-52
14. DUMBLETON M.J. & WEST G. The suction and strength remoulded soils as affected by composition. R.R.L. report LR 306, 1970.
15. SHERWOOD P.T. The reproduceability of the results of soil classification and compaction tests R.R.L. report LR339, 1970
16. SHERWOOD P.T. & RYLEY M.D. An examination of cone penetrometer methods for determining the liquid limit of soils. R.R.L. report LR 233, 1970.

ACKNOWLEDGEMENTS

This work was carried out as part of a continuing research programme by Ground Engineering Ltd. and financed by J. Laing & Son Ltd. The author wishes to thank the management of both companies for permission to publish the work, and particularly Mr. H.G. Clapham for his assistance and advice in preparing the paper.

W. J. DOHANEY, BSc, MSc, Associate Professor of Civil Engineering, University of New Brunswick, and M.C. FORDE, BEng, MSc, PhD, MIMunE, MIHE, Lecturer, University of Edinburgh

Maximum moisture contents of highway embankments

Maximum moisture contents of clayey soils used in highway embankment construction have always been of concern to highway engineers in England. Ministry of Transport specifications have not relieved this concern. A method of obtaining a workable upper limit moisture content for earthwork construction appears necessary. This paper illustrates one possible solution to the problem.

INTRODUCTION

Ministry of Transport (M.O.T.) specifications regarding maximum moisture content of soils used in earthwork construction have always been related to the Plastic Limit. In fact, even though the specifications have changed over the years, the specifications concerned with maximum moisture contents can be reduced to a common entity:

PLASTIC LIMIT x FACTOR

that is, the maximum moisture content of clay fill has been limited to the PLASTIC LIMIT multiplied by some factor.

Over this same period, engineers have found the specifications lacking. In many instances soils having moisture contents higher than specified were used economically in construction while in other instances the moisture content of a soil had to be reduced below that specified before it was used as a construction material. These anomalies were a function of the Plastic Limit rather than the multiplying factor which was continually changed in an effort to find a workable solution to the problem. These anomalies were a function of the quantity of clay particles and the mineralogy of the clay particles in the particular soil. Each soil reacted with moisture in a different manner dependent upon the quantity and mineralogy of the clay particles present in the soil. Hence there will never be accord between specifications and practice as long as the former is based on the Plastic Limit.

CRITERIA FOR SUITABILITY

The criteria as to a material's suitability for embankment construction have been previously outlined; Rodin, (ref. 3); Snedker, (ref. 5). These criteria are:

a) Can construction equipment work on the soil?
b) Will the embankment be stable?
c) Will future settlement of the embankment be excessive?

To answer criteria a) and b) the shear strength variations with change in moisture content of the soil must be related. For criterion c) the relationship of settlement with change in moisture content of the soil must be developed.

The previous researchers related these shear strength and settlement variations to the Plastic Limit thereby retaining the M.O.T. specification with its inherent anomalies rather than opting for a new criterion for the specification. Neither proposed method was successful because of the Plastic Limit base. Snedker (ref. 5) did point out the limitations of his proposed solution because of its Plastic Limit base and further mentioned that because of the difficulties encountered therewith it would be logical to substitute a shear strength criterion in the specifications to deliniate the suitability of a material for earthwork construction.

ANALYSIS OF CRITERIA

Embankment Stability

Stability of embankments is considered either as a short-term stability or a long-term stability problem. In the short-term stability problem, the solution is accomplished by a total stress analysis using unconsolidated-undrained shear strength parameters. In the long-term stability problem, an effective stress analysis using peak effective shear strength parameters from consolidated-drained tests is most often carried out. There has been some change in thought regarding the choice of parameters to use in the latter case. Some soils appear to weaken with time and consolidation. Slope design in such soils should be accomplished using residual shear strength parameters rather than peak strength values. The designer thus has three choices of parameters for design purposes. This should not present any difficulties however, since each set of parameters can be evaluated and the critical values selected and included in the specifications.

Embankment Settlement

Earlier work by Forde (ref 2) demonstrated that settlement is a function of embankment height, moisture content, air voids after compaction and the compressibility characteristics of the soil. The conclusion drawn was that no universal absolute value of settlement could be arrived at for embankments in general or even

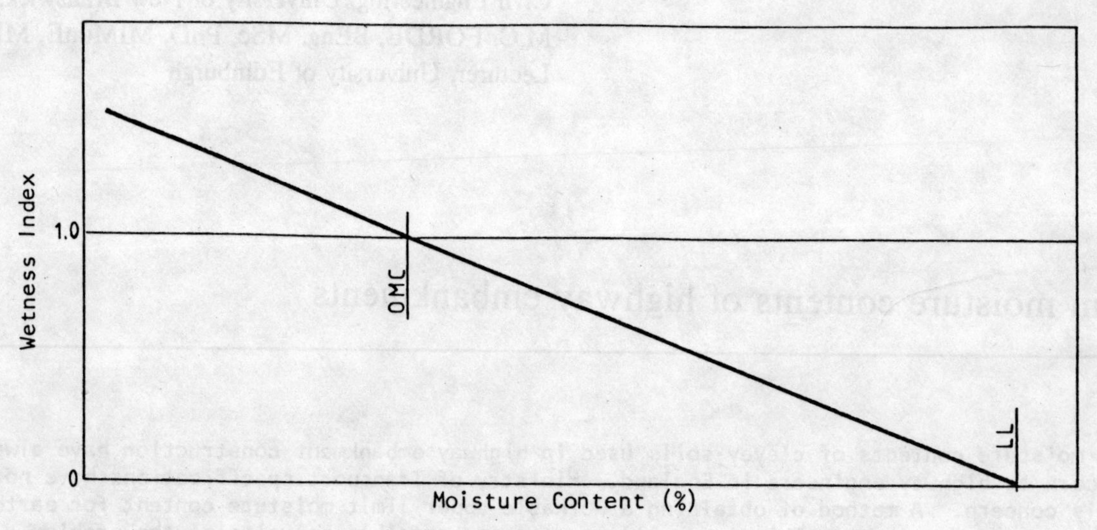

Fig. 1. Typical wetness index versus moisture content relationship

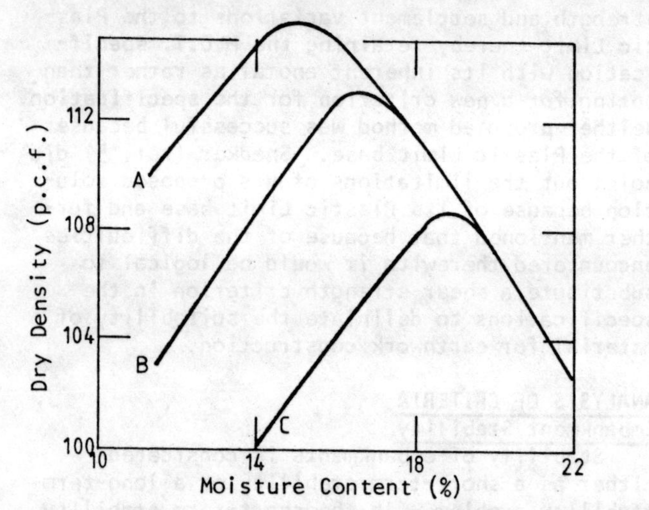

Fig. 2(a). Moisture-density relationship. After Seed, Mitchell and Chan (ref. 4)

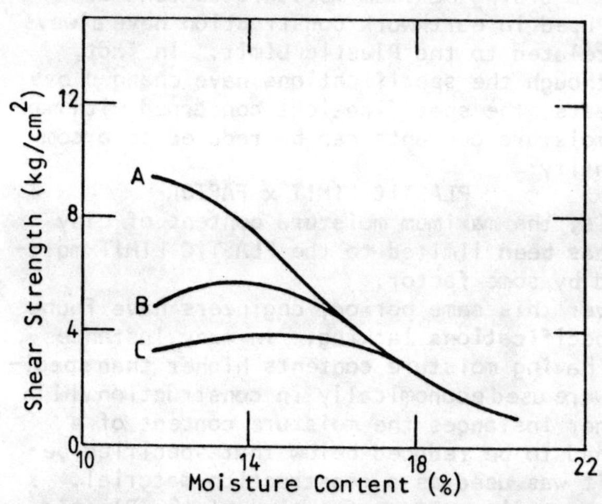

Fig. 2(b). Moisture-strength relationship

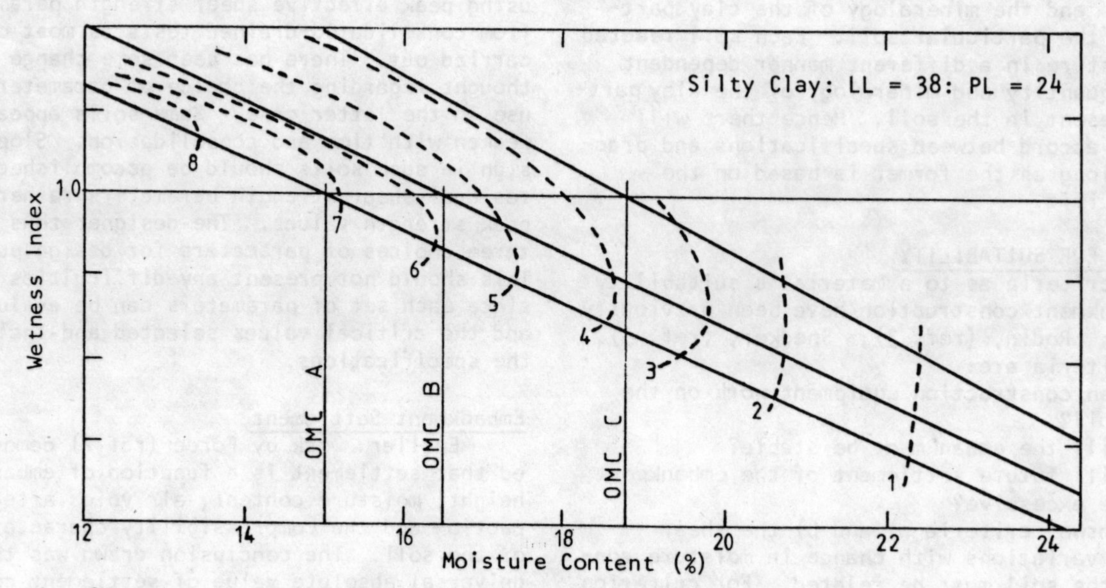

Fig. 3. Wetness index-moisture content-shear strength relationship

for one embankment if it were of varying cross-section over some distance. However for most highway projects the settlement criteria could be simplified to not more than four of five values. Thus it can be appreciated that a soil which has too high a moisture content to be placed in one location might be entirely suitable at another location on the project.

Plant Operation

Plant operation is undoubtedly a critical criterion in embankment construction. It is undeniably linked to construction costs. The shear strength of the soil in the plant operation is akin to the short-term stability condition. Since shear strength is a function of moisture content, provided a good relationship can be accomplished, better control of plant operation could be achieved.

NEW APPROACH TO PROBLEM

The wetness index concept, based on work by Birch (ref. 1), relates the natural moisture content of a soil to the range of moisture contents between the Liquid Limit of the soil and the Optimum Moisture Content of the soil when the soil is compacted at one energy level:

$$W.I. = \frac{L.L. - m.c.}{L.L. - O.M.C.} \quad (1)$$

where L.L. = moisture content at Liquid Limit.
m.c. = natural moisture content.
O.M.C. = optimum moisture content for soil compaction at a particular energy level.

The choice of this index to relate shear strength and settlements to moisture content was made for the following reasons:

a) The moisture content range (L.L. - O.M.C.) is the range of moisture contents at which soils defined as "wet fill" will generally be placed in an embankment.
b) The troublesome and inaccurate parameter of Plastic Limit is eliminated.
c) The concept introduces density, another factor related to shear strength and settlement and in so doing retains Forde's (ref. 2) parameter of air voids that was shown to affect settlements.
d) The elements of the concept, L.L., m.c. and O.M.C. are determined by standardized tests and as such operator bias is lessened and parameter reproduceability is greater.

The wetness index - moisture content relationship for a soil compacted at a particular energy level is illustrated in Fig. 1. There will be as many wetness index versus moisture content relationships for a soil as there are levels of energy used to compact the soil.

RELATIONSHIP OF WETNESS INDEX, MOISTURE CONTENT AND SHEAR STRENGTH

To illustrate this relationship, the author has borrowed data from Seed, Mitchell, Chan (ref. 4). This data is reproduced in Fig. 2(a) which illustrates the moisture content-density relationship for three different compactive efforts and in Fig. 2(b) which illustrates the unconsolidated-undrained shear strength-moisture content relationship for the as compacted conditions for the same soil.

Plotting the wetness index-moisture content relationships of these compactive efforts gives the linear curves of Fig. 3. Superimposed on these linear curves are the family of shear strength curves produced from the data in Fig. 2(b).

RELATIONSHIP OF WETNESS INDEX, MOISTURE CONTENT AND SETTLEMENT

Fig. 2(a) illustrates the moisture-density relationships of a soil compacted at different levels of energy. Over the moisture content range (L.L. - O.M.C.) there is very little change in air voids regardless of the magnitude of the compaction energy. Thus in terms of settlement we are dealing with a soil that is at essentially one air void content when it is placed in an embankment with its moisture content in the L.L. - O.M.C. range.

Forde (ref. 2), demonstrated that for a given soil at a given percent air voids, the amount of settlement increases as the wetness index decreases. This is illustrated in Fig. 4 where the calculated settlements for a ten meter high embankment constructed of Harmondsworth Brickearth are plotted.

USE OF RELATIONSHIPS AND CHOICE OF LIMITS

To ascertain the limiting moisture contents for placement of wet fill, one requires relationship between moisture content and shear strength and between moisture content and settlement. These relationships were always known to exist and as shown can be graphically related through the use of the wetness index concept.

Control of Plant Operation

Shown in Fig. 5 is a hypothetical relationship of wetness index-moisture content and shear strength for a wet fill. The chart covers the moisture content range that would encompass limiting moisture conditions. The heavy wetness index-moisture content line represents British Standard Compaction for this soil. Provided one can establish the correct limiting shear strength for plant mobility or uses some empirical standard such as Rodin's (ref. 3) one-fifth tyre inflation pressure or Snedker's (ref. 5) 5 - 7 p.s.i. limit one can then find this limiting shear strength curve among the family of such curves (dark line on chart) and by dropping an ordinate from the intersection of the two curves determine the limiting moisture content for plant operation.

Control of Embankment Stability

As pointed out, the limiting shear strength for stability would have to be decided upon between the limits of short-term analysis with unconsolidated-undrained parameters and the long-term analysis with effective peak or effective residual strength parameters.

Having determined the appropriate limiting strength one could find its location in the family of strength curves and again the abscissa coordinate of the intersection of the limiting strength curve and appropriate wetness index-moisture content curve would indicate the limiting moisture content for placement of the mat-

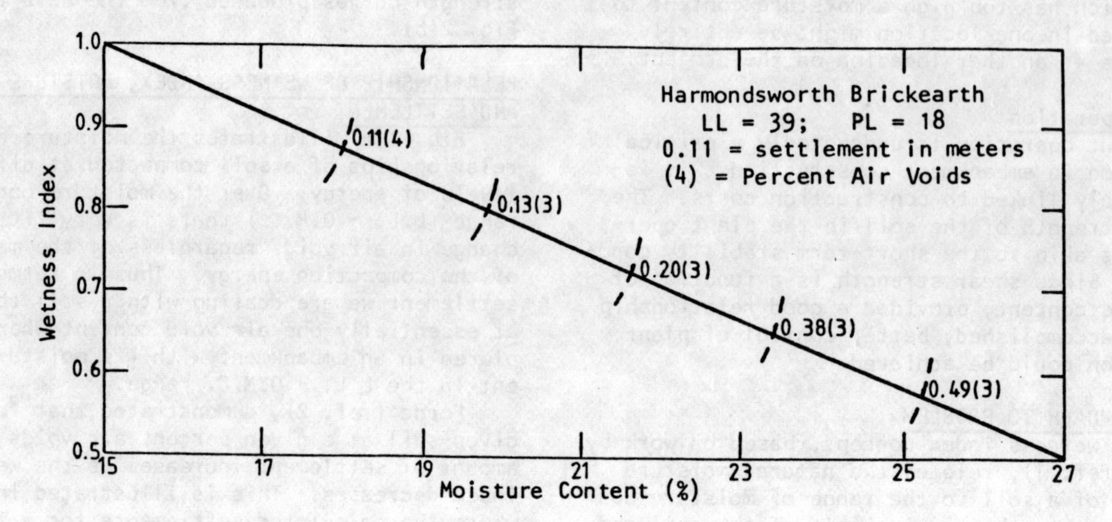

Fig. 4. Wetness index - settlement relationship. After Forde (ref. 2)

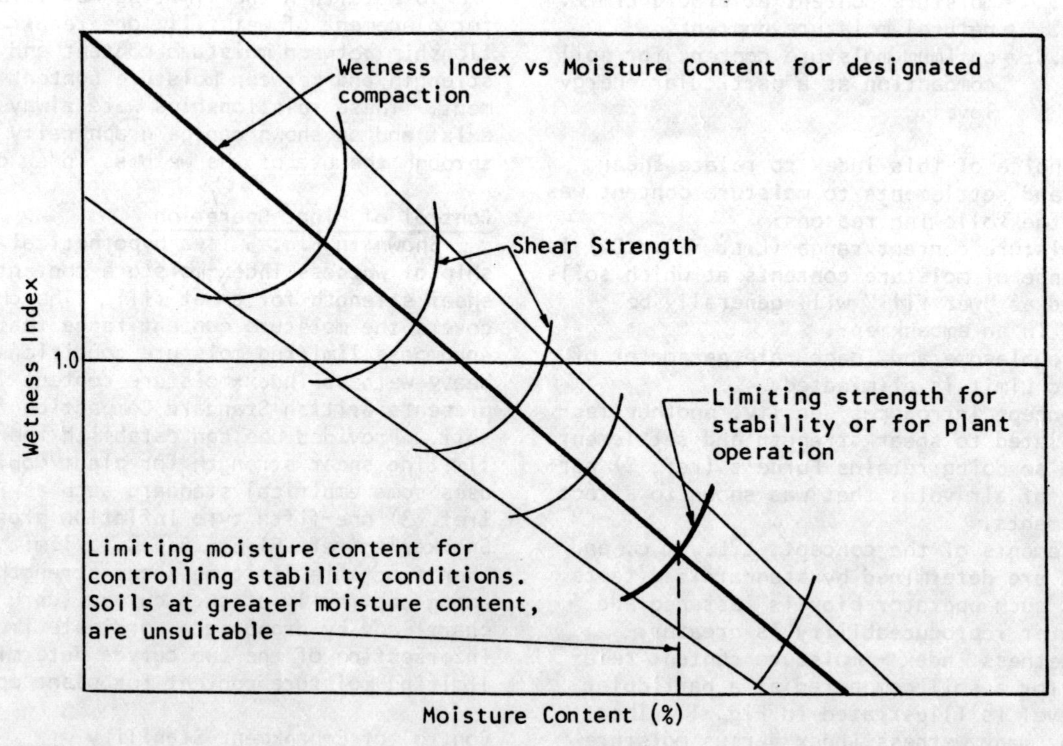

Fig. 5. Typical wetness index-moisture content control diagram for stability

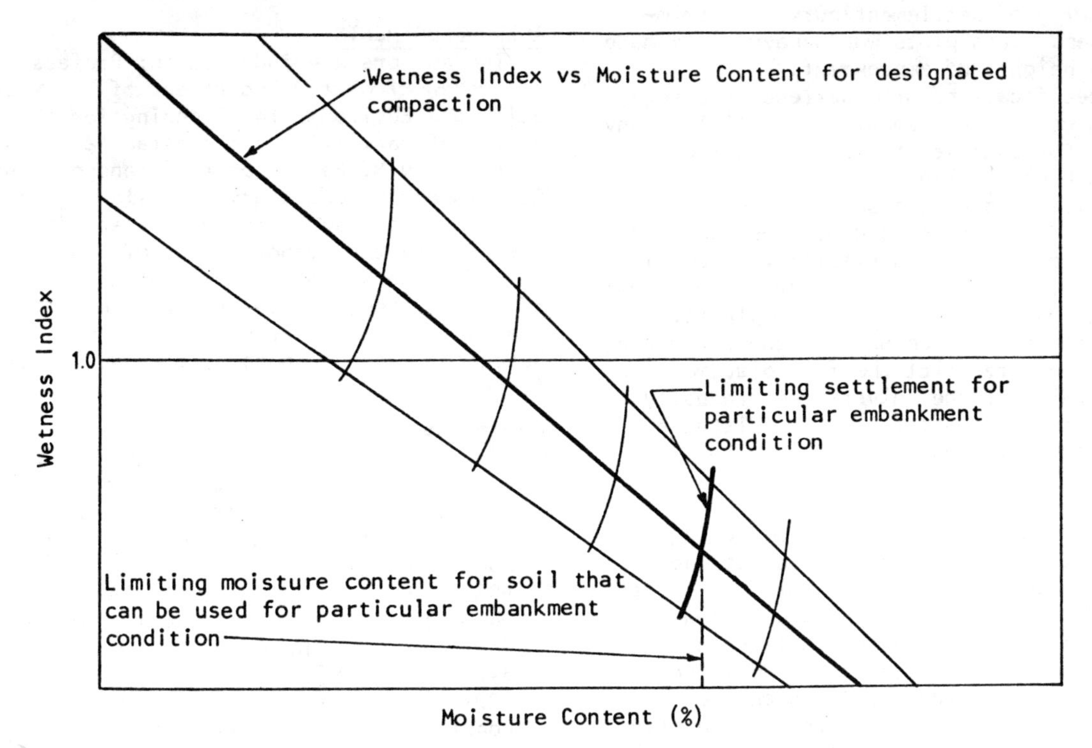

Fig. 6. Typical wetness index-moisture content control diagram for settlement

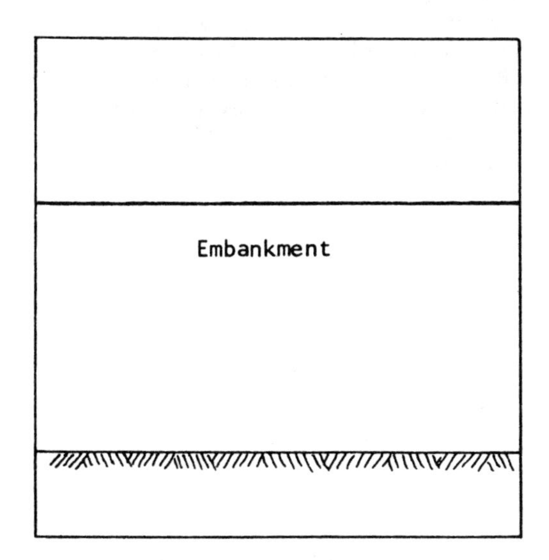

Fig. 7. Embankment of uniform height

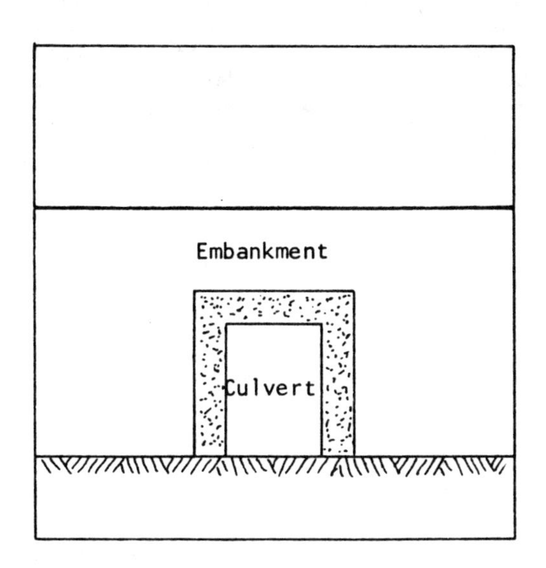

Fig. 8. Embankment over buried structure

erial in an embankment.

Control of Embankment Settlement

Fig. 6 depicts the same hypothetical wetness index-moisture content relationship as Fig. 5 with the family of settlement curves superimposed on them. Such plots would have to be made for various heights of embankment.

With specifications or experience the engineer could control placement of wet fill in any situation. For example, the case of uniform height embankment in Fig. 7.

If the limiting condition as shown on Fig. 6 did not exceed the limiting conditions pertaining to plant operation and stability as developed from Fig. 5, then the engineer could allow fill at high moisture content to be placed, whereas in the case of an underground structure, Fig. 8, he could restrict the fill placement to dryer material near the culvert and gradually relax the moisture content limit as distance from the culvert increases.

DISCUSSION

With a set or sets of figures depicting wetness index-moisture content-shear strength relationships and wetness index-moisture content-settlement relationships an engineer through a series of field moisture determinations can control embankment construction by deciding which is the controlling factor, plant operation, stability or settlement, in any one situation.

With the same set of figures, the contractor could control his operation on a day to day basis in a more economic manner.

The designer, with such figures could provide more realistic and workable specifications for construction, writing in limiting moisture specifications to govern any construction situation envisaged. In this respect, the charts would become part of the tender document, so accurate test results would be required. Since most of these tests are normally part of the soil exploration stage little extra expense is foreseen in developing the necessary curves. Any extra expense incurred at this stage would surely be offset by less field testing and better construction progress plus a possible cheaper tender from the contractors.

ACKNOWLEDGEMENTS

The authors are indebted to: Professor J. Kolbuszewski of the Department of Transportation and Environmental Planning for the provision of facilities and constant encouragement, Mr. D.M. Farrar of the Transport and Road Research Laboratory for allowing access to data, Professor I.M. Beattie and Professor A.W. Hendry the authors heads of Department at the Universities of New Brunswick and Edinburgh, the Science Research Council, the Rees Jeffreys Road Fund and the I.C.E. Radley Research Studentship for the provision of finance.

References

1. BIRCH, N. A laboratory study of the engineering properties of Kenper Marl. M.Sc. Thesis, Dept. of Transportation and Environmental Planning, The University of Birmingham, 1964.
2. FORDE, M.C. Wet fill for highway embankments. Ph.D. Thesis, Dept. of Transportation and Environmental Planning, The University of Birmingham, 1975.
3. RODIN, S. Ability of clay fill to support construction plant - Part 1. Civil Engineering and Public Works Review. 1965, 60, February, 197-202 and Part 2. Civil Engineering and Public Works Review. 1965, 60, March, 343-345.
4. SEED, H.B.; MITCHELL, J.K.; CHAN, C.K. The strength of compacted cohesive soils. Proceedings of Research Conference on Shear Strength of Cohesive Soils, American Society of Civil Engineers, 1960, June, 877-964.
5. SNEDKER, E.A. Choice of an upper limit of moisture content for highway earthworks, Highways Design and Construction, 1973, January.

D. M. FARRAR, MSc, MInstP, Transport and Road Research Laboratory

Settlement and pore-water pressure dissipation within an embankment built of London Clay

In the construction of earthworks for roads, economies will usually be achieved by making the maximum use of cohesive soils already on site and avoiding the expensive importation of fill. However, the engineer must ensure that embankments constructed using such soils are stable and do not settle excessively.

This paper describes observations made during and after construction of an embankment built of London Clay. Laboratory tests were carried out to enable predictions to be made of the rate of settlement and dissipation of pore-water pressure within the embankment. The predicted rates were then compared with those measured during the first four years after completion of the embankment.

It was found that care was needed in the interpretation of laboratory consolidation tests, both because the consolidation of the specimens departed somewhat from theory and because of the variability of the results. The dissipation of pore-water pressure predicted from these results using a finite difference technique was found to be in reasonable agreement with that measured by piezometers installed in the embankment.

The settlement within the embankment was found to be small in relation to that of the foundation soil and in reasonable agreement with that estimated using compressibility data derived from laboratory consolidation tests carried out on samples of the fill at the equivalent field moisture contents.

INTRODUCTION

1. In the construction of earthworks for roads economies will usually be achieved by making the maximum use of cohesive soils occurring on the site, and avoiding the expensive importation of higher quality fills. The feasibility of using the soils as fills is determined in part by the difficulty of earth moving[1,2] but an important consideration is that the engineer must ensure that a stable embankment can be built during the proposed construction period and that excessive settlement will not take place after construction has been completed. This paper describes observations made during and after construction of an embankment built of London clay. Laboratory tests were carried out to enable predictions to be made of the rate of settlement and dissipation of pore-water pressure within the embankment. The predicted rates were then compared with those measured during the first 4 years after completion of the embankment.

DESCRIPTION OF SITE

2. The site is immediately west of the Denham interchange on the M40 Gerrards Cross by-pass. The embankment is 12m high and accommodates a 3 lane carriageway and slip road. A transverse profile is shown in Fig. 1.

The subsoil consists of 1-2m of clay of intermediate plasticity overlying gravel, with a permanent water table at a depth of 1-1.5m. To allow time for dissipation of pore-water pressure in the subsoil, the embankment was built in 4 stages. Construction commenced in October 1971 with 150mm of ash being spread to provide a working platform for construction of the first stage, and the road was opened to traffic in August 1973. It was originally anticipated that the fill would be placed at a high moisture content, but in fact much of the fill dried appreciably during placement in the summer months. All the fill was compacted to the then current specification[3] by towed vibrating rollers. At the lower elevations, the cohesive soil was contained between granular fill wedges to enhance the stability of the embankment. The motorway pavement consists of 150mm thick soil cement sub-base overlaid by 275mm of unreinforced concrete.

SOIL TESTS

3. All soil tests were carried out in accordance with BS1377[4]. The density of the fill, measured by the sand replacement method, is shown in Fig. 2. The results of the soil classification tests are summarised in Table 1.

4. Consolidation tests were carried out on undisturbed specimens obtained from the fill and also from the cut from which the fill was obtained. For the stony fill used in the first stage of construction, however, it was necessary to test soil compacted into the oedometer ring at the same density as the undisturbed fill.

Clay fills. Institution of Civil Engineers, London, 1978, 101-106

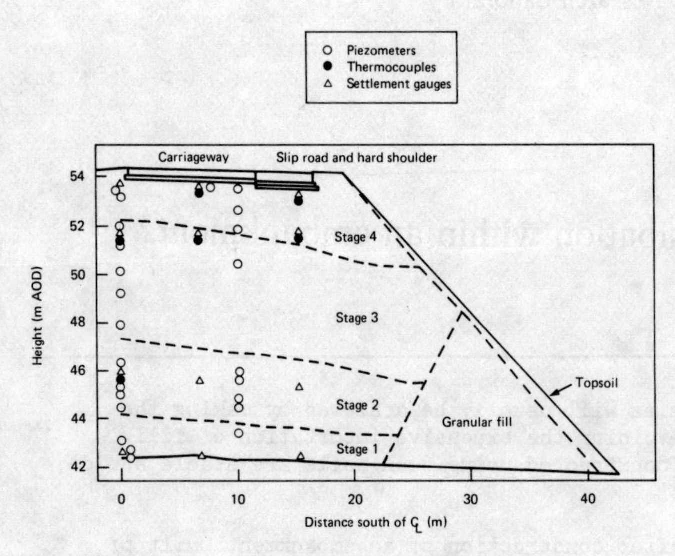

Fig. 1. Cross-section of embankment on M40 at Denham

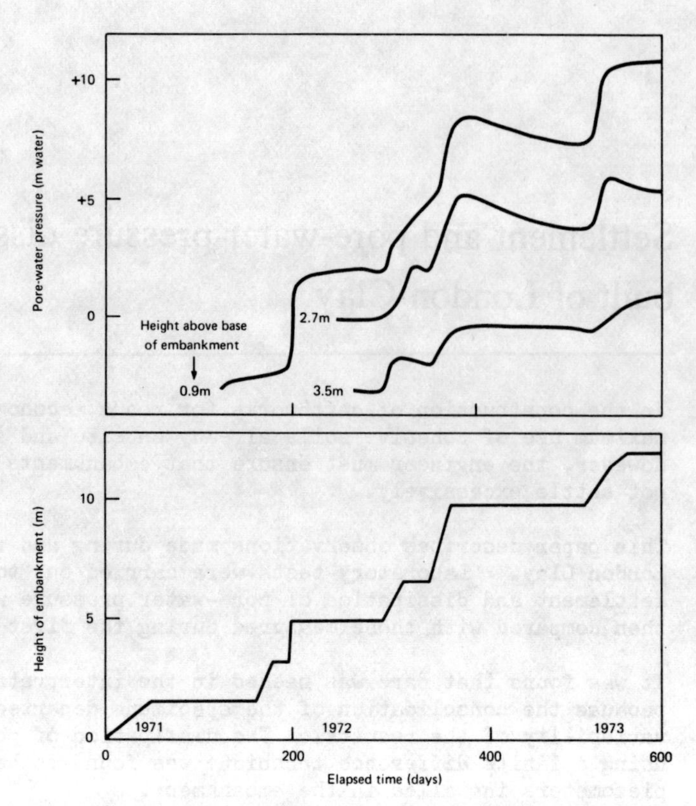

Fig. 3. Change in pore-water pressure with height of embankment during construction

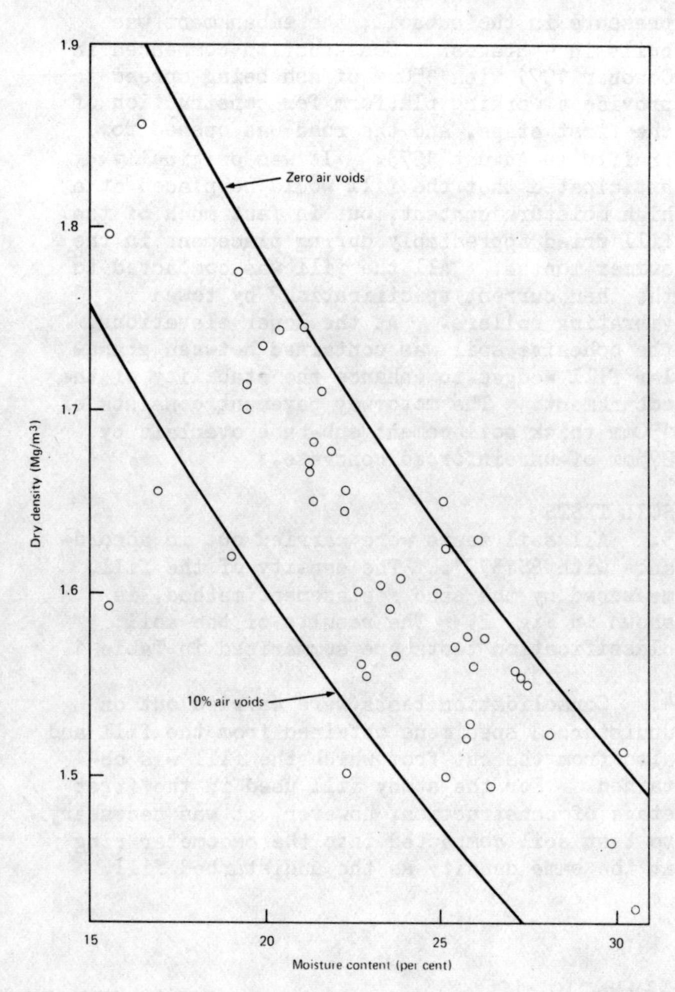

Fig. 2. Density test results on M40 at Denham

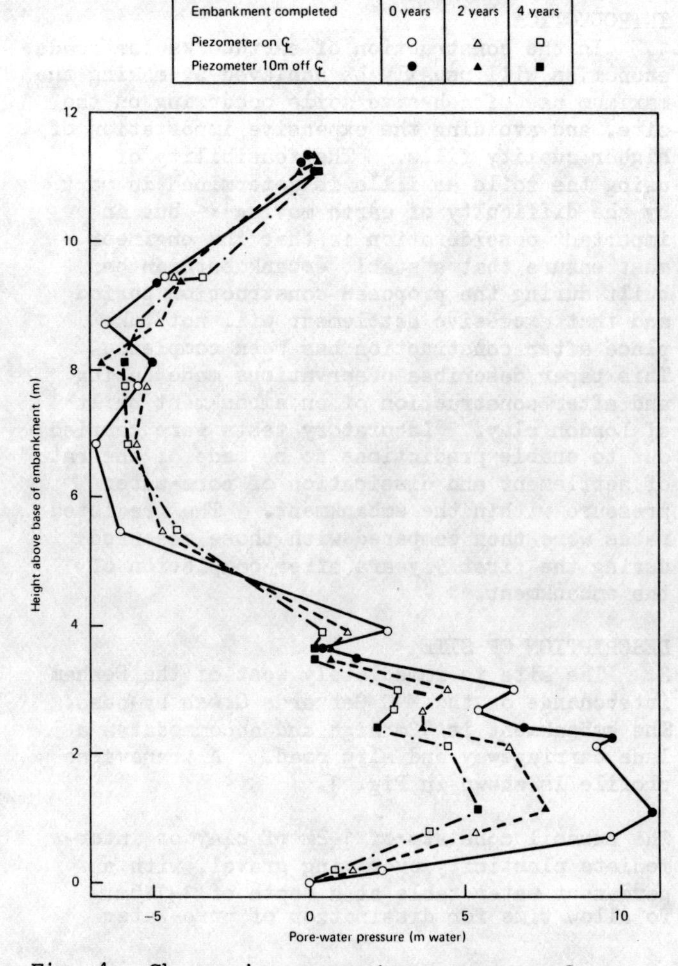

Fig. 4. Change in pore-water pressure after construction of embankment

Table 1. Results of classification tests on fill

Construction stage	1	2	3	4
Moisture content (%)	22.7	23.6	23.3	24.8
Liquid limit	70	74	73	73
Plastic limit	22	25	26	23
Clay and silt (%)	56	73	95	81
BS Compaction (2.5 kg)				
Max.dry density (Mg/m^3)	1.69			
Opt. moisture content (%)	21			

5. Preliminary tests with a large (150mm diameter) oedometer incorporating a miniature piezometer had shown that the relation between excess pore-water pressure and time was flatter than that calculated from consolidation theory for saturated soil. Clearly the application of unsaturated consolidation theory[5] would be appropriate in these circumstances. However, considerations of the relatively limited accuracy required for highway embankment purposes, and of the variability of the fill, led to the conclusion that the additional complexity of this method might not be justified. The simpler conventional approach was therefore used. It was found in all tests that the square-root-of-time ($\sqrt{t}$) curve fitting method[4] did not give a good fit to the experimental results, and gave a coefficient of consolidation about 50 per cent higher than that given by the logarithm-of-time (log t) fitting method. The log t method only was therefore used to evaluate consolidation test results.

6. The results of all tests on undisturbed samples from the fill are summarised in Table 2.

Table 2. Measured coefficients of consolidation (C_v), and swelling (C_s) of fill

	Pressure increment (kPa) 107-215	215-429	429-858
C_v (m^2/yr)	0.56	0.37	0.35
Number of tests	25	33	16
C.o.V.* (%)	56	46	36
	Swelling of samples under load of 54 or 107 kPa		
C_s (m^2/yr)	0.38		
Number of tests	12		
C.o.V.* (%)	95		

*Coeff. of variation (C.o.V.)= $\frac{\text{Standard deviation}}{\text{mean}}$

7. There was no significant difference in the consolidation characteristics of the fill placed in the different stages. The method of sample preparation also did not affect the results for this soil, as shown in the example in Table 3.

8. The measured coefficients of volume compressibility of the fill were in the range 0.08-0.19 m^2/MN. The sub soil and ash layer immediately beneath the embankment were highly

Table 3. Effect of method of sample preparation for fill (pressure increment 215-429 kPa)

	Undisturbed samples From fill	From cut	Remoulded samples
C_v (m^2/yr)	0.37	0.41	0.39
Number of tests	33	10	14
C.o.V. (%)	46	24	30

permeable (C_v of 5 m^2/yr and over 10 m^2/yr respectively).

SITE INSTRUMENTATION AND RESULTS OBTAINED

9. Piezometers. These were installed in two vertical profiles as shown in Fig. 1. Each consists of a porous tip of high air entry value ceramic connected through two polythene coated nylon tubes to a mercury manometer in an adjacent gauge hut. The piezometers were installed by pushing them into an augered hole of the same diameter, and carefully backfilling the hole with clay from the site and a 'plug' of bentonite. This procedure had the disadvantage of producing a poor response time (2-7 days), but was necessary to ensure that negative pore-water pressures could be measured.

10. Fig. 3 shows typical examples of the change in pore-water pressure during construction of the embankment. Positive pressures were developed in the fill placed in Stages 1 and 2 of construction, but pressures in the fill placed in Stage 3 remained negative. Fig. 4 shows the distribution of pore-water pressure at the end of construction of the embankment, and the subsequent changes. Pressures measured in the two upper piezometers on the central profile showed seasonal variations and are not plotted.

11. Thermocouples. These were installed as shown in Fig. 1. The mean temperature of the fill during 1973-5 was 11°C, with an annual fluctuation of ±5°C at a depth of 3m.

12. Settlement gauges. Four three-point mercury gauges[6] were placed at different elevations in the fill as shown in Fig. 1. The gauges were installed by cutting a shallow trench transversely across the formation during a pause in construction, mounting each gauge point in a block of concrete cast in situ, and carefully backfilling the trench.

13. To provide a datum level, a bench mark was installed at the boundary of the site. This consisted of a 32mm diameter steel rod inside a lining tube driven to refusal into the gravel underlying the site.

14. The measured rates of settlement at the base and near the top of the embankment, after completion of construction, are shown in Fig. 5. There was about 140mm settlement at the base of the embankment during construction, and some further settlement is taking place at a reduced rate. Settlement within the embankment is

still continuing, and is shown in more detail in Table 4 below.

Table 4. Measured and predicted settlement (-) or swelling (+) within the embankment

Height of layer of fill above base (m)	Gauge point	Time after completion of embankment (years) 0	1	2	4	Predicted total (see Para 20)
		Settlement (-) or swelling (+) within layer (mm)*				
0- 3.3	1	-19	-31	-38	-42	-60
	2	-17	-26	-33	-41	
	3	-11	-20	-28	-43	
3.3- 9.3	1	0	-4	-5	-12	+30
	2	0	-10	-8	-10	
	3	0	-12	-15	-12	
9.3-11.2	1	0	+12	+13	+15	+70
	2	0	+12	+14	+19	
	3	0	+5	+11	+7	

*ie difference between settlement gauges at upper and lower boundary

15. Road surface settlement. Small steel studs, with a recess to locate the point in the foot of a levelling stave, were fixed with epoxy resin to the surface of the road slabs, 300mm in from the rearside edge. A settlement of 50mm has been measured in the carriageway above the settlement gauges. There has been a corresponding swelling of 40mm on the carriageway in an adjacent cutting 3m deep.

MEASURED AND PREDICTED PORE-WATER PRESSURE DISSIPATION

16. The dissipation of excess pore-water pressure is usually estimated using Terzaghi's theory of consolidation. When the initial distribution of pore pressure with depth is not uniform, as in the present case, the consolidation equation may be expressed in a finite-difference form[7] suitable for solution using a desk calculator. The accuracy of this procedure, when applied to the unsaturated fill in the embankment at Denham, was separately tested for the lower and upper parts of the embankment.

17. Lower part of embankment. For this case, the coefficient of consolidation was assumed to be a uniform 0.47 m^2/yr (Table 2) corrected to the temperature of the fill (11°C). It was also assumed that the base of the embankment is free draining and that consolidation is one-dimensional. Using these assumptions, the change in pore-water pressure one year and 4 years after completion of construction was calculated and is compared in Figure 6 with the measured dissipation.

18. Figure 6 shows that the measured dissipation of pore pressure is on average 30 per cent lower than that calculated. Whether a discrepancy of this magnitude would be significant will depend upon the accuracy to which the coefficient of consolidation (C_v) can be determined in practice. In the present study, the coefficient of variation of C_v was about 50 per cent, but because a large number of consolidation tests were carried out it is possible to conclude that a 30 per cent discrepancy is statistically significant. On most motorway construction sites, however, it is likely that a much smaller number of tests will have been carried out. A 30 per cent discrepancy would not be significant, and the agreement between calculated and measured values would probably be adequate for practical purposes.

19. Upper part of embankment. The coefficient of swelling in the dry fill in this part of the embankment was assumed to be 0.38 m^2/yr (Table 2), 2), corrected to the temperature of the fill. Consolidation was again assumed to be one-dimensional and two boundary conditions were tested; (a) free water at formation level and (b) formation covered by impermeable layer. The increase in pore-water pressure four years after completion of the embankment, calculated using these assumptions, is compared in Figure 7 with the measured increase.

20. Unfortunately, the changes in pore-water pressure are small, and this experiment does not provide a very sensitive test of the accuracy of the calculation. The most that can be concluded is that the calculated change is of the correct order of magnitude. It can also be concluded, both from the comparison in Figure 7 and from the measured pore-water pressures shown in Figure 4, that it is correct to assume free water at formation level beneath the road pavement.

MEASURED AND PREDICTED SETTLEMENT

21. The total settlement or swelling within an embankment built of clay fill may be predicted from the results of a published laboratory study[8]. The settlement may alternatively be estimated from the coefficients of compressibility determined in the present tests, from which a similar result is obtained.

22. The predicted total settlement is compared in Table 4 with the measured settlement which has so far taken place. Considering each layer of fill separately:-

23. 0-3.3m above base. About 60 per cent of the excess pore pressure in this part of the embankment has now dissipated (Figure 4), giving a settlement of about 40mm. It is therefore likely that the final total settlement will be in reasonable agreement with the predicted 60mm.

24. 3.3-9.3m above base. There has been little change so far in the negative pore pressures in this part of the embankment, and it is not possible to determine from the present measurements if there will be a small total swelling as predicted.

25. 9.3-11.3m above base. Some 20 per cent of the negative pore pressure has now dissipated, to give a swelling of 10-20mm. The final

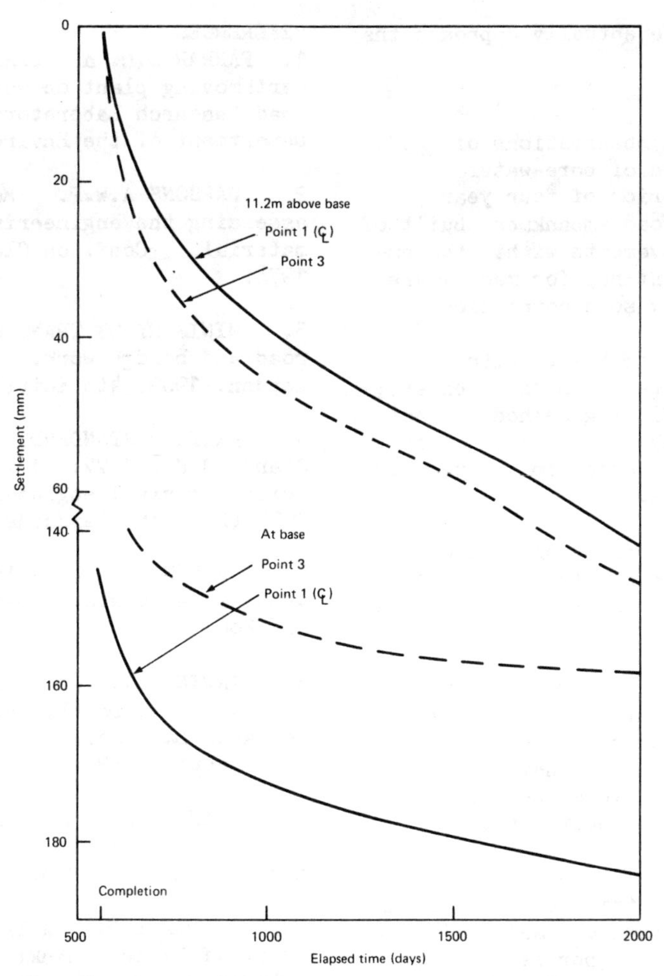

Fig. 5. Settlement near top and at base of embankment, after construction of embankment

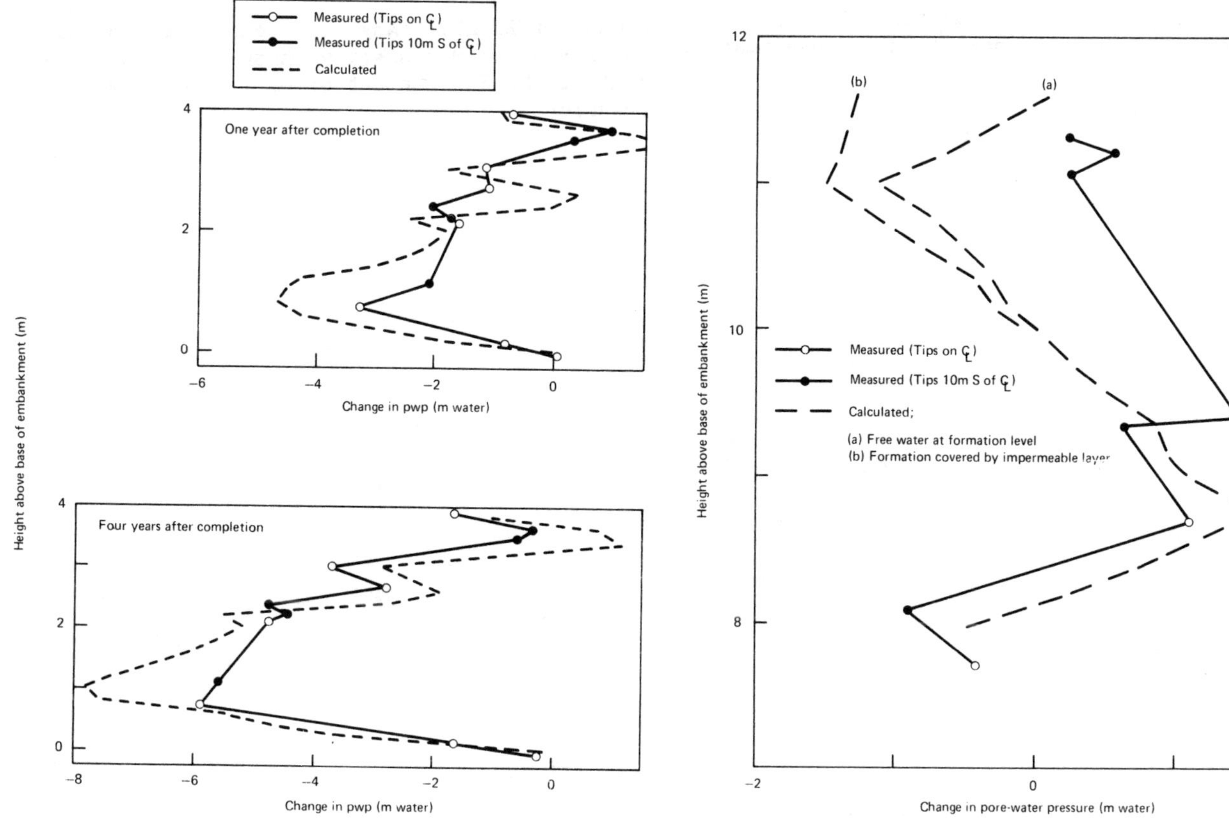

Fig. 6. Measured and calculated changes in pore-water pressure in lower part of embankment after completion

Fig. 7. Measured and calculated changes in pore-water pressure in upper part of embankment four years after completion

swelling could therefore eventually approach the predicted value of 70mm.

CONCLUSIONS

26. This paper describes observations of settlement and dissipation of pore-water pressure made during a period of four years after construction of a road embankment built of heavy clay. Moisture movements within the embankment are likely to continue for many years, but it is possible to draw some conclusions.

(1) Care is needed in using the results of laboratory consolidation tests in this context; the square-root-of-time fitting method may not be appropriate, and it may not be possible to attach a high degree of accuracy to the results of a small number of tests.

(2) The settlement and dissipation of excess pore-water pressure within the lower parts of the fill are in reasonable agreement with those predicted from theory.

(3) The changes of pore-water pressure in the upper part of the embankment are of the same order of magnitude as those predicted from theory. The measurements are consistent with pore pressures at formation level beneath the road surface remaining at or near zero.

ACKNOWLEDGEMENTS

27. The work described in this paper forms part of the programme of the Transport and Road Research Laboratory and the paper is published by permission of the Director. Thanks are due to the Director of the Eastern Road Construction Unit for permission to carry out measurements at this site, and to the staff at the R.C.U. and the contractors for their co-operation and assistance.

REFERENCES

1. FARRAR D.M. and DARLEY P. The operation of earthmoving plant on wet fill. Transport and Road Research Laboratory, Crowthorne, 1975. Department of the Environment, TRRL Report LR668.

2. PARSONS A.W.P. Moisture condition test for assessing the engineering behaviour of earthwork material. Conf. on Clay Fills, London, November, 1978.

3. MINISTRY OF TRANSPORT. Specifications for road and bridge work. HM Stationery Office, London, 1969, 4th Edition.

4. BRITISH STANDARDS INSTITUTION. British Standard No. 1377: 1975. Methods of testing soils for civil engineering purposes. London, 1975 (British Standards Institution).

5. BARDEN L. Consolidation of compacted and unsaturated clays. Geotechnique 1975, 15(3) 267-286.

6. IRWIN M.J. A mercury-filled gauge for measuring the settlement of foundations. Road Research Laboratory, Crowthorne, 1970. Ministry of Transport, RRL Report LR344.

7. SMITH G.N. Elements of soil mechanics for civil and mining engineers. Crosby Lockwood and Son Ltd, London 1973, 3rd Edn.

8. FARRAR D.M. A laboratory study of the use of wet fill in embankments. Road Research Laboratory, Crowthorne, 1971. Department of the Environment, RRL Report LR406.

M. C. FORDE, BEng, MSc, PhD, MIMunE, MIHE, Lecturer, University of Edinburgh, and A. G. DAVIS, BSc, MSc, PhD, Head of Soils and Foundation Research Division, CEBTP, Paris

Trafficability studies on wet clay

A laboratory study was undertaken upon two clays, London Clay and Cheshire Clay, at a range of typical moisture contents encountered in the U.K. Soil-rubber and soil-soil shear tests were undertaken using a direct shear box. The results of these tests are related to work undertaken on soil-rubber wheel shear using a simple experimental test facility in which the soil was moved into the instrumented wheel. Conclusions have been drawn regarding the mobility characteristics of wheels on London Clay and the leaner Cheshire Clay.

INTRODUCTION

1. Earthmoving problems can be due to one or both of two reasons - wet weather or low shear strength soil. Satisfactory material may become unsuitable due to an increase in moisture content from rain or it may be unsuitable due to its high initial moisture content at the time of excavation.

Military Engineering Approach

2. An enormous research effort has been put into off-the-road locomotion by the Military Engineers. Much of this work has been spearheaded by the various U.S. military research establishments (Ref.1,2,3). Although representing an ingeneous line of attack in the years immediately following World War 2, the approach suffers from the limitation of being based on a semi-empirical curve fitting technique extrapolated from the early work of Taylor(Ref. 4.).

The British Total Force Approach

3. The so called British Total Force approach was initiated by Micklethwaite (Ref.5.), also during World War 2. This work was later developed by Uffelmann (Ref.6.) and Nottingham University (Ref. 7, 8.).

The Civil Engineers Approach

4. Apart from the work at Nottingham University in collaboration with the military engineers very little work has been published by Civil Engineers. Early work by CERA (later CIRIA) on Earthworks in Wet Weather (Ref.9.) was aimed at Earthmoving on the macro-time scale. That is productivity over an earthmoving season rather than on a particular soil in a particular condition at one instance in time. One of the move ingeneous approaches by a Civil Engineer is that due to Rodin (Ref.10). This work is limited by the partial invalidity of the work of Black (Ref.11.) upon which much of Rodin's effort is based.

EXPERIMENTAL APPROACH ADOPTED

5. The approach adopted in this work was along the total force/conventional Civil Engineering Soil Mechanics Approach.

6. Soil-soil and soil-rubber shear tests were undertaken in a 6 cm x 6 cm square shear box. For the soil-rubber shear tests a block of rubber was placed in the lower half of the box flush with the preformed slip plane.

7. The soil rubber wheel experiments were undertaken in a simplified mobility rig experimental test facility (Ref.15) - see flow chart Fig.1. A smooth rubber wheel and a treaded rubber wheel were used - Fig.2.

THE DIRECT SHEAR TESTS

The results of the direct shear tests have been reported in detail elsewhere (Ref.12). Figs. 3 and 4 summarise the trend lines of the relationships between shear strength and consistency index, I_c, for London Clay, Keuper Marl and Cheshire Clay. The classification data of the soils are included in Table 1.

From the curves in Figs. 3 and 4 it will be noted that soil rubber values tend to be on a curve below the soil-soil values and also the experimental values from Skempton and Northey (Ref.13.). The reason why the soil-soil values from this project do not coincide with Skempton and Northeys is that the latter's soils were saturated whilst the authors' soils were compacted.

8. At low consistency indices, less than approximately 0.7, soil-soil and soil-rubber shear are of the same order. This would indicate that at these consistency indices soil-soil shear equals soil-rubber shear or more probably that the shear is taking place within the soil. At higher consistency indices soil-rubber shear is taking place and is of a lower value than soil-soil shear.

9. An explanation of the above behaviour is assisted by considering Fountaine's (Ref.14) model for an idealised relatively wet fine soil - Fig.5. In this model the water film between the plate of foreign material and the

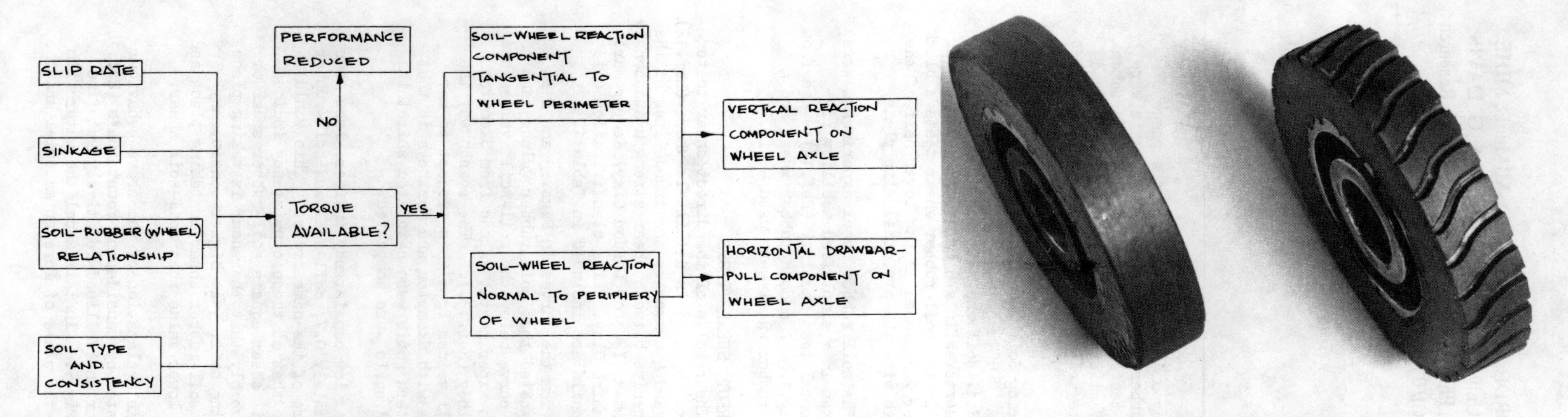

Fig. 1. Principle of mobility rig experimental test facility

Fig. 2. Smooth and treaded rubber wheels

Fig. 3. Trend lines soil-soil/soil-rubber shear - 172 kN/m^2

Table 1. Soil classification data

	LONDON CLAY	KEUPER MARL	CHESHIRE CLAY (Soil-Wheel Shear Tests)	CHESHIRE CLAY (Direct Shear Tests)
LIQUID LIMIT (%)	69	50	28	25
PLASTIC LIMIT (%)	29	28	19	15
BRITISH STANDARD COMPACTION TEST 4.5 kg Rammer				
MAX. DRY DENSITY (Mg/m^3)	1.75	1.79	1.97	2.09
O.M.C. (%)	17	16	14	9
S.G.	2.72	2.72	2.71	2.70

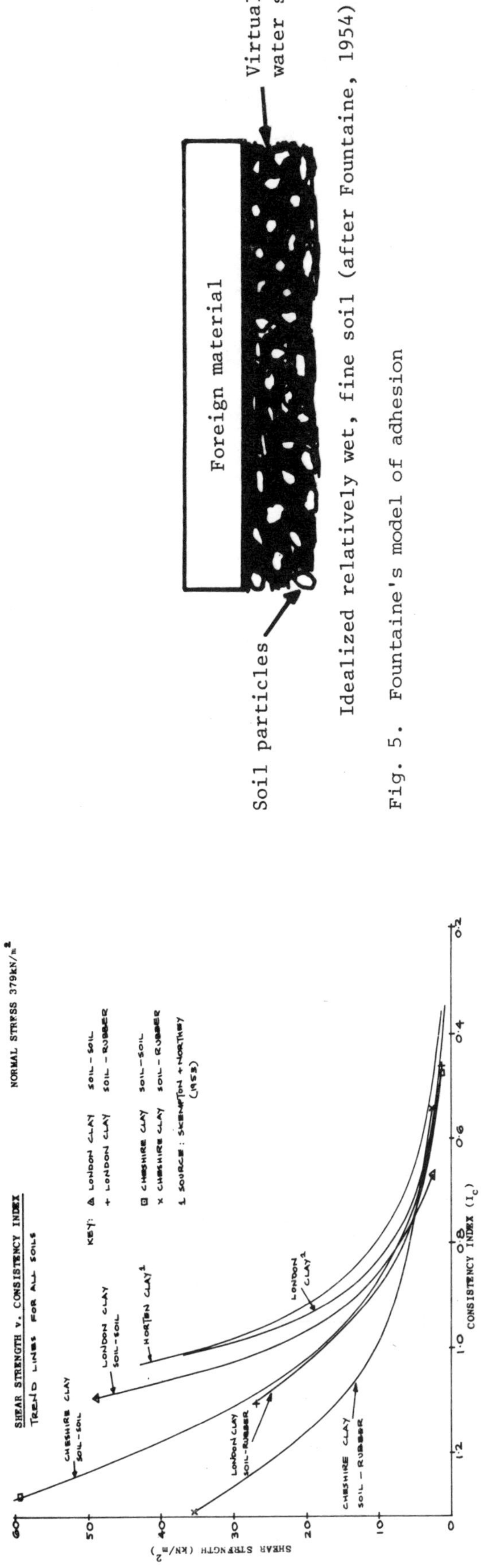

Fig. 4. Trend lines for soil-soil/soil-rubber shear - 379 kN/m^2

Fig. 5. Fountaine's model of adhesion

Fig. 6. Particle size distributions

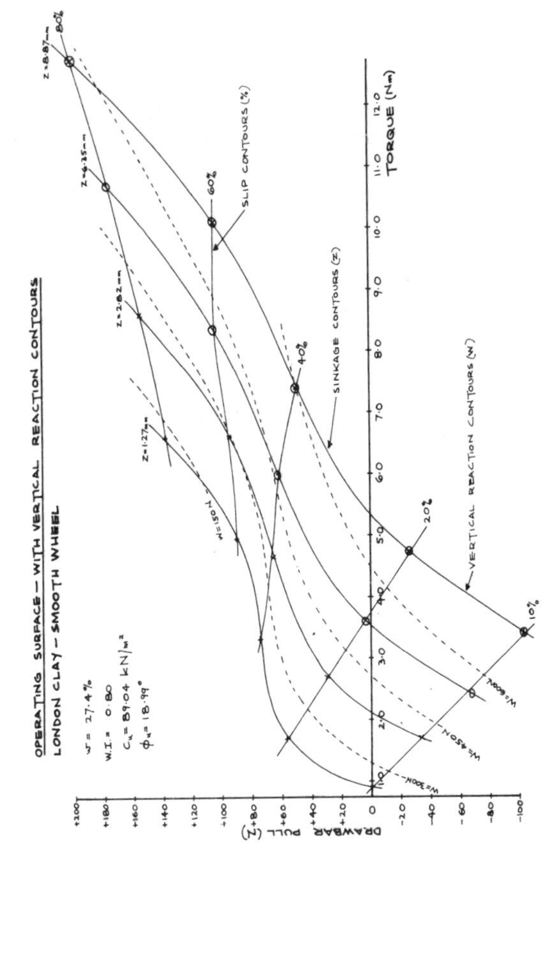

Fig. 7. Operating surface - London Clay

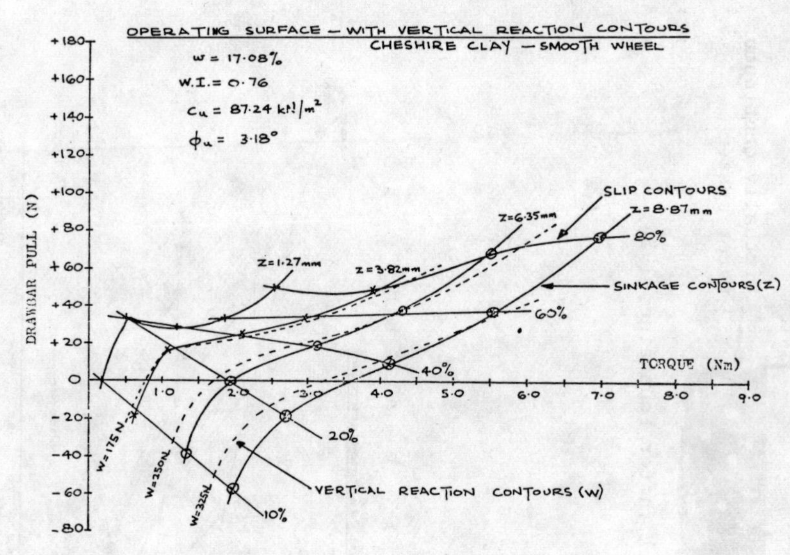

Fig. 8. Operating surface - Cheshire Clay

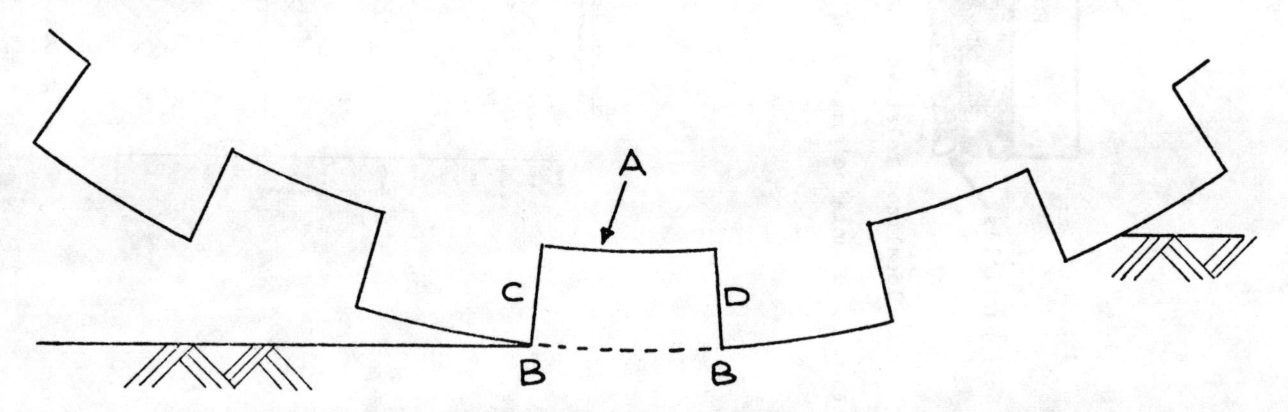

Fig. 9. Wheel tread cross section

Table 2. Soil consistencies for mobility rig experiments

LONDON CLAY.					CHESHIRE CLAY.				
w_L = 69% w_p = 29% OMC = 17%					w_L = 28% w_p = 19% OMC = 14%				
w	I_w	I_c	$\frac{w}{w_p}$	V_a	w	I_w	I_c	$\frac{w}{w_p}$	V_a
36.6	0.63	0.81	1.26	6.35	20.6	0.55	0.82	1.05	2.36
24.4	0.67	0.87	1.19	5.24	17.08	0.75	1.21	0.90	5.0
27.4	0.80	1.04	0.94	0.75	16.14	0.85	1.32	0.85	0.3

soil is connected with the bulk of the soil water - i.e. the pore water. If the system were allowed to come to equilibrium the pore water would move until it was under the same tension throughout. If the plate were pulled off quickly, before the pore water was given a chance to equilibrate, this would result in the air/water meniscus in the vicinity changing in such a way as to increase the pore water tension locally and thus increase adhesion. Before it may be considered directly relevant it must be appreciated that there has to be sufficient pore water available at the soil surface to wet the foreign surface and thus allow adhesion to occur. If there is sufficient soil surface moisture to permit adhesion and the soil shear strength, or suction, is sufficienly low then failure will occur within the soil mass. If the soil suction is sufficiently high then failure will occur at the rubber soil interf ace. This modification of Fountaine's model takes account of the observation that at low consistency indices soil-soil and soil-rubber shear are of the same order, that is the shear is taking place within the soil. Also it accounts for the fact that at consistency indices greater than approximately 0.7 shear takes place at the soil rubber interface.

10. The overall conclusion to be drawn is that above a consistency index of approximately 0.7 soil-soil shear is greater than soil-rubber shear. Thus if a tyre were able to present a soil face, for example by having a clogged tread pattern, then greater potential traction would be available at a given water content.

MOBILITY RIG EXPERIMENTS

11. In order to investigate the effect of tread pattern upon the soil-wheel shear process, London Clay and Cheshire Clay were each mixed to three different consistencies and a smooth wheel and treaded wheel used for the experiments. The soil consistencies are given in Table 2 and the particle size distribution in Fig.6.

Principle of Operation of the Mobility Rig.

12. The detailed design of the mobility rig is related elsewhere (Ref. 15). The basic operation of the mobility rig is illustrated in Fig.1. The principle of operation was that the test wheel was fixed in space on an instrumented frame. The soil under test was contained in a trolley and propelled into contact with the test wheel. Both the test wheel and the trolley were propelled by separate motors, thus enabling varying rates of slip to be achieved. An ultra-violet oscillograph recorded five separate items: test wheel speed, trolley speed, torque input to test wheel axle, vertical reaction on test wheel axle and horizontal reaction on the test wheel axle. A varying sinkage was achieved by setting the trolley to a slight inclination.

The Operating Surfaces

13. Each run of the rig was performed at a fixed rate of slip on a particular soil at a given moisture content with the trolley at a slight inclination. From the continuous trace of the oscillograph it was possible to plot graphs of torque v. slip, drawbar pull v. slip and rolling reistance v. slip. Since only one input variable, torque input, was being measured a previous worker compounded the results in the form of an operating surface (Ref.16.). The operating surface was a plot of drawbar pull against torque input upon which could be drawn contours of slip, sinkage. vertical reaction and efficiency. The detailed results of this work are reported elsewhere (Ref.17). Typical operating surfaces are shown in Figs. 7 and 8.

14. The observation which was drawn from the operating surfaces was that the slope of the sinkage contours of London Clay, to the horizontal, were greater than for Cheshire Clay. The consequence of this was that at a given wetness index

$$\left(I_w = \frac{w_L - w}{w_L - OMC}\right)$$

more drawbar pull is available from London Clay than from Cheshire Clay, thus reinforcing the trends observed in the direct shear tests.

15. A second observation made was that at approximately fifty per cent slip the slope of the slip contours changes from negative to positive. It was noted that for the Cheshire Clay the positive slope achieved tends to be lower than for London Clay.

16. As observed with respect to drawbar pull and slip the operating surface displays drawbar pull values which are negative. This was due to the design of the rig, forced wheel sinkage occurring under forced slip conditions.

TYRE TREAD CLOGGING

17 From the runs with the treaded wheel it was noted that the tyre clogged at slip rates exceeding 33%. At the higher slip rates the mechanism of clogging was possibly that : on the internal face, A in Fig.9, of a tread recess there is soil/rubber adhesion and that for the soil to be removed from a recess this adhesion on face A must be overcome and in addition the soil/rubber friction on faces C and D must be overcome. At low rates of slip the soil shear strength along the dotted line B-B may be sufficient to overcome the above resistances but as the slip rate increases soil/soil shearing starts to occur along B-B therefore weakening the soil shear strength at this point to such an extent that the above resistance cannot be overcome.

MOBILITY CHARACTERISTICS

18. As a result of the investigation into soil-rubber shear using both the shear box and the mobility rig three sets of guidelines for wheel tyre tread configuration have been put forward for clay soils (Ref.12.).

Case (A) - Hard Subsoil covered by a Thin Weak Layer

19. In this case maximum traction would be developed if the tyre could come into contact with the firm subsoil. An example of this situation could be a haul road on a construction site or an access route for military vehicles where the foundation of the road is basically sound but covered by slurry. The best tyre pattern would be one with a wide self cleaning tread pattern which can cut through the slurry to the firm surface below.

Case (B) - Homogeneous Weak Soil

20. This case could be the surface of an embankment being constructed with wet weak clay fill. Two soil consistencies can be identified. First at consistency indices $I_c > 0.7$ soil-soil shear strength is greater than soil-rubber shear strength. Second at consistency indices $I_c < 0.7$ soil rubber shear strength is greater than soil-soil shear strength. In the first instance a treaded tyre which is not self cleaning would give the greatest drawbar pull. The best tread pattern would be the one which not only clogs up but is also capable of presenting the largest soil surface for traction purposes. In the second instance when $I_c > 0.7$ a smooth rubber wheel would be most effective.

Case (C) - Hard Subsoil Covered by a Thick Layer of Weak Cohesive Soil.

21. In this case it is assumed that the weak layer is too deep for the tyre tread pattern to cut through. Effectively this case is the same as case (B).

CONCLUDING REMARKS

22. From a review of the literature it was evident that no work to date exists which can be directly applied to earthmoving by the Civil Engineer.

23. It has been concluded from the work with the direct shear box and the mobility rig that greater traction (or drawbar pull) can be achieved on the higher plasticity London Clay than on the lower plasticity Cheshire Clay at a given consistency index.

24. Also, an insight has been obtained into a rational approach to tyre tread design for three different examples of soil surface.

ACKNOWLEDGEMENTS

25. The authors are grateful to Professor J. Kolbuszewski of the Department of Transportation and Environmental Planning, University of Birmingham for the provision of laboratory facilities and his encouragement throughout this project. The authors are also grateful to Professor A.W. Hendry of the Department of Civil Engineering and Building Science, University of Edinburgh and the Director of CEBTP, Paris for their continuing encouragement. The assistance of Nottingham University and Dunlop Ltd., is acknowledged as also is the financial assistance of the Science Research Council, The Rees Jeffreys Road Fund and the Institution of Civil Engineers.

REFERENCES

1. BEKKER M.G. Theory of Land Locomotion. The University of Michigan Press, Ann Arbor, 1956.
2. BEKKER M.G. Off-the-Road Locomotion - Research and Development in Terramechanics. University of Michigan Press, Ann Arbor, 1960.
3. BEKKER M.G. Introduction to terrain - vehicle systems. The University of Michigan Press, Ann Arbor, 1969.
4. TAYLOR, D.W. Fundamentals of Soil Mechanics. John Wiley, N.Y., 1948.
5. MICHLETHWAITE E.W.E. Soil Mechanics in relation to fighting vehicles. Military College of Science, Chobham Lane, Chertsey, 1944.
6. UFFELMANN F.L. The performance of rigid cylindrical wheels on clay soil. Proceedings 1st Int. Conf. on Mechanics of Soil-Vehicle Systems, Turin, 1961.
7. MAYFIELD B. The performance of a rigid wheel moving in a circular path through clay. Ph.D. Thesis, University of Nottingham, 1963.
8. CULLEN R.M. Forces on a rigid wheel moving in an artifical clay. Ph.D. thesis, University of Nottingham, 1966.
9. NORMAN R. The effect of wet weather on the construction of earthworks. CIRIA Research Report No. 3., 1965.
10. RODIN S. Ability of clay fill to support construction plant. Parts I and II, Civil Engineering and Public Works Review, 60, 2 and 3, 1965.
11. BLACK W.P.M. A method of estimating the California Bearing Ratio of Cohesive soil from Plasticity Data. Geotechnique XII, 1962 4, pp. 271-282.
12. FORDE M.C. and A.G. DAVIS, The Tyre Terrain System. Proceedings of 5th Int. Conf. of I.S.T.V.S. Vol.1, June 1975.
13. SKEMPTON A.W. and R.D. NORTHEY, The Sensitivity of Clays. Geotechnique, Vol.3, 1953.
14. FOUNTAINE E.R. Investigation into the mechanisms of soil adhesion. Journal of Soil Science, Vol. 5, No.2, 1954.
15. FORDE M.C. A simpler test facility for laboratory wheel-terrain evaluation of wet clay. Proceedings of 6th Int. Conf. of I.S.T.V.S. Vienna, 1978.
16. THOROGOOD R.P. The steering of wheeled vehicles moving over soft ground. Ph.D. Thesis, University of Nottingham, 1969.
17. FORDE M.C. An investigation into Rubber Wheel Mobility on London Clay and Cheshire Clay. Proceedings of 6th Int. Conf. of I.S.T.V.S., Vienna, 1978.

H. GRACE, SM, MSc, FICE, and P. A. GREEN, ACGI, BSc, DIC, FGS, FICE, Scott Wilson Kirkpatrick and Partners

The use of wet fill for the construction of embankments for motorways

Attention is drawn to the need to specify the widest range of materials which will perform satisfactorily in motorway embankments. Reference is made to the different properties of fill material which are required in different parts of an embankment. A description is given of a trial embankment on the M6 Motorway which showed that fill material much wetter than that normally permitted by current specifications could be safely incorporated in selected zones of the M6 Motorway embankments if suitable drainage measures were provided. Attention is also drawn to the large financial savings which resulted from the use of wet fill on the Lancaster Penrith Motorway and mention is made of the application of this technique to other works.

INTRODUCTION

1. Until recently it has been common practice when preparing specifications for civil engineering works to specify only the best material available.

2. Recently however more thought has been given by engineers both at home and abroad to specifying the widest range of materials which would perform satisfactorily in the finished works.

3. Nowhere is this approach more appropriate than in motorway works where a major proportion of the cost relates to earth works and where appreciable proportions of the earthworks have often been classed as unsuitable, not because the solid element of the soil was in any way unsuitable but because its moisture content was higher than that recommended by the current Ministry of Transport Specification (ref. 1).

4. The problem of wet soil combined with adverse climatic conditions was particularly acute on the Lancaster Penrith sections of the M6 Motorway which reaches a height of over 1,000 ft where it passes over Shap summit.

5. This lead the authors to propose to the then Ministry of Transport that a trial embankment should be constructed using the prevalent boulder clay at a moisture content which would normally render it unsuitable according to the current specification. The object of the trial was to ascertain if boulder clay classed as unsuitable could be used satisfactorily in selected parts of the earthworks and if the provision of horizontal drainage layers would accelerate the dissipation of pore pressures and the corresponding settlements.

6. Drainage layers have been used extensively in embankment dams to accelerate the dissipations or pore pressures and so increase the shear strength to provide stability at crucial stages such as the end of construction and during rapid draw down. (ref. 2) Stability of motorway embankments is an important but a relatively rare problem but the control of settlements to ensure that the finished road profile is acceptable to the road users is an ever present problem to the road engineer, if closures and expensive reconstruction of the finished work are to be avoided. This paper therefore is concerned with settlements and the measures which should be taken to enable materials normally classed as unsuitable to be used in embankment construction in such a way that the final work is satisfactory.

FUNCTION OF MOTORWAY EMBANKMENTS

7. The two top metres of a motorway embankment have to withstand the shearing forces induced by the wheel loads and resist the densification which can occur as the result of many millions of repitions of load. The surfacing, base and subbase are especially designed to resist these forces and the underlying fill to a depth of about two metres is usually compacted to not less than some specified density to ensure that deformation of the pavement does not occur due to the repetition of stresses due to traffic.

8. Beneath the top two metre layer the stresses in the embankment are governed primarily by its own weight and the stresses induced by traffic are relatively small.

Clay fills. Institution of Civil Engineers, London, 1978, 113-118

The stability of a motorway embankment is rarely a problem provided it is founded on a strong foundation but the consolidation characteristics of the compacted fill must meet certain exacting requirements to insure that unacceptable deformations of the finished pavement are avoided. This paper is confined to a study of the requirements of the lower part of the embankment which is relatively unaffected by traffic.

9. To define unacceptable deformations is no easy matter. Typical specifications lay down that the final level of the motorway surfacing shall not vary by more than 5 mm (3/16") when tested under a 3.05 m (10') straight edge. However, this specification does not adequtely define the acceptable limits of deformations of the pavement as a whole. This is a complex problem depending not only on the differential settlements of the pavement but on the characteristics of the suspension mechanism and the speed of the vehicles using the pavement. It can be readily appreciated that quite appreciable settlements of an embankment, say up to 10 cm or more are acceptable provided they are reasonably uniform whereas differential settlements of this magnitude over a relatively short distance would be quite unacceptable.

10. More information on the maximum vertical acceleration which can be tolerated by passengers and their vehicles is required. Until this information is available it is not possible to define adequately the limits of curvature and their frequency which are acceptable in the finished profile of a motorway pavement.

THE PROBLEM

11. The main problem for motorway embankments is usually one of limiting ultimate settlements and so reducing differential settlements to acceptable limits.

12. In order to construct satisfactory embankments it is necessary to reduce the void ratio of the compacted fill to a value which would ensure that only minor additional settlements will take place after the completion of the works. In the early stages of the British Motorway Programme this was affected by specifying an upper limit for the moisture content of the fill material and a specified dry density. Above this moisture content the fill material could not be compacted to give the required void ratio and in addition it was found impossible to operate the rubber tyred plant economically over such material.

13. Any material having a moisture content greater than that specified was classed as unsuitable. A certain proportion of this unsuitable soil was usually composed of peat or highly organic clay which was unsuitable to incorporate in the works regardless of its moisture content but by far the greater proportion of the unsuitable fill was unsuitable only on account of its excess water content. The problem is an important one, not only are many materials in their natural state unsuitable but the usual methods of excavation on a relatively horizontal surface exposes large areas to the elements and appreciable quantities of soil which are originally suitable become unsuitable during periods of cold and wet weather.

14. In many of the motorway contracts appreciable quantities of excavated material were tipped to spoil and replaced with other material. The resulting increase in cost was appreciable.

15. Unsightly spoil tips of unsuitable material alongside the motorway and the excavation of equally unsightly borrow areas for suitable material became necessary. The Engineer had great difficulty when preparing the contract documents in estimating the amounts of unsuitable material. He had not only to determine the quantities of insitu unsuitable materials in the cuttings which had to be excavated but he had also to make an estimate of the amount of suitable material which would become unsuitable during the construction period due to the weather and the operations of the contractors plant. It was not surprising that the estimated quantities of unsuitable material were usually greatly exceeded during construction, a point which contractors were not slow to recognise.

16. The problem was particularly acute in the Northern part of England and on the higher ground with its colder climate and shorter spells of warm drying weather.

17. The problem is of course not new and has been experienced in the construction of dam and railway embankments. In the case of dams, settlement can be provided for by building the crest of the dam higher at the end of construction. In the case of railway embankments settlements can be rectified after the completion of construction by packing up the rails.

18. Unfortunately neither of these methods of dealing with settlements are practical for motorways and it is not surprising that this problem did not attract widespread attention until the advent of the motorway programme with its exacting requirements regarding settlements.

METHOD OF TREATING WET FILL

19. Fill may either be wet in its insitu state in the borrow area or it may become

wet during placing; the construction of Logan Airport Boston (ref. 3) Kai Tak Airport (ref. 4) and the Plover Cove Dam in Hong Kong (ref. 5) are interesting examples of the methods which can be employed to produce satisfactory embankments when fill is deposited through water or transported hydraulically to the site.

20. This paper however is concerned with material which in its insitu state in the borrow area, is too wet to be compacted to the required dry density.

21. There are three principle methods of treating this wet material:

1) By stock piling
This is only applicable to self draining granular materials having a very small precentage of fines. Occasionally such material can be piled in heaps or spread in 15-20 cm layers on a surface having a crossfall of about 1 in 10 and after three or four days the water will have drained away sufficiently to enable satisfactory compaction to take place.

2) By aeration
This method was used successfully during the construction of the Sasumua Dam in Kenya (ref. 6). Granular and most plastic materials can be successfully treated in this way, the only requirements being that the material is sufficiently friable so that the soil will break down enabling it to dry out reasonably uniformly. Most soils can be broken down by agricultural equipment such as harrows and pulvimixers to form a suitable tilth providing this process is carried out when the soil is in a damp condition. However this method is time consuming and expensive and can only be used when climatic conditions are favourable. It is not generally applicable in the British Isles.

3) By consolidation with time, with or without surcharge
This method has been used successfully during the construction of many embankment dams to reduce the excess pore pressures and so increase the shear strength both of foundations and fill material. Sand drains and horizontal drainage layers are commonly used techniques to accelerate this process, many British Dams have been built in this manner. The main problem has been one of increasing the shear strength by reduction of pore pressures, the resulting settlement of the embankment dam being of secondary importance. It has also been used to accelerate the consolidation of soil beneath road embankments, the construction of Western Avenue, Lagos (ref. 7) being a typical example.

22. However the use of drainage layers combined with wet fill on the Lancaster Penrith Motorway is to the authors knowledge the first time that this technique has been used to control the settlement of an embankment.

TEST EMBANKMENT OF WET FILL

23. In the early 1960's when the authors' firm was appointed to report on the section of the M6 Motorway between Lancaster and Penrith few trial embankments for highways had been constructed. Those that had been constructed were usually concerned with the behaviour of the embankment foundation and few had been concerned with the performance of the fill material.

24. The authors' firm had been responsible for a section of the M6 in Cheshire where approximately 30% of the excavated material had been discarded as unsuitable. On sections of the M6 further north the percentage of unsuitable material had increased.

25. The section of the M6 between Lancaster and Penrith traverses the mountainous region between the Lakes and the Pennines where climatic conditions are highly adverse. To make matters worse the proposed line of the motorway crossed extensive areas of boulder clay. This material is primarily a gravel sand silt containing small percentages of clay. Although it is a very suitable fill material when adequately compacted at the right moisture content, it is highly moisture susceptible and when in a loose state can absorb additional water readily. A small increase in the moisture content above the optimum results in a very marked reduction in shear strength often turning it into a material having the consistency of wet concrete.

26. It was thought that an appreciable proportion of the insitu boulder clay would be classed as unsuitable and that a considerable volume of suitable material would become unsuitable when exposed to the cold and wet climate in this region. The only suitable material readily available for embankment construction was rock and the cost of constructing embankments of rock was prohibitive.

27. It was known that the boulder clay contained a high proportion of granular material a typical sample having 5% less than 2 microns in size and 30% passing the No 200 sieve. It was concluded that after compaction the ultimate settlement would not be large. Even at high moisture contents the material was relatively impervious due to its well graded properties.

It was thought therefore that to obtain a reasonably degree of consolidation a relatively long settlement period would be required for 5 - 10 metre high embankment unless drainage layers were provided to accelerate the dissipation of the pore pressures.

28. Based on the classical theory of consolidation a diagram was prepared showing the relationship between the time required for 90% consolidation, the permeability and the spacing of drainage layers assuming that the coefficient of consolidations Cv remained constant during the test. This diagram is shown in Fig. 1.

29. Theoretically the time required to reach 90% consolidation, which it was assumed was necessary to avoid unacceptable differential settlements, could be controlled by the spacing of horizontal drainage layers in the fill. This time required for consolidation was a vital matter because although it would be possible to specify a settlement period in the contract documents any additional time required for consolidation of the embankment before laying the pavement would disrupt the contract and inevitably result in claims from the contractor.

30. It was therefore decided that the theoretical findings should be checked by a full scale test embankment. The construction of this embankment would in addition provide useful information on the problem associated with the operations of plant on wet material.

31. The trial embankment was constructed during June and July 1966. The Road Research Laboratory agreed to undertake the installation of the instrumentation, the recording of all measurements and the writing up and publication of the findings. These are recorded in RRL Report LR238 (ref. 8).

32. The trial confirmed the opinions that:-

1) It would be possible to use boulder clay to form parts of embankments at a much higher moisture content than would normally be allowed by the current specification.

2) That the rate of dissipation of pore pressures could be controlled by the spacing of the drainage layers.

3) That the total settlement would be small owing to the high stone and sand content of the boulder clay.

33. The findings from this investigation produced a number of important consequences in the carrying out of the Lancaster Penrith Motorway which are listed below:

1) It was possible to incorporate boulder clay into the works having moisture contents of up to 18% compared with the maximum moisture content of 9 - 11 % permitted by the current M.O.T. specification.

2) This relaxation of the specification meant that extensive and unsightly spoil dumps and borrow pits were dispensed with along the line of the motorway.

3) The work was substantially accelerated.

4) There was an appreciable saving in the cost of the earthworks. It is difficult to assess the exact saving but it can be said with certainty that at least £500,000 was saved on 35 miles of motorway construction and it is roughly estimated that the actual saving was about £1 million.

CONSTRUCTION PROBLEMS WHEN USING WET FILL

34. The use of wet fill requires different placing and compacting techniques to those normally employed. The normal rubber tyred plant is unable to operate over this type of material. Excavation has to be carried out by excavators or front loaders, and the wet material conveyed to the site by truck or dumper. The wet fill is unable to support construction traffic and suitable haul roads have therefore to be provided. Spreading can be effected by bulldozers provided with extra large tracks. Due to the high water content compaction to a small percentage of air voids is no problem and one or two passes of the bulldozer is all that is required. Further compaction is counter productive, the only effect produced is to induce excess pore pressures in the fill with a consequent decrease in stability.

35. The effect of the bulldozer when operating over the wet fill was interesting. A large bow wave was produced which extended in front of and on each side of the machine. However, no permanent displacement or shearing of the material took place and the level of the fill reverted to its original level after the passage of the bulldozer.

36. The bearing capacity of fill of this type is negligible and the wet fills were so located that 6 or 8 feet of suitable fill could be placed over them to obtain a firm surface before any base or subbase was laid.

37. Wet fills were limited to a thickness of 8 - 10 feet and drainage layers of crushed rock were invariably located above and below the wet fill. After the placing of the overlying suitable fill a settlement period of 3 months was specified. However in practice settlement took place quicker than anticipated and the settlement period was reduced. The settlement period was useful as the pore pressures dissipated and the semi liquid wet fill gained in strength.

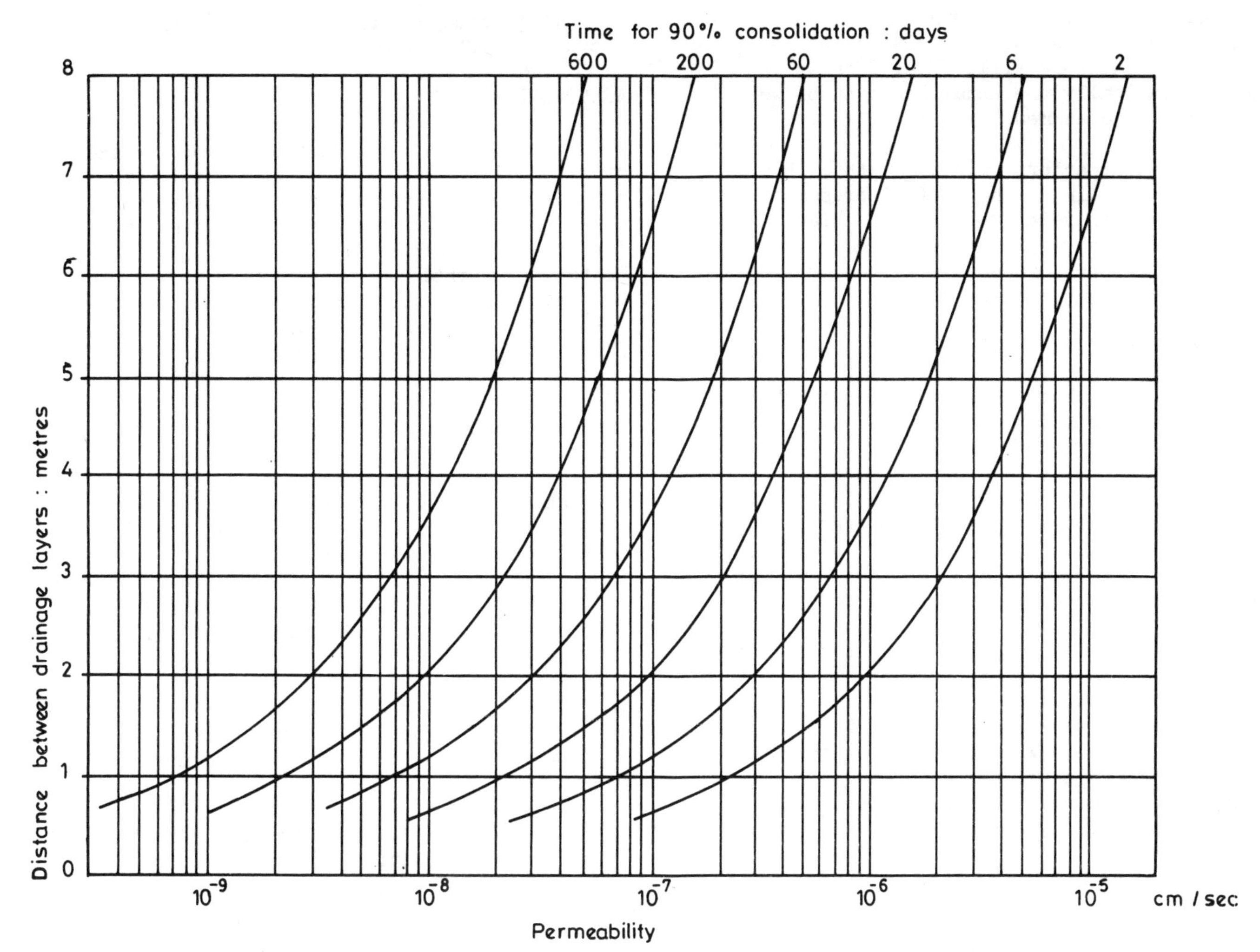

Fig. 1. Time required for 90% consolidation of layers of varying thickness and soils of varying permeability (assumes Mv = 0.186 m^2/MN

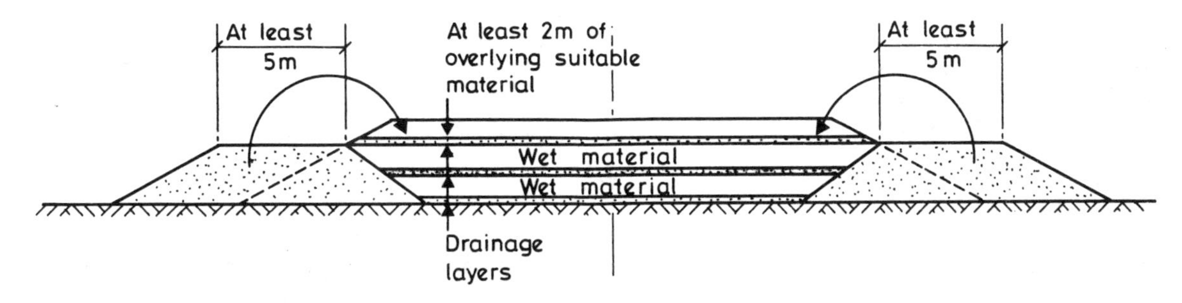

Fig. 2. Diagrammatic cross section of embankment showing location of wet material

38. It was usual to select relatively flat areas for the wet fills and they were normally contained by the construction of bunds of suitable fill placed on each side, the major part of these bunds being subsequently excavated and placed as overlying suitable fill. Fig. 2 shows diagrammatically the usual location of wet fill in an embankment.

39. The Contractors maintained that they preferred dealing with wet fill presumably because it was relatively easy to place and required negligible compaction. The wet fill has performed satisfactorily since the motorway was opened.

APPLICATION TO OTHER WORKS

40. Wet fills have also been used on the M2 Motorway in Northern Ireland. In this area the predominant soil is also a boulder clay but it contains a much higher proportion of clay and is therefore much less sensitive to small moisture content changes.

41. A trial embankment was also constructed in Northern Ireland which proved to the satisfaction of the Engineers that unsuitable material could be incorporated in the permanent works. The technique employed on the M2 was somewhat different to that used on the M6 in Westmorland. On the M2 a high proportion of the soil was only marginally too wet to comply with the specification. A technique was developed whereby layers of unsuitable soil were alternated with layers of suitable soil which may well have acted as drainage layers. It is regretted that results of the Northern Ireland work are not available to incorporate in this paper.

CONCLUSIONS

42. The trial embankments on the M6 in Westmorland and on the M2 in Northern Ireland have shown that fill wetter than that normally permitted by current specifications can be incorporated into the lower parts of motorway embankments and will subsequently perform satisfactorily.

43. Theoretical studies indicate that soils classed as unsuitable which have a permeability greater than 10^{-8} cm/sec (this includes a large proportion of soil present in the British Isles) can be used in the construction of motorway embankments provided drainage layers are incorporated at suitable intervals.

44. Further investigations and trial embankments will be necessary if other soil types are used and it will probably be necessary to investigate the shearing strength of the fill to check the stability of the higher embankments in addition to the consolidation characteristics.

45. The Transport and Road Research Laboratory published in 1971 a report on a laboratory study on the use of wet fill in embankments (ref. 9) which engineers would do well to study if they are contemplating the use of wet fill.

ACKNOWLEDGEMENTS

46. The authors are indebted to Prof. Bishop of Imperial College London who worked with them as their consultant when formulating their proposals for the use of wet fill and for the trial embankment. Mr. J. L. E. Sutton was jointly responsible for the proposals and prepared Fig. 1 showing the relationship between permeability, time for 90% consolidation and layer spacing.

47. The authors are also indebted to the Transport and Road Research Laboratory for installing the instrumentation in the M6 Trial Embankment and for recording and publishing the results in RRL Report LR238.

REFERENCES

1. MINISTRY OF TRANSPORT, Specification for Road and Bridge Works London 1969 (H.M. Stationery Office)

2. VAUGHAN P. R., LOVENBURY H. T. & HORSWILL P., The design construction and performance of Cow Green embankment dam. Geotechnique 25 No. 3 555-580

3. CASAGRANDE A., Soil Mechanics in the Design and Construction at Logan Airport. Journal Boston Soc. Civ. Engs. April 1949 pp 192 - 221

4. GRACE H. & HENRY J.K.M., The Planning and Design of the new Hong Kong Airport. Airport Paper No. 35 Proc ICE June 1957 Vol 7 pp 275 - 325

5. ELLIOTT S. G. & FORD S. E. H., Investigation and Design of the Plover Cove Water Scheme Proc ICE Oct 1965 Vol 32 pp 255 - 293

6. DIXON H. H., EDDINGTON G. A. & FITZGERALD E. P., The Chania Sasumua Water Supply for Nairobi Proc ICE April 1958 Vol 9 pp 345 - 368

7. JOHNSTON C. M., Western Avenue, Lagos; the design and construction of a soil cement pavement. Proc ICE Sept 1961 Vol 20 pp 107 - 140

8. McLAREN D., RRL Report LR238 1968 Trial Embankment at Killington

9. FARRER D.M., RRL Report LR406 1971 A Laboratory Study of the Use of Wet Fill in Embankments.

R. E. HARPSTER, MA, Senior Project Engineering Geologist, Woodward-Clyde Consultants, San Francisco

Selected clays used as core for a rockfill dam designed to cross a potentially active fault

Selected clay borrow materials were obtained from an old lake bed to form a flexible core in the construction of Cedar Springs Dam, a 215-foot- (66-meter) high zoned earth- and rock-filled structure, San Bernadino County, California, U.S.A. The dam is designed to withstand a maximum credible accident, which included 3 to 5 feet (0.9 to 1.5 meters) of lateral or vertical displacement that could possibly occur within the foundation of the dam during a strong earthquake.

The clay borrow was selected after an extensive investigation for plastic material that could deform without cracking or losing strength. Results of soil tests on materials placed in the dam differed somewhat from the results of soil test data used in design. However, based on the tests and upon the performance of the dam structure since 1971, the inplace material met the design requirement and is able to maintain its integrity as an impervious membrane. The design objective required the core to be flexible in nature and to be able to adapt to large deformations without cracking. The suitability of the inplace clay core was measured by its Atterberg limits and grain size gradation. Although the clay core was more clayey and more plastic inplace than predicted, it is believed to meet the design intent.

Factors leading to the differences between the design soil test results and the inplace material soil test results are principally attributed to differences in handling of the materials. The source of the selected clay fill was approximately 25 miles (40 kilometers) from the dam. The contractor's construction procedures required the stockpiling and moisture conditioning of the clay fill near the embankment before it was re-excavated and placed in the dam. There were measurable differences in test results with various amounts of handling of the clay materials. These differences have been noted by others and are generally anticipated and quantifed for design purposes by adjusting the testing program to simulate mechanical mixing of clay particles and methods of adding water anticipated during construction. In the final analysis, the laboratory program should serve as a model of the construction processes utilized.

INTRODUCTION

1. Cedar Springs Dam was designed and constructed by the California Department of Water Resources across the west fork of the Mojave River in San Bernardino County, California, U.S.A. (Figure 1). The construction was started in 1968 and completed in 1971. Most of the exploration and design work was done in the early to late 1960's. The dam is located about 10 miles (16 kilometers) north of the city of San Bernardino. It is a zoned earth- and rock-fill structure, about 215 feet (66 meters) high and about 2,200 feet (671 meters) long.

2. The dam was constructed using six zones of specific materials: central impervious core (zone 1); flanking transition zones (zone 2); flanking sand and gravel sections (zone 3); downstream shells of rock (zone 4, 4A and 5); and a protective rip rap zone (Figure 2). Lean clays excavated from lake bed deposits were found during design investigation to have physical properties suitable for the construction of the impervious core for Cedar Springs Dam. Approximately 1,160,000 cubic yards (887,400 cubic meters) of lake bed clay were needed. The uniqueness of the design of this dam is that it takes into consideration the so-called "maximum credible accident", which includes the concept that faults in the foundation may move during the life of the dam and cause a displacement of 3 to 5 feet (0.9 to 1.5 meters), (1).

3. The dam crosses a geologic fault judged active; that is, geologic evidence shows the fault has cut gravel deposits of Holocene age (the last 11,000 years).

4. The objectives of this paper are to: 1) relate the specific need for a clay material required for an impervious core expected to maintain its integrity during substantial deformation without excessive cracking; 2) describe the soil test requirements as stated in the Bid and Contract Specifications for the zone 1 dam embankment; 3) describe the soil test results from samples taken from the borrow source area during the design phase of the project; 4) describe the soil test results from samples taken from the zone 1 dam embankment; and 5) describe the differences in soil test results between the

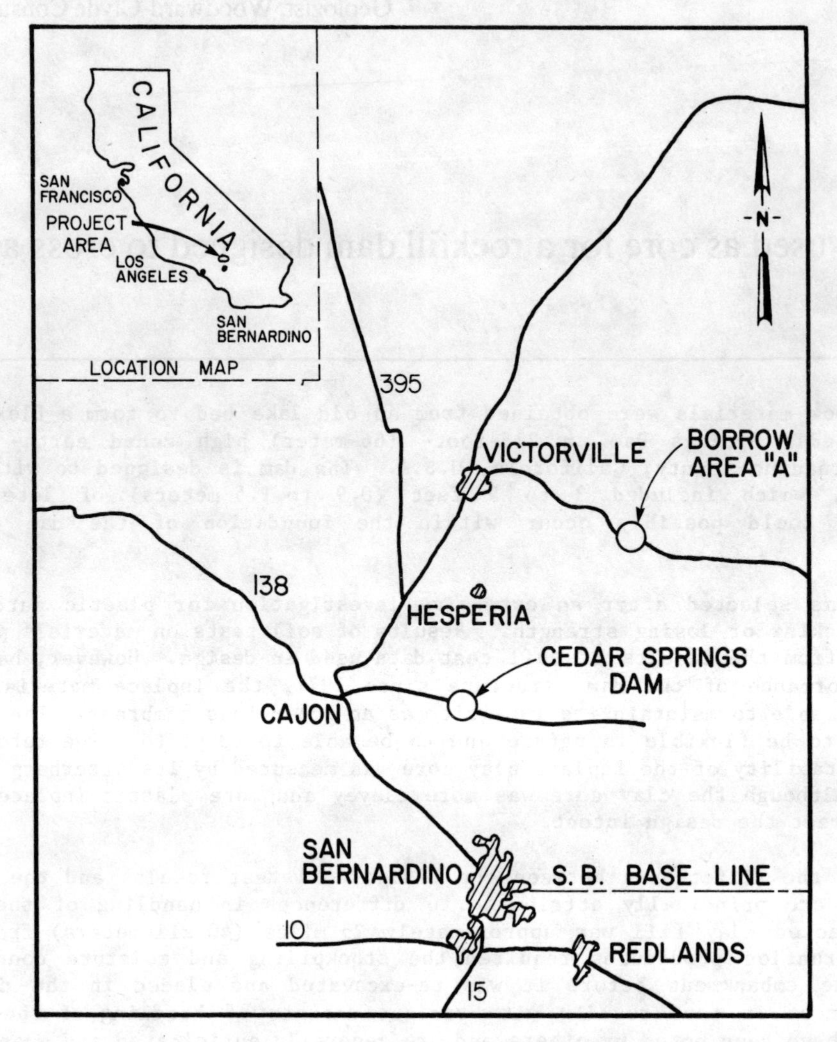

Fig. 1. Location map of dam site and borrow area A

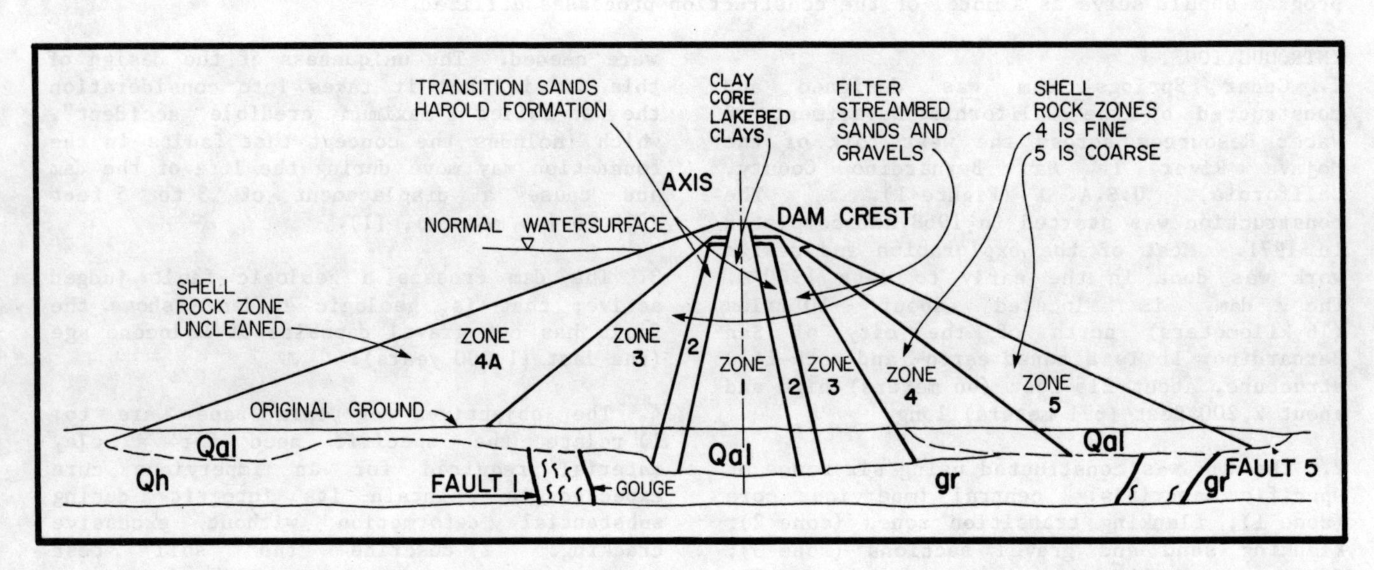

Fig. 2. Transverse dam section showing zones

samples taken in the borrow area during the design phase and the samples taken from the zone 1 dam embankment during construction.

AREA GEOLOGY

5. Cedar Springs Dam is located in the northwest portion of the San Bernardino Mountains within the Transverse Range geomorphic province. The Mojave Desert province lies to the north of the dam. The Transverse Range in this area is distinguished by the east-west-trending San Bernardino Mountain block, approximately 55 miles (88 kilometers) long and as wide as 30 miles (48 kilometers). The mountains are bounded on the south by the San Andreas fault and are less well defined by a series of subparallel faults on the northern boundary.

6. The oldest bedrock in the area is Precambrian gneiss and other associated metamorphic rock types. In the northwest section of the San Bernardino Mountains, which includes the dam location, Mesozoic intrusions predominate. They are composed of granitic and gneissic rock types. Fault troughs, some paralleling the San Andreas and others extending east-west to the northern boundary of the mountain range, contain younger terrestrial sediments of Miocene, Pliocence, and Holocene ages.

7. Paralleling the northern border of the San Bernardino Mountains are gently to steeply dipping younger terrestrial sediments of the Crowder Formation, Harold Formation, Shoemaker gravels, and Recent alluvium. These formations represent the remnants of a series of large fluvial and fan deposits created during time intervals when the north slope of the San Bernardino Mountains drained northward. Subsequently, Cajon Creek has incised the range, creating Cajon Pass by severely eroding the softer terrestrial sediments and crossing the San Andreas fault zone, as well as other parallel fault zones.

8. Approximately two-thirds of the foundation of the dam is in granitic rock. The granitic rock has been subjected to various degrees of metamorphism, and to some extent, exhibits characteristic black, white and pink banding of gneissic rock. The rock is moderately to slightly weathered. The more metamorphosed rock is considerably more friable and weak.

REGIONAL FAULTING

9. The San Andreas fault, striking about N70^{o}W, is mapped at the southern border of the San Bernardino Mountains, 7 miles (11.3 kilometers) south of the dam site. The San Jacinto fault, also striking in a northwestern direction, is about 10 miles (16 kilometers) southwest of the dam site. Within the reservoir area the design exploration verified the location of the east-west Cleghorn fault. The Cleghorn fault dips steeply to the north and displaces older alluvium in Cleghorn Canyon, about 2 miles (3 kilometers) south of the dam site.

10. During the exploration phase of the study, seven faults were mapped in the vicinity of the dam site. Three of these faults were in the proposed foundation area (Figure 3). An apparent vertical displacement of 5 feet (1.5 meters) on recent alluvium was exposed in a trench across fault 1 (Figure 4). Smaller displacements were observed in seven other trenches across fault 1 and in three trenches across fault 2. The dam was placed so as to avoid crossing these faults where evidence of Holocene faulting was observed. Faults 1 and 2 are probably structurally connected and regionally form the Harold Formation/granite contact. This contact can be observed to be at least 200 feet (61 meters) upstream from the dam axis location.

11. Fault 1 crosses the axis of the dam on the left abutment. The fault strikes N67^{o}W and dips 80°N, where it crosses the core trench. Generally, fault 1 is characterized by a band of gouge 1 foot to 9 feet thick (0.3 to 3 meters), with a zone of breccia and altered rock 10 to 50 feet (3 to 15 meters) wide. The fault is considerably more narrow where it crosses the dam axis. The fault crossing was covered by zone 1 core and blanket. Curtain and blanket grouting also cover the fault crossing. Grout holes were drilled at angles positioned to cross the fault and intersect both contacts of the fracture zone and the fault gouge. Piezometers were placed in the foundation on either side of the fault crossing.

CONSTRUCTION MATERIAL FOR ZONE 1

12. Because the geologic evidence indicated active faults in the foundation, prudent design dictated the need for a flexible clay core and blanket to meet the potential displacement requirements. The physical properties of materials from potential impervious borrow sources adjacent to the dam site were studied in detail, (2). The soil characteristics required by the designers were not found and, as a result, local clay sources were not selected. Instead, an unnamed dry lake north of California State Highway 18, about 25 miles (40 kilometers) from the dam site, was chosen as the source of dam impervious borrow. This more distant source was chosen because the clays possessed physical properties favorable for use in a potentially flexible clay core. Design exploration at this site showed that a sufficient quantity of lean clay with uniform characteristics existed within the lake bed, designated as borrow area A.

13. Placement control, as stated in the specifications for the required blended- and moisture-conditioned clays obtained from borrow area A, came from moisture content measurements on samples taken from zone 1 material in place. This control was exercised by the engineer in specifying that the moisture content required for compaction would be approximately 1 to 2 percent above optimum, as determined by soil tests on the borrow

area A material. Laboratory tests indicated the plastic behavior of zone 1 material varied with placement moisture content confining pressure. Guidelines were given from the design engineer to the construction personnel regarding the desired placement moisture content variation relative to increments of embankment elevation.

14. General requirements, within the specifications, required the clay material to be compacted at a uniform moisture content within 1 percent of that designated by the engineer. Detailed control of the embankment was based on a minimum testing frequency program of one test per 2,000 cubic yards (1530 cubic meters) placed.

15. Other general guidelines to construction personnel were developed to improve the quality of the embankment and to ensure that the design intent was achieved. These guidelines were not part of the construction contract; however, they served to relate the physical properties of the constructed embankment to those anticipated in the design. The most important design objective was for the impervious core of the dam to be plastic in nature and to adapt to rapid deformation without excessive cracking. Testing programs during the design phase confirmed that the most suitable materials available were the lean clays that were sampled from the dry lake beds in borrow area A. The designers required that material with liquid limits greater than about 50 or less than 30, and with a plasticity index less than 15, not be accepted for core material. Design gradation curves were also presented that showed the limits acceptable. It was required that most materials tested within borrow area A meet the acceptable range of gradation for the designed impervious core. Anomalous lenses or thin beds were expected to have little influence on the physical properties of the blended material and were to be considered acceptable if blended into the soil mass. Principal tests used to evaluate the suitability of borrow material in place and the quality of the impervious core were size gradation tests and moisture content tests. The more completely blended material appears to have a slightly higher optimum moisture content than the values measured on samples taken from the borrow area. The consistent testing program ensured the material placed was within the blended limits of size distribution and met the desired expectations for the clay materials. Within the recommended procedures for placement of the clay core was required that the first foot of embankment have additional moisture to allow the clay material to form a tight contact with the irregular surface of the foundation rock.

SOIL TEST DATA

Borrow Area A

16. Thirty-three bucket auger holes were dug, logged and sampled at 5-foot (1.5-meter) intervals in borrow area A, covering about 75 acres (30 hectares) of a dry lake bed. The holes were dug to a depth of 35 to 40 feet (10 to 12 meters). Tests made on the samples were mechanical analysis, hydrometer, liquid limit, plastic limit, specific gravity, in place moisture, and some compaction tests. The average values of the soil tests are shown in Table 1. The values vary depending upon whether they reflect areal relationships or varying thickness of the strata.

17. Weighting of the test values with the thickness of the strata more accurately reflects the geologic distribution of the

Table 1. Comparison of soil test results

Samples taken From Borrow Area A	Atterberg Limits		Size Analysis		Optimum Moisture Content		Maximum Dry Density	
	Plasticity Index	Liquid Limit	Number 200 Seive	5 Micron				
Averages Reflecting Area Relationships	21.0% 50 Tests	34.0% 50 Tests						
Averages Reflecting Thickness of Strata	23.0% 50 Tests	38.7% 50 Tests	86.7% 50 Tests	62.6% 50 Tests	18.1% Average of 3 Tests From 3 Holes on From Composite Samples	18.2% 1 Test on Composite Sample	110.3 Pounds Per Cubic Foot (1767 Kilograms per Cubic Meter) Average of 3 Tests on Composite Samples	111 Pounds Per Cubic Foot (1778.22 Kilograms per Cubic Meter) 1 Test on Composite Samples from 4 Holes
Samples Taken From Zone 1								
Averages of Tests	32% 474 Tests	47% 474 Tests	91% 467 Tests	66% 467 Tests	19.0% 498 Tests		110.0 Pounds Per Cubic Foot 1762.2 Kilograms Per Cubic Foot 498 Tests	

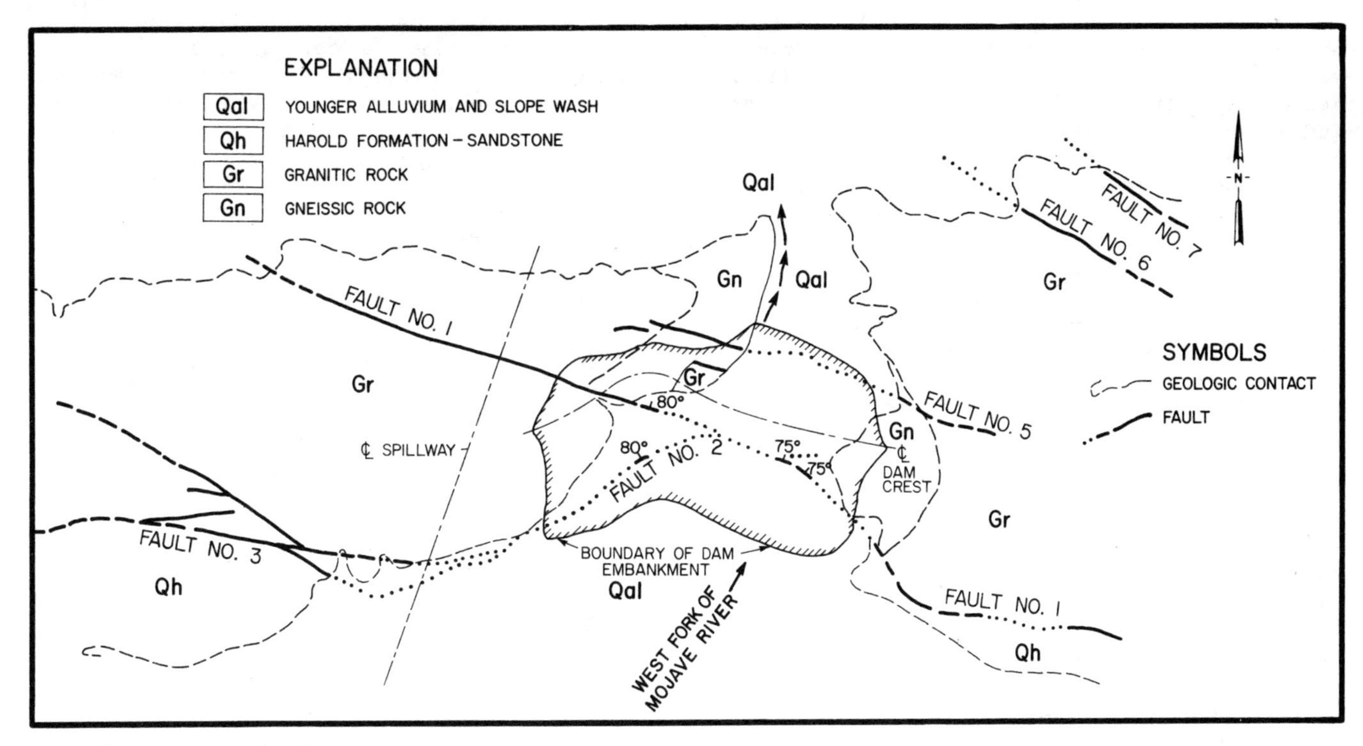

Fig. 3. Site geologic map

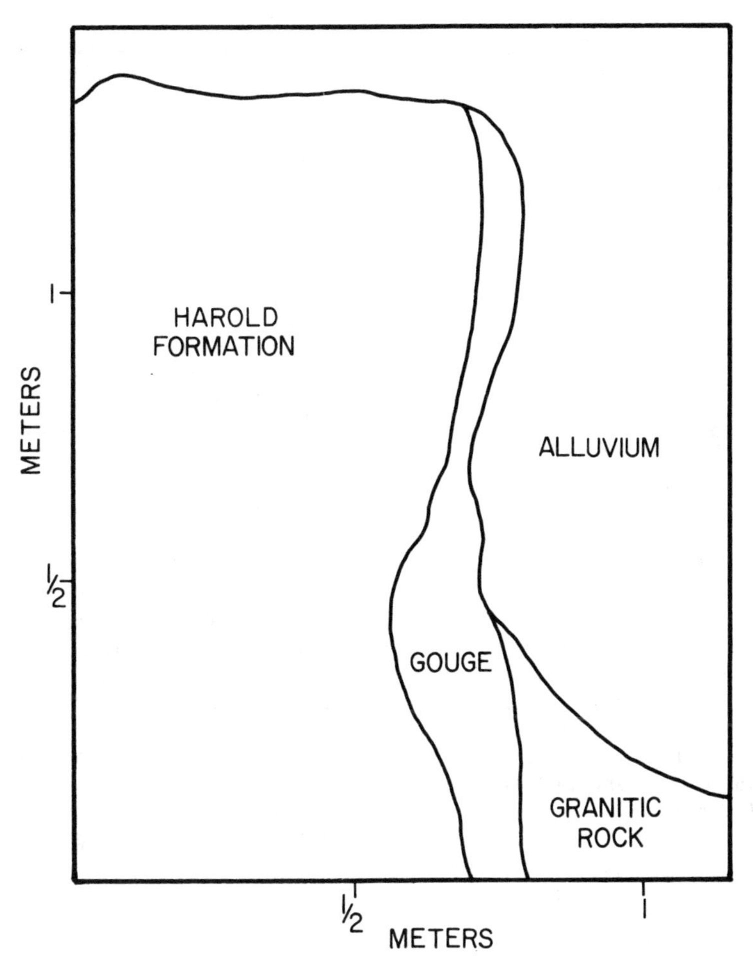

Fig. 4. Log of exploration trench across fault 1

various materials, and is also shown in Table 1. Accurately weighted values may not exactly reflect soil test results performed on blended samples because of the unknown effect of the physical interaction of the various materials. The California Department of Water Resources testing program reported each test individually. Assumptions and averaging procedures allow possible small errors and variations in average values to occur, as shown in Table 1; however, averaging test results is an acceptable procedure to arrive at design parameters.

18. The design soil test results shown in Table 1 indicate average values of liquid limits vary from 34.0 percent to 38.7 percent, depending upon the method of averaging. The plasticity indexes correspondly vary from 21.0 percent to 23.0 percent. Weighted average values calculated from size analysis tests show 86.7 percent material passing the 200 sieve size and 62.6 percent passing the 5 micron size. Results of compaction tests yield optimum moisture content results of 18.1 percent and 18.2 percent, with corresponding dry density values of 110.3 and 111.0 pounds per cubic foot (1767 and 1778.2 kilogram per cubic meter).

Zone 1

19. An estimated 1,160,000 cubic yards (887,400 cubic meters) of material excavated from borrow area A was needed to supply the clay for the core of the dam. The distant source of the clay material required the construction of a 25-mile- (40-kilometer) long haul road from the borrow area to a stockpile area in the reservoir. The dry lake materials were prewetted in the borrow area and excavated with a Staker Euclid belt loader powered by D-9 tractors. The final depth of excavation was approximately 20 feet (6 meters).

20. The excavated material was placed in the stockpile area in thin layers and sprinkled with water and worked with a disc to obtain the proper moisture content. When zone 1 material was needed, the clay in the stockpiles was excavated and placed in the dam. Further moisture conditioning and blending with the disc was necessary prior to compacting the clay on the embankment with tamping rollers.

21. The tests of zone 1 material after placement showed an average liquid limit of 47 percent and an average plasticity index of 32 percent. The average of sieve analysis tests showed 91 percent of the material passed the 200 size sieve and 66 percent of the material was minus 5 micron size.

22. The average optimum moisture was 19.0 percent and the average maximum dry density was 110.0 pounds per square foot (1762.2 kilograms per cubic meter).

Comparison of Soil Test Data

23. A comparison of weighted averages calculated from soil test data shows an increase in the minus 200 size from 86.7 percent from the borrow area A samples tested during design to 91 percent from the zone 1 samples. The minus 5 micron size increased from 62.6 percent to 66 percent. The Atterburg limit tests show a corresponding increase in the liquid limit and, therefore, in the plasticity index for test values taken from samples from the zone 1 embankment over the test values of samples from the borrow area A tested during design phase of work. These differences in test values represent the results from a large number of individual tests (Table 1). The data show consistency in differences and not a large scatter of data points. The differences, although small, represent a slight but real change in the physical properties of the clay borrow. However, the values still fall completely within the accepted range of values meeting the design objectives.

24. The difference between test values from samples taken from borrow area A during the design phase of work and the test values from samples taken from zone 1 during the construction phase of work may be attributed to any one or a combination of several factors, including: 1) Precision of test measurements; 2) Consistency of test procedures; 3) Consistency of sampling procedures; 4) Different degree of blending resulting from methods of initial excavation and amount of handling, such as hauling, stockpiling, discing, and reexcavation; and 5) Different quality of water used for moisture conditioning.

25. The test results and the physical procedures used to excavate and place the clay borrow indicate that different degrees of blending may be responsible for the differences in test values. It has been demonstrated through laboratory testing that grinding of clay minerals causes an increase in cation exchange capacity, (3). The multiple handling of the clay borrow may have had a grinding effect causing a breakdown in particle size of the stacks of crystal plates, resulting in an increased surface area for the clay minerals. The postulated increase in surface area could have created a slightly increased capacity for the clay to hold water. The soil test results show an increase in average liquid limit from 38.7 percent to 47 percent and a corresponding increase in percent finer than the 200 sieve size and clay size (minus 5 microns). The soil test results also show a slight increase in optimum moisture content with a decrease in maximum dry density. This, too, is consistant with a slight increase in the amount of the clay size fraction.

26. No testing was done to investigate whether or not cation exchange took place. However, if cation exchange occurred in the

handling process, slight changes in the water holding capacity of the clay could be expected.

CONCLUSIONS

27. The case history of design and construction of the clay core at Cedar Springs Dam is an excellent example of a detailed combination of specification requirements resulting from a carefully controlled design investigation together with a detailed construction testing program. These specifications accomodated the general design objectives and produced an impervious core with specific plastic characteristics consistent with the geologic environment at the site.

28. In this way, a structure has been constructed that can relate to its natural environment. The further development of such an approach to design and construction through a complete review of 1) soil test results and 2) field instrumentation results will help to define the application and limitation of design requirements. This will allow further use of clay materials for critical engineering structures.

REFERENCES

1. Department of Water Resources, 1967, Design Considerations for maximum credible accident, Cedar Springs Dam.
2. California Department of Water Resources, 1968, Geology and construction materials data, Cedar Springs Dam, Project Geology Report D-102.
3. Grimm, R. E., 1953, Clay minerology McGraw-Hill Book Company, Inc., New York, 138.

The author acknowledges with appreciation the technical reviews and helpful suggestions provided by Bernard B. Gordon, David J. Gross, and Janet L. Cluff.

M. INADA, Dr Professor of Civil Engineering Department, Tokai University, K. NISHINAKAMURA, Construction Project Consultants, Inc., T. KONDO, H. SHIMA and N. OGAWA, Japan Highway Public Corporation

On the long-term stability of an embankment of soft cohesive volcanic soil

The study in this paper involves a stability analysis during and after the construction of a motorway embankment with the height of 20 m using volcanic cohesive soil called "Ashitaka Loam". The soil is widely deposited around the southern foot of Mt. Fuji and has high natural water content ranging from 120 to 230 percent. The strength of the soil decreases very sensitively with remolding actions during the earthwork operations.
Focus is placed in this paper on the deformation and settlement of an embankment and the variation of slope stability with the passage of time. All slope stability analyses were made by the effective stress method on the basis of in-place pore pressure measurements.

1. PREFACE

Japan has many volcanos and the volcanic soil covers wide area of mountain foots and hilly plateaus. This volcanic soil has changed its characteristics by the meteorological conditions of much rainfall and high humidity and become a peculiar kind of cohesive soil with high water content. As the most important feature, the volcanic cohesive soil possesses relatively high strength in the state of nature before disturbance but reduces it remarkably by remolding during earthwork operations.

Tomei Motorway runs through the foot of Mt. Ashitaka located in the southeast of Mt. Fuji. The foot is thickly covered with the volcanic spouting deposits and the top of 10 to 20 m with the volcanic cohesive soil called "Ashitaka Loam". (Ashitaka Loam is very similar to Kanto Loam in its characteristics.) The deposit of Ashitaka Loam has been eroded remarkably and many valleys have been developed. Hence, the extremely up-and-down topography has been formed.

The distribution of Ashitaka Loam along the motorway extends about 20 km in length from the east end of Yoshiwara city to Susono town. The natural water content of the most of Ashitaka Loam ranges from 120 to 200% and has the sensitivity ratio of 3 to 10, and shows the worst engineering characteristics among the volcanic cohesive soils in Japan.

At the time of the construction of the motorway running through the mountain foot with this topography and soil, the followings were problems to be solved:

(1) The selection of suitable construction equipment and earthwork methods.

(2) The high embankment structure with the long-term stable slope and the smooth road surface.

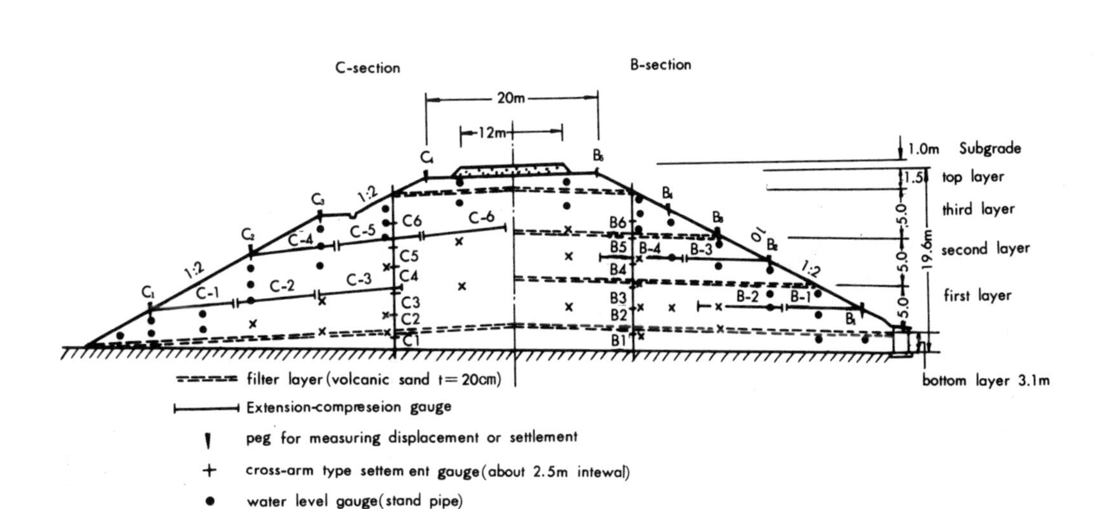

Fig. 1. Cross section of embankment and arrangement of instruments

Clay fills. Institution of Civil Engineers, London, 1978, 127-132

(3) The suitable subgrade structure on the cut ground and the embankment of Ashitaka Loam. In order to investigate above problems, a test embankment was constructed and its various behaviors and state of soil conditions were observed and tested for a long time before the main construction of the motorway began.
The authors report here the long-term stability of a high embankment with Ashitaka Loam focusing on term (2) of above problems.

2. THE EMBANKMENT STRUCTURE AND THE EARTHWORK

The test embankment of 100 m in length and 20 m in height was constructed, filling the V-shaped eroded valley with Ashitaka Loam. The cross-section is shown in Fig. 1. The slope gradient was set as 1:2 and for the purpose of comparing the effect of the consolidation acceleration and the stability of the embankment, filter layers were placed differently in test sections; 4 layers for B-section (5 m interval) and 2 layers for C-section (15 m interval) as shown in Fig. 1. The permeable volcanic sand material was applied to the filter layers and the thickness of the layer was 20 cm.
For hauling, spreading (second hauling) and compaction in the embankment field, marsh bulldozers were used. The number of passes for compaction by marsh bulldozers was 1 to 2 for every spreading thick lift of 30 cm.
On and in the embankment, as shown in Fig. 1, the extension- compression gauges, the pegs for measuring displacements, the cross-arm type settlement gauges, the water level gauges (stand pipes) and the pore-water pressure gauges (Bourdon type) were installed and used for measurements. And the changes of the state of the embankment with time were investigated during and after the construction by applying the cone penetration tests (cross sectional area = 3.23 and 10 cm^2, cone angle = 30° and 60°), the boring and sampling, and the other soil tests.

3. THE CHARACTERISTICS OF EMBANKMENT AND THE CHANGES WITH TIME

The characteristics of the embankment under the subgrade and their changes with time both in B-section and C-section are shown in Fig. 2. From this figure, the followings are verified:

(1) Immediately after the compaction the water content was 120 to 140%, the degree of saturation 94%, the unit weight about 1.32 t/m^3. The water content decreased by 10% on average from the natural water content before excavation. During about one month after the completion of the embankment, the water content hardly changed from that immediately after the compaction, but after about 32 months passing, it was about 10% lower than that at the compaction time. The rate of the decrease of water content in B-section was more remarkable than that in C-section.

(2) The cone penetration indices measured on each layer immediately after compaction were 2 to 3 kgf/cm^2 {200 to 300 kN/m^2}. The cone penetration index increases remarkably with time after the completion and the rate of increase, as seen in Fig. 2, in the lower part of the embankment and in B-section were higher than those in both the upper part and C-section. The increase of cone penetration index near the filter layer was remarkable.

(3) The cohesion, Cu, by the unconsolidated - undrained shear test on the soil sampled immediately after the compaction of each layer is 0.14 kgf/cm^2 {13.7 kN/m^2} on average and it

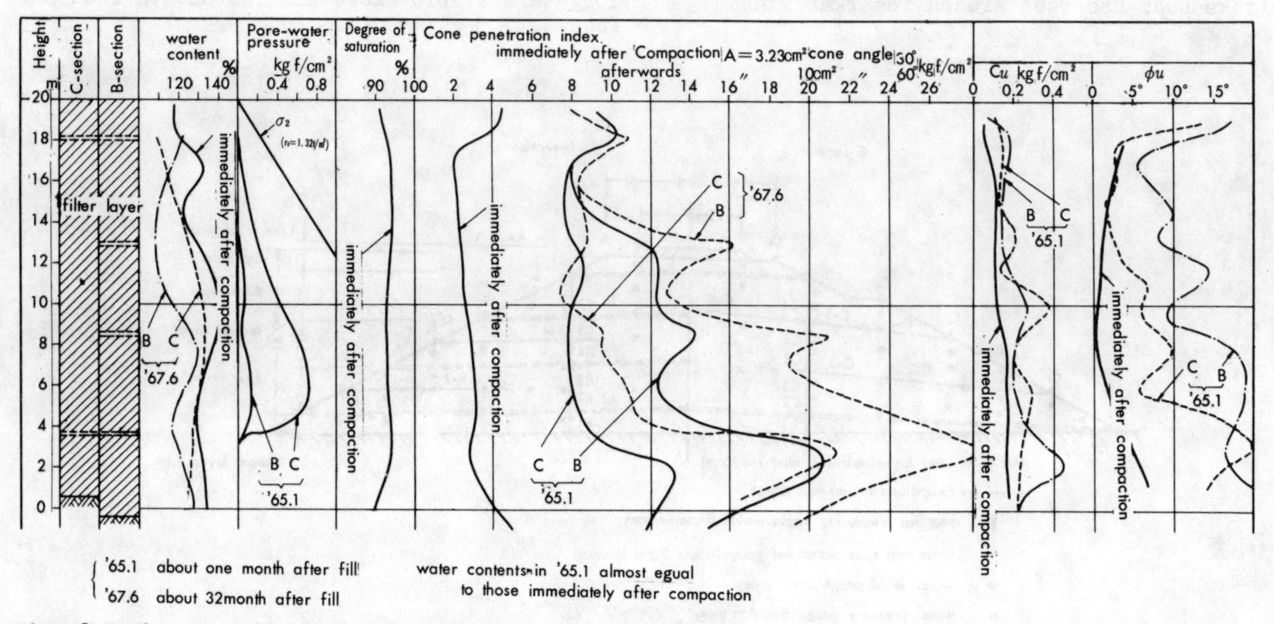

Fig. 2. Changes of embankment soil characteristics

was higher in the lower part of the embankment. Cu at about one month after the completion of the embankment had increased considerably but the difference between B and C-sections was not clear.

On the other hand, the angle of shearing resistance, ϕ_u, by the unconsolidated - undrained shear test on the soil immediately after the compaction ranged from 0° to 6° and increased at about one month after the completion of the embankment. Especially, the increase of ϕ_u of the soil in B-section was higher than that in C-section and the increase of ϕ_u in the lower part of the embankment was remarkable. But the effect of the filter layer both on Cu and ϕ_u was not clear.

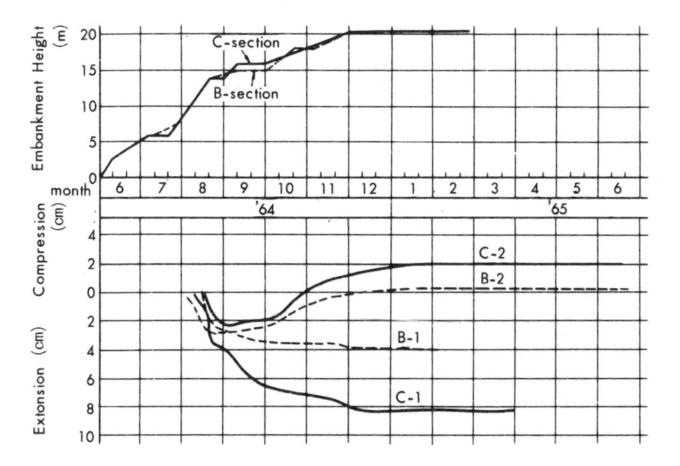

Fig. 3. Horizontal deformation at lower part of embankment

(4) For an example of the pore-water pressure distribution in the embankment, the water pressure under the shoulder at about one month after the completion of the embankment is shown in Fig. 2. Compared with C-section, the pore-water pressure in B-section was remarkably low and the effect of the filter layer was obvious. The maximum value of the pore-water pressure exists at about 6 meter upper from the bottom of the embankment.

4. THE DEFORMATION AND THE COMPRESSION OF EMBANKMENT

Fig. 3 shows the result of the lateral deformation measured by extension-compression gauges installed in the lower layer of B and C-sections versus the height of the embankment. From these results, the interesting facts were that the central part of the embankment extended remarkably at the first stage and after that gradually changed in the state of the compression, but the soil near the slope surface continued to extend constantly and never being compressed. The deformation in B-section with many filter layers is about one half of that in C-section with few filter layers and, hence, the effect of the filter layers was clear.

Fig. 4 shows both the sum of the layer compressions measured by using cross-arm type settlement gauges since the beginning of the embankment construction and the settlement with time measured with the pegs, B_5 and C_4, set on the shoulder after the completion of the embankment. The compressions of the soil during the construction were 4.0 and 3.5% of the height of the embankment for B and C-sections, respectively. The compressions three years after the completion of the embankment were 1.2 and 1.0% for B and C-sections, respectively. Both the

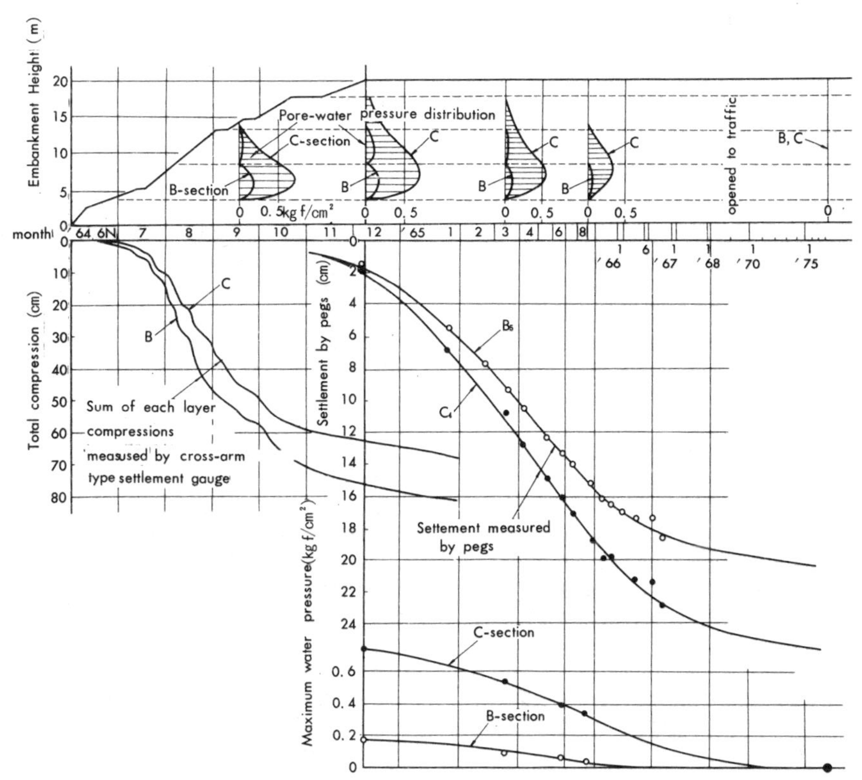

Fig. 4. Pore-water changes and settlement with time

compressions during and after the construction of the embankment for B-section with many filter layers were more than those for C-section with few filter layers.
The settlement of the embankment had almost finished in 2 or 3 years after the completion of the embankment and after that only the secondary settlement has slightly continued.

5. THE STABILITY OF THE EMBANKMENT

1) The pore-water pressure in the embankment
Fig. 5 shows the pore-water pressure measured at the various portions in the embankment during the construction versus the overburdon soil thickness at the same points.
Since the unit weight of the embankment was averaged as 1.32 t/m^3, the maximum value of the pore-water pressure in C-section was about 75% of the overburdon earth pressure. On the other hand, the maximum value of that in B-section was only about 30% of the overburdon earth pressure. The ratio of the maximum water pressure to the overburdon earth pressure decreased with the progress of fill and the passage of time, and it was shown that the water was drained through the filter layers during the construction of the embankment.

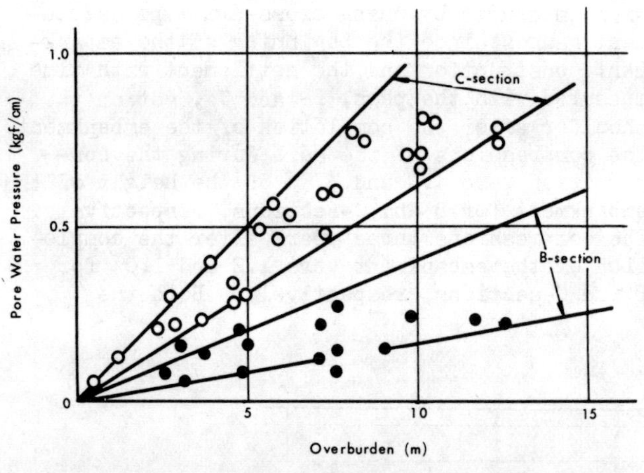

Fig. 5. Pore-water pressure rise due to overburden

From the iso-water pressure lines drawn by measured values at various points in the embankment, the pore-water pressure distributions in depth under the shoulder for each section were obtained and their changes with the passage of time were shown in Fig. 4. The maximum water pressure changes are also shown in Fig. 4. From these figures and field observations the followings were clarified:

(1) The drainage through the filter layers was very effective and the water pressures in B-section with many filter layers were very low compared with those in C-section with few filter layers.

(2) At the time of rainfall the increase of the water pressure could be slightly noticed (maximum of 20 to 120 cm).

(3) At 2 to 3 years after the completion of the embankment, the pore-water pressure in the embankment was almost zero. The rate of the water pressure decrease in C-section was much slower than that in B-section.

2) The strength of the embankment
It has been clarified in Fig. 2 that the cone penetration indices of the embankment increase remarkably with the passage of time. The values of Cu, ϕ_u by the unconsolidated - undrained shear test (UU-test) on the soil samples obtained from the each layer immediately after the compaction and immediately after the completion of the fill are shown in Fig. 2.
The $\sigma - \tau_f$ relationships obtained from UU-test results previously shown in Fig. 2 and the other consolidated - undrained shear test (CU-test and CU-test accompanied with the pore-water pressure measurements) are shown and also contrasted in Fig. 6.

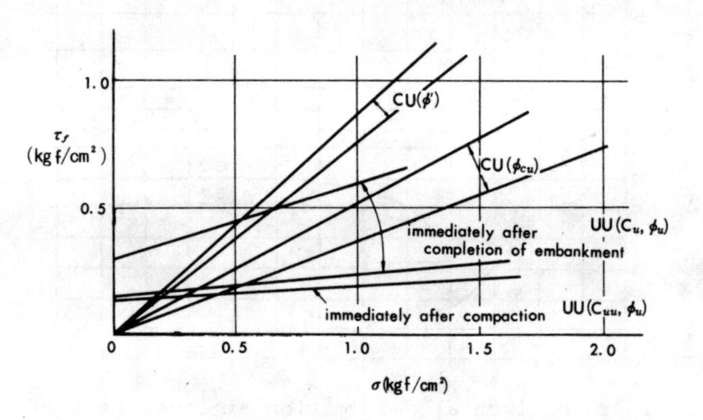

Fig. 6. Relationship between UU and CU-test results

With the comparison in Fig. 6, although there are differences in strength between B and C-section, or the upper and lower part of the embankment, it is clear that the soil strength increases remarkably with the increase of overburdon pressure by filling the upper part of the embankment and the passage of time.
It can be presumed that the increase of the strength is due to the increase of the density with the dispersion of the pore-water pressure raised during the construction, and the effect of thixotoropy commonly observed in the cohesive volcanic soil.
In order to investigate the effect of the embankment weight and the pore-water pressure on the strength of the soil, the shearing strength, τ_f, on the envelope of UU-test corresponding to the overburdon earth pressure, p_v , was obtained and the value is plotted versus the effective overburdon earth pressure, p_v-u , and the relationship is shown in Fig. 7. From the relationship in the figure, the shearing strength increased nearly in proportion to the effective overburdon earth pressure and, comparing with the CU-test results shown in the same figure, it lay in the range of ϕ_{cu} lines. From this relationship the state of stress in the embankment can be roughly assumed to have

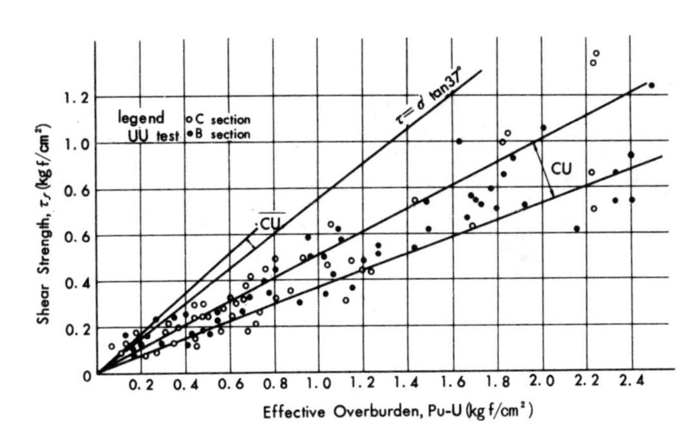

Fig. 7. Relationship between shear strength and effective overburden pressure

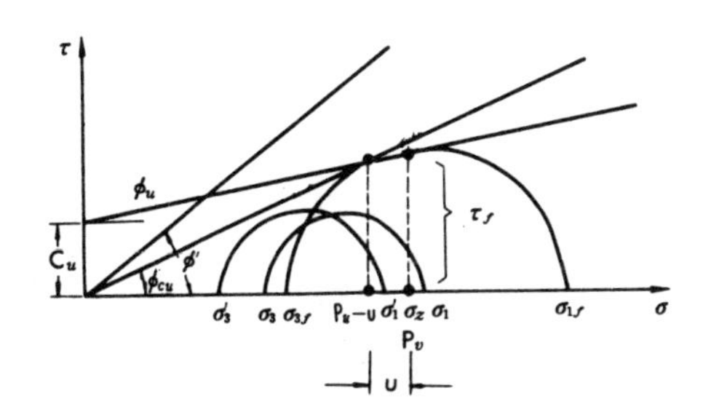

Fig. 8. State of stress relationship in embankment

the relationship shown in Fig. 8.

3) The stability of the embankment slope

The factors of safety for the slope during and after the construction of the embankment were analysed by using the following equation:

$$F_s = \frac{1}{\Sigma W \sin\alpha} \cdot \Sigma \frac{C'b + (W - u.b)\tan\phi'}{\left(1 + \frac{\tan\alpha . \tan\phi'}{F_s}\right)\cos\alpha}$$

where;

- W: the wet weight of the slice,
- b: the width of the slice,
- α: the angle between the failure surface and the horizontal,
- u: the pore-water pressure over the failure surface,
- C': the effective cohesion of the soil, (taken as C'=0), and
- ϕ: the effective angle of shearing resistance (taken as $\phi'=37^{\circ}$)

The pore-water pressures over the failure surface were obtained by drawing iso-water pressure lines with the measured values at the various points in the embankment and determing the basic values at the intersecting points of the failure surface and the iso-water pressure lines.

Fig. 9 shows the results of the calculated safety factors of the slope in B and C-sections at the completion of the embankment. Also Fig. 10 shows the calculated minimum safety factors assuming the conditions in the case of rainfall at the time of the completion of the embankment and adding the maximum increased values of the water pressure measured with the water level gauges in the shoulder at the time of rainfall.

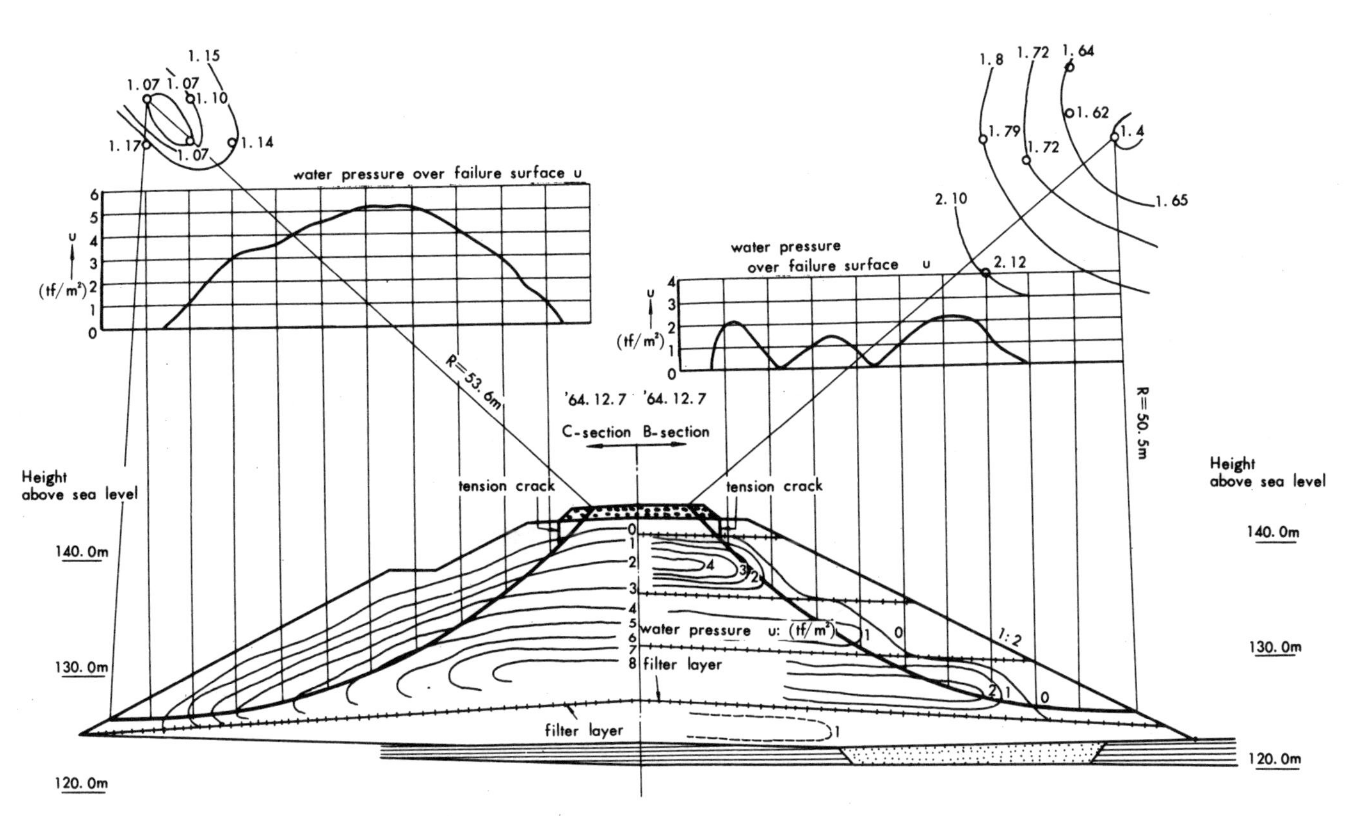

Fig. 9. Stability analysis immediately after completion of embankment

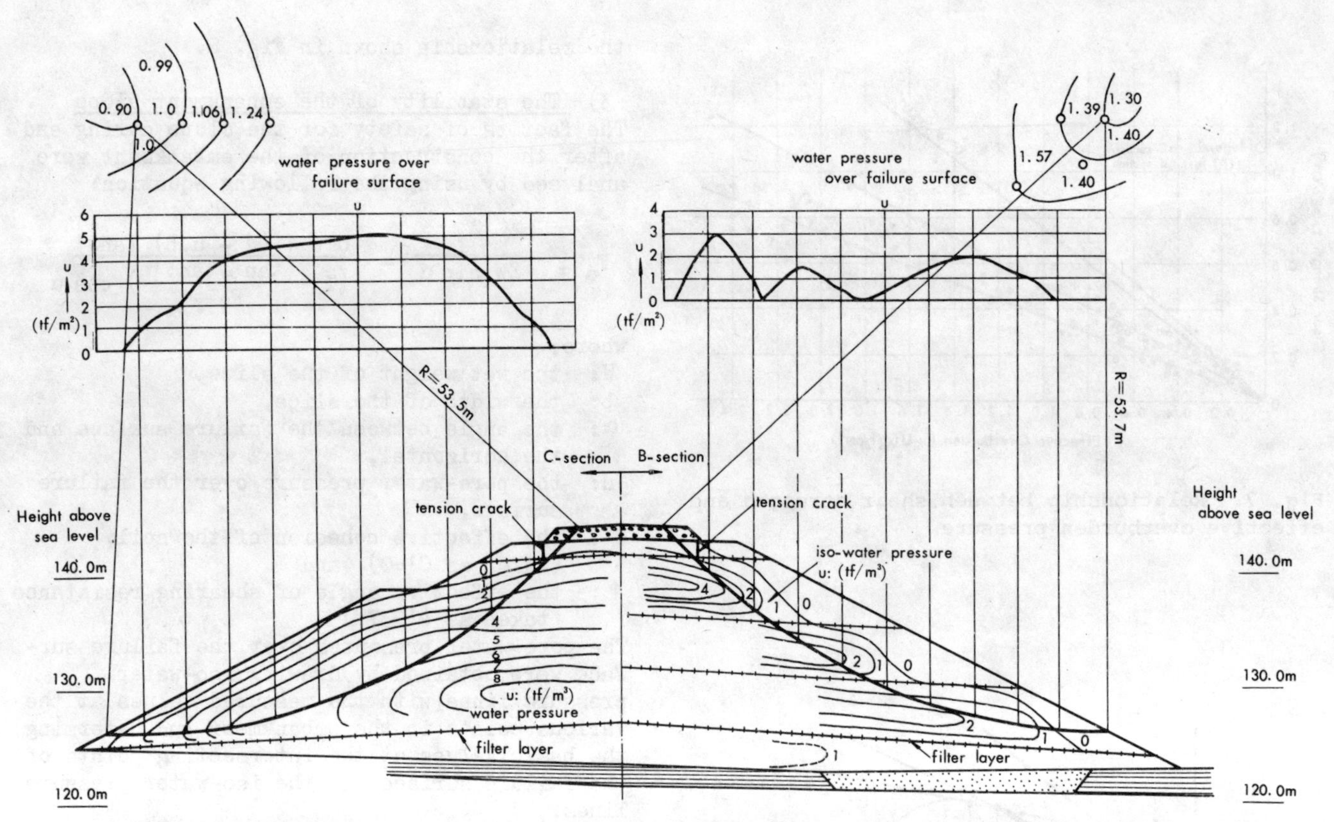

Fig. 10. Stability analysis after completion of embankment (during rainfall)

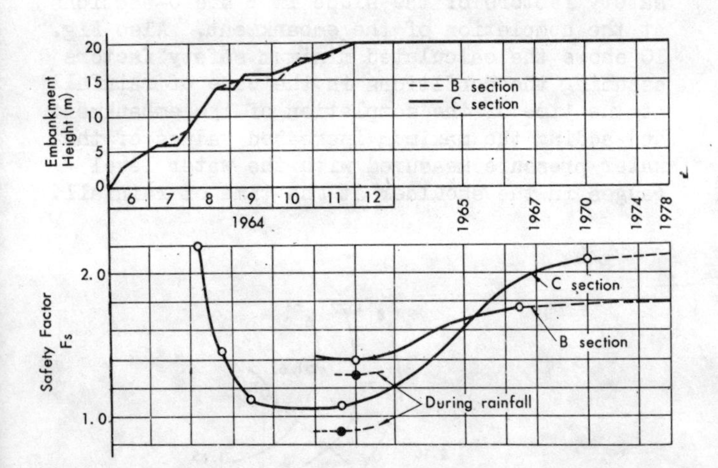

Fig. 11. Variation of safety factor with time

The safety factors of the slopes in the case of the pore water pressure equal to zero at 2 to 3 years after the completion of the embankment are 1.76 and 2.08 for B and C-sections respectively.
The relationship shown in Fig. 11 was obtained by plotting the safety factors determined in the stability analyses versus the height of the embankment. From this result the slope of C-section was in the considerably critical state when the embankment was completed. It could be assumed that the failure would have occurred if the big rainfall had continued. On the other hand, it was clear that the safety factor of the slope in B-section was high enough against the failure at the completion of the embankment.

6. POSTSCRIPT

In order to investigate the long-term stability of a high embankment with the soft cohesive volcanic soil of Ashitaka Loam as a filling material, the embankments with the filter layers of 5 m interval and 15 m interval were compared regarding the soil characteristics, the deformation and the stability during and after the construction of the embankment. From these analyses it was clarified that the embankment with many filter layers became stable at the earlier stage and remained in good conditions for a long time.

The authors wish to acknowledge the useful comments on the draft of this paper by Dr. Goro Kuno, Chuo University, Mr. Reiji Shinoki, and other members of Japan Highway Public Corporation.

T. S. INGOLD, BSc, MSc, DIC, MICE, MIHE, MASCE, FGS, Chief Engineer, and C. R. I. CLAYTON, BSc, MSc, DIC, PhD, MICE, FGS, Research Engineer, Ground Engineering Ltd

Some observations on the performance of a composite fill embankment

When certain configurations of suitable granular fill and suitable cohesive fill are placed without mixing in the same cross-section of a highway embankment the performance of the resulting composite fill is not always satisfactory. This paper reports an investigation into an embankment with a history of slope instability, and demonstrates the possible effects of the injudicious use of materials of widely different permeability.

Embankment Geometry

1. The section of embankment subject to slope instability and softening extended approximately 400 m from the low point on a sag curve at a river crossing in the east to a transition point between embankment and cutting in the west. A typical cross-section of the embankment showing the composition of the fill and drainage details is shown in Fig. 1.

2. As can be seen the lower 2.5m, approximately of the embankment was formed of gravel becoming clay and gravel. This lower granular layer was overlain by approximately 5m of silty clay, W_L 58, W_p 22, which in turn was overlain by up to 1.5m of coarse gravel and clay with extensive gravel and sand lenses. Results of insitu density tests indicated that the cohesive fill had been compacted to an average dry density of 1.54Mg/m^3 at a moisture content of 19%, this represents compaction to 95% of the maximum dry density determined using B.S.heavy compaction. At this average moisture content of 19% the clay fill compacted to a very stiff consistency with an undrained shear strength of approximately 200 kN/m^2.

Slope Failure and Softening

3. In the period January to March 1975 five slips occurred in the batters of the section of embankment described, two occurring in the northern batter and three in the southern batter In the course of the remedial works to the fifth slip, excavation revealed the line of a slip surface which in the cohesive fill took the form of a very softened layer approximately 100mm thick near the crest of the embankment and 300mm thick at the toe. Several shear vane strength determinations in this layer gave an average undrained shear strength of 18kN/m^2. The location of the slip surface is shown in Fig.2 together with a detailed description of the fill material exposed.

4. Six determinations of undrained shear strength were made in each of the six layers of clay fill. The average value had dropped from 200kN/m^2 to 75kN/m^2, the average undrained shear strength for each layer is shown in Fig.2. This drop in shear strength was associated with a corresponding increase in average moisture content of 9% in the clay, with the moisture contents ranging from 24% to 32% compared with an average placement moisture content of 19%. During excavation for the remedial works seepage was observed in the granular layers Nos.9 & 10 and the clay and gravel in layer No.8 was extremely wet. The topsoil was a well compacted silty clay with traces of gravel that varied in thickness from 150mm at the crest to 550mm at the toe of the embankment. A stability analysis was carried out assuming $\phi'=27^o$ and 35° for the clay and granular fill respectively with zero cohesion for both materials. A calculated factor of safety of 1.25 was obtained with $r_u=0$ for layers Nos.1 to 10. It was found that with an $r_u=0.5$ for the granular layers Nos.8 to 10 the factor of safety dropped from 1.25 to 0.95. A general view of the slip is shown in Fig.3.

The Source of Seepage

5. It was apparant from the state of the cohesive fill that seepage had occurred giving rise to general softening and the generation of high porewater pressure at the toe of the embankment which resulted in slope instability. Excavation in the virgin soil at the toe of the embankment and inspection of the toe ditch indicated that it was unlikely that water had entered the lower granular fill from the embankment foundation. The only other likely source of seepage was thought to be the median french drain. To check the possibility of such seepage a dye trace test was carried out during a period of prolonged heavy rainfall. Diluted Fluorescein Sodium was introduced into the median drain catch pit upstream of the slip and storm water run-off was sampled from the northern toe ditch down stream of the slip. Prior to the introduction of the dye a control sample of water was taken from the same point in the ditch. The samples were analysed using two methods. Firstly the test and control samples plus a sample of distilled water and distilled

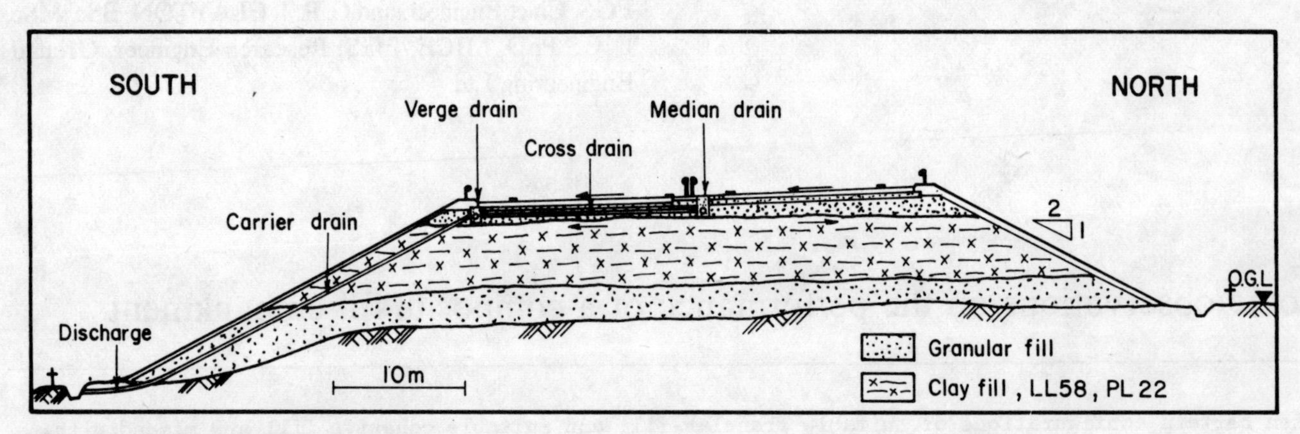

Fig. 1. Typical cross-section

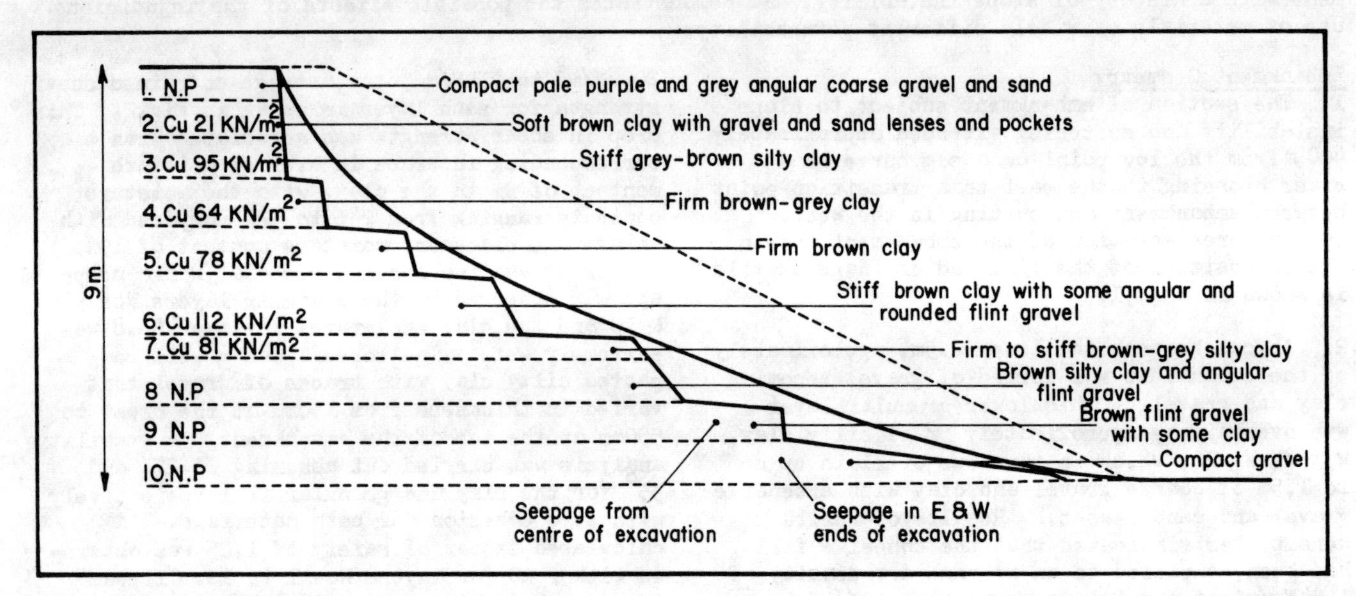

Fig. 2. Details of fill and slip surface location

Fig. 3. A general view of the slip

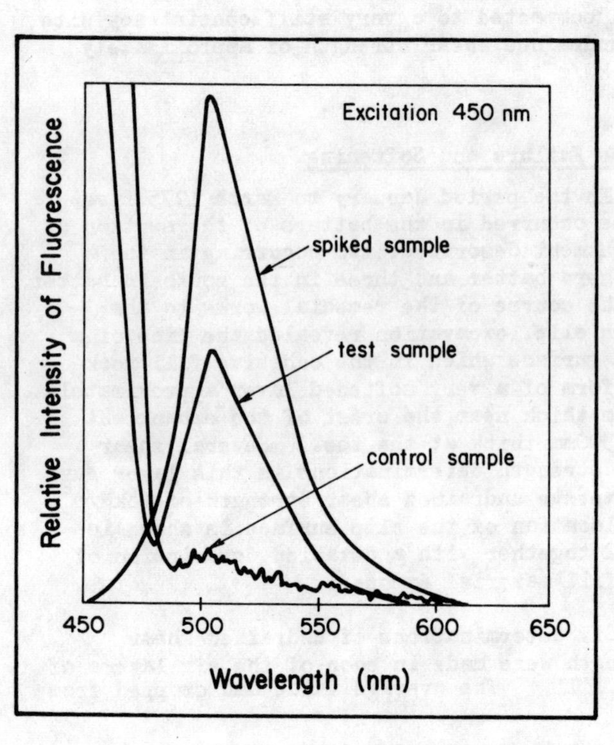

Fig. 4. Flourescence emission spectra

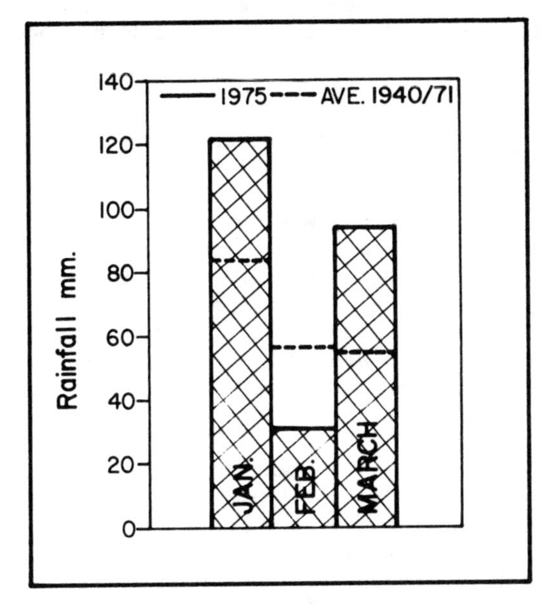

Fig. 5. Rainfall data

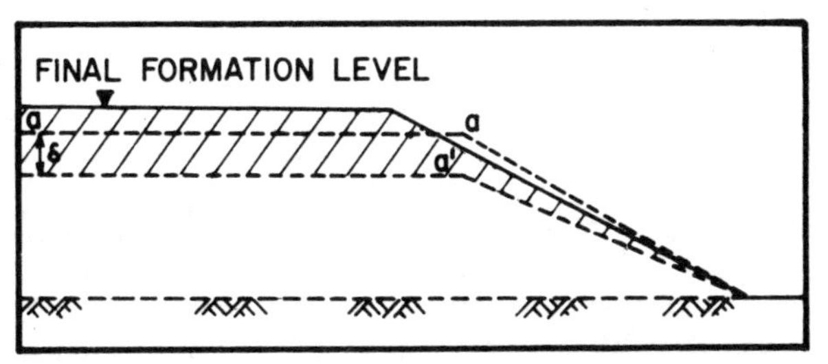

Fig. 6. Change in embankment geometry

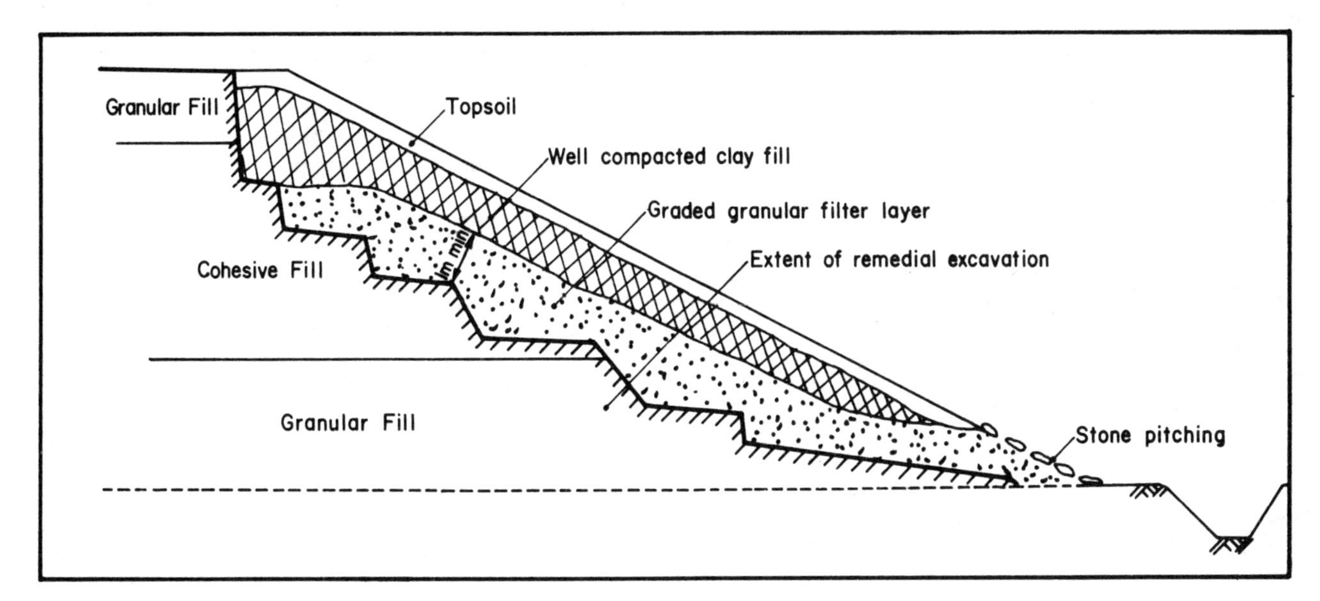

Fig. 7. Remedial works

water spiked with Fluorescein were analysed using using a Turner 111 Filter Fluorometer at the standard wave lengths for Fluoroscein analysis. The results shown in Table 1 show clearly that the test and spiked samples have high readings that indicate the presence of Fluoroscein.

Table 1

SAMPLE	FLUOROMETER READING
Test sample	56
Control sample	8
Spiked sample	73
Distilled water	1

6. In the second analysis an Aminco Bowman Ratio Spectrofluorometer was used to determine the emission spectra of the test sample, control sample and a spiked sample. The spiked sample was excited at a wave-length of 450nm and showed a maximum relative intensity of fluorecence at an emission wave-length of 500nm. These excitation and emission wave-lengths are characteristic of Fluorescein Sodium. The emission spectra for the test and control samples, as well as the spiked sample are shown in Fig.4. The high values of intensity of emitted light for the test and control samples at the excitation wave-length of 540nm are due to scattering of light through the optical system, however, at the Fluorescein Sodium characteristic emission of 500nm it can be seen that the relative intensity of fluorescence of the test sample was much higher than that of the control sample thus again confirming the presence of Fluorescein in the test sample. The results of the two types of analyses showed conclusive evidence of seepage from the median drain into the north batter.

The Seepage Mechanism

7. The exact mechanism of seepage from the median drain into the lower granular fill layer is not known. However, two possible modes are postulated. The first and most obvious mechanism is that of lateral flow of water from the median french drain along the upper layer of granular fill associated with gravitational flow from the granular layer down through the clay fill into the lower granular layer. It is highly possible that a sufficiently high head was generated at the median drain to cause such lateral flow since the section of drain in question was at a slack gradient which was fed from a much steeper gradient, also rainfall during the period of the failures was particularly heavy. Motorway storm water drains are designed for one year storm i.e. a storm that gives a certain intensity of rainfall once every year. Reference to Fig.5 shows that for the months of January and March 1975 the monthly rainfall exceeded the average monthly rainfall for 1940/71 by 45% and 71% respectively Although amount of rainfall and intensity cannot be directly equated it is a reasonable assumption that during this period of heavy rainfall the median drain would at times have been severely surcharged.

8. A second possible mode of seepage involves a series of granular lenses near the surface of the batter connecting the upper and lower granular layers. During excavation for the remedial works several lenses of granular material were observed orientated approximately parallel to and near the surface of the batter. Generally the thickness of the lens reduced down slope.

9. A possible mechanism for the formation of such a feature is shown very simplistically in Fig.6. During embankment construction it is common practice to overfill the batter and then trim back to the design batter which is then formed in fill that was compacted during the main filling operation. When this construction method is used the position of the toe is located in the design position by a batter board thus to achieve overfilling the batter must be constructed at a steeper slope than the design slope. This situation is shown in Fig.6 for some intermediate formation level a-a.

10. If the embankment undergoes a settlement then the formation level a-a will reduce and the current crest level will reduce from a to a'. This settlement effectively induces a rotation of the batter about the toe of the embankment causing the batter to slacker; in certain circumstances this rotation can be sufficient to make the constructed batter slacker than the design slope. When the embankment is taken up to final formation level and to the design side slope the final layer of fill, shown attached in Fig.6, extends some way down the previously constructed batter thus bringing the slope of this batter up to the design slope. If this upper layer of fill is granular then obviously there would be a tendency to form a granular layer or lens at the surface of the final batter.

Remedial Works

11. The remedial works were designed to prevent softening of the fill close to the surface of the batter and the generation of high pore water pressures in the lower granular fill layer. It was not economically viable to prevent seepage from the median drain itself.

12. The embankment was cut back in benches as shown and a granular filter drain was laid to a minimum thickness of one metre, Fig.7. Examination of the grading curves for the cohesive fill and granular fill indicated that the filter drain could be conveniently formed of material with a Type 1 subbase grading. The filter was laid to within one metre of the underside of the upper granular fill layer and extended down slope to the toe. The remainder of the embankment was taken up to profile with well compacted clay fill which was designed to seal the filter drain and the upper granular fill layer. The slope was topsoiled and seeded ensuring that the topsoil only extended to the top of the filter drain where it passed out into the toe of the embankment. The remedial works have given satisfactory performance from May 1975 to date.

H. C. INGOLDBY, Transport and Road Research Laboratory

The classification of chalk for embankment construction

The paper describes an investigation of chalk as a freshly placed fill material and the development of a new classification for chalk in relation to its stability during construction. From measurements made on eleven chalk earthworks sites, the onset of unstable conditions in the fill has been related to values of moisture content and degree of crushing. The new classification is based on the prediction of these two parameters and depends on the measurement of saturation moisture content and the "chalk crushing value" (using a newly developed test) of the chalk. The classification includes recommendations on the selection, when necessary, of the methods of earthwork construction which minimise the risk of instability in the freshly placed fill material, while at the same time using a degree of compaction compatible with a satisfactory long term performance of the embankment.

INTRODUCTION

1. The behaviour of freshly placed fill material resulting from chalk cuttings varies considerably depending on the strength and moisture content of the chalk being used. Fills resulting from soft chalk at high moisture contents can become unstable during the formation of embankments, with high positive pore water pressures making compaction impractical, and work sometimes stopping altogether due to the inability of the fill material to support the construction plant. Delays, disruption of the construction programme, and greatly increased costs can result. Conversely, with less serious results, very hard chalk can be difficult to excavate and the coarse particle size of the resulting material also causes compaction problems.

2. This wide variation in the performance of chalk as a fill material gives rise to the need for methods of predicting, at the site investigation stage, the potential behaviour of the material during construction, and particularly the means of identifying the more troublesome softer chalk. This paper describes the development of such a predictive procedure and provides guidance on methods of earthwork construction to use on chalk. A Report describing the work in detail has already been published (ref. 1).

FACTORS AFFECTING THE STABILITY OF CHALK FILL

3. Chalk is a porous limestone of widely varying strength which consists almost entirely of the remains of calcareous marine organisms. The natural moisture content of the chalk remains at, or very near to, the saturation moisture content of the individual lumps (values of the saturation moisture content varied from as low as 8 per cent to as high as 36 per cent in the chalk encountered in the recent research). During earthworks the excavation and compaction processes partly break down the natural rock structure of the chalk, releasing some of the contained water. Chalk fill therefore consists of a mixture of chalk lumps, chalk fines and water, the latter two forming a slurry or "putty chalk" at higher values of moisture content. A fill material which is temporarily weak and unstable results if the proportion of putty chalk is high enough to control the behaviour of the whole. The stability of the freshly placed fill depends, therefore, on its moisture content and its fines content.

4. A programme of research was implemented to provide information from which an appropriate chalk classification could be formulated. Measurements were made of the degree of crushing and the moisture content of the compacted fill at eleven road construction sites where earthworks in chalk were in progress. For simplicity the degree of crushing was defined as the percentage passing the 20 mm BS sieve.

5. The relation between the moisture content and the degree of crushing of the compacted fill in both stable and unstable conditions is shown in Fig. 1. Unstable conditions were defined as those where ruts were formed under the construction plant. The line dividing points representing stable and unstable conditions can be considered to be the relation between moisture content and degree of crushing at the onset of unstable conditions. The results indicate that stable conditions are always likely to occur at moisture contents below 23 per cent, and, as the moisture content increases above 23 per cent, stable conditions can be maintained by progressively reducing the degree of crushing of

Clay fills. Institution of Civil Engineers, London, 1978, 137-142

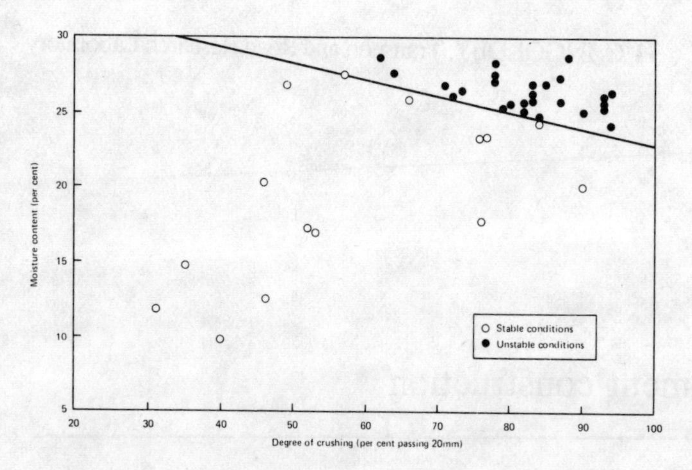

Fig. 1. The relation between moisture content and degree of crushing of compacted chalk fill for both stable and unstable conditions

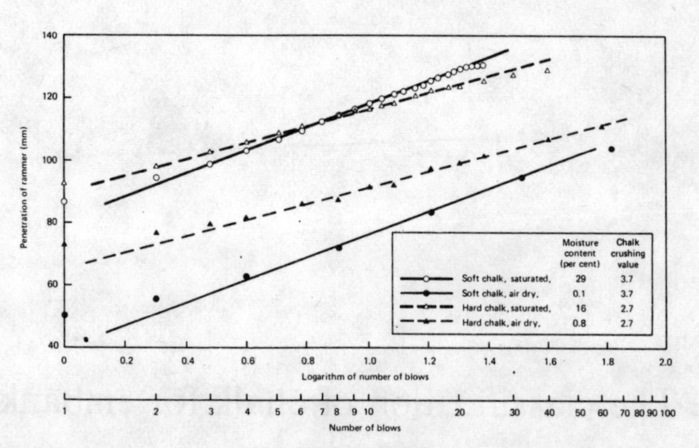

Fig. 3. Relations between the penetration of the rammer and the logarithm of the number of blows obtained in impact crushing tests with typical samples of soft and hard chalk in both saturated and air dry states

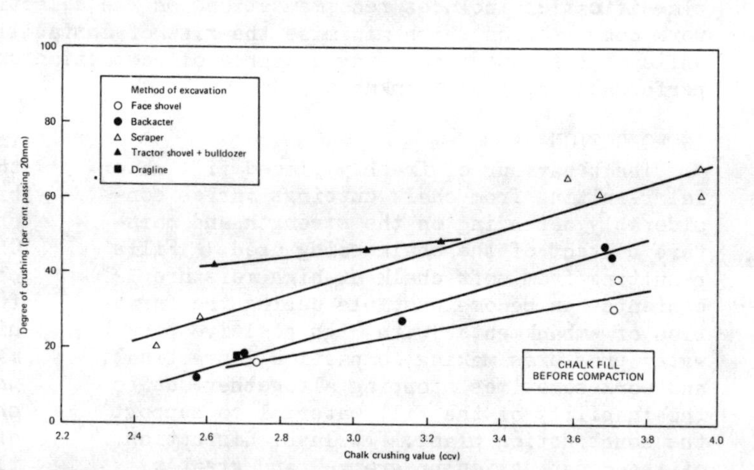

Fig. 4. Relations between the degree of crushing of the fill material before compaction and the chalk crushing value

Fig. 2

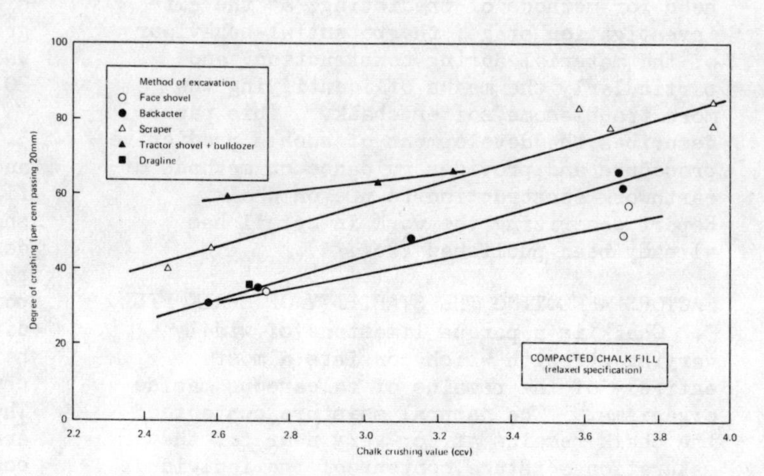

Fig. 5. Relations between the degree of crushing of the fill material after compaction to a relaxed specification and the chalk crushing value

the fill material. Thus a classification of chalk related to its behaviour as a fill material must be based on the prediction of both the degree of crushing and the moisture content, especially with soft chalk with moisture contents in excess of 23 per cent.

PREDICTION OF THE DEGREE OF CRUSHING

The impact crushing test

6. The degree of crushing of the fill material is dependent on the susceptibility to crushing of the chalk being used and the crushing action to which it is subjected during the earthwork construction processes. The prediction of the degree of crushing in the chalk fill, therefore, requires the measurement of the susceptibility of chalk to crushing and the establishment of relations between such measurements and the degree of crushing produced by various methods of earthworks construction currently in use. Thus a rapid and simple crushing test was required which would be applicable to chalk samples available at the site investigation stage.

7. It was thought that none of the existing methods of measuring the strength of soils or rock, if applied to chalk, would accurately represent its susceptibility to crushing. The vibrocrushing test (ref. 2), developed in France, was being considered for the purpose (refs. 3 and 4), but it was felt that a less complicated test, less susceptible to possible operator variability, would be more appropriate.

8. An apparatus (Fig. 2) was under development at the Laboratory for the rapid measurement of the moisture condition of soils (ref. 5). It was found that this apparatus could also be used to test the susceptibility to crushing of chalk samples and fulfilled the requirements of speed and simplicity of use. Susceptibility to crushing is taken to be the rate at which a 1 kg sample of single sized chalk lumps (passing 20 mm and retained on 10 mm BS sieves), contained in a mould 100 mm in diameter, crushes under impacts from a 7 kg rammer falling freely through a height of 250 mm. The amount of crushing during the test is determined by measuring the penetration of the rammer into the mould. The rate of crushing or "chalk crushing value" (CCV) of a sample is defined as one tenth of the slope of the straight section of the relation between the penetration (mm) and the logarithm of the number of blows. A steep slope indicates a fast rate of crushing and therefore a soft chalk. CCVs varied during the recent research from about 4.2 for a very soft chalk to about 2.4 for a very hard chalk.

9. CCV was found to be independent of the moisture content of the test sample. Fig. 3 shows the relations between the penetration of the rammer and the logarithm of the number of blows for a soft and a hard chalk in both saturated and air dry states. The relations are displaced with variations in moisture content but the slopes of the lines, and hence the CCVs, are effectively the same for samples of the same chalk.

On site investigations

10. The type of excavation plant observed during the site visits were divided into four main categories:

(1) Face shovels
(2) Backacters
(3) Scrapers
(4) Tractor shovels with bulldozer assistance.

On one site a dragline was used as a supplementary item of excavation plant.

11. The methods of compaction observed were divided into two main categories:

(1) Compaction to Clause 609 of the specification for Road and Bridge Works (ref. 6) (normal compaction specification).
(2) Compaction to a special specification which generally restricted the type of compactor to smooth wheeled or grid rollers with a mass of 2700-5400 kg per metre width of roll, reduced the number of passes to three, and increased the depth of compacted layer to 450-550 mm (relaxed compaction specification).

12. Relations were established between the degree of crushing of the fill material achieved by each of the categories of excavation plant observed and the CCV (Figs. 4 to 6). Fig. 4 shows the relations obtained for the loose uncompacted chalk, Fig. 5 for the material after the application of a relaxed compaction specification, and Fig. 6 after application of the normal compaction specification. For a given method of excavation, the degree of crushing increased with increase in CCV at all stages, but the different types of excavation plant produced significantly different relations. Tractor shovels and scrapers caused the greatest crushing of the chalk and backacters and face shovels the least.

13. A comparison of the values of the degree of crushing before compaction (Fig. 4) with those after compaction (Figs. 5 and 6) demonstrates the increase in crushing caused by the compaction process over that already achieved by the excavation process. With soft chalk it is particularly noticeable that the greater proportion of the crushing was attributable to the excavation process.

14. The additional degree of crushing caused by the normal compaction specification over that by a relaxed compaction specification is shown by a comparison of Figs. 5 and 6. For soft chalk this also is small in relation to the degree of crushing caused by the excavation process.

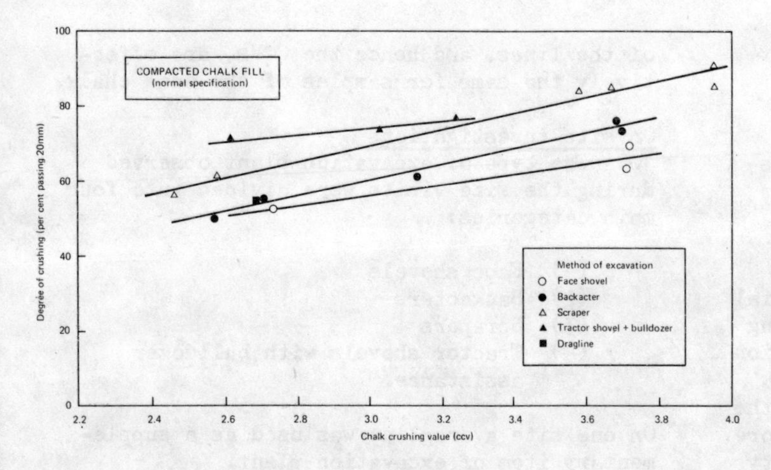

Fig. 6. Relations between the degree of crushing of the fill material after compaction to the normal specification and the chalk crushing value

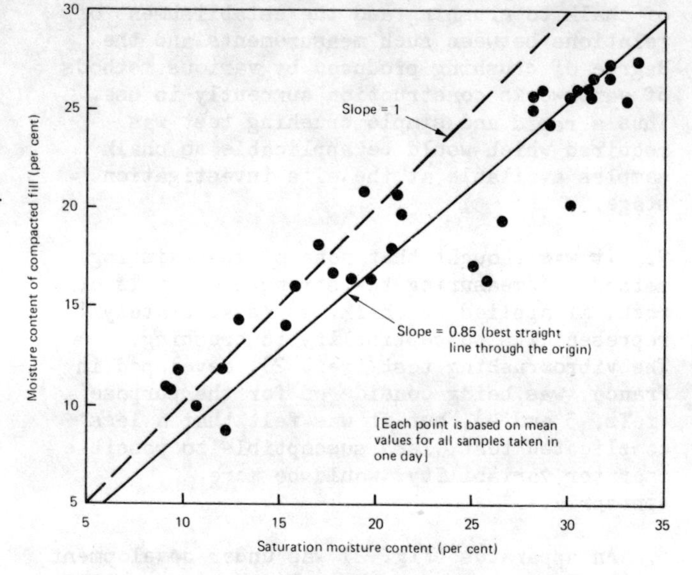

Fig. 7. The relation between the moisture content of the compacted fill and the saturation moisture content of the intact chalk lumps

Fig. 8. Chalk classification with measures required to avoid or minimise instability

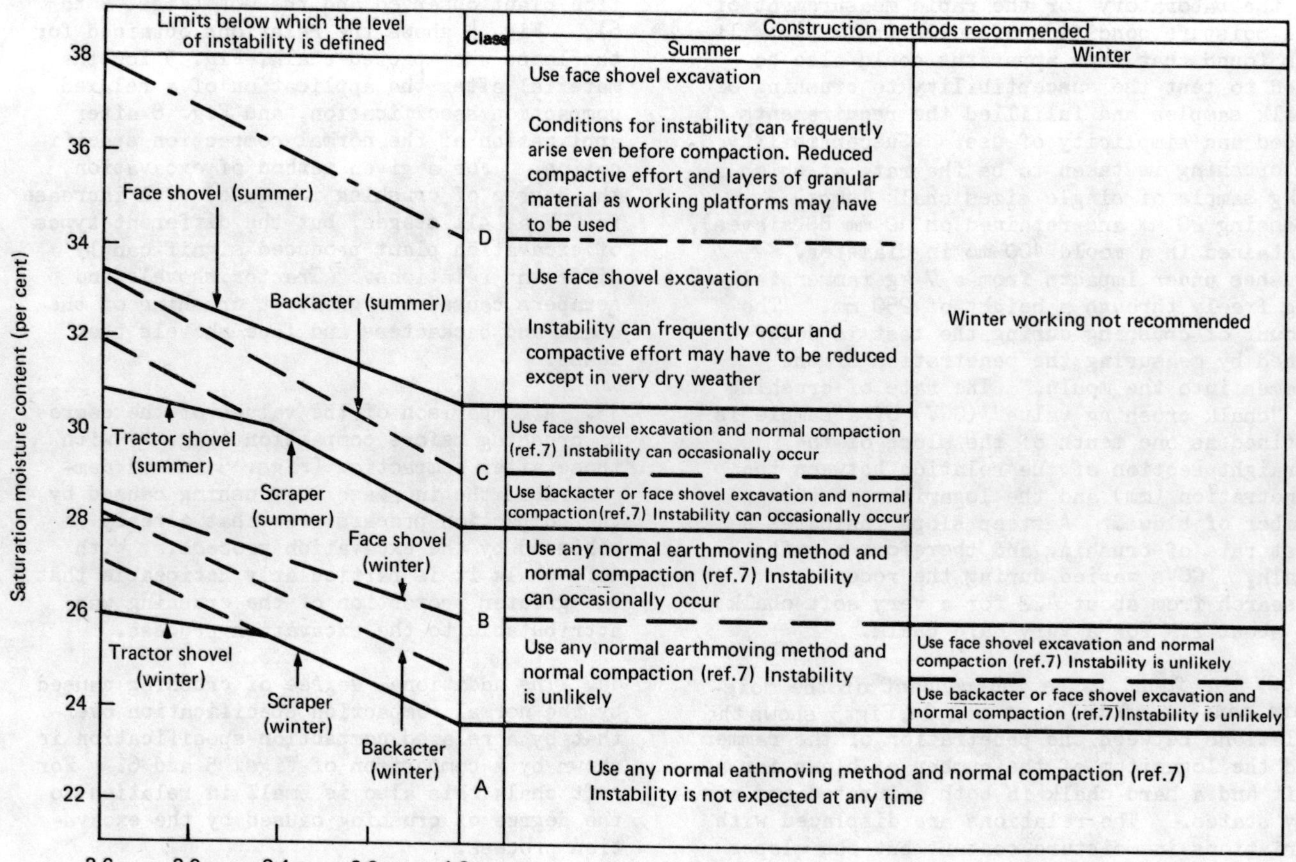

Limits below which the level of instability is defined

Saturation moisture content (per cent)

Chalk crushing value (ccv)

Face shovel (summer)

Backacter (summer)

Tractor shovel (summer)

Scraper (summer)

Face shovel (winter)

Tractor shovel (winter)

Scraper (winter)

Backacter (winter)

Class	Construction methods recommended: Summer	Construction methods recommended: Winter
D	Use face shovel excavation Conditions for instability can frequently arise even before compaction. Reduced compactive effort and layers of granular material as working platforms may have to be used	Winter working not recommended
	Use face shovel excavation Instability can frequently occur and compactive effort may have to be reduced except in very dry weather	
C	Use face shovel excavation and normal compaction (ref.7) Instability can occasionally occur	
	Use backacter or face shovel excavation and normal compaction(ref.7) Instability can occasionally occur	
B	Use any normal earthmoving method and normal compaction (ref.7) Instability can occasionally occur	
	Use any normal earthmoving method and normal compaction (ref.7) Instability is unlikely	Use face shovel excavation and normal compaction (ref.7) Instability is unlikely Use backacter or face shovel excavation and normal compaction (ref.7)Instability is unlikely
A	Use any normal eathmoving method and normal compaction (ref.7) Instability is not expected at any time	

15. The opportunity to control the degree of crushing of the softer chalk, and therefore its behaviour, clearly lies more with the choice of excavation plant than with the compaction specification. Predictions of degrees of crushing likely in chalk of different CCVs can be made from Figs. 4, 5 and 6.

Compaction of chalk fill

16. The selection of a particular method of excavation for the purpose of reducing the degree of crushing would result in a coarser fill material which would be more difficult to compact to a low air content than would be the case had it undergone more severe crushing. Additionally, the possibility of very dry weather conditions, similar to those experienced during the summer of 1976, could lead to the chalk being at a moisture content at which a satisfactory state of compaction would be more difficult than normal to produce. It is likely that any relaxation of compaction, particularly by increasing the depth of compacted layer, could, in many instances, be detrimental to the subsequent performance of the embankment. Furthermore, the difference in the degree of crushing associated with compaction to the normal specification with that of a relaxed compaction specification is shown to be relatively small in the softer chalk (see para 14). In terms of maintaining stability in the freshly placed fill, therefore, the advantages of relaxed compaction are limited and will usually be outweighed by the risk of increased settlement within the embankment.

17. It was therefore considered that a classification of chalk, in relation to its behaviour as a freshly placed fill material, should be based on maintaining stable conditions, whenever possible, with compaction to the full requirements of Clause 609 of the Specification for Road and Bridge Works (ref. 7). For this purpose the prediction of the degree of crushing in the compacted fill was made from Fig. 6. The circumstances in which relaxation of compaction can be of benefit are described later (paras 22 and 23).

PREDICTION OF THE MOISTURE CONTENT

18. Determinations of the saturation moisture content of intact chalk lumps were made on samples from the excavation areas in use during each site visit. The test used was an extension of the British Standard test for the determination of the dry density of soil by the "immersion in water" method (Test 15E BS 1377:1975 (ref. 8)). The porosity, and hence the saturation moisture content, was calculated from the measured dry density assuming the specific gravity of chalk solids to be 2.70.

19. Fig. 7 shows the relation established between the moisture content of the compacted fill (determined by the oven drying method (ref. 8)) and the saturation moisture content. During the period of the research, the moisture content of the compacted fill was, on average, 0.85 of the saturation moisture content. For the purpose of predicting the moisture content of the compacted fill it was concluded that the saturation moisture content could be used to indicate the highest possible value and 0.85 of the saturation moisture content the mean summer level.

THE CLASSIFICATION OF CHALK AS A FILL MATERIAL

The classification chart

20. The chalk classification chart (Fig. 8) was derived from the information contained in Figs. 1, 6 and 7. Values of the degree of crushing, which each of the four methods of excavation would be likely to achieve on chalk of different CCVs, were read from Fig. 6. These values were related to critical values of moisture content, at which instability is likely to occur, by an inspection of Fig. 1. These moisture content values were in turn related to saturation moisture content values by reference to Fig. 7. Assumptions were that, in winter, the moisture content of the compacted fill would be equal to the saturation moisture content and, in summer, to 0.85 of the saturation moisture content. The classification chart thus relates levels of saturation moisture content and CCV at which the onset of unstable conditions is likely when each of the four categories of excavation plant is in use, and normal compaction is applied.

21. Unstable conditions are not expected to occur with chalk in Class A, but chalk in Classes B and C (especially at higher values of saturation moisture content) can produce unstable conditions occasionally, even when the recommended category of excavation plant is in use. Unstable conditions can occur frequently with chalk in Class D.

22. Methods of compaction when instability occurs. When instability occurs during construction using chalk in Classes B and C compactive effort may be temporarily relaxed. As weather conditions can fluctuate considerably, and chalk properties may also vary, it is advisable that this relaxation should be on the basis of day to day decisions on site. Such relaxation should follow the advice given in Section NG 608 of Notes for Guidance on the Specification for Road and Bridge Works (ref. 9), whereby the number of passes of the compaction plant or the size of the compactor is reduced.

23. With chalk in Class D, where instability is likely to occur more frequently, it may be preferable to make provision for relaxation of compaction from the outset. In extreme cases, Class D chalk can be in a potentially unstable condition even before compaction. These extreme cases are possible when the coordinates of the degree of crushing associated with face shovel excavation before compaction (Fig. 4), and moisture content (equal to 0.85 of the saturation moisture content), indicate an unstable condition (Fig. 1). An additional line has been included in Fig. 8 to simplify the recognition of the likelihood of this

condition arising. If the use of such chalk is unavoidable, it may be necessary to introduce layers of stable granular material into the embankment to act as working platforms for the construction plant.

Alternative chalk classifications

24. The two main assumptions which contribute to the classification in Fig. 8 are that, in summer, the moisture content in the compacted fill will be 0.85 of the saturation moisture content, and that normal compaction, ie to the requirements of the Specification for Road and Bridge Works (ref. 7), will be used. Alternative chalk classifications, to suit particular circumstances, may be deduced using the procedure described in para 20, but with alternative data. For instance, if a classification is required assuming a relaxed compaction specification as described in para 11, Fig. 5 can be used to estimate the degree of crushing involved instead of Fig. 6. If it is considered that the moisture content of the compacted fill in summer is more likely to be 0.90 of the saturation moisture content than 0.85, the last step of the procedure can be changed to allow for this; in this case the resulting classification will have the summer working lines closer to the winter working lines, the latter being in the same position as in Fig. 8.

CONCLUSIONS

25. The onset of instability under construction plant operating on chalk fills has been shown to depend on the moisture content and the degree of crushing of the material. These parameters have been related respectively to the saturation moisture content and the chalk crushing value (CCV). A method of classifying chalk in relation to its behaviour as a freshly placed fill material, based on saturation moisture content and CCV, has been developed. The classification includes recommendations on the selection, when necessary, of methods of earthwork construction which minimise the risk of instability in the freshly placed fill material, while at the same time using a degree of compaction compatible with the long term performance of the embankment. Data on which to base alternative classifications to suit particular requirements has been provided.

ACKNOWLEDGEMENTS

The work described in this paper forms part of the programme of the Transport and Road Research Laboratory and the paper is published by permission of the Director. The cooperation is gratefully acknowledged of the Directors of RCUs, County Surveyors, Resident Engineers and their staffs, on whose sites the investigations were made.

REFERENCES

1. INGOLDBY H.C and PARSONS A.W. The classification of chalk for use as a fill material. Department of the Environment, Department of Transport, TRRL Report 806. Crowthorne 1977 (Transport and Road Research Laboratory).

2. STRUILLOU R. Study by vibrational grinding of the aptitude of chalks for compaction. (Etude par vibrobroyage de l'aptitude des craies au compactage). La Craie. Bulletin de Liaison des Laboratoires des Ponts et Chaussées, Spécial V. Paris, 1973 (Laboratoire Central des Ponts et Chaussées), 99-110.

3. CLAYTON C.R.I. Chalk in earthworks, performance and prediction. The Highway Engineer, 1977, XXIV(2), 14-20.

4. CLARKE R.H. Earthworks in soft chalk: performance and prediction. The Highway Engineer, 1977, XXIV(3), 18-21.

5. PARSONS A.W. The rapid measurement of the moisture condition of earthwork material. Department of the Environment, TRRL Laboratory Report 750. Crowthorne, 1976 (Transport and Road Research Laboratory).

6. MINISTRY OF TRANSPORT. Specification for road and bridge works. London, 1969 (HM Stationery Office).

7. DEPARTMENT OF TRANSPORT. Specification for road and bridge works. London, 1976 (HM Stationery Office).

8. BRITISH STANDARDS INSTITUTION. Methods of test for soils for civil engineering purposes. British Standard BS 1377:1975. London, 1975 (The Institution).

9. DEPARTMENT OF TRANSPORT. Notes for guidance on the specification for road and bridge works. London, 1976 (HM Stationery Office).

M. F. KENNARD, BSc(Eng), FICE, FIWES, MASCE, FGS, MConsE, Partner, H. T. LOVENBURY, BSc(Eng), DIC, PhD, MICE, Senior Geotechnical Engineer, F. R. D. CHARTRES, BSc, MICE, AMIMM, FGS, Geotechnical Engineer, and C. G. HOSKINS, BSc(Eng), MICE, Geotechnical Engineer, Rofe, Kennard and Lapworth

Shear strength specification for clay fills

In recent years, the placing of clay in clay cores of several embankment dams has been controlled by the use of a specification based on an acceptable range of undrained shear strength. The method has enabled the required properties of a non-cracking and plastic clay core to be readily tested. Measures to increase the water content of clays have been developed by contractors at different sites to achieve the required shear strength. The experience gained, especially with Boulder Clay at Cow Green Dam; and Wadhurst Clay at Bewl Bridge and Ardingly Dams are described. The problems and limitations of the method are discussed.

INTRODUCTION

1. The evolution of rolled clay as a core material for embankment dams in the United Kingdom is illustrated by the dams constructed for the then Tees Valley and Cleveland Water Board between 1955 and 1970, namely Selset, Balderhead and Cow Green, and elsewhere including Derwent Dam. With the development of modern compaction plant and increased wages, the labour-intensive method of puddle clay has been replaced by the more economical mechanised placement of clay fill.

2. Specifications based on optimum water content and/or maximum density, following the work of Proctor (ref.1) and as described in B.S.1377 (Methods of Testing Soils, 1948) have been used extensively. These normally take the form of a percentage of the maximum dry density, often combined, in the case of embankment dam cores, with water contents within specified percentages above or below the optimum value.

3. In recent years the undrained shear strength method, in conjunction with a compaction criterion based either on density or air voids ratio, has frequently been adopted for specifying placement of clay fill. This approach has the advantages of being fairly quick to perform and of controlling a major design parameter directly.

SOME DEVELOPMENTS LEADING TO A SHEAR STRENGTH SPECIFICATION

4. One of the last major dams to be constructed in this country with a puddle clay core was Selset Dam (1955-60) on a tributary of the River Tees. The shoulders of the dam were formed of boulder clay fill (ref.2). Control of the puddle clay was by water content, shear strength and permeability tests. Placement of the shoulders was controlled by density, the fill being required to be compacted to a dry density not less than 95% of the Proctor optimum density, after correction for the stone content. This was also a water content specification by definition. The compaction was achieved by a D8 bulldozer, loaded rubber-tyred dump trucks and a 15-ton smooth roller.

5. At Derwent Dam (1960-1965) on a tributary of the River Tyne, the original design included a puddle clay core, with a shear strength specification, but the design was changed during the contract to a rolled clay core with a specified shear strength of 90 ± 20 kN/m^2 (ref.3).

6. As a result of the experience with the shoulders at Selset Dam, a core of rolled boulder clay was selected for Balderhead Dam. This dam was constructed with shale fill shoulders, between 1961 and 1965 in the next valley to Selset. Placement was controlled by water content and density. The clay was originally specified to be placed between 1-3% wet of Proctor optimum water content. However, when it was found that the optimum water content itself varied over a range of 4%, the placement water content was related to the plastic limit (ref.4). The plastic limit was approximately 2% higher than the optimum water content, but could be determined much more readily. The compaction criterion was that the in-situ density (corrected for stone content) should be not less than 98% of the density obtained in the standard compaction test on the clay at the same water content. The water content of the boulder clay was increased by about 3 to 5% during placement in 150mm thick layers. Compaction was by a taper foot roller.

7. At the 25m high Cow Green Dam (1967-1970), on the River Tees, a central rolled boulder clay core between gravel fill shoulders was used for the embankment section. Placement was originally controlled by water content limits relative to the Proctor optimum water content; this method however quickly proved to be cumbersome due to delays in establishing the optimum water content. The specification was then developed

further to relate placement directly to undrained shear strength, with the knowledge of experience gained at Derwent Dam and after tests had established a reasonably consistent relationship between undrained shear strength and water content relative to the optimum (ref.5). The specification was altered by defining a range of undrained strength equivalent to the original water content limits as shown in Fig.No.1. The compaction criteria was the same as that used at Balderhead Dam.

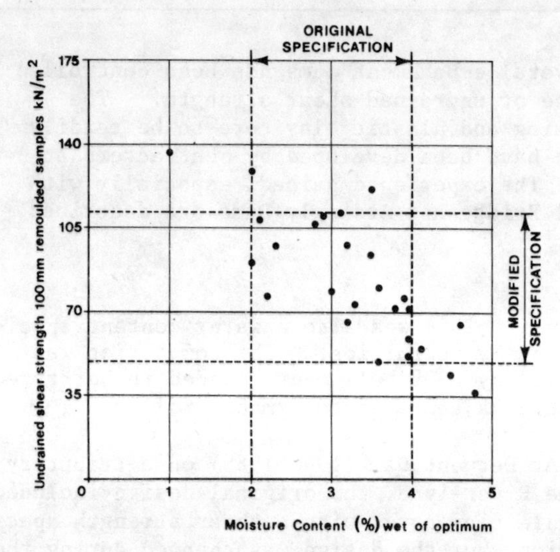

Fig. 1. Specification of core at Cow Green Dam

8. Some examples using the method of shear strength specification from the authors' experience are presented below. A similar method has also been adopted at several other recent embankment dams in the United Kingdom.

EXPERIENCE OF THE SHEAR STRENGTH SPECIFICATION IN USE

Cow Green Dam, Co. Durham/Cumbria

9. The original placement water content range was from 2% to 4% above optimum water content. The substituted specification required the undrained shear strength of the compacted clay to be between 48 kN/m^2 and 110 kN/m^2 with the additional proviso that two-thirds of all test results should fall within the specified range. The compaction criterion was unchanged with the clay placed in 150mm thick layers and compacted by the construction plant (which included loaded dump trucks) and a 7.3 tonne smooth-driven vibrating roller. Some watering of the clay was necessary. The undrained shear strengths were obtained from triaxial compression tests on 100mm diameter remoulded specimens at a cell pressure of 620 kN/m^2 and at 25% strain. It was possible to use remoulded samples since previous investigation had shown that the remoulded and undisturbed strengths of the boulder clay were equivalent. With this method the number of tests for each layer placed was increased because the results could be obtained from the triaxial tests within 1-2 hours whereas the Proctor tests had taken between 1-2 days to complete. Over 90% of the test results lay within limits of the specification, (ref.6).

Bewl Bridge Dam, Kent/East Sussex

10. This 30m high dam constructed between 1973-75 has a central core and upstream blanket of Wadhurst Clay fill supported by shoulders of Lower Tunbridge Wells Sands fill. The original specification was based on water content and shear strength as well as density. Placement was to be in 300mm thick layers at a water content between 0-2% wet or Proctor optimum to give a shear strength in the range 55-85 kN/m^2. The compacted dry density was to be not less than 95% of the Proctor optimum density. The shear strength was determined from undrained triaxial tests on 200mm diameter specimens prepared from U100 samples taken from the placed layer. The cell pressure was 200 kN/m^2 and the strain rate 2% per minute.

11. Fill placement was carried out for only one month in the 1973 season. The results of the shear strength tests performed together with those from a preliminary trial bank, showed considerable scatter. 37% of all the results lay above the upper specified limit, the mean value being 84 kN/m^2. Samples taken from trial pits in the placed clay suggested that some drying out of the clay had occurred after placement. 70% of the results of 17 tests were higher than 85 kN/m^2 and the mean value was 114 kN/m^2.

12. The clay is a shaly clay containing hard mudstone layers. The optimum water content was between 15-16% whereas the in-situ water content in the borrow pit was 16-17%. Tests indicated, however, that water contents in the range 18-22% were required to achieve the specified strength limits, a net increase in water content of 2-6%. Discussions were held with the Contractor as to how this could be achieved and the action taken is described later. In addition, the rather narrow strength limits were amended to 55-110 kN/m^2 with the layer thickness reduced to 150mm. A further stipulation was that at least 80% of all test results should fall within these limits with the running mean value lying within the range 70-100 kN/m^2.

13. In the 1974 season, 76% of over five hundred test results obtained lay within the overall strength limits, with only 5% above. The overall mean value was 74 kN/m^2, which was in fact within the original limits.

14. Using this method of control, each test took about 30 minutes to perform. A check on the water content was obtained by drying the specimen in a microwave oven. Tests showed that the water content obtained using a microwave oven was within $\pm$ 0.1% of that determined using the conventional British Standard drying method.

Ardingly Dam, Sussex

15. This 17m high dam, constructed between 1976-78, has a central core and shoulders of Wadhurst Clay. The fill, a finely brecciated shaly clay, with similarities to that used at Bewl Bridge Dam. The core specification required placement in 150mm thick layers at such a water content so as to give a shear strength in the range 60-100 kN/m^2 and a dry density of at least 98% of that obtained when compacted under standard Proctor effort (2.5 kg rammer method) at the same water content. The undrained shear strength was to be determined by laboratory triaxial tests with a confining pressure on 200mm long by 100mm diameter specimens taken from the compacted clay. For any area tested, it was required that 70% of all the strength test results should fall within this specified range and the running average of 5 consecutive tests should lie within the range 70-90 kN/m^2. In addition a minimum shear strength provision of 80 kN/m^2 was placed on the shoulder fill.

16. Tests carried out on re-compacted samples of clay from boreholes in the borrow area at the design stage indicated that placement water contents in the ranges 16-24% were required to satisfy the strength limits. These represented water content increases over the in-situ value varying from 1 to 6% or more.

17. In practice, the placed water contents generally ranged from 18 to 22%, an increase of 0 to 4% over the average insitu water content. The optimum water content was around 19%. During the course of the contract over one hundred triaxial tests were carried out on placed core material; of these 51% were within the specified limits, although only 15% were above. The overall mean was found to be 75 kN/m^2 which was within the specification. 75% of the results lay within the strength range of 50 to 110 kN/m^2. The results are not wholly representative of the placed fill, since the sample locations were frequently confined to doubtful areas where visual observations suggested that testing was required. A more random method of sampling would have given a higher proportion of results within the specified strength range.

Cadney Carrs Embankment, Lincolnshire

18. This 6m high bunded reservoir forming part of the River Ancholme Scheme, was constructed between 1972-73. The central core was constructed of soft laminated clay and some weathered Oxford Clay with shoulders of unweathered Oxford Clay. The core fill generally required placement in 150mm layers to give a shear strength in the range of 45-110 kN/m^2 and a dry density of at least 97% of that obtained when compacted under standard Proctor effort (2.5 kg rammer method) at the same water content. The undrained shear strength was to be determined on 200mm long by 100mm diameter specimens remoulded in the laboratory from samples taken from the compacted layers at the same water content as the placed clay. Some triaxial tests were initially carried out to obtain experience of the clay and to check the suitability of hand vane testing. Subsequently vane testing was carried out in the field, as the method of control testing.

19. The water contents of the fill were often about 25%, several per cent above the optimum, yet 66% of all the tests were within the specification. The overall mean value was 56 kN/m^2.

METHODS OF INCREASING THE WATER CONTENT OF CLAY FILL

20. Previous experience with boulder clays at Pennine sites had shown that the water content could be increased sufficiently, if necessary, by spraying with water from hoses during spreading. At Balderhead Dam, where a fairly large increase in water content was necessary, some difficulties were experienced with this method of spraying in achieving systematic watering. At Cow Green Dam, the clay was watered from hoses attached to standpipes on a steel main alongside the filter. The main was subsequently abandoned and the clay was watered whilst being spread by dozer. It was found that applying water to the blade was the only effective way to get water into the boulder clay.

21. However, with the over-consolidated shaly clays encountered at Empingham and Bewl Bridge Dams, where large amounts of water had to be introduced, more effective techniques had to be developed. In particular, attention had to be given to methods of working the water into the material. At Bewl Bridge Dam, the contractor (Gleeson Civil Engineering Ltd.) drew on his experience gained with the Lias Clay at Empingham Dam.

Bewl Bridge Dam

22. The Wadhurst Clay was won with motor-scrapers from the valley side within the reservoir basin, generally above the ground water table because of the type of plant used, and spread out by the boxes during unloading. Initially, both a pumped spray irrigation system through pipes laid on the fill (output of 150 gallons/minute) and two 500 gallon bowsers towed in tandem with a rear spray bar were tried, but with little success. This cumbersome system produced an uneven application, tending to over-wet the areas near the discharge points and not reach the extremities of the core. The bowsers were an improvement but of insufficient capacity to do more than offset the high rates of evaporation experienced during transit and placement caused by very hot weather and strong drying winds.

23. The method finally adopted was to use a specially converted motor scraper carrying a 3000 gallon tank inside the box with two pumped spray outlets at the tail. After spreading, the clay was worked by ripping with agricultural tynes followed by power-harrowing to reduce the size of the material to 60mm maximum. The clay was then watered and re-harrowed, the operation being repeated if necessary. Compaction was by a Cat 825B compactor, the surface being sealed with a smooth wheeled roller. After the first

season, when problems with drying out of the clay blanket were experienced, the watering operation was generally carried out in the early evening so as to allow time for the material to absorb the water. The fill was tested in the morning and re-worked locally if necessary and then kept watered during the day to arrest surface evaporation. In dry weather, evaporation can lead to a loss in water content of the order of 0.5% per hour. The method managed to increase the water content of the fill by the required 4-5% and in places even higher increases were achieved. However, strict inspection was necessary to ensure that a uniform consistency was achieved. An attempt was also made the previous winter to increase the water content of the clay in the borrow pit by stripping the area and ripping the surface but this had little effect.

Ardingly Dam

24. The Wadhurst Clay was excavated from the valley bottom and lower slopes of the reservoir basin using hydraulic backacters and transported to the dam in dump-trucks. Initially, the face of the borrow pit was sluiced using water pumped from the stream, a method that had been used by the Contractor (Shephard Hill Ltd.) previously with good results at Farmoor Reservoir. Once the excavation was below the water table, the pit was allowed to fill with water and the excavators mixed the clay at source prior to loading. The material was spread and rolled using a Cat 815 compactor. The surface was continually watered using a converted 6000 gallon road tanker with pumped rear spray bar to stop excessive drying out.

25. The method proved successful from the start and during the summer of 1976, when the temperatures were in excess of 25°C on most days, it was able to cope with raising the water contents of the fill by up to 8% and to counteract the high evaporation rate. On the rare occasions when a layer was too dry it was removed and incorporated in the downstream shoulder. Conversely, wet layers were re-worked and allowed to dry.

Cadney Carrs Reservoir

26. The weathered and unweathered Oxford Clay was excavated by motor-scrapers. Occasional watering was required. This was carried out with a bowser fitted with a pump and a hose. Spraying from a sparge pipe on the bowser was not effective, and hosing into the clay as it was spread by a dozer or hosing into the scrapers as they discharged, were found to be suitable methods.

DISCUSSION

27. The use of a specification based on shear strength has certain advantages over one based on optimum water content but the site engineer must be fully conversant with the design concept if the specification is to be applied successfully.

28. The principal advantages are that the measured strengths relate directly to the design fill performance and that the results, as well as being fairly quick to obtain, are independant of any variation in optimum water content of the material. The minimum shear strength can be readily related to a stability analysis of the embankment. It is considered that the method is more precise than only a relative density requirement as with modern plant there is usually no difficulty in achieving satisfactory compaction. In specifying the range of strength to be attained, the lower bound is usually governed by the limits of plant trafficability not stability considerations, and the upper bound by the need to ensure a plastic core.

29. Table No. 1 shows the range of specified shear strengths for the clay cores of several recent embankment dams in England. This shows that the range commonly in use is approximately 55-100 kN/m^2.

Table 1. Range of specified clay core shear strength at several recent dams

Dam	Specified range
Ardingly	60 - 100 kN/m^2
Ardleigh	55 - 110
Bewl Bridge	55 - 85
	55 - 110 (amended)
Cadney Carrs	45 - 110
Covenham	62 - 103
Cow Green	48 - 110
Derwent	70 - 110
Draycote	42 - 103
Empingham	60 - 90
	50 - 100 (amended)

30. There is a danger in specifying too narrow a range as this may be very difficult to achieve in practice, because of the nature of the material available. Both the Empingham and Bewl Bridge specifications had to be amended for this reason. There is a need for more testing of the material at the design stage to ensure that the shear strength limits are compatible with the materials available without an excessive amount of water having to be added. The authors consider it is easy to include figures in a specification, but it is much more difficult to achieve them in practice. The contractor's methods of excavation, working, watering, and compaction (and the tender rates) could lead to contractors achieving different results.

31. It is considered prudent to define a wider range overall, whilst placing tighter limits on the mean value, and this was done at Ardingly. This flexibility can then take account of the natural variations of the material, which cannot be ironed out by uniform watering, and will thus be reflected in the strength data obtained.

32. Although it is possible to gain some idea of the likely increase in water content required to satisfy the design limits beforehand from laboratory tests on re-compacted borehole samples, it is unwise to specify actual values since the amount may well have to be revised when the borrow pit is opened up. The contractor should be warned that the natural water content will need to be raised to achieve the specified strength range. He should be allowed to quote separately in the bill of quantities for the provision of water, either in terms of volume added or on a percentage increase in water content basis, with a further item to cover the additional working involved. The fill placement rate should, however, include for any watering required to prevent the placed clay from drying out.

33. It is considered essential that both the contractor and the engineer should regard control testing as an integral part of the placement operation. For the method to be effective, the results must be obtained as quickly as possible and the trend of results reviewed continuously. It is not satisfactory if the results are not available from the site laboratory until the next day. Until the placement technique has been mastered by the contractor as many tests as possible should be carried out. This is also essential where the variability of the clay, especially when inclusions are present, can affect the results of individual tests. Therefore, a larger number of tests in the early stages of a contract is desirable. Thereafter, control can depend to a greater extent on the site staff's observational experience of and feel for the behaviour of the material, with the main testing centred on the more doubtful areas. For example, the depth of rutting, or its absence, provides a visual guide to the lower limit of acceptable shear strength. Hand vane testing can also serve as a guide to placed strength but must be used with discretion because of the very small size of sample tested. Routine testing must still be continued so that results obtained according to the specified testing method are available in case of disagreement.

34. The speed of testing has been improved by the use of the microwave oven for rapid sample drying out and the automatic plotting of data from triaxial testing. The latter method can be justified on large contracts. An alternative technique is in-situ plate load testing since this has the advantage of testing a larger and therefore a more representative sample of fill in one test.

35. A shear strength method specification is of particular value where doubtful materials exist (or where there is a change of circumstances) since it enables a decision to be more readily made on whether or not to incorporate them in the fill. Sometimes, the specification can be met by mixing a softer material with a firmer one.

36. At one site weathered clays, from the previous overburden and now lying in worked out brick pits, were subsequently used in forming embankments up to 20 metres in height. The natural water content of this material was 10-15% in excess of its original water content which was close to Proctor optimum. It is considered a shear strength specification enabled selection of suitable materials whereas a relative density specification would have shown the material to be outside normally acceptable limits.

37. The authors believe that a shear strength specification, combined with a compaction requirement, for controlling and placing clay materials, principally in rolled clay cores of embankment dams, although not perfect, is considered to be an advance on previous methods based exclusively on optimum density and moisture content.

REFERENCES

1. PROCTOR R.R. (1933). Fundamental principles of soil compaction, and Description of field and laboratory methods. Engineering News - Record 111. 245-248 and 286-289.

2. KENNARD J, KENNARD M.F. (1962) Selset Reservoir: Design and Construction. Proceedings of the Institution of Civil Engineers, 1962 21, Feb. 277-304.

3. BUCHANAN N. (1970). Derwent Dam - Construction. Proceedings of the Institution of Civil Engineers, Part 1, 1970 48, March 401-422.

4. KENNARD M.F., PENMAN A.D.M., VAUGHAN P.R. (1967) Stress and Strain measurements in the clay core at Balderhead Dam. International Commission on Large Dams, Ninth Congress, Istanbul, 34, 129-131.

5. VAUGHAN P.R., LOVENBURY H.T., HORSWILL P. (1975) The Design, Construction and Performance of Cow Green Embankment Dam. Geotechnique 25, No.3, 555-580.

6. KENNARD M.F., READER R.A. (1975) Cow Green Dam and Reservoir. Proceedings of the Institution of Civil Engineers, Part I, 1975 58, May 147-175.

G. KUNO, Dr Ing, Professor of Civil Engineering Department, Chuo University, Tokyo, R. SHINOKI, Technical Adviser, T. KONDO, Head, and C. TSUCHIYA, Engineer, Japan Highway Public Corporation

On the construction methods of a motorway embankment by a sensitive volcanic clay

This paper covers construction methods that were developed during the construction of motorway embankment using a weathered volcanic ash deposit called "Kanto Loam" which prevails widely in the eastern Japan and has high natural water content ranging from 60% to 180%. Two fundamental problems should be considered in using this soil as an embankment material, which were as follows:
(1) Sensitive decrease of the strength with remolding.
(2) Low trafficability against heavy equipments in the field.
To cope with these problems, special techniques were developed as to reduce the disturbance of soil during earthwork operations to the minimum and to preserve a required strength for maintaining sufficient trafficability or stability of embankment. Some of these techniques were to construct a hauling path within the work area for assuring the trafficability of dump trucks and to make the best of crawler-mounted dozers with low ground contact pressure such as bulldozers or the improved scrapedozers, depending on the haul distance. In view of the meteorological conditions in Japan expressed by high humidity and severe rainfall, special considerations were also paid on the drainage system during construction, which were effective to minimize the suspended days after rainfall. As for the compaction control standard of this soil, the degree of saturation (85-98%) or the air void ratio (10-1%) was adopted on the basis of the facts that the maximum dry density was hardly determined in the laboratory and that an effort to obtain the specified density resulted in the phenomenon of overcompaction.

1. GENERAL PROPERTIES OF KANTO LOAM SOIL

Because of a number of volcanos distributed in Japan, soils originated from their eruptions prevail widely on hilly countries and volcanic mountainous areas covering fairly large part of the Japanese islands. Most of these soils are completely decomposed to clayey soils, together with meteorological condition of high-rain, high-humidity and from volcanic soils having high water content. So called Kanto loam soil covering terraces around Tokyo and the western part of hilly countries is a representative of those soils. The properties of Kanto loam are summarized in Table 1. In Table 1, two soils of the same kind as Kanto loam are also shown in comparison. Similar kind of soils exist also in Papua New Guinea (ref. 1) and in Indonesia (ref. 2).

Although these soils exhibit fairly high strength at their natural undisturbed condition (100-230 kN/m^2 in unconfined compressive strength), when they are remolded they show characteristics of decreasing in their strengths to a considerable extent as discussed in the following Chapter.

Table 1. Properties of representative volcanic cohesive soils in Japan

Soils	Distance from Volcanos (km)	Natural Water Content (%)	Soil Gradation				Consistency		Sensitivity	Engineering Properties			Dominant Clay Mineral
			Gravel (%)	Sand (%)	Silt (%)	Clay (%)	Liquid Limit	Plasticity Index		$qu^{1)}$ (kgf/cm^2)	$E^{2)}$ (kgf/cm^2)	$qc^{3)}$	
Kanto Loam	60-100	100-140	0	8	58	34	136	56	Very High	0.2-0.6	5-10	2-5	Allophane, Halloysite
	40-60	80-130	2	28	35	35	135	59	High	0.2-0.8	5-10	2-7	Allophane, Hydrated Halloysite
	15-40	60-90	16	37	25	19	111	550	None	1.2-1.6	80	15	Halloysite, Hydrated Halloysite
	5-10	130-180	0	55	40	5	228	114	Very High	0.2-0.6	5-10	2-4	Allophane
Kanuma Soil		150-240	60-65	30-35	2-5	0	105 -150	-	Very High	-	-	0.2-5	Allophane
Volcanic Ash		50-60	0	19	47	34	55	24	Very High	0-0.6	-	2-5	Halloysite

1) unconfined compressive strength
2) modulus of deformation
3) cone index

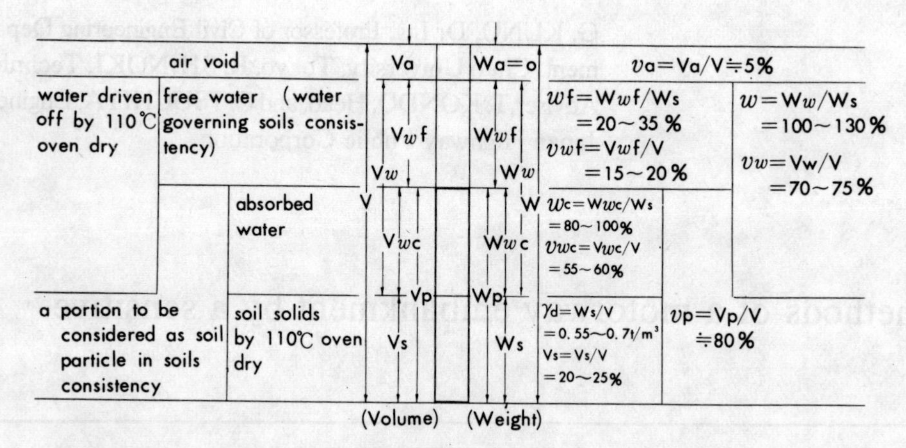

Fig. 1. Illustrative constitution of Kanto loam soil

It is also noticeable that the strength decreasing tendency is not always related with natural water content of soils. The water driven off from Kanto loam under 110°C oven dry may be divided into free water which has an intimate relation with engineering characteristics of soil such as strength and deformation, and the water which is absorbed on the surface of soil structures. The rate of these waters among soils is shown in Fig. 1 in which the amount of free water is defined as the water driven off from a soil when subjected to a centrifugal force equivalent to pF = 4.2. The threshold value was determined on the basis of the fact that dehydration of natural Kanto loam was not almost observed at pF = 4.2 or more (ref. 3).

As is clear in Fig. 1, a fairly large amount of absorbed water exists at natural state of Kanto loam. Further researches on this subject show that some of absorbed water turn into free state (or normal water condition) when the soil is remolded and its soil structure is disturbed. Furthermore, it is verified through many examples that the strength characteristics of the soil does not depend on a total amount of water in soil but depends highly on the amount of free water.

2. VARIATION IN STRENGTH CHARACTERISTICS DUE TO REMOLDING

(1) Laboratory investigation

A penetration test by the Portable Cone Penetrometer is popularly used as a test to evaluate the trafficability of construction equipment as well as to know the strength characteristics of soil both in the field and the laboratory. The cone has an angle of 30 degrees and a shadow area of 3.24 cm^2 similar to the WES cone for cohesive soil and soil strength is expressed by cone index which is obtained by dividing the penetrating resistance by the area.

It is a common procedure in laboratory that number of rammer blows in the compaction test is varied to control a extent of remolding of a sample. When natural Kanto loam is compacted in a 10 cm or 15 cm diameter mold by using 2.5 kgf (24.5 N) rammer having 30 cm free drop height, the strength expressed by cone index decreases with an increase in the number of blows as seen in Fig. 2. Measurement of the cone index is performed by penetrating the cone into the molded specimen at a speed of 1 cm/second.

It is indicated in Fig. 2 that application of compactive effort to a soil having high natural water content is not effective in increasing soils strength but simply results in an increase in the degree of saturation, that is, overcompaction state in which soils structure is disturbed by the rammer blows causing an increase in a free state water, thus a softening of soil.

A research committee on Kanto loam in Japanese Society of Soil Mechanics and Foundation Engineering proposed a cone index testing method to classify Kanto loam (ref. 4). According to this method, fresh soils having the same water content are compacted in a 10 cm diameter steel cylinder by a 2.5 kgf (24.5 N) rammer applying 10, 25, 50 and 90 blows each layer to make three compacted layers. The resistance of penetrating the cone penetrometer into the specimen at depths of 5, 7.5 and 10 cm is measured and the mean value of three readings is defined as a cone index. By plotting the cone index against the number of rammer blows, two types of soils are defined based on the shape of curves. They are, type

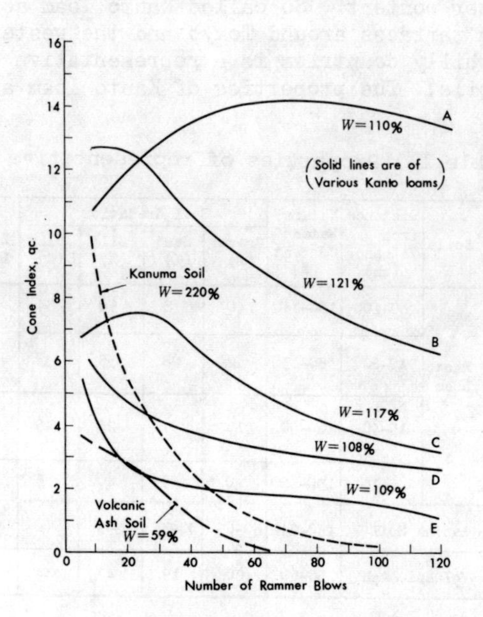

Fig. 2. Relationship between cone index and number of rammer blows

I which has a peak in curve as seen in Fig. 2, (see curves A, B and C) and type 2 which does not have a peak simply showing a decreasing tendency (see curves D and E). The shape of curves associated with three informations, the maximum cone index, cone index at 55 rammer blows and the difference in cone indices between at 55 and 90 rammer blows are tried to classify the Kanto loam soil. During actual operations, it has been found that about eighty percent of soils which showed good workability belongs to type I soil and the maximum cone index exceeds nine. On the contrary, those soils that showed poor workability belong to type II soil and the cone index at 55 rammer blows is mostly less than 2.5.

In measuring the cone index briefly in the laboratory, soil specimen is compacted in a 15 cm diameter mold by a 2.5 kgf (24.5 N) rammer applying 56 blows to each layer making three compacted layers.
No significant difference is observed for type II soils between the cone index values obtained at the above 55 and 56 rammer blows, whereas for type I soil, the preceding maximum cone index (qc max) tends to show higher values, which may attribute to the confining effect by the mold.

(2) Observation in the field
During pushing or spreading work of Kanto loam, soils are remolded considerably by the crawler line or the blade of bulldozer resulting in the decrease of strength. Fig. 3 and 5 show the strength decreasing tendency during operation in terms of cone index values.
Fig. 3, an example at the Ashitaka test fill, shows that soil is remolded during the works by a blade and reduces its strength. Generally the cone index at fill surface tends to decrease with increasing number of equipment passes as shown in Fig. 5, which is contrary to a normal notion that equipment passes compact

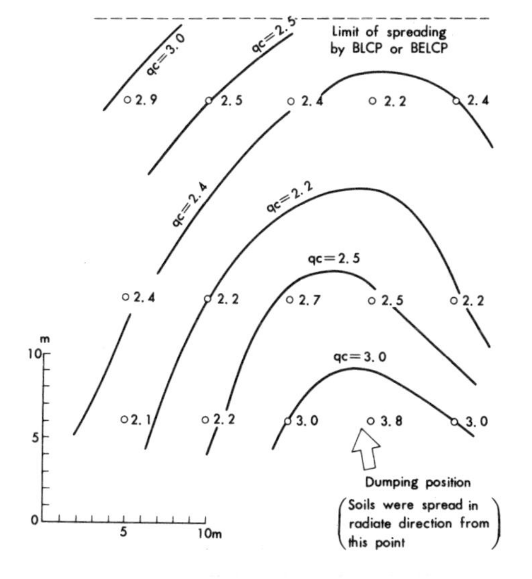

Fig. 3. Contour of strength at the surface of embankment

Table 2. Relationship between trafficability of equipments and cone index

Type of Equipment	Cone Index qc	Remarks
Bulldozer with extra low contact pressure	1.5-2	BELCP
Bulldozer with low contact pressure	2-4	BLCP
Ordinary bulldozer	4-7	
Scrape dozer (4m3)	4-5	low contact pressure
Towed scraper	5-7	tawed by bulldozer with low contact pressure
Self-loading scraper	at least 10	
Twin-engine scraper	4-5	self-powered, front and rear drive
Dump truck	at least 10	

soils and increase their strength. A relationship between a number of cone index values measured in the field and in the laboratory reveals a state of soil expressed by the minimum cone index in the laboratory at which equipment can pass the same track by 2 to 4 times. The minimum cone index value for each equipment is shown in Table 2, which provides a guide for selecting appropriate equipment (ref. 5).

3. CARES AND MEASURES FOR BETTER OPERATIONS
Fundamental cares and countermeasures that have been taken during earthwork operation of high natural water content clay materials such as Kanto loam are summarized as follows;
a) prevention of strength decrease due to remolding,
b) prevention of the increase in water content, or reduction of it if possible,
c) best use of strength recovering property of soil, and
d) treatment of material by additives
In the following paragraph, concrete countermeasures are described in details introducing the experiences during the motorway construction.

(1) Use of equipment with low contact pressures
It is a liable situation that an equipment loses its mobility after several passes on Kanto loam fills because of deep rutting. Fig. 4 indicates the cone index values showing the required strength of fills for equipments to pass 2 to 4 times on the same paths. The data in Fig. 4 were obtained through the test fill at Ashitaka. Fig. 5 also shows that the cone index at the surface of a volcanic ash soil embankment decreases with number of passes of bulldozers. Specifications of these bulldozers are summarized in Table 3 in which it is indicated that a bulldozer with low contact pressure will be effective in reducing the remolding of soils due to crawling, thus providing higher trafficability and efficiency

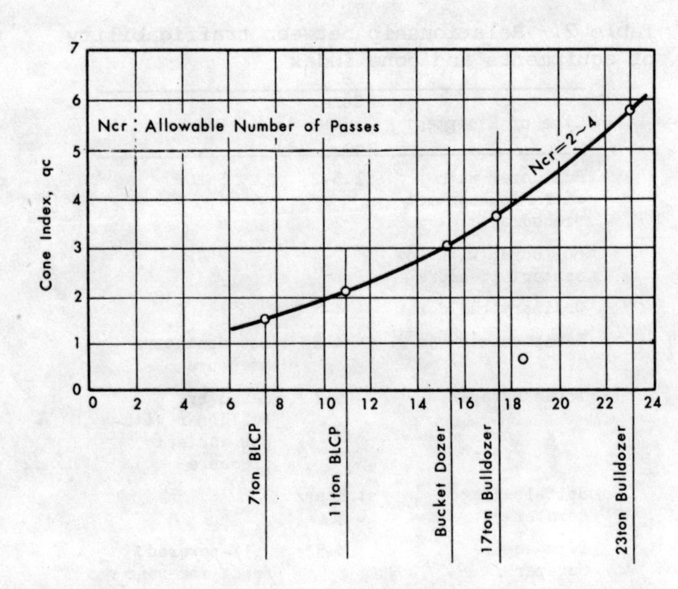

Fig. 4. Relationship between cone index and the trafficability of various bulldozers

of work. Dozing work in the earthwork of Kanto loam is mostly done by bulldozers with low contact pressures or scrapedozers which have lower contact pressures by widening the width of crawler link.

(2) Selection of methods to reduce remolding

a) Bulldozer work system: Bulldozers are efficiently used for excavating, hauling and spreading in the Kanto loam area when a haul distance is approximately less than 80 m. The use of ordinary bulldozers having normal contact pressures is however, difficult even when soil condition is fairly good. Fundamentally bulldozers having lower contact pressures are more efficiently used. For a Kanto loam with medium soil condition, it is a common practice that the use of ordinary bulldozers is restricted at the cut section only and a bulldozer with low contact pressure (BLCP) does all the works at the fill section. When the soil condition becomes poorer, BLCPs do the works in the cut from excavating to hauling and a bulldozer with extra low contact pressure (BELCP) is adopted for the works at the fill such as secondary hauling from the cut-fill boundary into the fill section and spreading. A bucket-dozer which is intalled with a bucket instead of a blade was developed in an effort to reduce the remolding of soils by blade works during hauling. The bucket-dozer has not been widely used since its hauling capacity is fairly small (1.7 m^3) and it has difficulty in balancing when it is loaded.

b) Scrapedozer work system: Towed scrapers are used for earth-moving in the Kanto loam terrain when a haul distance is within 300 m. Their use are, however, restricted to the cases where soil condition or working condition is favorable such as during summer when aeration of soils is effective in reducing water content, etc.. Since a turning of scraper at the fill area accompanies much difficulties, unloading is done at the cut-fill boundary from which the secondary hauling and spreading works are done by BLCP.

Scrapedozers were effectively used even under a fairly bad soil condition, which may attribute to those merits of scrapedozers as relatively low contact pressure and no use of turning after unloading. In general, self-loading scrapers are not used in the earthwork operation of Kanto loam. Twin-engine scrapers

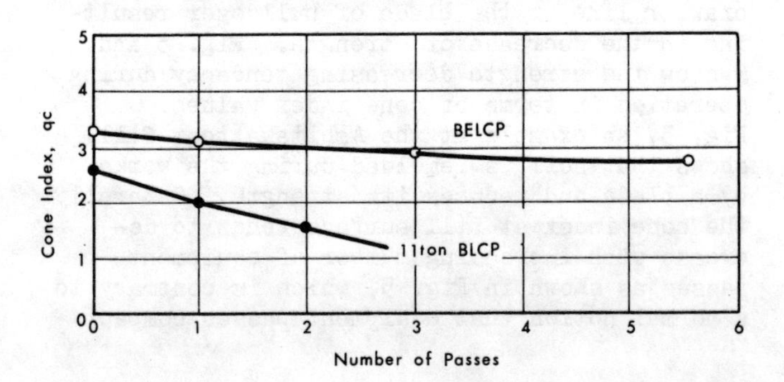

Fig. 5. Reduction of cone index with number of bulldozer passes

Table 3. Specifications of various bulldozers

Type of Equipment	Weight (kgf)	Length (mm)	Width (mm)	Contact Length of Crawler Line (mm)	Width of Crawler Line (mm)	Contact[1)] Pressure (kgf/cm²)
Ordinary Bulldozers						
23 ton class	23,200	5,970	4,110	2,840	560	0.73
17 ton class	17,550	5,510	3,880	2,525	510	0.68
BLCP						
11 ton class	11,000	4,830	3,190	2,580	800	0.26
7 ton class	7,600	3,920	2,900	2,100	762	0.24
BELCP	9,700	4,290	3,550	2,600	1,050	0.165
Bucket dozer	15,710	5,950	3,460	2,985	860	0.31
Scrape dozer	15,800	4,865[2)]	2,980[2)]	3,050	600	0.43

1) without load
2) blade is not installed

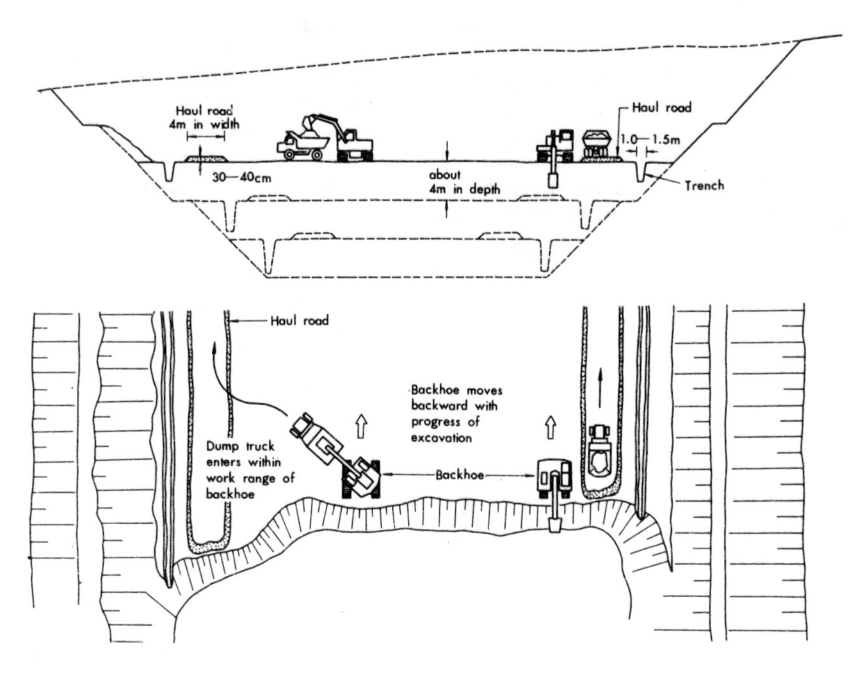

Fig. 6. Backhoe-dump truck system and the arrangement of haul roads at the cut

are often used in preparing a large housing site although their use is limited at highway construction sites.

c) <u>Shovel-dump truck system</u>: Shovel-dump truck system is adopted when a haul distance exceeds 300 m or when excavating and hauling can not be effectively done by scrapers or bulldozers. In this system, it is required as a matter of course to construct haul roads using a selected sandy soil both at the cut and the fill and to adopt BLCP for spreading work at the filling area. When progressing the cut works as shown in Fig. 6, the shovel-dump truck system will be often employed but there arise problems associated with maintaining the haul roads constructed on disturbed materials and keeping the excavating area well-drained condition as the excavation progresses. Otherwise the condition at loading area becomes worse by wetting and mixing soils with rain or underground water.
To cope with these problems, the best method was found to adopt a dragshovel (or a backhoe) or a dragline system which excavates backwards at the cut as seen in Fig. 6 providing that topographical or site conditions are favorable.
In this method, an excavating equipment can work always on undisturbed or less disturbed, well-drained ground and it is not required to construct a haul road on loose ground after excavating, which makes the maintenance of haul roads far easier.

Among highway construction sites in the Kanto loam terrain, backhoes or draglines are adopted at the same rate as power shovels as an excavator of long-distance earth moving operations. A wheel-type loader was not popularly adopted because the remolding of soils during excavating or loading work was found to be remarkable.

The haul roads measure from 4 to 5 m in width and are normally constructed by crushed-stones or well graded gravels having thickness of 30 to 40 cm at the cut and 50 to 60 cm at the fill. When the forwarding of these selected materials costs highly, steel mat, synthetic fiber or cloth is effectively used in reducing the thickness of the haul roads by placing under the selected material. Since rutting progress rapidly once the road surface was disturbed, prompt supply of repairing materials was required to smooth the road surface and its repairing frequency reached two or three times a day at the high season of earth moving. Soils dropped on the road surface often deteriorated the surface conditions of the road and their removal whenever they were found was

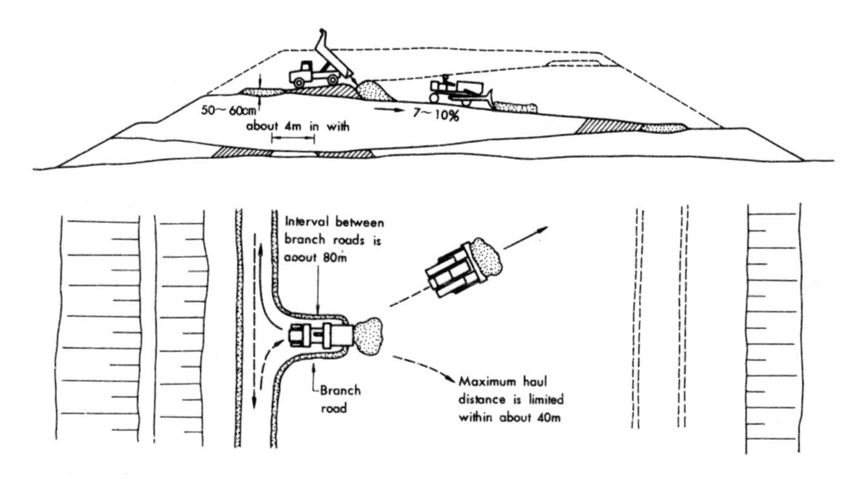

Fig. 7. Hauling and spreading work and the arrangement of haul roads at the fill

found to be one of the most important cares for maintaining the road in good condition. A steel mat as haul road was found unsuitable except as a temporary expedient since remolded soils on the mat produced a very slippery surface.

Typical array of the haul roads at the cut and the fill is shown in Fig. 6 and 7 respectively. As seen in Fig. 7, many branch roads at the fill are required for effective dumping in the fill area. The arrangement of the branch roads at the fill is made so that secondary hauling distance by BLCPs does not exceed 20 to 40 m, above which soils are remolded considerably resulting in a decrease in efficiency. Besides, when it is apprehended that the haul roads collect and lead penetrated water into the fill, they are excavated and mixed with adjacent fill materials after their missions are fulfilled.

(3) Countermeasures against rainfall

During fill works of Kanto loam, rainfall usually affects the progress of work seriously by suspending one to three days after each rainfall exceeding 10 mm per day. Various attempts are made to reduce the suspended days to the minimum. In the cut section, longitudinal trenches are placed along the edge of cut slopes preceding to excavation in an effort to prevent water from flowing into work areas (see Fig. 6). These trenches are also useful for draining underground water and for lowering the water content of soils to be excavated. Another effort is to place a ditch horizontally along the cut-fill boundary to drain water that tends to flow from the cut section into the fill area.

A superelevation is kept fairly steep ranging from 4 to 6 percent downgrade from the center of the fill even when haulroads are not used. Especially when a heavy rainfall is forecasted, the trace of crawler line such as by bulldozers at fill surface is smoothed and compacted by a light rubber tired compactor to drain surface water smoothly. Water proof cloths having a size of 10 m by 10 m are also useful to prevent water from penetrating into the fill by spreading over important haul roads and the areas where spreading work is scheduled.

An increase in water content at the fill surface due to rainfalls can be limited within 10 cm in depth provided the surface is compacted well and has smooth drainage. After rainfall, heavy equipments should avoid to enter the field unless sufficient bearing capacity for them is ensured, otherwise the surface is remolded and softened considerably.

(4) Measures to reduce the water content of soils

Aforementioned measures such as longitudinal trenches, horizontal trenches and drainage layers (or filter layers) placed in a high embankment are found effective in draining underground water as well as in reducing the water content of Kanto loam soil itself. Another more active measure may be the drying of spread Kanto loam through aeration by use of a disk harrow. Fig. 8 is an example of this method showing a significant effect in reducing the water content. However, the effect is expected only under the best meteorological condition such as a high temperature during summer. During winter, disk harrow work is suspected to reduce the strength of soils by remolding rather than to dry them.

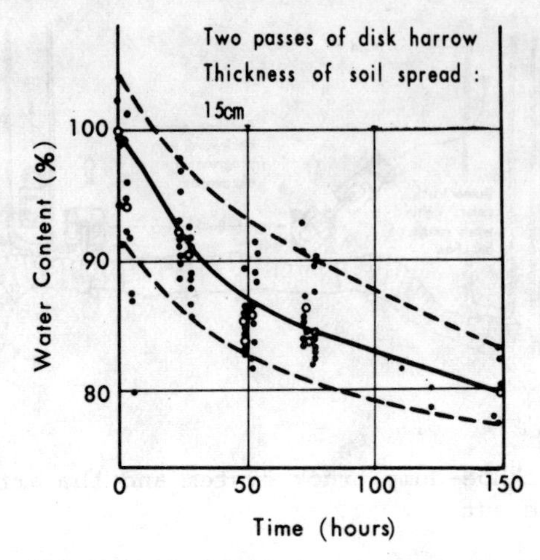

Fig. 8. Aeration effect on the water content of soil

It is concluded that aeration is effective in drying soils during summer only when higher temperature is expected in spite of high humidity. During winter, the drying effect becomes negligible coupled with thawing although a long spell of fine weather with lower humidity is expected around Tokyo.

(5) Strength recovering characteristic of Kanto loam after operation

Fig. 9 shows the attitude of strength recovery of soils inside the embankment, which was recognized by using a double tubed cone penetrometer at the Ashitaka test fill. In Fig. 9,

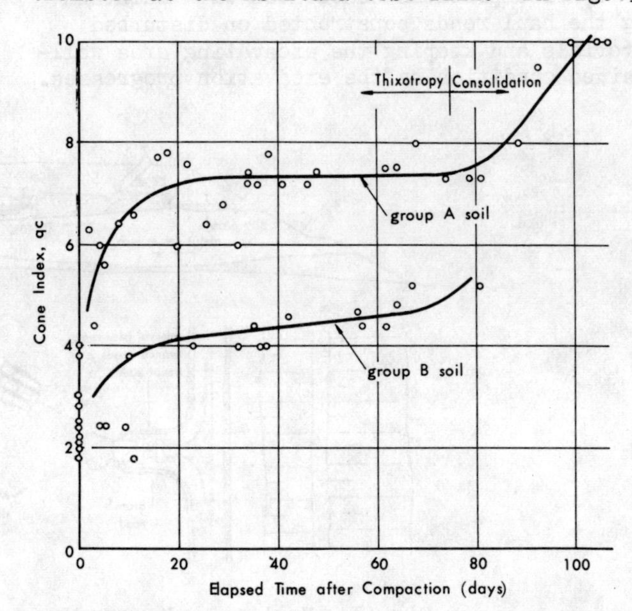

Fig. 9. Strength recovery of soils inside embankment

group A soil has a comparatively good quality placed at lower part of the embankment with about 15 meters overburden, whereas group B soil has poorer quality placed upper part of the embankment. In this example, it is seen that strength recovery in the first twenty days reaches about 2.5 times and twice of the original strengths for group A and B soil respectively. These strength gains may attribute not only to thixotropical effect but to consolidation by dissipating the excess pore water pressure in soils.

It is because of the strength recovering characteristic of Kanto loam, even if it is remolded during operation, that high embankments having over 10 m in height and 1.5 to 1.8 (in cotangent) in slope gradient keep their stability in spite of the low initial strengths after compaction such as 2 in qc value which is approximately equivalent to 0.2 kgf/cm^2 (19.6 kN/m^2) in the unconfined compressive strength. So much as strength increase attributes to consolidation, the compression of embankment becomes large. According to the measurement on seven motorway embankments having a order of 20 m in height, their compressions amounted three to five percent of their heights. Residual settlement after completion of the filling is, however, small 0.5 to 0.8 percent, indicating that the settlement progresses comparatively fast.

(6) Soil stabilization

Generally a compacted loam does not have sufficient bearing capacity as subgrade. Although a selected subgrade soil such as sandy soil has been borrowed from outside roadway, recent economical and environmental situations require to make the best of in-place materials by stabilization especially when a borrowing material is not economical because of a large amount of surplus soil to be wasted or a long haul distance, and when environmental disruption around construction site or disturbance in the existing traffic due to excavation or hauling of borrow materials is legally restricted.

In the treatment of Kanto loam with additives, a quick lime is generally preferable to a portland cement or a bituminous material since uniform mixing of additive with clayey soils having high water content such as Kanto loam is difficult often leading to softening of the mixture by mixing excessively, and also the addition of liquid materials will lead to the same result. Although hydrated lime is sometimes used, quick lime is more preferably adopted expecting to lower their water content by the reaction heat and the dehydration effect during mixing.

As for mixing methods, in-place mixing method or pre-mixing method at the cut is popularly used whereas central mixing method is rarely adopted. In pre-mixing method, trenches or holes are dug at excavating area in which quick lime is filled. Excavation starts one or two days later expecting the water content of soils adjacent to the holes or trenches to be reduced to a certain extent and the pre-mixing is completed during excavating, hauling and spreading works. In an example of this method, 15,000 m^3 of Kanto loam was ameliorated and used as subgrade soils by reducing the water content from 108% to 70%, the plasticity index from 50 to 14 and increasing the California Bearing Ratio (CBR) from 2 to 18 (ref. 6). Trenches having about 1.5 m in depth were dug by trencher or backhoe at regular intervals so that the lime mixing content becomes approximately twenty percent of soils on the dry-weight basis.

Meanwhile, a simple compulsive drying method is experimentally adopted at many construction sites including fill-type dams since soil characteristics of Kanto loam are obviously improved by reducing its water content. The compulsive drying method includes two categories, one of which is to reduce the water content of whole soils by 20 to 30 percent and the other is to dry a part of soils to the maximum, which are used to dehydrate natural soils by mixing or by placing between natural soil layers. The former category is practically adopted in highway construction site. In an example of treating 10,000 m^3 of motorway subgrade soils by use of an asphalt plant dryer, natural Kanto loam having 95 to 110% in natural water content was dried up to 80 to 90% and the strength was improved from less than 2.2 in CBR values at natural state to 33 to 50 by adding 10 to 13% of quick lime content (ref. 6).

(7) Earthwork operation records

Work rates of various equipments during the

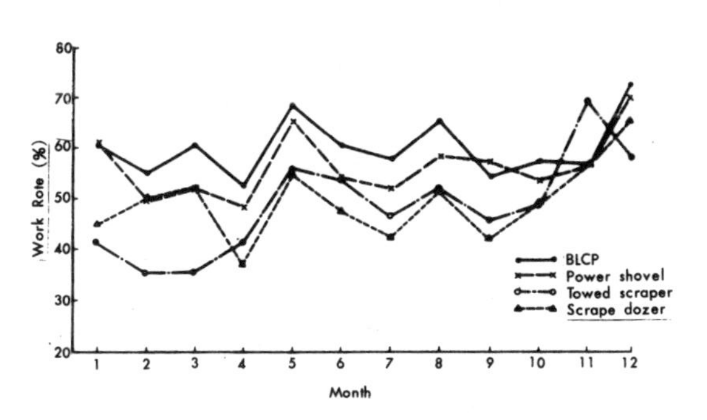

Fig. 10(a). Work rates of various equipments (1965-1967)

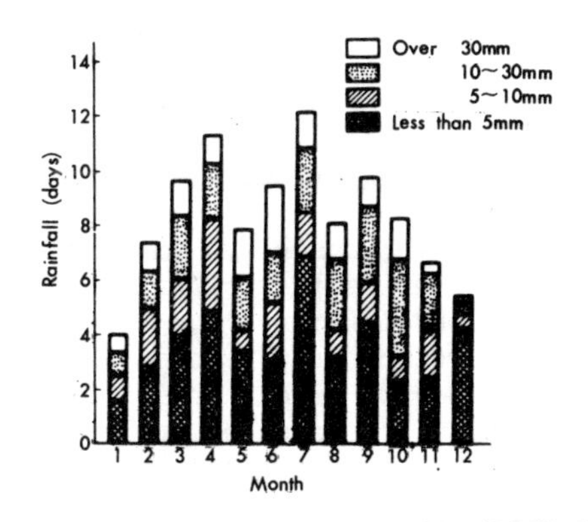

Fig. 10(b). Record of rainfall (1965-1967)

motorway construction at the Kanto loam terrain are shown in Fig. 10 (a) on the basis of actual operations from 1965 to 1967 while earthwork operation was at its zenith recording about 17 millions cubic meters earth moving volumes (ref. 6). The daily rainfall record during these three years are also shown in Fig. 10 (b).

It is seen in Fig. 10 (a) that the average work rate of bulldozers remained 60.2% (51 to 73%) whereas the rate reached to 68.3% (55 to 77%) in the area where no volcanic clay as Kanto loam prevails. The difference in work rates indicates how the handling of Kanto loam is difficult. The work rate of scrapers went down during winter (December and January), which was explained by the fact that the data included the operations in the area where soil conditions were very poor and the rate was below 20%. There can be seen in Fig. 10 (a) some months during summer when the work rate of scrapedozers is lower than that of towed scrapers, which was due to the situation that scrapedozers were newly induced at that time and their operations in the Kanto loam terrain was inexperienced resulting in frequent troubles. Since then, the work rate of scrapedozers kept constantly high throughout the year.

Power shovel showed almost constant work rate throughout the year regardless of soil conditions, which was attributed to those efforts ensuring the haul road constructed by selected materials as stated above and it is understood that the assistance by BLCPs was considerable showing high workability for any type of loams from good to poor quality.

4. SPECIFICATIONS FOR COMPACTED SOILS

A criterion for compaction based on the maximum dry density (γd max) and the optimum water content (Wopt) in the standard compaction test is widely used in Japan. The criterion is not, however, applicable to volcanic soils having high natural water content such as Kanto loam because of the following reasons:

a) Since the natural water content at operation often exceeds the minimum value to give the designated dry density, it is required to reduce the water content drastically by some measures which may be almost impossible practically.

b) It is usually observed during the standard compaction test of cohesive volcanic soils as Kanto loam that γd max or Wopt varies with the initial water content of a sample before test under the same compactive effort. This makes the standard compaction test infirm of significance as a standard method (ref. 7).

Since there exist various clays in Japan except cohesive volcanic soils that show the same situation as stated in item a), a specification in which the degree of compaction is evaluated in terms of the degree of saturation or the air void ratio is popularly adopted in Japan. That is, it is specified in the earthwork operation of Kanto loam that the degree of saturation or the air void ratio of compacted soils is to be within 85 to 95% or 10 to 1% respectively, limiting also the maximum water content of soils above which sufficient strength can not be ensured for the mobility of equipment to be used (for upper embankment within one meter below subgrade, water content at compaction is specified so that the submerged CBR value of compacted soil is more than 2.5).

Almost all Kanto loams are successfully handled by BLCPs satisfying the aforementioned compaction criteria by two or three passes of a 11 to 13 ton bulldozer and additional passes of a roller will lead to the condition of overcompaction where soils strength decreases on the contrary. Although BLCPs are not considered as a compactive equipment in general, compaction of Kanto loam soil is regarded to be completed by approximately four uniform passes of them. Especially for a Kanto loam having good quality less susceptible to remolding, a light 7 to 10 ton towed roller is occasionally used but other type of roller is not adopted.

As discussed in the previous sections, a remolded Kanto loam recovers its strength in the course of time. Making the best of this characteristic of Kanto loam, an expedient measure for ensuring the trafficability of equipments is to delay the succeeding work on the finished layer as possible. It may also be concluded that Kanto loam is not applicable as a material for filling rapidly at narrow areas.

Acknowledgement

The authors express many thanks to Prof. M. Inada for invaluable suggestions and also to those who provided many data in preparing this paper.

References

1) Wallace, K.B., Structural behaviour of residual soils of the continuary wet high lands of Papua New Guinea, Geotechnique, vol. 23, No. 2, 1973, P.203-P.218.

2) Wesley, L.D., Discussion on the paper submitted by Wallace, K.B., Geotechnique, Vol. 24, No. 1, 1974, P.101-P.105.

3) G. Kuno and S. Iitake, Water in soil and its effect on the compaction. Bulletin of the faculty of science and engineering, Chuo University, Tokyo, No. 12, 1969. (in Japanese)

4) Research Committee on Kanto loam, A proposed testing method for classifying Kanto loam. Symposium on Kanto loam (cohesive volcanic soils), Japanese Society of Soil Mechanics and Foundation Engineering, 1970. (in Japanese)

5) Japan Highway Public Corporation, Design manuals, Part 1 earthwork, 1970. (in Japanese)

6) The Expressway Research Foundation, Report of research committee on Kanto loam, 1972. (in Japanese)

7) G. Kuno and M. Yabe, On the compaction of wet cohesive soils. Proceedings, 2nd Asian Regional Conference on Soil Mechanics and Foundation Engineering, 1963, P.276-P.279.

I. P. LIESZKOWSZKY, BASc, Chief Engineer, Dominion Soil Investigation Inc., Toronto

Fort Creek Dam—impervious clay core

Experience with the construction of an impervious clay core for an earth fill dam are related. For the construction of the impervious core, locally available highly plastic and sensitive glacial lake clay deposit was used. The clay was borrowed at moisture contents 16 to 18% above the optimum and placed at moisture contents 6 to 10% above optimum. The placement of the fill was controlled by checking its moisture content and the undrained shear strength both by laboratory compression tests and field vane tests. Good correlation exists between the relative density, moisture content and shear strength of the compacted material. Construction control included also the monitoring of piezometers installed in the lower half of the clay core and the foundation material. The pore pressures in these piezometers were recorded as the placing of the fill progressed and are presented in this paper.

INTRODUCTION

Fort Creek Dam, the construction of which is described in this paper, was built during the summers of 1968, 1969 and 1970 in Northern Ontario, Canada. The primary function of the dam was the control of the flood waters of the water shed of Fort Creek, which before the construction of the dam caused serious damages in parts of the City of Sault Ste. Marie. The secondary function of the dam was the creation of a recreational lake.

The earth works consisted of two zoned earth fill dams founded on a deep deposit of firm, varved clay, and an emergency spillway. The main earth dam is about 150 m long and 15 m high. The secondary dam is much smaller.

For the construction of the impervious clay core, locally available, highly plastic and sensitive, varved clay was used. The natural moisture content of the clay was well above the optimum and to attain the shear strength properties assumed in the design, the moisture content had to be reduced considerably before compaction. Unfortunately, the summers when the construction of the dam was carried out, were very wet and the drying of the clay was difficult to achieve, causing serious delays in construction. Further delays were caused by high pore pressures set up in the foundation material, requiring the shut down of the construction for periods of time.

This paper describes briefly the project, the geology of the site and the subsurface conditions. It discusses in more detail the properties of the clay used for the impervious core, both in its undisturbed and remoulded state; it describes the method of borrowing and placement and the results of the quality control tests. Piezometers were installed in the lower half of the clay core as well as in the foundation material. These piezometers were monitored throughout the construction and the results of some of these piezometers are also presented.

SITE AND GEOLOGY

The City of Sault Ste. Marie is located on the north shore of Lake Superior and the St. Mary's River, which discharges the head waters of Lake Superior into Lake Huron through a drop of about 6 m. The dam site is located in the northern outskirts of the City. The Fort Creek at this point follows a slightly meandering course through the base of an approximately 200 m wide steep walled valley.

The geology of the site is typical of the Canadian pre-cambrian shield. A high degree of glaciation is in evidence, sound igneous or sedimentary bedrock, metamorphosed in places, persist throughout. Rock outcrops, practically void of drift, occur frequently, but within short distances the rock could dip to great depths below the ground surface. These deep valleys were eroded and gouged out by the glaciers. As the last Wisconsin ice sheet retreated, the waning glaciers exposed the basin of the present Great Lake system which was then occupied by a complex series of lakes. The size, shape and distribution of the post glacier lakes varied greatly with fluctuations in the ice front, the uncovering of new outlets and the result of differential uplift of the northern part of the lake basins. The hole of the Lake Huron basin may have been uncovered by ice over 9000 years ago, at which time the lake level was estimated to be at about Elevation 190 to 200 m. Differential uplift finally brought about the gradual lowering of the lake level, going through intermediate stages referred to as the Nipissing Great Lakes and Algoma Lake stage and, finally, in the not too distant past, the Huron level dropped to that of the present lake level at about Elevation 183 m.

The remnants of this past lacustrine environment are the deep, varved clay and the fine beach sand deposits encountered around the fringe of the Great Lakes. The floor of these clay beds is a sandy ground moraine which overlies the uneven surface of the bedrock.

SUBSURFACE CONDITIONS

At the dam site, the typical soil profile consists of an approximately 3 m thick recent alluvial deposit, consisting of layered sandy silt with some organic matter. The alluvium is underlain by an approximately 15 m thick deposit of sensitive reddish varved clay. The clay has a firm to stiff consistency with undrained shear strength values of about 40 to 60 kN/m^2. The sensitivity is about 5. The clay in turn is underlain by very dense bouldery till with a matrix of clayey silt or silty sand. A typical borehole log at the dam site is shown on Figure 1.

The borrow pit for the impervious material was located immediately adjacent to the dam site and included some of the material excavated at the site of the emergency spillway and the access road to the dam. The location of the borrow area in relation to the dam site is shown on Figure 5. Also shown on this Figure are the locations of the exploratory boreholes put down during the site investigation and a summary of the soil profiles encountered in these boreholes.

In the borrow area, the red varved clay was exposed at the ground surface or was covered by less than 300 mm of organic topsoil. The clay was desiccated and had a stiff consistency to an average depth of about 1.5 m. The underlying material had a higher moisture content and a firm consistency. For the construction of the clay core, this upper dryer crust was considered to be suitable and the approximate boundary between the suitable and underlying unsuitable material is shown on the borehole logs, given on Figure 5.

PROPERTIES OF BORROW MATERIAL

The clay found in the borrow area, and at the dam site, is typical of the glacial lake clays found in the Sault Ste. Marie area. It is a red clay with grey silt laminations. The deposit exhibits a definite banded (varved) structure. Darker red coloured clay bands alternate with coarser textured light coloured silty bands. At the site, the clay bands dominate and are typically 25 mm thick while the silt bands are about 5 mm thick. The principal constituent minerals are quartz, kaolinite and illite, with a small percentage of chlorite and vermiculite (Ref. 1). The distinctive reddish colour of the clay is due to the presence of hematite. The clay fraction, i.e. particles smaller than 2 microns, varies between 45 and 65%. From this and the plasticity index, the activity of the clay was found to be between 0.43 and 0.7, indicating an inactive clay as defined by Skempton (Ref.2).

The clay is highly plastic. The liquid limit ranges between 40 and 85% with an average value of about 54%. The plasticity index is typically between 25 and 30. The natural moisture content within the desiccated crust ranges between 30 and 36% and below the crust moisture contents are typically in the 43 to 50% range. The index properties of the clay are summarized in Table 1.

Table 1. Average index properties of borrow material

Percentage passing No. 200 sieve	- 95%
Percentage smaller than 2 micron	- 45%
Liquid limit	- 57%
Plastic limit	- 26%
Natural moisture content	
within crust	- 35%
below crust	- 43%
Specific gravity	- 2.72

The compaction characteristics of the clay were determined in the laboratory, using the Standard Proctor compaction test. It was found that the maximum dry density is obtained at an optimum moisture content of 24% and about 80% saturation. The maximum dry density at optimum moisture content ranged between 14.7 and 15 kN/m^3. Unconfined compression tests were performed on samples of the clay compacted at various moisture contents. The variations of the undrained shear strengths with moisture content during compaction are shown on Figure 4. Consolidated undrained triaxial compression tests with pore pressure measurements were performed to determine the effective shear strength parameters of the compacted clay. The samples were compacted at a moisture content of 33%. The degree of saturation at this moisture content was calculated to be about 97%. The effective angle of shearing resistance from these tests was determined to be 23.5 degrees with a cohesion intercept of about 20 kN/m^2. Consolidation tests were also performed on samples compacted at a moisture content of about 35%. Beyond a consolidation pressure of about 100 kN/m^2, the clay behaved like a normally consolidated clay of medium compressibility. The compression index in the 100 to 1000 kN/m^2 pressure range is about 0.3. The coefficient of permeability was estimated from the consolidation test data and was found to be 7 x 10^{-8}cm/sec. for pressures less than 100 kN/m^2. At higher pressures, the coefficient of permeability decreased. The engineering properties of the compacted clay material are summarized in Table 2.

Table 2. Average engineering properties of compacted borrow material

Maximum dry density (Standard Proctor)	15 kN/m^3
Optimum moisture content	24%
Undrained shear strength at 30% moisture	100 kN/m^2

Table 2 (continued)

at 35% moisture	50 kN/m^2
at 40% moisture	25 kN/m^2
Effective angle of friction	23.5 degrees
Cohesion intercept	20 kN/m^2
Compression index	0.3
Coefficient of consolidation	0.03m/day
Coefficient of permeability	7x10^{-8} cm/sec.

DESIGN

In the early stages of the design it was recognized that the construction of a homogeneous earth fill dam using the varved clay as construction material would be difficult and probably uneconomical. If this material was to be used, side slopes possibly of the order of 5 horizontal in 1 vertical would be required and suitable borrow material in these quantitites would unlikely be available. A zoned earth filled dam of the cross section shown on Figure 6 was, therefore, adopted.

The stability of the proposed dam cross section has been analyzed, assuming both circular and plain failure surfaces. The stability of the dam at various stages of construction has been computed. The stability at the end of construction was analyzed in terms of total stresses, using the undrained shear strength of the soil and, for this case, the undrained shear strength of the foundation material as well as the compacted clay core has been taken as 50 kN/m^2. Secondly, the stability has been analyzed when the dam has been completed and the reservoir is first filled, but before consolidation of the underlying soil takes place, i.e. the induced pore water pressures in the subsoil have not been dissipated. Again, this case has been analyzed using a total stress analysis because of the inherent difficulties of estimating the pore water pressures at this stage. Finally, the long term stability of the dam with the reservoir full has been investigated, using the effective stress analysis with the equilibrium pore water pressures which will be governed by the steady seepage conditions. Stability calculations were performed using a computer. Due to the precarious location of the dam in a heavily populated area, a safety factor of 1.5 was chosen. The selection of this rather high safety factor resulted in the need for two stabilizing berms of 35 and 20 m length at the downstream and upstream sides respectively.

SPECIFICATIONS

To be acceptable, the clay core material had to possess the properties that were assigned to it in the design. The undrained shear strength had to be not less than 50 kN/m^2, which based on the laboratory test data shown on Figure 4 could be realized if the material was placed at moisture contents less than 35%. Assuming a fully saturated material at this moisture content, the maximum degree of compaction that could be obtained is 92% of the Standard Proctor maximum dry density. It was also felt that if the clay was compacted at moisture contents of 30 to 35%, the material will have a good texture and a coefficient of permeability of 10^{-7}cm/sec. or less.

To achieve these objectives, the specification called for an average of 97% compaction with no value less than 92%. It was also specified that the moisture content during placing shall be not less than 26% and not more than 35%. It was specified that the maximum thickness of a single lift shall not exceed 200 mm and that the material shall be compacted with a sheeps-foot roller. Areas inaccessible to a sheeps-foot roller shall be compacted by other approved mechanical or pneumatic equipment.

The moisture content of the borrow material shall be determined in the borrow area before delivered to the dam site. Where necessary, the clay should be dried in the borrow area by plowing and discing to give maximum aeration. After the material has been sufficiently dried and approved, the dried layer shall be stripped off and placed in the core. The specification also included a clause that if the surface of the already placed clay core in the engineer's opinion is too dry or too wet, or otherwise damaged by frost or construction equipment, the contractor shall remove, recompact or rework this material so that the homogenity and the bond between consecutive lifts is maintained.

CONSTRUCTION

The contract was awarded in the spring of 1968. The contractor moved to the site in May 1968 and after site preparations such as grubbing and stripping, started to work on the construction of the drop inlet structure and the outlet conduit which was intended to serve as a by-pass for the creek during construction. Progress, however, was very slow mostly because of poor weather conditions. Some work had been done on the stripping of the clay borrow area, but this too progressed slowly due to the wet ground conditions. The excavation of the clay core area was scheduled to start in July but, because of almost continuous rains, was never carried out in 1968. No construction was carried out during the winter and work was resumed in May 1969. However, rainy weather again interferred and it was only in July of 1969 that the excavation of the cut-off trench could be carried out. The excavation of the cut-off trench to the dimensions shown on Figure 6 were carried out with a drag line, starting at the west abutment of the dam.

The placing of the clay core was began on July 21, 1969, starting from the east abutment. Because of the soft and plastic nature of the subgrade material, the following procedure had to be adopted at the beginning of the placing of fill. A layer of loose fill material, approximately 300 mm thick, was spread over about a 15 m length of the base of the cut-off trench and this material was compacted as much as possible, using a light bulldozer equipped with caterpillar tracks. It was impossible to place a heavy dozer or a roller over this thin layer and tracking was continued until the material began to wave. Another thin layer

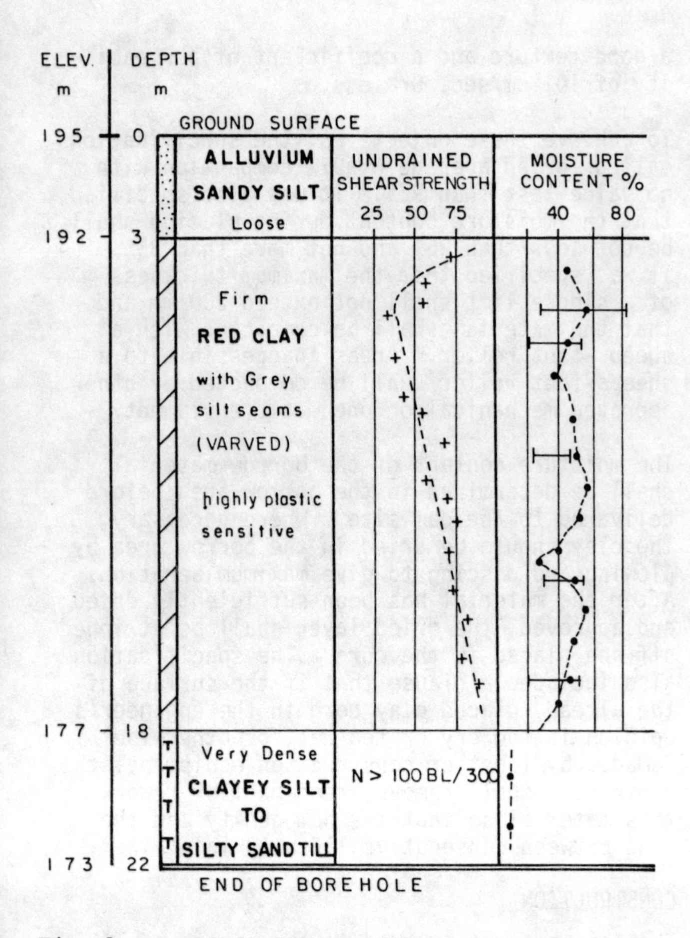

Fig. 1. Borehole log (dam site)

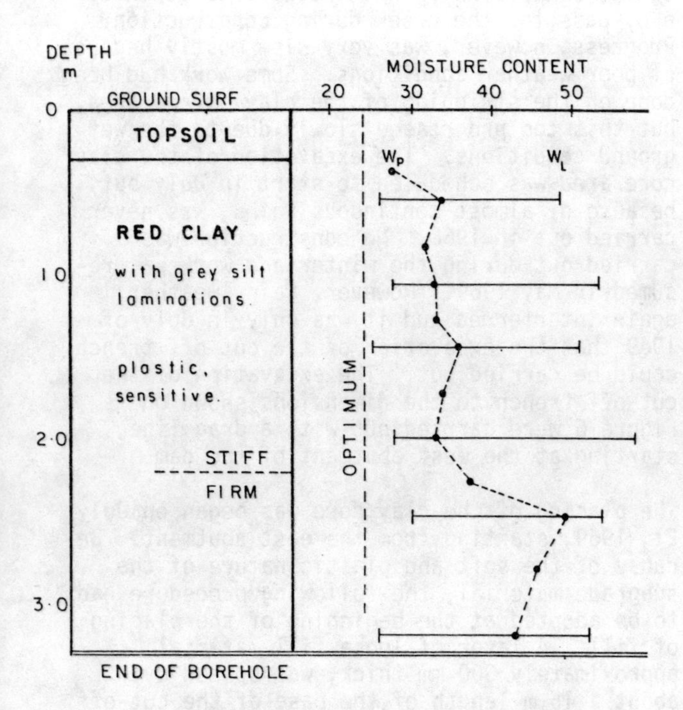

Fig. 2. Borehole log (borrow area)

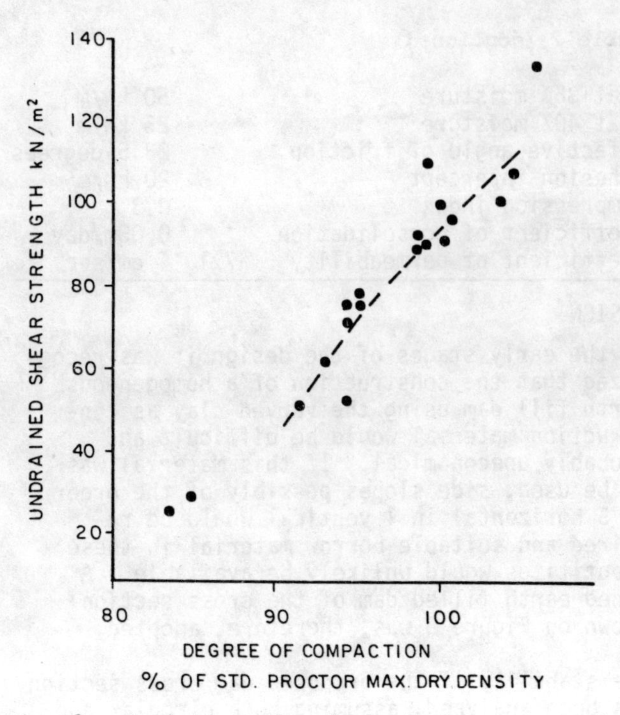

Fig. 3. Variation of undrained shear strength with degree of compaction

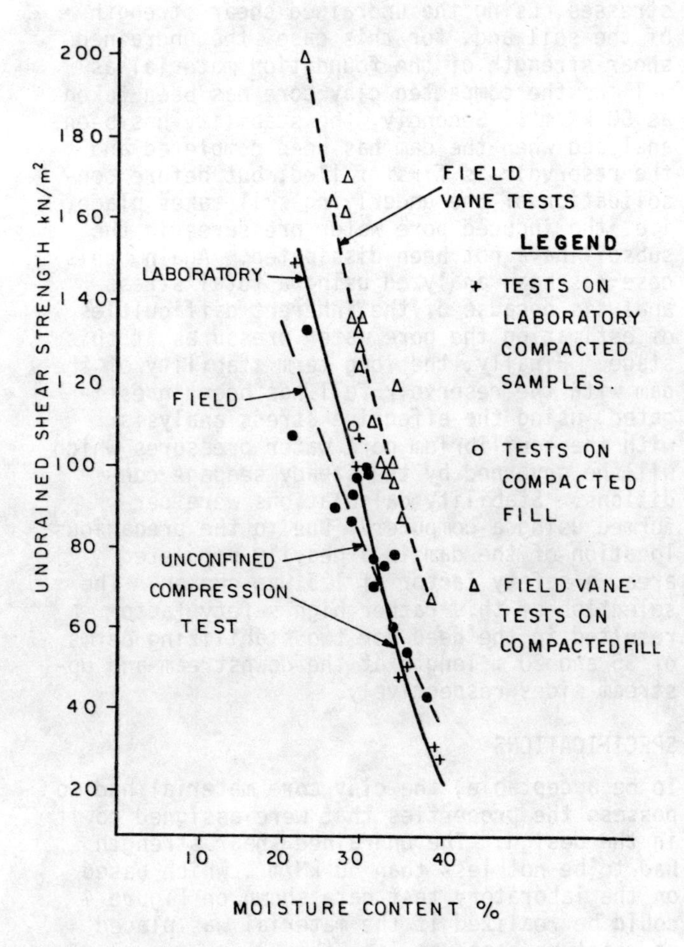

Fig. 4. Variation of shear strength with moisture content (material in place)

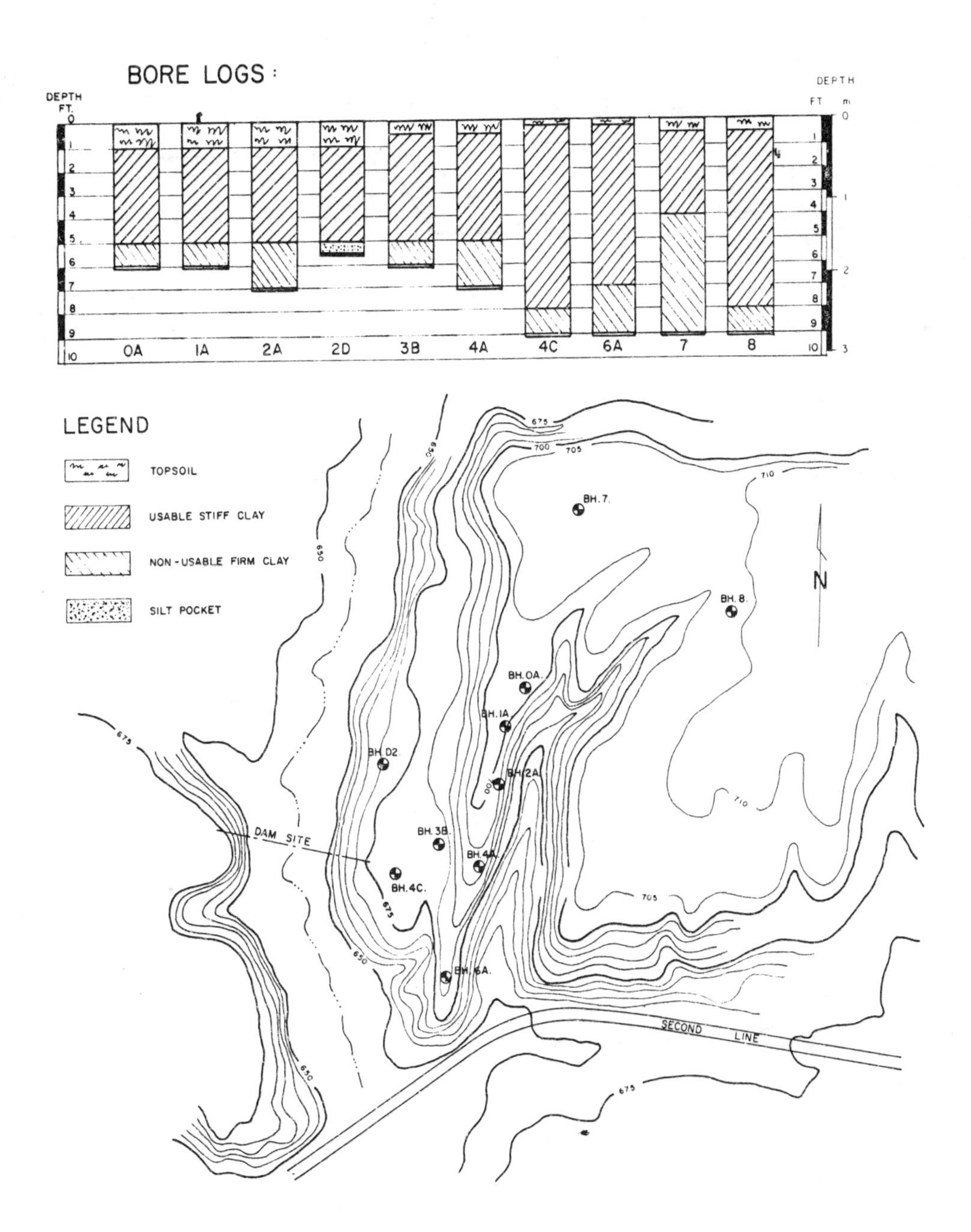

Fig. 5. Bore logs and location of boreholes (borrow area)

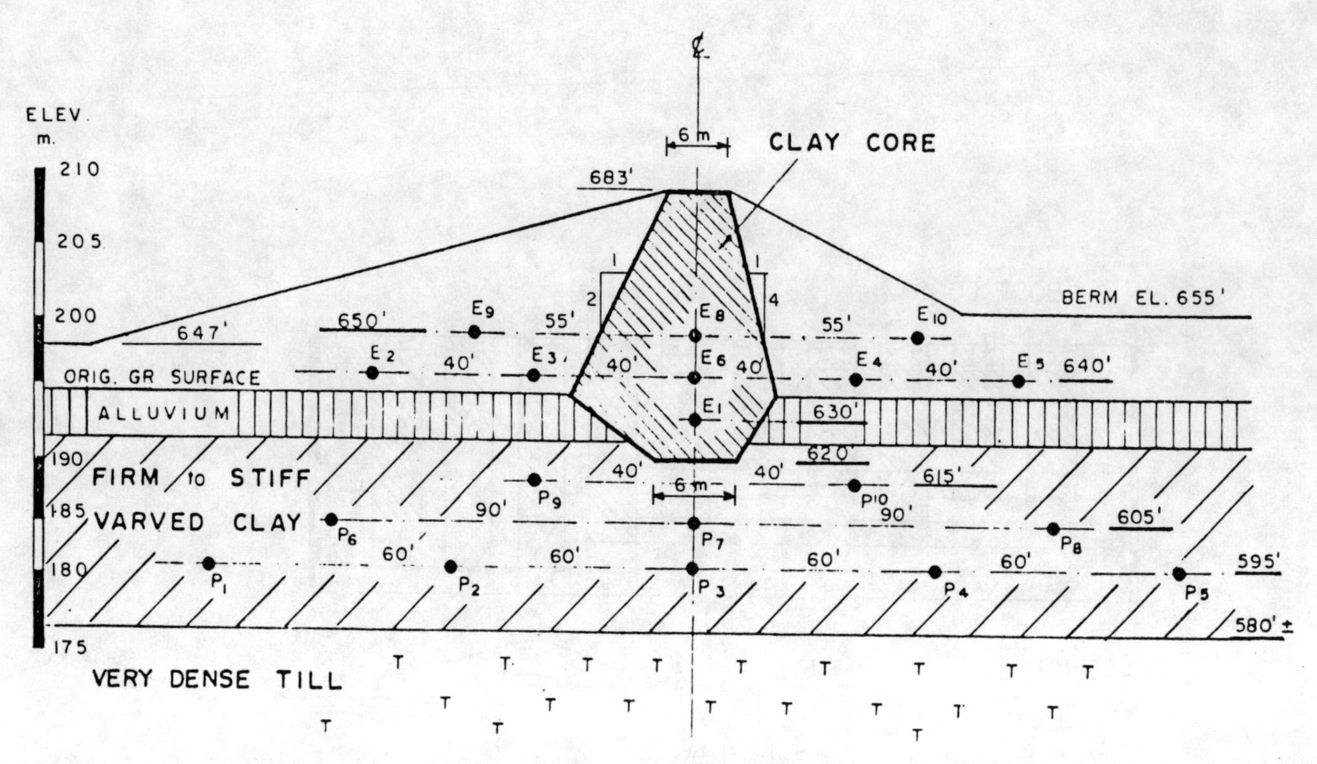

Fig. 6. Cross section of dam showing piezometer installations

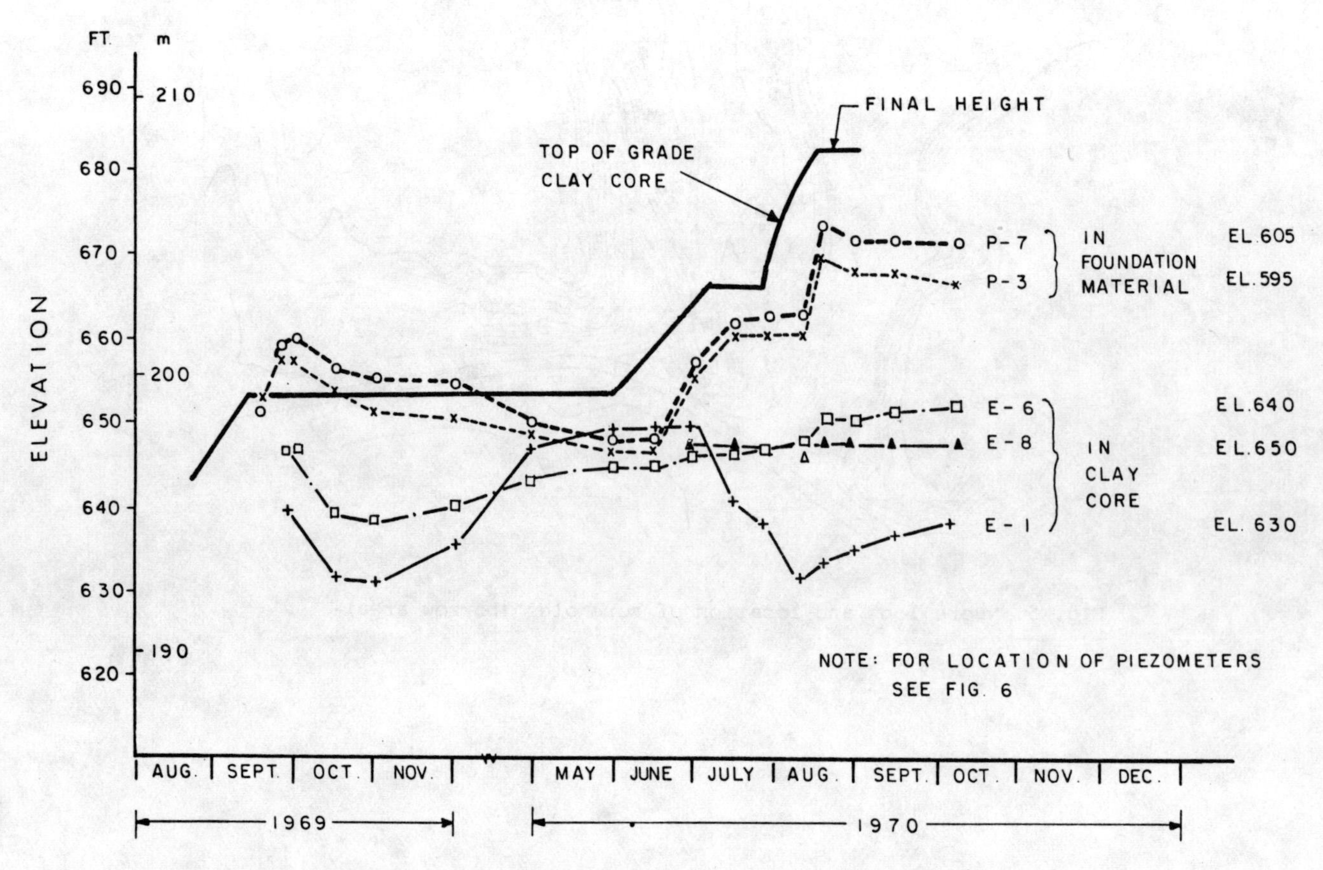

Fig. 7. Piezometer readings

was then placed over the top of the first layer and the procedure was repeated until a layer approximately 0.6 to 1 m thick was built up at which stage it was possible to place a small sheeps-foot roller over the area which has been filled. A field density test carried out on this material indicated 83% of compaction and a shear strength of 25 kN/m^2, considerably lower than the 50 kN/m^2 assumed in design. When a reasonable working base had been established at the east end of the cut-off trench, the drag lines were employed to clean off the base of the cut-off trench to the west of the advancing front of the fill and the same process of establishing a working base was carried through until the western abutment was met. It was at the end of the second day that a relatively uniform base had been established right through from the east to the west abutment. At this time, the thickness of fill placed was at an average about 1 m and the base was rigid enough to support a D7 dozer pulling a large sheeps-foot roller. At the edges of the excavation of the cut-off trench, there was a zone about 150 mm thick of material which had become desiccated due to excessive drying out. This layer was removed with a small bulldozer and the sheeps-foot roller was worked tight against the face so that the fill and natural subsoil were well knit together.

Initially, the roller was directed to make 12 passes over the entire surface at which stage the fill could be seen to be visibly knit together and to be stiff. Field density tests carried out on the first well compacted lift indicated between 98 and 101% degree of compaction and the natural moisture content was between 23 and 26%. Field vane tests indicated undrained shear strength values of 130 to 170 kN/m^2. From these tests it was then inferred that approximately 10 passes will be sufficient to achieve the desired degree of compaction. As the placing of the fill progressed, it was noted that the density was increasing noticeably and this was attributed to the fact that an increasingly hard base was being built up, giving something against which one could compact. Vane tests carried out in the fill indicated undrained shear strength values of 150 to 200 kN/m^2. Following these observations it was felt, that the number of passes could be reduced to eight, which was then maintained as a minimum throughout the construction.

During the placing of the fill, it was noted that in general the material coming from the borrow area was becoming more silty. A check of the borrow area indicated that occasional larger lenses of silt are present and an inspector was placed in the borrow area to assure that this material is removed and wasted and is not included in the fill delivered to the dam site.

Construction progressed well and by August 4, 1969, the clay core was built up to Elevation 195 ft. or about 1.5 m above the original grade. At this stage, the placing of further fill was temporarily halted to allow the installation of the control piezometers. A total of 20 piezometers were placed; ten into the firm varved clay foundation material, three in the clay core, and seven in the granular shell. The location of the piezometers is shown in Figure 6. The installation of the piezometers was completed in early September and the clay core was raised to Elevation 199 m by the second half of September. This was followed by the placing of the granular shell and by the time further clay core could be placed, bad weather set in, preventing the placing of more fill before early frost and cold weather necessitated the shut-down of the construction for the winter.

Work on the earth fill was resumed again on June 1, 1970. An inspection of the condition of the clay core indicated that the top 300 mm of the compacted clay fill had deteriorated during the winter. Moisture contents of 35 to 42% were measured with corresponding shear strength values of 30 to 40 kN/m^2. The clay below this disturbed zone was surprisingly intact and showed no signs of damage during the winter. It was decided to remove the affected zone, as its recompaction in place due to the high moisture content was not considered to be possible. Fill was placed throughout the entire month of June and by the end of the month the clay core was at Elevation 202.5 m. At this time, however, the piezometers installed in the foundation material registered high pore pressures, especially near the toe of the dam. A computer check on the stability of the dam using the effective shear strength parameters indicated a safety factor of about 1.05 and since the contractor was able to keep active elsewhere on the project, it was decided to allow for some pore pressure dissipation befor placing more fill. By July 28, the pore pressures under the granular shell had dropped and work on the dam was continued. Because of good sunny weather, construction progressed rapidly, and by August 15 the clay core had reached its final level at Elevation 208.2 m. The placing of the granular shell, however, once more had to be interrupted due to a further significant increase in the pore pressures and a theoretical safety factor of 1.02. The piezometers were monitored daily and by September 21 the pore pressures were at a sufficiently safe level that work on the granular shell and the placing of the rip-rap could be resumed. All work on the dam was completed by December 1970.

For the work in the borrow area, the contractor employed two tractors pulling a farmer's disc and harrow to rip and break up the clay for drying. The depth penetrated was about 150 to 200 mm and during one single warm and sunny day the moisture content of this layer dropped sufficiently from an in-situ value of about 40% to a moisture content of about 30 to 34%. The dried out material was then removed by a bulldozer assisted scraper which carried the material either to the dam site for placing and compaction or to a stock pile where the dried out material was stored for future use.

QUALITY CONTROL

Quality control was carried out at the borrow pit area and at the dam site. At the borrow pit, the exposed material was visually inspected and if it was found to be too silty, it was removed and wasted. At the beginning of the contract, moisture contents were frequently taken to determine the in-situ moisture content of the material. At the beginning, the moisture content of the material in the crust was found to be of the order of 35% but, as the desiccated crust was depleated, the underlying intact material, which was originally condemned, was also used when weather conditions permitted the drying out of this material. At the beginning of the contract, the moisture contents were frequently checked, but with time the inspector developed a feel for the clay and was able to determine visually the moisture content of the clay within 1 or 2%.

At the dam site, in addition to visual inspection and observations, control tests consisting of in-situ density tests and field vane tests were performed. Undisturbed samples were also recovered for unconfined compression tests in a field laboratory set up at the site. At the beginning of the contract, tests were made frequently and later, when already sufficient data was available, the quality control consisted mainly of visual and tactile examination of the material and in-situ vane tests. Since there was good correlation between moisture content and vane tests as well as the degree of compaction and undrained shear strength, it was possible to estimate from the vane test results the moisture content of the clay as well as the degree of compaction. Representative test results are plotted on Figures 3 and 4. Nearly 200 tests were performed which gave an average dry density of 14.65 kN/m^3 in place at an average moisture content of 30%. The unconfined compression tests gave an average undrained shear strength of 85.3 kN/m^2 and the field vane tests indicated shear strengths generally 50% higher. The average degree of compaction was 98%. Test pits dug in the clay core indicated a good homogeneous structure without any sign of layering.

Piezometers were installed both in the foundation material and the embankment in August 1969 and these were monitored on a daily basis when fill was placed and at least once a week during the winter when no work was carried out. While it is not the intention of this paper to discuss the piezometer installations and readings in detail, Figure 7 shows typical piezometer readings carried out in five piezometers. Three of these piezometers (E-1, E-6, E-8) were installed in the lower half of the clay core. Two other piezometers shown on this figure (P-3, P-7) were installed in the foundation material, the firm red clay below the core. A reference to Figure 7 indicates that the piezometers installed in the foundation material responded immediately to construction activities. Usually, the placing of the fill was followed by an immediate and proportionate rise in the pore pressures. When construction was stopped, the piezometers registered an immediate pore pressure dissipation. In contrast, piezometers installed in the clay core showed a pore pressure change which was inversely proportionate to the placing of the fill. Immediately after the fill was placed, the pore pressures dropped significantly and increased steadily during prolonged inactive periods of construction. Particularly piezometer E-1 behaved in this manner, while the other piezometers in the clay core showed a lesser response to construction activity. The observations made in piezometer E-1 are consistent with the behaviour of non-saturated clays in which shear deformations result in the development of negative pore pressures. A subsequent increase in the pore pressure signifies the dissipation of these negative pore pressures and the adjustment of the clay to the long term stress conditions.

CONCLUSION

Experience related in this paper indicates that highly plastic, sensitive clay materials found in Northern Ontario can be successfully compacted at moisture contents 6 to 10% above the optimum. The compacted material will exhibit homogeneous texture, sufficient strength and low permeability for dam core construction. Good correlation exists between moisture content - degree of compaction - shear strength - of this soil type and the quality of the fill can be controlled by simple and quick in-situ vane tests.

ACKNOWLEDGEMENT

The author would like to extend his thanks to the owner, the Sault Ste. Marie Conservation Authority, and Proctor and Redfern Limited, consulting engineers for this project, for their permission to publish this paper.

REFERENCES

1. WU T.H. Geotechnical Properties of Glacial Lake Clays. Journal of the Soil Mechanics and Foundations Divisions, A.S.C.E. 84 SM3 Aug. 1958
2. SKEMPTON A.W. The Collodial Activity of Clays. Proc. 3rd. Intern. Conf. on Soil Mech. and Found. Eng. 1953, I. 57-61

M. P. MOSELEY, BSc, FGS, Director, and
B. C. SLOCOMBE, MSc, Engineer, Ground Engineering Division, GKN Keller Foundations

In-situ treatment of clay fills

Uncompacted clay fills present very difficult and expensive foundation problems and this condition is a common feature on sites formerly used for brick pits or some similar form of mineral extraction. The paper describes two sites where clay fills have been treated by geotechnical processes in order to achieve acceptable bearing capacity and settlement criteria. Soil and load test data are included to define the soil conditions and show the effect of the ground treatment. In particular, the paper describes the application of Dynamic Consolidation on a large site in the English Midlands, where an abandoned clay pit had been filled without significant compaction control. The area was successfully treated by Dynamic Consolidation prior to a major local authority housing development.

GROUND IMPROVEMENT

To function properly, every structure must be provided with suitable foundations and over the years various construction techniques have been evolved to meet this basic requirement. The essential problem in foundation design is to select the most economical method of supporting applied loads consistent with adequate safety margins and acceptable settlements. In most cases, the optimum solution occurs when the ground conditions permit some form of shallow foundation arrangement. However, in cases where sites are covered by fills or soft or loose natural soils, shallow foundations become impractical and in the past the only solution in such cases has been to adopt the costly alternative of transferring the applied loads down to strong soils either by piling or deep excavation.

During the past decade, ground improvement techniques have been increasingly used to solve this problem; many poor quality soils and fill materials may be treated in-situ to improve their bearing capacity and settlement characteristics, thus enabling the use of conventional lightly reinforced shallow foundations.

When considering clay fills for the application of ground improvement techniques it is vital that attention be paid to the age of the material, its past history and the presence of the ground water table, in addition to the normal geotechnical properties. The time dependent nature of self weight settlement in clay fills is well known but little documented, and more data is urgently required. Similar observations would apply to ground water table movements. The need for accurate and comprehensive site investigation work prior to assessing clay fill sites cannot be over emphasized.

GROUND IMPROVEMENT BY VIBRO COMPACTION

Vibro techniques were first evolved in the early 1930s to compact loose sands in Germany and were first introduced to the United Kingdom about 20 years ago. The basic tool is essentially a high powered poker-vibrator where vibrations are produced in a horizontal plane by an eccentric weight assembly enclosed in a heavy tubular casing. The nose of the vibrator is tapered to aid penetration of the ground and jet nozzles emerge from the main housing just above the vibrator tip. Depending on the prevailing ground conditions, air (Vibro Replacement) or water (Vibro Compaction) may be used as the jetting medium.

Detailed description of the use of Vibro Compaction techniques may be found in a paper by Greenwood (1970) whilst useful case histories have been presented by West (1975) and Engelhardt, Flynn and Bayuk (1974).

In the treatment of clay fills the vibrator is used as a means of forming a hole through the weak material. After reaching the required depth it is withdrawn and a small quantity of graded stone aggregate is tipped into the hole. The vibrator is lowered again to compact the infill and interlock it tightly with the surrounding soils. This cycle is repeated until a compact stone column is built up to ground level. When soils conditions require the use of the water jetting medium or "wet" process the vibrator is rapidly surged up and down to allow the high pressure jetting action to remove the soft materials prior to the construction of the stone column.

Treatment by Vibro techniques in granular soils results in physical compaction between the positions of the stone columns. This is not the case for clay materials since the fine-grained nature of the soils constituents

Clay fills. Institution of Civil Engineers, London, 1978, 165-168

inhibits and dampens the action of the horizontal vibrations. Here, treatment results in reinforcement of the clay-type soils, the clay providing passive resistance to permit the stone column to develop its high bearing capacity relative to the surrounding soil. This composite structure provides an overall increase in bearing capacity and at the same time the stiffness of the columns reinforces the stabilised clay, markedly reducing the amount of consolidation movement that can occur.

Modern warehouse systems require the provision of large floor areas capable of supporting high average loads and yet with low settlement tolerances. Recently Vibro Replacement techniques were used to improve the ground conditions beneath a warehouse with an average sustained floor load of 55 kN/m^2. Treatment was also performed beneath the structural foundations; column loads were in the range of 25 to 50 tonnes.

The site investigation boreholes revealed two conditions, namely, natural soils composed of firm and stiff clays or backfilled opencast workings resting on weak shale. The backfill to the workings, which was at least 20 years old, was composed of a variable mixture of silty clay, with sandstone, shale, and coal fragments, all generally soft to firm and firm in-situ but locally firm to stiff and very friable in places. This fill material extended to depths of up to 8.0 m below ground level, and allowing for side slope batters during coal excavation occupied about one-half of the building area. The index property tests showed the clay fill to have liquid limits in the range 35 to 54 and plastic limits 16 to 20 (Casagrande Symbol CI to CH). Moisture contents varied between 16 and 35%. Values of coefficient of volume decrease of about 0.0003 sq m/kN were obtained from consolidation tests, and undrained triaxial compression tests gave values of c_u in the range 39 to 68 kN/m^2.

Settlement control was a prime criterion in the design of this distribution depot, and it was evident that the existing soil conditions would result in high total and differential settlements of both floor slab and structure. In these circumstances, ground improvement of the fill material by Vibro Replacement was adopted. Stone columns were installed beneath the floor slabs at centres varying between 2.0 and 2.25 m, dependent on the depth of fill being treated, whilst closer centres were used beneath the structural foundations. Treatment was provided to the full depth of the fill materials.

In addition to normal testing, a large scale zone loading test was performed on a base 4.0 m x 4.0 m in plan. The test position was selected to examine the worst site condition, and the base was loaded to three times working load (ie a pressure transmitted to the soil surface of about 165 kN/m^2.) The result of this test, together with a site plan, and typical borehole logs, are presented in Fig 1. In particular, the settlement of the base at working load was measured at 10 mm, the rate of settlement over the final five hours of load application falling to 0.02 mm per hour. This value of settlement indicates a marked degree of improvement when compared to an estimated pretreatment value for a similar base of about 50mm.

GROUND IMPROVEMENT BY DYNAMIC CONSOLIDATION

Dynamic Consolidation was developed in France by Techniques Louis Menard and introduced to this country in 1973. In this system the ground is compacted by repeated surface tamping using heavy steel and concrete weights. Typically the tamper weighs between 10 and 20 tonnes and is dropped in free fall from heights of up to 25 m to provide treatment to considerable depth. Although the system is particularly effective in granular soils it can be applied to a wide range of soil classifications.

In low permeability silty soils the high stresses induced by the high energy tamping cause a compression of the voids within the soil structure simultaneously generating high water pressures. In this state of partial liquefaction preferential drainage paths are created and these allow the excess pore water pressures to dissipate much more rapidly than would be the case under static loading. Consolidation takes place as the pore water pressures dissipate and therefore by carrying out the tamping in a number of carefully phased passes it is possible to compact and strengthen many fine-grained low permeability soils as well as virtually all coarse granular materials.

Detailed description of the use of Dynamic Consolidation techniques may be found in a paper by Menard (1972) whilst useful case histories have been presented by West (1975) and West and Slocombe (1973).

Housing developments on filled sites require careful attention to their foundation design. Ground improvement techniques can frequently provide the most economic solution to this type of foundation problem. On a housing site in the Midlands, Dynamic Consolidation techniques were used recently to improve the engineering properties of freshly placed and old fill materials.

The 3ha site was occupied by an old clay pit up to 6.0 m deep. The base of the pit consisted of irregular "dumplings" of clay waste with a works railway embankment running down the centre of the site. The pit was backfilled by end-tipping and lightly tracking in materials obtained from a nearby road excavation, after first having pumped away any water that had collected in the lower areas. The material used to backfill the pit was composed of a highly variable mixture of clays, shales, limestone, ash, sand and pottery fragments. A 500 mm thick carpet of granular material was provided to ensure continuity of operations and for use in backfilling the large craters induced by the localised very heavy tamping.

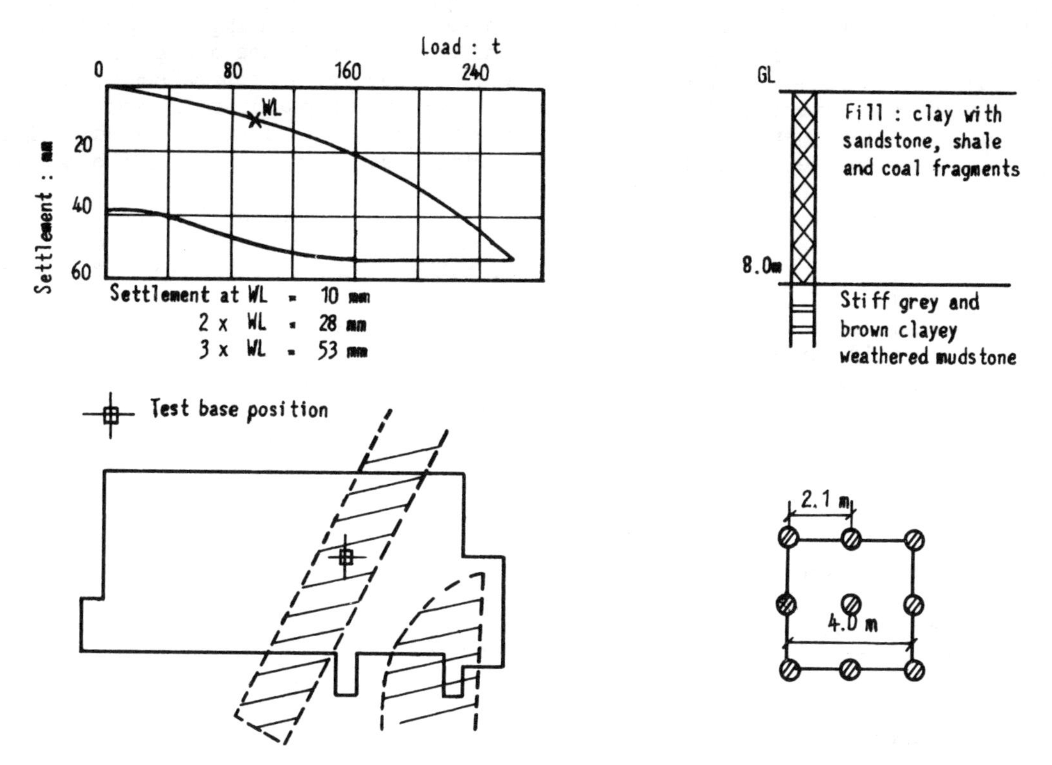

Fig. 1. Distribution depot: plan, soils conditions and load test results

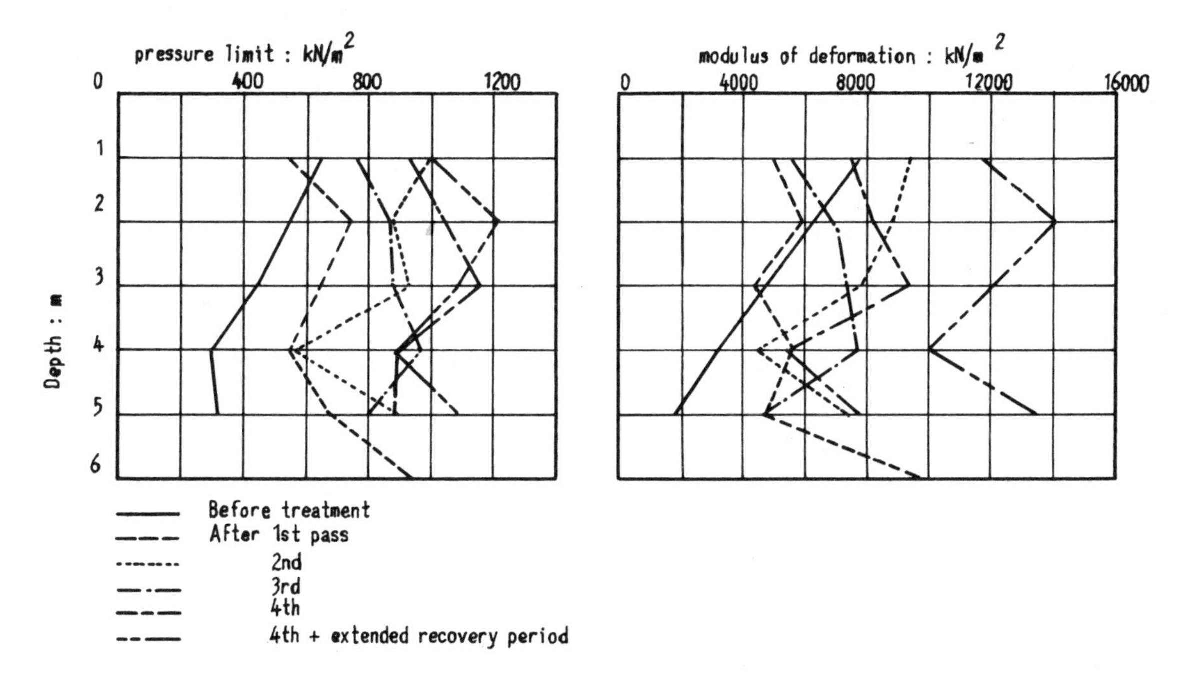

Fig. 2. Housing development; comparison of improvement to successive 1.0 m layers with consecutive tamping passes

Laboratory tests on the clay fill materials revealed liquid limits in the range 27 to 37, plastic limits 16 to 18 (Casagrande Symbol CL and CI) whilst moisture contents were found to lie in the range of 14 to 18%. Pressuremeter tests performed before treatment to determine the properties of the soils in-situ revealed pressure limits ranging between 130 and 700 kN/m^2 with moduli ranging between 1400 and 10400 kN/m^2, the soils generally becoming weaker with depth.

Trial pit examination of the soils revealed the old clay fill materials to be dessicated at the surface and to locally contain large limestone boulders. This investigation also revealed the fill in the lowest part of the site near to a pond to be of a significantly lower quality where ground water had softened the clay infill.

The general area is to be subject in the near future to underground mining and medium mining rafts designed to 110 kN/m^2 were specified. The layout of the proposed 2-storey local authority housing consisted of long irregularly shaped blocks. Dynamic consolidation was performed to permit the construction of these blocks anywhere within the treatment area, irrespective of the areas of new or old fills, and restricting foundation settlements to less than 25 mm.

In view of the predominantly fine-grained nature of the soils, treatment was performed by several tamping passes, each carefully phased to permit dissipation of the high pore water pressured induced by the localised heavy tamping before performing the next tamping pass. In view of the poor quality fill materials in the low-lying area some of the tamping passes were repeated or divided in order to provide a satisfactory engineering performance of the treated clay fills.

Continual monitoring of the reaction of the soils was performed during the treatment operations including intensive pressuremeter testing. A comparison between the results obtained before treatment and after an extended recovery period revealed an improvement by a factor of at least 2 for both the pressure limits and moduli.

In addition, an analysis was performed to provide an indication of the improvement of soils properties to successive 1.0 m layers with consecutive tamping passes. This analysis was independent of the recovery period which would have a significant effect on the results, in particular the pressuremeter modulus, due to thixotropic effects. The results of this analysis are given graphically in Fig 2 which shows a particular improvement in the properties of the soils at depth as a direct result of the first two tamping passes. These two tamping passes had been designed specifically to improve the soils at depth, the third to further improve these together with improving the middle layers whilst the fourth was designed to improve the surface layers, it can be seen from Fig 2 that this was readily achieved. The improvement of the soils with time is also readily illustrated, particularly with respect to the modulus.

The Dynamic Consolidation treatment induced a surface settlement of the site of about 250 mm or 6.25% of the average depth of fill materials.

A zone test was performed on a base 2.0 m x 2.0 m in plan, the base being loaded to 1.7 times working load (ie a pressure transmitted to the soil surface of about 185 kN/m^2). The settlement of the base at this pressure was measured at 3.9 mm, the rate of settlement over the final 23 hours of load application falling to 0.015 mm per hour and remaining steady for 2 hours.

It is of interest to note that the pressuremeter test results in the area of the test base had predicted a total settlement of 4.0 mm. Calculations using the modulus results obtained from the pressuremeter tests that were performed after allowing sufficient time for the soils to recover predicted a maximum total long-term settlement of foundations of the order of 10 mm.

CONCLUSION

The two case histories presented in this paper show that Ground Improvement techniques can be used to improve clay fill materials. Further research allied to measurement of real soil conditions is needed however to further extend our understanding of the processes and to lead to more rigorous design methods.

REFERENCES

ENGELHARDT, K. FLYNN, W.A. and BAYUK, A.A. (1974). A method to strengthen cohesive soils in situ. American Society of Civil Engineers National Structural Engineering Meeting, Cincinnati, Ohio, Meeting Preprint 2281.

GREENWOOD, D.A. (1970). Mechanical improvement of soils below ground surface. Proc. Conf. Ground Engineering, Institution of Civil Engineers, London 11 - 22.

MENARD, L. (1972). The dynamic consolidation of recently placed fills and compressible soils. Application to maritime works. Travaux No 452.

WEST, J.M. (1975). The role of ground improvement in foundation engineering. Proc. Conf. Ground treatment by deep compaction, Institution of Civil Engineers, London 71 - 78.

WEST, J.M. and SLOCOMBE, B.C. (1973). Dynamic Consolidation as an alternative foundation. Ground Engineering 6, No 6, 52 - 54.

A. W. PARSONS, Transport and Road Research Laboratory

Moisture condition test for assessing the engineering behaviour of earthwork material

A test has been developed for the rapid measurement of a "moisture condition value" (MCV) of earthwork material. It is intended as a means of assessing the suitability of earthwork materials for use in embankments while avoiding the measurement of moisture content with its associated delays. It has been found that the performance of earthmoving plant can be related to MCV, and relations between undrained shear strength of remoulded soil and MCV have been established. There are strong possibilities that CBR can be predicted from MCV and that a new method of soil classification can be based on the characteristics of the moisture condition calibration line (the straight line relation between moisture content and MCV). The possible applications of the moisture condition test in the prediction of the relative effects of wet weather on different soils and the degrees of difficulty of compacting different soils are also discussed.

INTRODUCTION

1. This Paper describes the principles and methods of measuring a "moisture condition value" (MCV) of soil. The test provides a rapid and reproducible means of assessing the suitability of earthwork material in relation to specified limits of moisture content.

2. The control of the quality of earthwork material in highway construction has often proved difficult because the parameters normally used for such control (refs. 1 and 2) (moisture content, moisture content to plastic limit ratio, moisture content minus BS optimum moisture content (2.5 kg rammer method) (ref. 3)) do not relate uniquely to the engineering quality of the material within the variations of soil type encountered on an earthwork construction site. Adding to this problem, further difficulties arise from the delay associated with the measurement of moisture content and, when applicable, the lack of reproducibility of the plastic limit test (ref 4). The moisture condition test has been developed, therefore, with the objectives of providing an immediate result, of being applicable to a wide range of soil types, of being usable on site or in the laboratory, of eliminating as far as possible the effect of the human factor (operator error), and of utilising a sufficiently large sample to minimise variability associated with sampling. Evidence is put forward to suggest that MCV can be a useful indicator of the engineering quality of the soil and of other aspects of its behaviour.

PRINCIPLE OF THE TEST METHOD AND THE APPARATUS

3. The test basically consists of determining the compactive effort necessary, in terms of the number of blows of a rammer, to fully compact a sample of soil (ref. 5). It is well known that relations between density and moisture content produced by different compactive efforts tend to converge as the moisture content increases (Fig 1). At any given moisture content a comparison of the densities produced by two compactive efforts will indicate whether the lower compactive effort was sufficient to produce full compaction. For example (Fig. 1), compactive effort A is sufficient to produce full compaction at moisture content A′, as no further increase in bulk density can be achieved when using compactive efforts B and C. Similarly, compactive effort B is sufficient to produce full compaction at moisture content B′, as no further increase in bulk density can be achieved when using compactive effort C. The higher the moisture content of the sample the lower the compactive effort (eg. the number of blows of a rammer) beyond which no further increase in density occurs. The apparatus developed for the test has been designed so that the penetration of a rammer into the mould can be measured, thus avoiding the determination of density.

4. The apparatus (Fig 2) (ref 5) has a 100 mm internal diameter mould with a detachable base, a free falling rammer with a mass of 7 kg and a diameter of 97 mm, and an automatic release mechanism with adjustable height of drop. A sample of soil of 1.5 kg is normally used, together with a height of drop of 250 mm. The penetration of the rammer into the mould can be determined from a scale attached to the rammer. The height of drop is kept constant with the aid of a vernier scale, one half of which is attached to one of the lower guides of the rammer and the other half to a rod connected to the adjustable cross member containing the rammer release mechanism. Pollution of the rammer and extrusion of soil between the rammer and the sides of the mould are avoided by placing a lightweight disc of laminated phenolic sheet, 99 mm in diameter, on top of the soil in the mould. To eliminate, as far as possible, the effects of using the apparatus in varying conditions, eg. on soft soil on site or on the

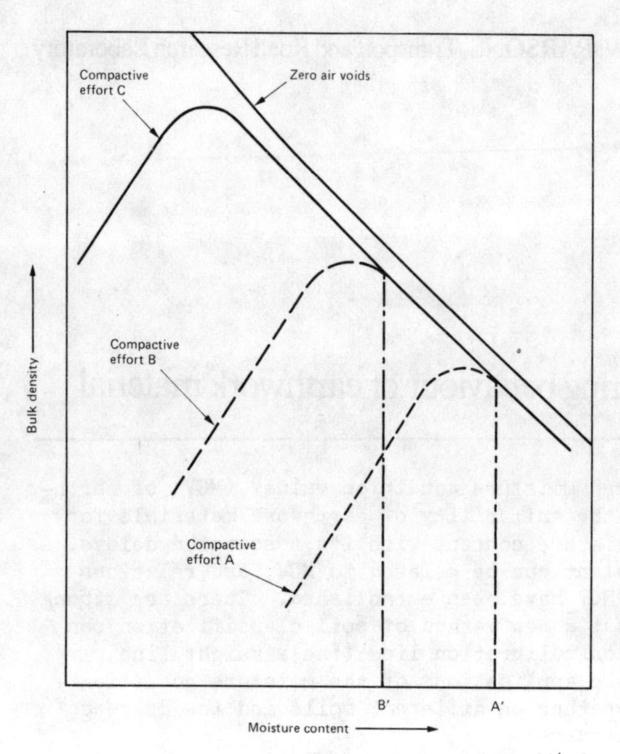

Fig. 1. Idealised relations between bulk density and moisture content

Soil: Heavy clay   Moisture content: 26.3 per cent

Number of blows of rammer 'n'	Penetration of rammer into mould (mm)	Change in penetration between 'n' and '4n' blows of rammer (mm)
1	41	33.5
2	57.5	33
3	67	33.5
4	74.5	26.5
6	84	17
8	90.5	10.5
12	100.5	0.5
16	101	
24	101	
32	101	
48	101	

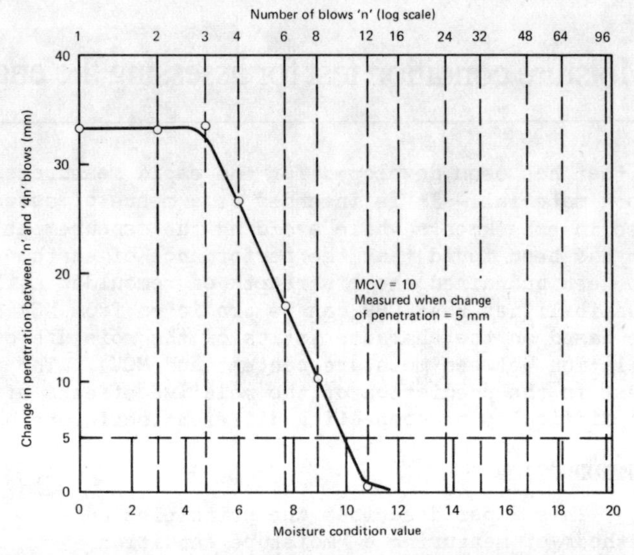

Fig. 3. The determination of the moisture condition value of a sample of heavy clay

Fig. 2

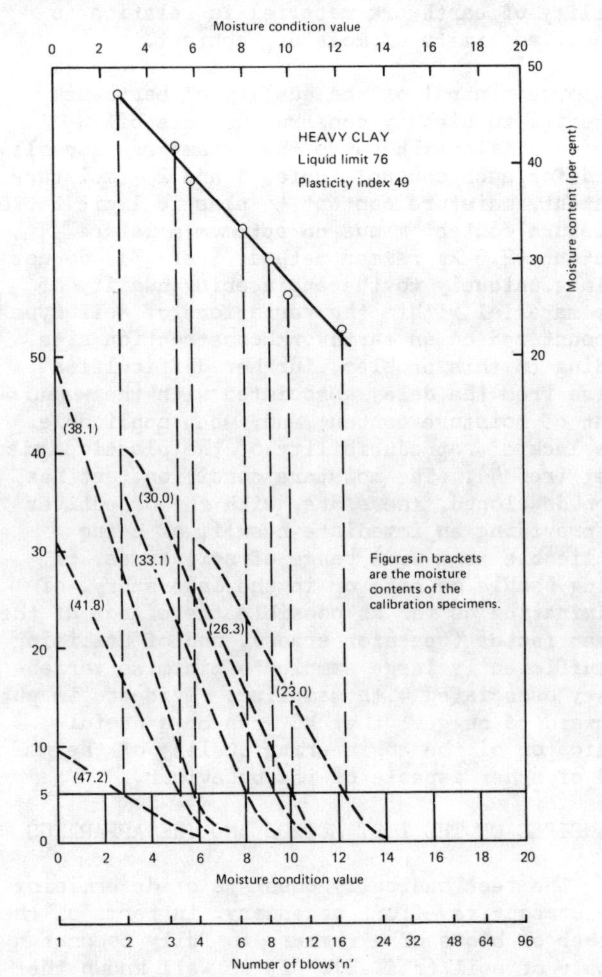

Fig. 4. Moisture condition calibration of a heavy clay soil

concrete floor of a laboratory, the base of the apparatus has been made as heavy as possible while maintaining overall portability. Thus, the total mass of the apparatus is about 50 kg, of which 31 kg is contributed by the base.

CALIBRATION AND USE OF THE MOISTURE CONDITION TEST FOR THE ASSESSMENT OF SUITABILITY OF EARTHWORK MATERIAL

5. To determine the MCV of a sample of soil the penetration of the rammer into the mould is measured at various stages of the compaction of the soil (Fig 3). The penetration of the rammer at any given number of blows is compared with the penetration for four times as many blows and the difference in penetration determined. This "change in penetration" is plotted against the lower number of blows in each case (Fig 3), the number of blows being on a logarithmic scale. To avoid predicting the point in the plotted relation when the change in penetration reaches zero, a change in penetration of 5 mm has been arbitrarily selected as indicating the point beyond which no significant change in density occurs. The moisture condition value (MCV) is defined as 10 times the logarithm (to the base 10) of the number of blows corresponding to a change in penetration of 5 mm on the plotted curve. Charts can be prepared from which the MCV can be read directly (Fig. 3).

6. To carry out a moisture condition calibration the MCV is determined at various moisture contents and the relation between moisture content and MCV determined (Fig 4). Over a substantial range of moisture contents this relation is a straight line. The limit or limits of moisture content that have been determined for the particular soil may, therefore, be translated to a MCV. For instance, for highway construction an upper limit of moisture content could be determined using soil tests to measure undrained shear strength and compressibility, taking account of the relevant features of the embankment design. Calibrations of soil types ranging from heavy clay to gravel-sand-clay (ref 5) indicate that standards normally applied for the limits of suitability of such soils in highway construction in this country are equivalent to a MCV of about 8. Thus, it is possible that a constant MCV could be used as a criterion of suitability for a range of soil types encountered on any one earthwork construction site.

7. To assess the suitability of material being used during earthwork construction the MCV of various samples can be measured as described above and the values compared with that equivalent to the designed limit of suitability. Alternatively, an even more rapid method can be used which simply determines the change in penetration of the rammer between the number of blows equivalent to the MCV of a specified limit and four times as many blows. For example, the specified upper limit of suitability commonly used in highway construction is equivalent to a MCV of 8, corresponding to 6 blows of the rammer. The change in penetration between 6 blows and a total of 24 blows would, therefore, be measured. If the change in penetration exceeds 5 mm the material would be suitable, if less than 5 mm the material would be unsuitable. The degree of suitability or unsuitability would not, however, be given by this method.

CLASSIFICATION OF SOIL

8. The moisture condition calibration of a soil produces a straight line (Fig 4) with an equation of the form

$$w = a - b\ (MCV)$$

where w is the moisture content (per cent)
a is the moisture content (per cent) at MCV = 0
and b is the slope of the line.

Thus "a" and "b" are parameters that could be indicative of the soil type. The factor "b" is, inter alia, indicative of the sensitivity of the soil to changes in moisture content, whilst "a" is an arbitrary low-strength moisture content value which could be used in the same context as the liquid limit of the soil.

9. The relations between liquidity index and MCV for a heavy clay and a sandy clay soil are shown in Fig. 5. Providing other homogeneous clay soils show similar relations the values of "a" and "b" could be used to derive indicative values of liquid limit, plastic limit and plasticity index of the soil. Clearly, the linking of the characteristics of the moisture condition calibration to soil classification is an interesting area for future research. If such research is successful, a very practical aid to classifying soil could be provided. As an indication of how such a classification might appear some general soil types have been assigned within areas of "a" and "b" co-ordinates in Fig. 6. This classification is very tentative and has been based on a limited number of calibrations. The term silt as a soil description is used to define the M-soils in the Casagrande plasticity chart (ref 6).

RELATION OF MCV WITH SOIL STRENGTH

10. The relation between undrained shear strength and liquidity index has been established by Skempton and Northey (ref. 7). The good correlation between liquidity index and MCV shown in Fig 5 suggests, therefore, that MCV would also be related to soil strength. Thus it is of interest to obtain an indication of the relation between MCV and soil strength parameters often used in embankment and pavement design.

11. During the course of laboratory work involving moisture condition calibrations of a wide range of soil types, remoulded undrained shear strength measurements were made, whenever possible, using a small hand vane in the fully compacted specimen at the completion of the determination of MCV. (The hand vane was 20 mm diameter and 25 mm long; it was pushed into the compacted specimen to a depth of about 55 mm and

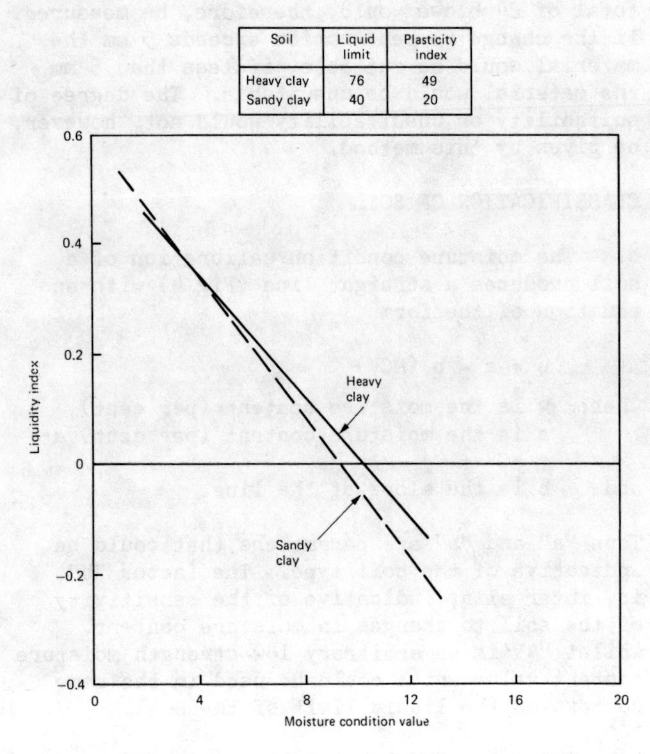

Fig. 5. Relations between liquidity index and moisture condition value

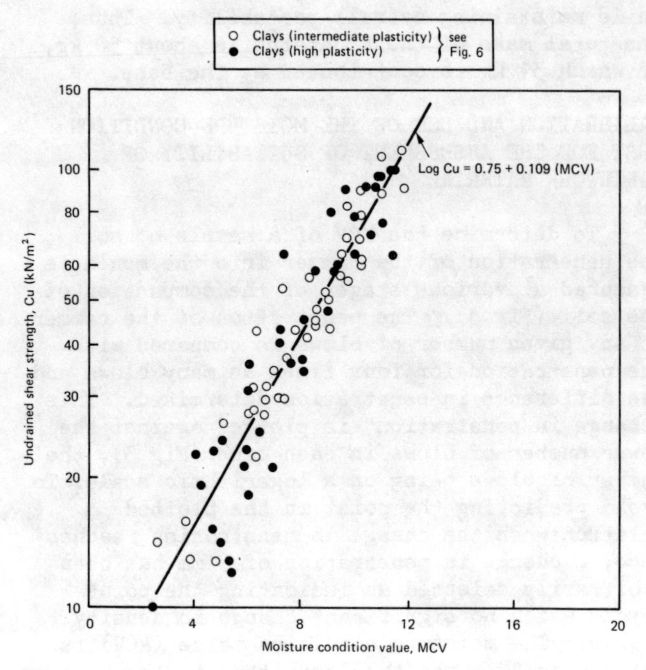

Fig. 7. Relation between undrained shear strength, determined by hand vane, and moisture condition value for clays of intermediate and high plasticity

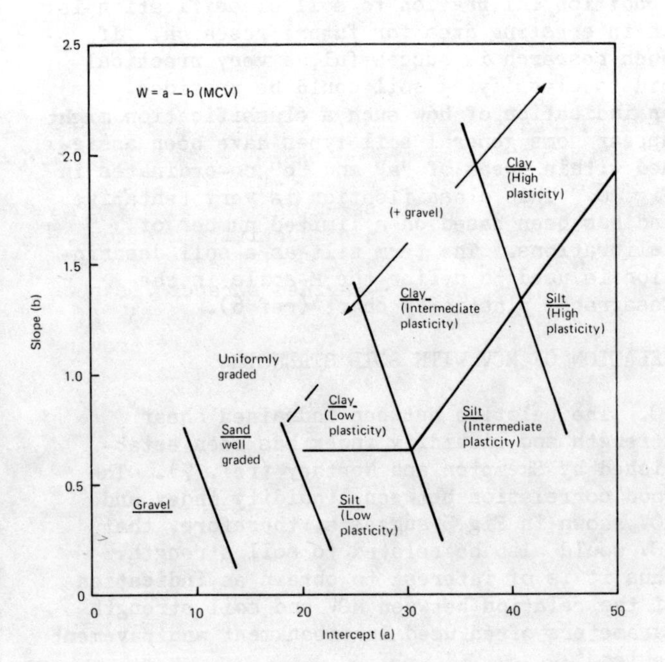

Fig. 6. Tentative soil classification based on the slope and intercept of the moisture condition calibration

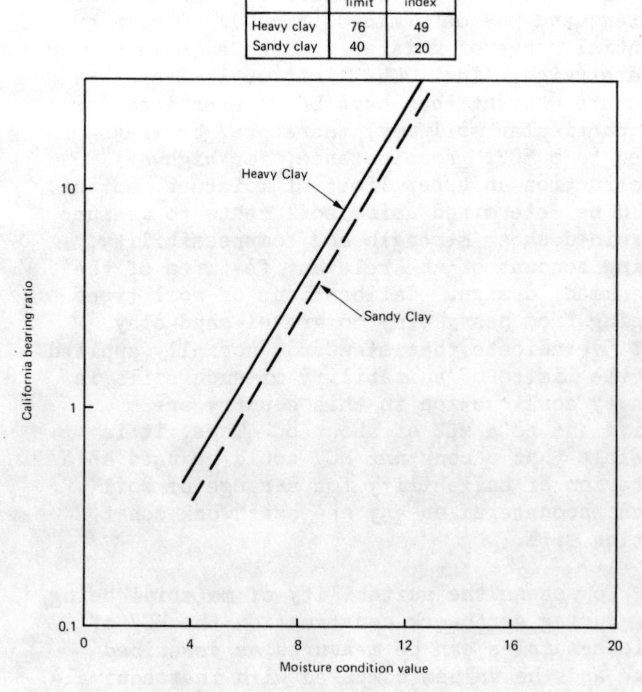

Fig. 8. Theoretical relations between CBR and moisture condition value. The CBR's were estimated from plasticity data[2]

rotated at a rate of approximately 6°/sec.) A range of measured values of the undrained shear strength was obtained at a given MCV, but good relations were obtained between the two parameters for each general soil type as determined from the slope and intercept of the moisture condition calibration line (see Fig. 6). The results for intermediate and high plasticity clays are shown in Fig. 7. A straight line relation was obtained when the undrained shear strength was plotted to a logarithmic scale. The undrained shear strength varied from about 70 kN/m^2 at a MCV of 10 to about 15 kN/m^2 at a MCV of 4 (Fig 7). The results obtained for all classes of soil are summarised in Table 1.

Table 1. Correlations of remoulded undrained shear strength (determined by hand vane) with moisture condition value

Soil type (Fig 6)	Number of results	Equation of relation *	Correlation coefficient
Clay - high plasticity	40	Log C_u = 0.74 + 0.111 (MCV)	+ 0.94
Clay - intermediate plasticity	44	Log C_u = 0.77 + 0.107 (MCV)	+ 0.96
Clay - low plasticity	14	Log C_u = 0.91 + 0.112 (MCV)	+ 0.89
Silt - high plasticity	15	Log C_u = 0.70 + 0.105 (MCV)	+ 0.97
Silt - intermediate plasticity	2	Log C_u = 0.80 + 0.100 (MCV)	-
Silt - low plasticity	4	Log C_u = 0.91 + 0.120 (MCV)	+ 0.96

* C_u = undrained shear strength (kN/m^2)

12. The results in Table 1 indicate that straight line relations between the logarithm of the undrained shear strength and MCV probably exist for each soil type amenable to measurement of shear strength. However, in the case of measurement by shear vane, at least, different relations exist for different soil types. It is possible that, with further experience, the relation between undrained shear strength for remoulded soil and the MCV could be predicted from a knowledge of the characteristics of the moisture condition calibration line (relation between moisture content and MCV).

13. With the two clay soils that were calibrated during the development of the moisture condition test (ref 5), California bearing ratios (CBR) have been predicted from a knowledge of their plasticity characteristics and moisture contents (ref 8). The relations between the predicted CBR and the MCV for the heavy clay and sandy clay are shown in Fig. 8. It appears from this that it should be possible to correlate CBR with MCV and, therefore, with sufficient experience, to use MCV as a guide to the potential CBR of remoulded soils.

PREDICTION OF THE PERFORMANCE OF EARTHMOVING PLANT

14. If the moisture condition test becomes widely adopted as a means of assessing the suitability of earthwork material, then the measurement of MCV will become routine during site investigations. It is important, therefore, to make maximum use of the results of such measurements, especially as such a wide range of soil types can be included in the test. One aspect of earthwork construction on which there is very little published data is the relation between earthmoving plant performance and the site conditions. The economics of making the maximum use of on-site soils are critically dependent on the selection of the appropriate plant for the soil conditions. However, there it little or no guidance in the literature on which the contractor can base his selection of the proper type or size of plant for particular conditions and, consequently, little guidance for the engineer in considering the contractor's proposed methods. Current research at TRRL is aimed at providing such information. An exploratory study has shown the concept to be feasible, and that MCV can be used as a measure of soil conditions

15. Figure 9 shows relations with MCV of single-pass rut depth, speed of travel on the haul road, and losses in productivity due to "bogging down" in the cut and fill areas for single-engined and twin-engined scrapers (struck capacity between 15 and 20m^3). Information of a similar nature for a wide range of earthmoving plant is being gathered at the present time.

16. A summary of this information, given in Table 2, lists limiting values (minimum) of MCV below which the two types of plant (Fig 9) would be expected to produce ruts, to operate at an "uneconomical" haul speed, and to cease to operate in the fill area. The first of these figures would indicate whether plant is likely to damage prepared formations and other graded areas, the second would indicate whether specially prepared haul roads might be required, and the last would be the prime factor in the selection of the type of plant appropriate to the soil conditions likely to be encountered. Of course, the MCV at the time of construction may be different from that measured at the site investigation. Some interpretation of the measured values may be necessary, therefore. and this aspect is discussed in para. 17. It is envisaged that relations such as those given

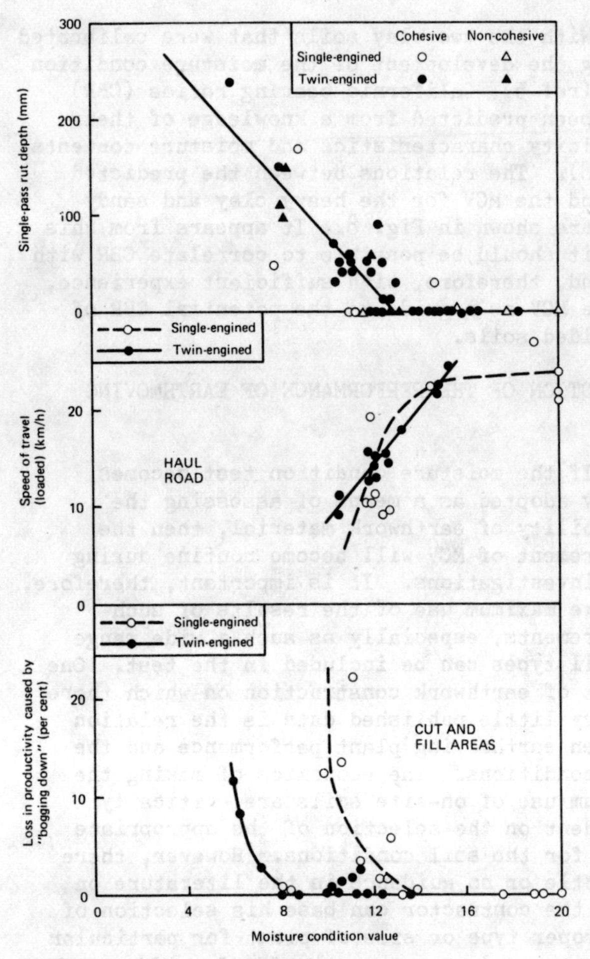

Fig. 9. Factors affecting operations with medium scrapers (15-20 m^3 struck capacity) related to moisture condition value

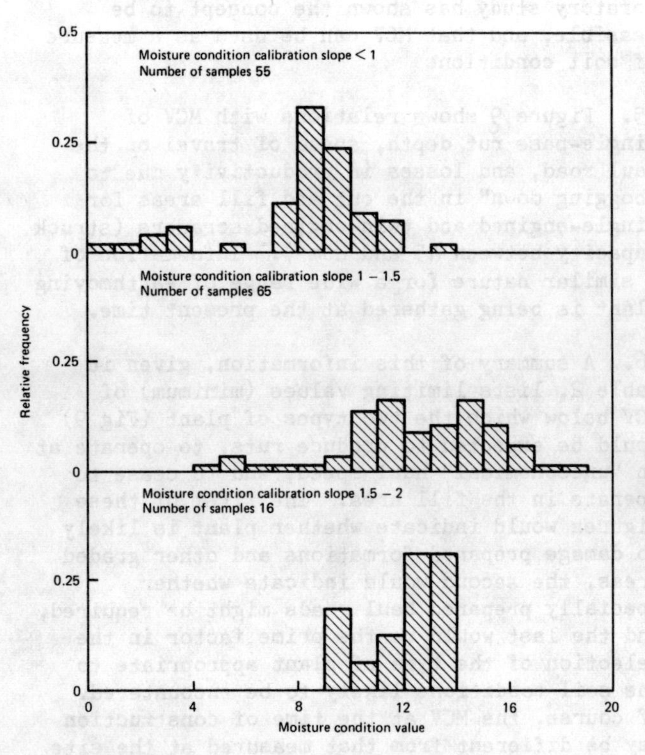

Fig. 10. Frequency distributions of moisture condition values of freshly placed fill material, classified according to the slope of the moisture condition calibration line

in Fig. 9 would find considerable application in the formulation of schemes intended to make the maximum economic use of on-site materials, particularly at the wet end of the range of soil conditions.

Table 2. Limiting moisture condition values for different criteria (single-engined and twin-engined scrapers, 15 to 20 m^3 struck capacity)

Criterion	Minimum MCV of soil: Single-engined scraper	Minimum MCV of soil: Twin-engined scraper
No ruts are formed by a single pass of a machine	13	13
Machines operate at an economical speed of travel on the haul road (not less than half of the maximum speed recorded during the studies)	12*	11*
Machines do not "bog down" in the fill area	10+	6+

* Refers to the average MCV along the haul road

+ Refers to the MCV of the fill material being placed

BEHAVIOUR OF SOIL DURING EARTHWORK CONSTRUCTION

17. The potential of the moisture condition test for providing information that can be related to soil strength and plant performance has already been discussed. Reference has already been made (para. 16) to the need for interpreting MCVs measured at site investigation stage so that potential changes in MCV before and during construction can be allowed for. Work is currently proceeding with a view to quantifying the sensitivity of soil to wet weather in terms of the slope of the moisture condition calibration line (the parameter "b", para 8). This slope is indicative of the change in strength for a given change in moisture content. A low value of "b" will indicate a high rate of change in strength with change in moisture content and a high value of "b" would be associated with a low rate of change in strength. A knowledge of the type of soil (possibly indicated by the parameters "a" and "b" taken together, paras 8 and 9) will allow an assessment of its drainage and permeability properties, thus providing a means, overall, of assessing the likely effects of wet weather on the soil behaviour. Results of recent research indicate that at the earthwork construction stage, soils with low values of "b" have the

lowest moisture condition values (Fig 10) and, therefore, are most likely to constitute "wet fills".

18. Another factor that is often of concern is the difficulty of compacting some soils which are encountered in a fairly dry condition. This is particularly important when a method specification is used for compaction (ref. 1), ie the number of passes and thickness of layer to be employed with various types of compaction plant are specified, and there is no specific requirement for the measurement of the state of compaction being achieved. The moisture condition test effectively determines the difficulty of compacting a soil sample and, therefore, might be a useful guide as to whether a compactive effort higher than that specified may be necessary in certain circumstances. Conversely, where weak soils that are easily overstressed are encountered, reduced compaction might be applied depending on moisture condition values of the soil. If the moisture condition test is adopted as a routine means of assessing the suitability of earthwork material the estimate of compactive effort required would not involve any additional testing.

CONCLUSIONS

19. Evidence has been put forward to indicate that a newly developed test, the moisture condition test, can be used as a rapid means of assessing the quality of material for earthwork construction. Because of the good reproducibility of the test and its applicability to a wide range of soil types, it has been shown to have potential in many other aspects of earthwork design and construction.

20. Thus the planning and execution of earthmoving operations can be aided by the relation of plant performance to the moisture condition value (MCV), and the designer may be assisted by the relations of undrained shear strength and CBR of remoulded soil with MCV. There are indications that a soil classification could be based on the characteristics of the moisture condition calibration, although a considerable amount of further work would be required in this area. Lastly, the possibility of predicting the effects of wet weather on a soil and the possible use of MCV to indicate where the attainment of high states of compaction may be difficult, or where there is a need for reduced compaction to avoid overstressing, have been highlighted.

21. It is clear that many areas of research can be explored, aimed at relating the MCV parameters to these factors. The acceptance of the moisture condition test as a means of control for assessing suitability of earthwork material is, therefore, of prime importance if results of such research are to be fully implemented.

ACKNOWLEDGEMENTS

22. The work described in this paper forms part of the programme of the Transport and Road Research Laboratory and the paper is published by permission of the Director. The author would like to acknowledge the assistance of his colleagues at TRRL who provided the results on which the paper is based. Thanks are also due to the Directors of the RCUs for permission to visit various road construction sites, and to the Resident Engineers, Contractors, and the members of their staff for their co-operation.

REFERENCES

1. DEPARTMENT OF TRANSPORT. Specification for road and bridge works. HM Stationery Office, London, 1976.
2. DEPARTMENT OF TRANSPORT. Notes for guidance on the specification for road and bridge works. HM Stationery Office, London, 1976.
3. BRITISH STANDARDS INSTITUTION. Methods of test for soils for civil engineering purposes. British Standard BS 1377:1975, The Institution, London, 1975.
4. SHERWOOD, P T. The reproducibility of the results of soil classification and compaction tests. Ministry of Transport, RRL Report LR 339, Road Research Laboratory, Crowthorne, 1971.
5. PARSONS, A W. The rapid measurement of the moisture condition of earthwork material. Department of the Environment, TRRL Laboratory Report 750, Transport and Road Research Laboratory, Crowthorne, 1976.
6. BRITISH STANDARDS INSTITUTION. British Standard Code of Practice CP 2001 1957. Site Investigations. British Standards Institution, London, 1957.
7. SKEMPTON, A W and R D NORTHEY. The sensitivity of clays. Geotechnique, 1952, III (1), 30-53.
8. BLACK, W P M. A method of estimating the California bearing ratio of cohesive soils from plasticity data. Geotechnique, 1962, XII (4), 271-282.

A. D. M. PENMAN, DSc, FICE, Building Research Station

Construction pore pressures in two earth dams

The shear failures at Chingford and Muirhead dams in 1937 and 1941 could be attributed to high pore pressures. Rates of construction had increased and there had been insufficient time for the dissipation of construction pore pressures. Measurements of pore pressures were made with standpipes at Knockendon: a dam similar to Muirhead being built at about the same time. During the designs of embankment dams on the rivers Usk and Daer consideration was given to pore pressures and twin tube hydraulic piezometers were installed during construction in 1951-3. The dams had similar sections and were both built from boulder clay, but while negligible pore pressures were measured at Daer, values greater than the overburden pressure were observed in the fill of the Usk dam. Tests showed that there were marked differences in the values of c_v and water content of the two fills.

The slip surface at Chingford passed through a layer of soft clay in the foundation. A layer of silt was found in the foundation at Usk and to ensure adequate dissipation of pore pressure from it to ensure stability when the dam was built, a simple system of sand drains was used.

INTRODUCTION

Investigations were made of shear failures that occurred during the construction of dams at Chingford and Muirhead in 1937 and 1941 by the Soil Mechanics Section of the Building Research Station. These have been described by Cooling and Golder (1942) and Banks (1948). Total stress analysis was used in both cases, not because a method of analysis using effective stresses was not available at that time, but because there was no pore pressure data.

Chingford dam had a central puddled clay core and was designed to be 10.4 m high. It was built with caterpillar tractors pulling tracked tipping waggons: a new method that permitted rapid construction. Shear failure occurred when it was almost 8 m high, the slip surface passing through the core and a thin layer of soft yellow clay in the foundation. The undrained strengths of the core and yellow clay, measured in the field by the portable unconfined compression apparatus and on samples returned to the laboratory, had values of about 14 kN/m^2. The clay layer was almost horizontal and about 0.9 m thick. The vertical load applied to it by the weight of the bank was about 144 kN/m^2 and drained tests made in a shear box showed that the shear strength of the clay under this load could be much higher. Skempton used the consolidation theory of Terzaghi and Fröhlich (1936) to calculate values of pore pressures at the centre of the layer for various times after application of the load. During the month before failure the bank was raised from 4 m to 8 m and it remained for a further 8 months before it was rebuilt. At this stage the undrained strength of the clay was found to be 36 kN/m^2, an increase of 22 kN/m^2 since failure of the bank. The coefficient of consolidation found from oedometer tests was 0.42 m^2/yr. Skempton's calculations gave values of pore pressure of 115 and 5 kN/m^2 after 37 and 277 days, ie an increase in effective stress of 110 kN/m^2. The drained tests indicated a corresponding strength increase of 30 kN/m^2 which compared favourably with the increase found in the field during the 8 month period. Cooling and Golder (1942) acknowledged Skempton's contribution and remarked that with a more permeable foundation layer, such as silt, the method of calculation could be used to control the rate of construction so that the strength of the foundation material should at no time be exceeded.

The dam at Muirhead used boulder clay fill for the shoulders. The method of construction had been changed when the bank was partly built. The lower layers were placed slowly with little compaction but to advance completion new Diesel powered earth moving machinery was brought in. Shear failure occurred in the lower fill before the bank reached full height. This could be attributed to high pore pressures, induced by the weight of the dense new fill and given insufficient time to dissipate by the rapid construction.

At the time of this failure, a sister bank was being built nearby at Knockendon from a similar boulder clay. Both dams were designed, by Babtie Shaw and Morton, for a maximum height of 27 m and they had similar cross-sections, both with a central puddled clay core. At the time of the shear failure at Muirhead the fill at Knockendon was about $^1/_5$th full height (Banks, 1952) and the section was modified by adding toe weighting to the upstream shoulder and

Clay fills. Institution of Civil Engineers, London, 1978, 177-187

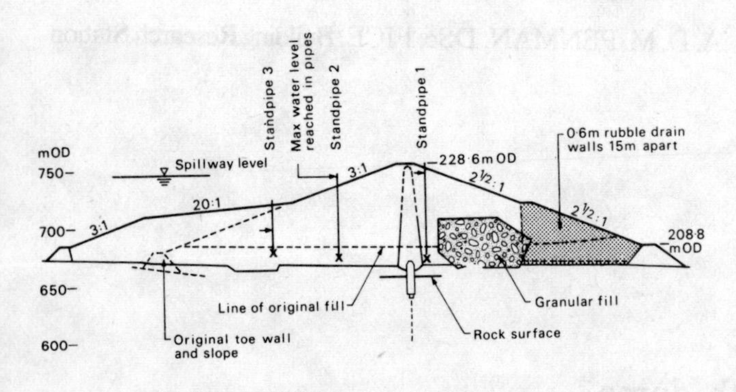

Fig. 1. Cross-section of Knockendon dam

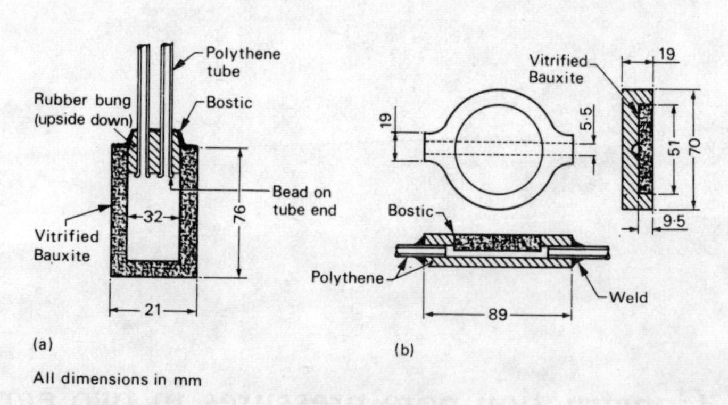

Fig. 4. Piezometer tips

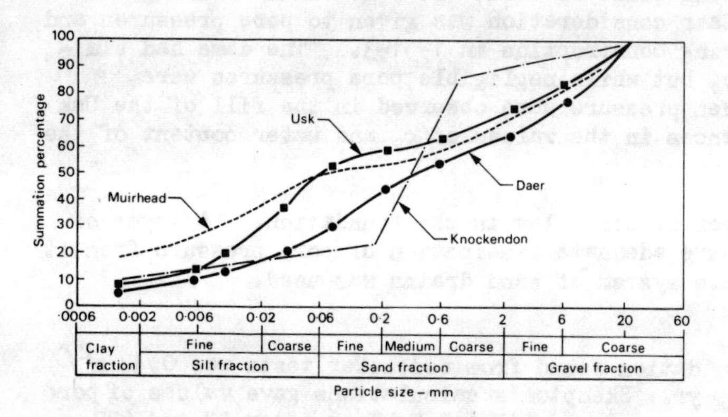

Fig. 2. Grading curves from samples of fill

Fig. 5. Piezometer tips at Daer

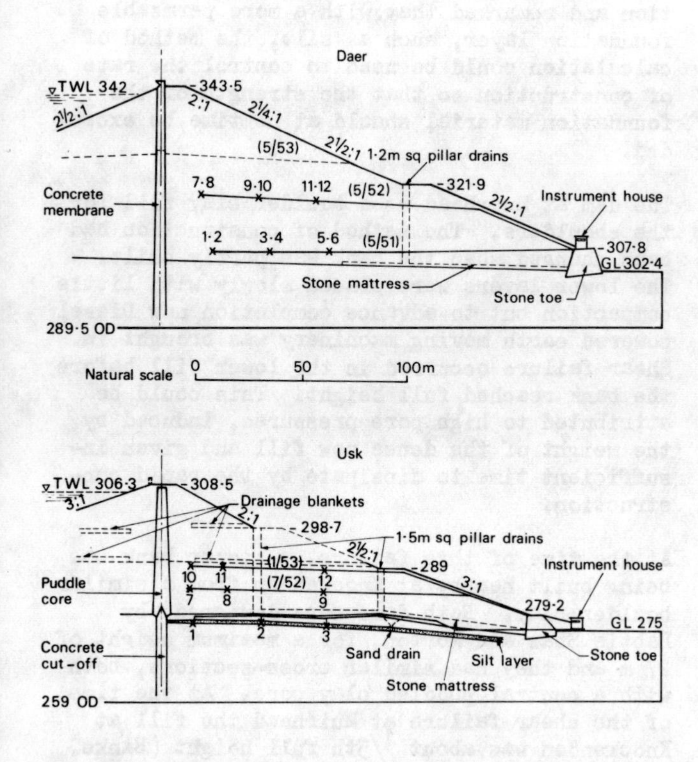

Fig. 3. Daer and Usk dams: major sections downstream

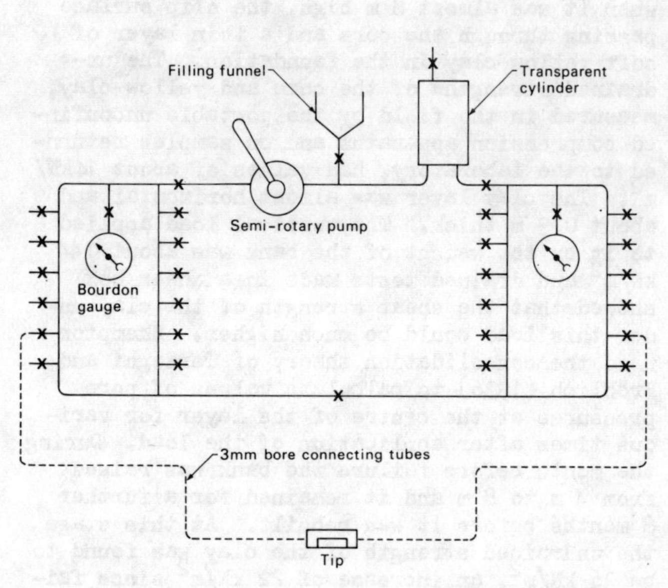

Fig. 6. Manifold system

including a zone of granular fill in the downstream shoulder as indicated by Fig 1.

When the fill had reached about ½ full height, ten boreholes were made to find out about the condition of the lower fill and see if it was going to be strong enough to support the full height of the dam. The casing of one of these holes happened to be left protruding about 0.6 m above fill level overnight and next day was found to be overflowing with water. Golder (1953) thought this was the first observation in Britain to reveal water level standing above the surface of a fill. In order to be able to continue checking on the pore pressure in the lower fill during the remaining construction four standpipe piezometers were installed.

They were made from 75 mm diameter steel tubes fitted with pointed shoes and perforated with 6 mm holes over their bottom lengths of 30 to 45 cms. They were driven into the fill at two positions in the upstream shoulder as indicated by Fig 1 and at two positions on different sections just downstream of the core. The upper half of the fill was placed at an average rate of about 0.38 m/month during the period 1944-46 and the maximum water levels measured in the standpipes are indicated in Fig 1. The maximum pore pressure at standpipe no 2 was 55 per cent of the overburden pressure.

Grading curves for the fills of the two dams, shown in Fig 2, indicate that the boulder clay at Muirhead contained much more material of the silt and clay size particle than did the fill used at Knockendon. Samples of the Knockendon fill had a Proctor optimum water content of 15 per cent and samples taken from the lower fill had an average water content of 20.1 per cent. The first fill was placed and compacted in 0.3 m layers sloping down towards the core. After the Muirhead experience, the slope was changed to be down away from the core to assist run-off and care was taken to avoid any ponding of rain water on the surface of the fill. Drainage was provided for in the original design by a stone mattress of quarry debris 0.6 m thick under half of the downstream shoulder with vertical wall drains 0.6 m thick at 15 m spacing built over it like buttresses. The two stone toe walls also provided some drainage from the fill. In addition, when the granular fill was included in the downstream shoulder, holes were drilled through the culvert to ensure that it was drained.

The values of measured pore pressure were used to check on stability through the results of drained shear box tests. The effective stress in the fill was assumed to be the overburden less the pore pressure and was used to obtain values for the shear strength of the fill from the shear box tests. These were applied to the total stress analysis. Banks (1952) remarked that when material such as boulder clay is used as fill, it might be prudent to place some limitation on the rate of construction afforded by modern earth-moving plant. He also suggested that this solution may rarely be practicable due to economic and other pressures and indicated that there should be more concentration on field measurement of pore pressure.

Cooling (1953) in response, pointed out that the pore pressure measurements made at Knockendon were probably the first of their kind in Britain and that pore pressure measurement in dams was part of the programme of the Building Research Station.

THE DAER AND USK DAMS

The next major embankment dams to be built in Britain were on the rivers Daer and Usk. Both were designed by Binnie Deacon and Gourley (now Binnie and Partners) who invited the Building Research Station to install piezometers in them during construction. They were to be about 35 m high with boulder clay fills (see grading curves, Fig 2) and had similar cross-sections as indicated by Fig 3. Usk used a traditional central puddled clay core while Daer had a central articulated concrete wall. Stability analysis based on effective stresses had indicated that for a satisfactory factor of safety the pore pressure at Daer should not be allowed to rise to more than 40 per cent of overburden pressure.

A twin-tube hydraulic piezometer apparatus, based on that used by the USBR in the Anderson Ranch dam was constructed at BRS. The tips were made with a larger intake area. They consisted of a plastic moulding, as shown by Fig 4b, containing a disc of vitrified bauxite (grindstone material) 50 mm dia x 10 mm thick: the discs at Anderson Ranch were 22 mm dia and 6 mm thick.

The first six tips were placed in the fill at Daer during the second half of May 1951 at positions shown by Fig 3. They were placed in pairs, as shown by Fig 5, on either side of a trench dug in the fill when it had reached the required level. They were connected by black pvc tubing of 3 mm bore and 1 mm wall thickness to Bourdon gauges in an instrument house built on the downstream toe wall. A manifold system, as indicated by Fig 6, was used to connect all the tips to the Bourdon gauges as required and a semi-rotary pump was provided to circulate de-aired water through the tips. The maximum length of the pvc tube manufactured was only 46 m so that joints had to be made in the connections to all the tips. The trench in the fill was dug by hand to a depth and width of about 0.5 m. Fill passing a 6 mm screen was used under the tubes and placed over them to protect them from stones in the fill. The backfill was placed in layers and compacted by hand-held power rammer. The level of these first tips was about 2 m below the filling funnel in the instrument house so there was not enough overburden to allow a pressure to be built up at the tip sufficient to return water and they were not de-aired at that stage.

At Usk, during excavation for outlet works and a stilling basin, a layer of silt was exposed at a depth of about 3 m below ground surface. Despite the reputation that silt had at that time of unstable behaviour, the engineers were keen to apply the principles outlined by

Cooling and Golder (1942) to ensure the stability of the silt in place and avoid having to remove it. Further site investigation showed the layer to extend under the downstream shoulder over the width of the valley floor and peter out just upstream of the cut-off. The results of oedometer tests made at BRS during October 1950 on samples of the silt carefully taken from an exposure, were used to design a sand drain system by the method given by Barron (1948). The boulder clay was difficult to bore so as an economy the silt layer was drained by a single row of sand drains placed across the valley floor under the position for the first berm. A trench was dug across the valley floor to within about 15 cms of the silt at the position shown by Fig 3 and 20 cm diameter holes at 3 m spacing were bored by hand auger from the floor of the trench to pass through the silt layer. They were backfilled with fine to medium sand that was brought up to form a layer 30 cms thick in the trench. The rest of the trench was filled with stone rubble in contact with the stone mattress drain later laid over the stripped ground.

To check on the behaviour of the drains, six piezometer tips were placed in the silt layer between the line of the drains and the central cut-off to record any undesirable pore pressures which might develop. Tips of the type shown by Fig 4a were placed in boreholes to lie near the mid-thickness of the silt during the last week of July 1951. They were surrounded by sand in the silt and the boreholes were backfilled with compacted clay above the silt. The black pvc tubes were carried in trenches to an instrument house built on the ground just below the downstream toe, as shown by Fig 3, to reduce the head difference between the tips and the Bourdon gauges. Another manifold system, the same as that used at Daer (Fig 6) was made in BRS workshops. It also had provision for 12 tips with the idea that at a later stage, 6 tips might be installed in the fill to find out what sorts of pore pressures were developed in a typical dam fill. This formed part of the BRS research programme and permission was obtained from the owners, through the consultants for these extra observations to be made. This use of sand drains to stabilize the foundations of the dam has been described by Sheppard and Little (1955) who claim that it was one of the first applications in Britain.

CONSTRUCTION AT USK

The contractor opted not to place fill during the winter months at Usk and it was decided to place three tips at about mid-thickness of the first seasons fill forming the downstream shoulder. The puddled clay core was regarded as a central water stop that should not be violated by passing plastic tubes through it and with provision for only 6 tips, measurements in the upstream shoulder were not considered.

When the fill had reached the desired level a trench about 0.6 m deep and 0.5 m wide was dug by hand from the positions for the tips to the top of the downstream stone toe wall adjacent to the instrument house. It was lined with puddled clay except at the positions for the tips (Fig 4b) which were placed on the fill, porous disc down, and covered with some fill that had passed a 6 mm sieve. The black pvc connecting tubes were laid on, and covered by more, puddled clay before the trench was backfilled. The tubes were taken through a steel pipe from the top of the stone toe into the top of the instrument house. This detail later caused trouble due to freezing. The tips, numbered 7, 8 and 9, were placed at positions shown by Fig 3, during the third week of July 1952.

The semi-rotary pump (see Fig 6) caused large pressure fluctuations and was not satisfactory for circulating air-free water through the long lengths of small bore tubing, so it was replaced by two transparent plastic cylinders. One of these contained a football bladder so that when it was filled with air-free water, it could be pressurised by pumping air into the bladder from a car tyre footpump. The other cylinder had a hand suction pump attached to it so that the air pressure in it could be lowered. The difference in pressure between the two cylinders could be kept fairly steady and was used to circulate air-free water through the piezometers. This modification was also made to the piezometer apparatus at Daer.

The fill at Usk was being placed by 11.5 m^3 scrapers that were push-loaded by D8 dozers in the borrow pits. In general the material was spread in 0.3 m layers and rolled by a 15 tonne wobbly wheel rubber tyred roller. The specification limited the rate of placement to 0.9 m/month but because of a winter shutdown of about 6 months, the contractor sought and was granted permission to double this rate during the placing season. A smooth 4 tonne roller was also used, particularly to smooth the fill surface to throw off rain. The bank was constructed with slopes down from the puddle core of about 1 on 25. The work has been described by Sheppard and Aylen (1957) who quote values of field dry density of 1.96 Mg/m^3 and water content of 12.2 per cent with Proctor optimum water content 9.8 per cent for the fill placed in 1952.

CONSTRUCTION AT DAER

Fill at the borrow pits was dug by face shovels and loaded into 9.2 m^3 bottom discharge Enclid articulated waggons which dumped the material on the bank in wind-rows. It was bulldozed into 0.6 m thick layers that were compacted by the Enclid trucks which were each directed into a new track by an efficient 'bank manager'. Flat tracked caterpillar tractors were used to flatten the surface to encourage run-off on the 1 on 20 slope and the placement working area was kept to a minimum so that if rain came it could be quickly smoothed. It was found best to stop work immediately when rain began falling because it was then possible to resume work as soon as it stopped, whereas if work continued during rain, the resulting quagmire was difficult to overcome. The construction of the Daer dam has been described by Kerr and Lockett (1957). Little (1957) has given histograms based on 336 observations showing a fill dry

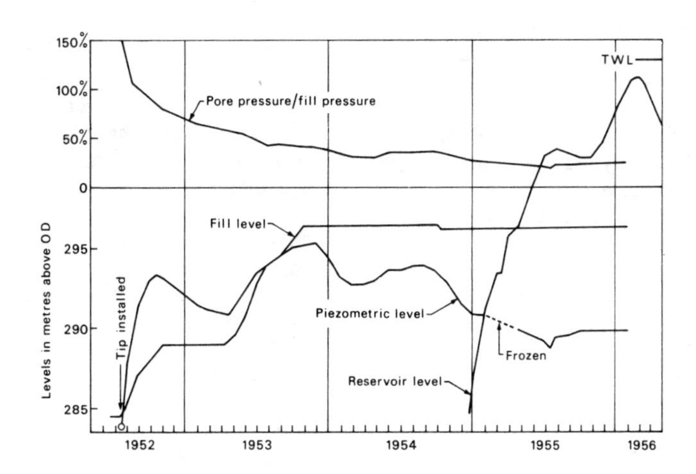

Fig. 7. Results from tip no 9 at Usk

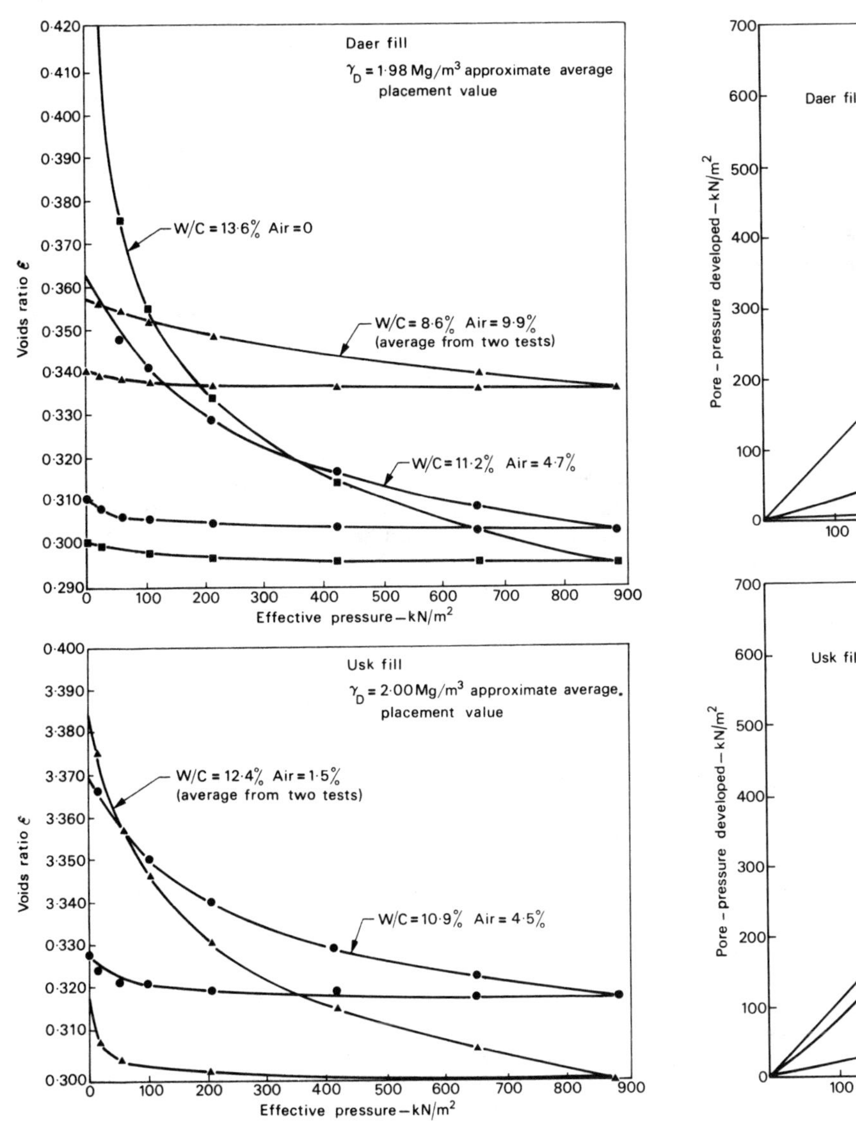

Fig. 8. Compressibilities of the fills from oedometer tests

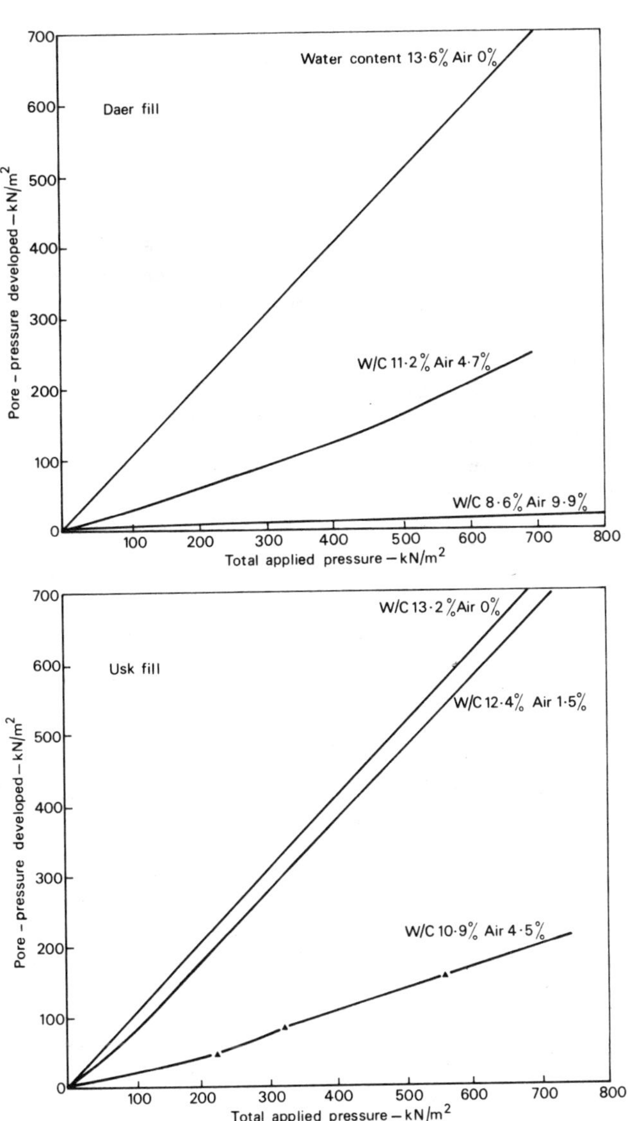

Fig. 9. Calculated pore pressures for various water contents

density of 1.97 Mg/m^3 and a water content of 7.5 per cent. Proctor tests gave an optimum water content of 9.6 per cent.

The upper set of six tips were placed in the fill at the positions shown by Fig 3 during the third week of May 1952. There was then 11.6 m height of fill over the lower six tips.

INITIAL PORE PRESSURES

Reading on the lower tips at Daer taken when the upper tips were placed, showed negative pressures of 0.06 to 0.3 m head of water. When the fill height over them had increased to 26 m the pore pressures had become positive and on dam completion, with 36.6 m over the lower tips, they recorded pore pressures of 0.2 to 0.9 m of water. The upper tips, with 25 m of fill over them showed pressures of 0.7 to 1.1 m head of water.

At Usk no significant pore pressures were measured by the tips in the silt layer, indicating that drainage was effective.

After tips 7, 8 and 9 had been placed during July 1952, filling continued as before, at a rate of about 1.3 m/month. Only one tip at a time could be connected to the Bourdon gauges and the response time was so slow that each tip had to be left connected for a few days in turn. Calculation of the initial pore pressure from the gauge readings showed them to be greater than the ever increasing overburden pressure and there was some concern about the accuracies involved and the validity of the readings. The Bourdon gauges reached their limit of a pressure of 14 m head of water from tip 9 on 1 Sept 1952 and new gauges had to be fitted. The results from tip 9 are given in Fig 7. Fill placement was stopped for the winter at the end of October, when the pore pressure/fill pressure ratio had already fallen to 80 per cent. The question then remained of how much dissipation would occur during the winter shut down, how much pore pressure would build up during remaining construction and what maximum values could be tolerated without endangering stability?

LABORATORY WORK

The pore pressure developed in an undrained element of soil under an applied total pressure is governed by the relative compressibility of the soil skeleton and the pore fluid. The compressibility of water is small compared with that of a soil skeleton so that a saturated soil usually developes a pore pressure equal to the applied pressure. Even a stiff soil such as a well graded clean sand develops almost 100 per cent pore pressure when fully saturated, as shown by Bishop and Eldon (1950).

With unsaturated soils, the compressibility of the pore fluid depends on the proportion of gas in the pores. When an external pressure is applied, both the soil skeleton and the gas compress. The pore pressure is affected by the gas pressure and if it becomes great enough the gas may all dissolve in the water phase of the pore fluid. The relationships between gas volume and pressure and solubility in water have been expressed by the clasical theories of Boyle and Henry. They have been used by Brahtz et al (1939), Hamilton (1939) and Hilf (1948) to calculate the pore pressures that will develop in an undrained element of unsaturated fill compressed by an external pressure.

This approach was used to demonstrate the difference between the observed pore pressures at the two dams. Oedometer tests were made on samples of the fills to measure their compressibility at different water contents and the results are given by Fig 8. Material passing a 19 mm sieve was compacted into an oedometer to form samples of 100 mm diameter and height. The large height was chosen to accommodate the coarse particles and adequate time was allowed for drainage during the test, although friction between the sample and the brass walls of the oedometer cylinder may have affected the results. An increase of water content increased the compressibility of the drained fills and the two materials showed comparable behaviour.

The compressibility of the pore fluid was calculated from the air and water content of the samples, derived from measured water content and density. The pore pressures calculated for applied total pressures up to 700 kN/m^2 for three water contents for each fill are given in Fig 9. Clearly the wetter the fill the greater the pore pressure response to total pressure until when the water content was high enough to exclude all air the ratio of pore pressure/total pressure became unity.

To confirm this behaviour, direct measurements of pore pressure were made on samples subjected to total pressures in a triaxial apparatus. Samples of the fills passing a 6 mm sieve were prepared at various water contents. Each was compacted into a rubber sheath mounted on a cell base, supported by a 3-piece mould, to form a sample 38 mm diamter and 89 mm high. Pore pressure was measured through a porous disc at the base of the sample by a no-volume-change device that has been described by Penman (1953). The results, given by Fig 10, show a pore pressure response for each water content similar to that obtained from the oedometer tests. Differences could be expected because of the even smaller fraction of fill used and because the samples were subjected to an equal all round pressure instead of the oedometer pressure distribution.

Both results showed clearly the effect of placement water content on the pore pressures developed in the undrained samples. They indicate that fill placed wet of optimum, as at Usk, would develop much higher pore pressures than fill placed dry of optimum, as occurred at Daer.

The undrained pore pressures developed in the dams could be different from the values given by these simple tests because of different stress conditions, because the tested fill contained only the smaller particles and because of differences of bulk density and water contents. Whatever the potential undrained values,

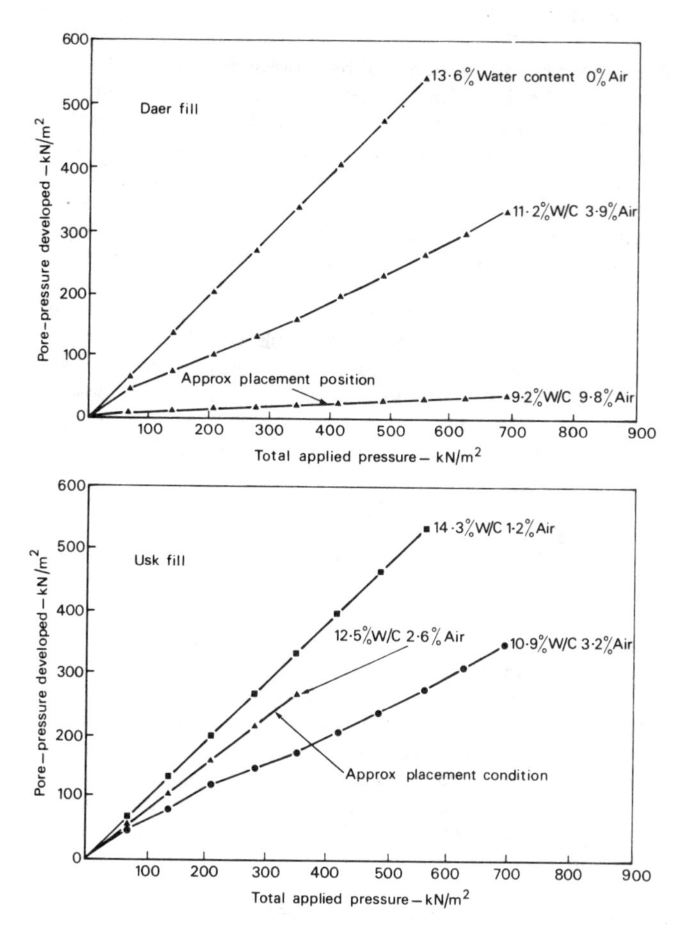

Fig. 10. Measured pore pressures for various water contents

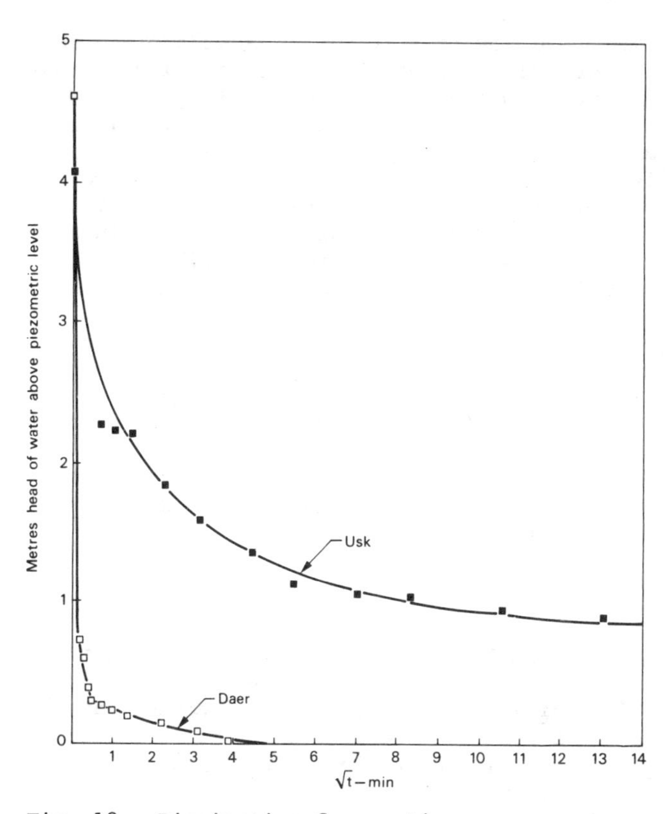

Fig. 12. Dissipation from a tip

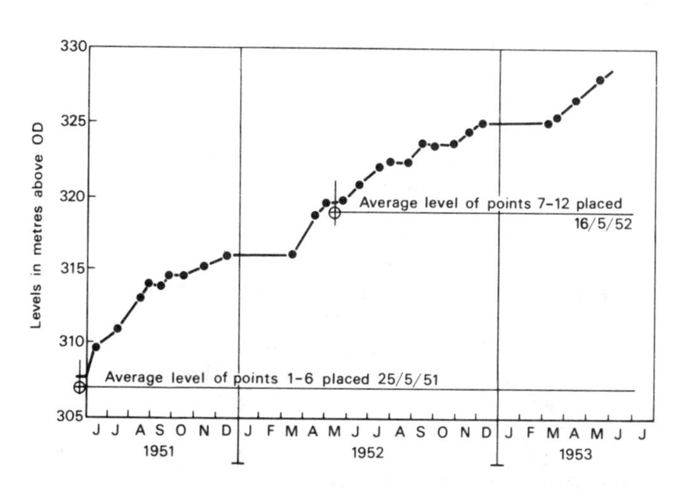

Fig. 11. Rate of construction of Daer dam

Fig. 13. Tip with polyethylene tubing at Usk

however, the actual pore pressures that could develop in the dams were governed by the freedom of the fill to drain and the rates of construction.

Values for the coefficient of consolidation were obtained from the oedometer test results by the method discussed by Cooling and Skempton (1941). For the Usk fill they ranged from 11 to 21 with an average of 15 m^2/yr under the pressures likely to be found in the dam. The rate of pore pressure dissipation measured at tips nos 7, 8 and 9 was used by Skempton (1957) to obtain a value of 12.3 m^2/year. The Daer fill had a much higher value of 334 m^2/yr.

Consolidation theory indicates that the pore pressure at the centre of a saturated layer of fill free to drain to its upper and lower surfaces would have dissipated to about 10 per cent of its initial value when the time factor T = 1. The relationship between the time factor T, the coefficient of consolidation c_v, the drainage path length d and elapsed time t is given by

$$T = \frac{c_v t}{d^2}$$

At Daer the lower set of tips were approximately at the centre of the first 4.3 m height of fill. If it had a coefficient of consolidation of 334 m^2/yr then the pore pressure at the tips would have dissipated to 10 per cent of its initial value in less than a week. The rate of construction of the dam, shown by Fig 11, seldom exceeded 2 m/month so that considerable dissipation could occur during construction.

Tip 9 at Usk was about at the mid-thickness of the 9.8 m of fill placed before November 1952. If the average coefficient of consolidation was 15 m^2/yr, the time for 90 per cent dissipation would be 1.6 years. Unfortunately, during construction, the length of the drainage path to the fill surface would continue to increase and seriously delay dissipation.

FIELD TESTS

Further indications of the marked difference of behaviour of the fills of the two dams were given by ad hoc tests on site. Just after some of the tips had been placed, when the cover over them was about 0.6 m, measurements were made of the pore pressure response to the pressure applied to the fill by placing machinery.

At Daer, just after the upper tips had been placed, readings were taken as a loaded Enclid was driven over them. The tyre pressure of the machine was 414 k N/m^2 and the maximum pore pressure increase observed was 3.4 kN/m^2.

A similar test was made at Usk just after Tips 10 to 14 had been placed. Readings were taken as a D8 bulldozer worked over three of them. Its weight, if distributed over the whole area of its tracks, caused a pressure of 68 kN/m^2. The maximum increases of pore pressure measured at the tips was 35, 27 and 22 kN/m^2. Values of the ratio of pore pressure/overburden pressure were 51, 85 and 29 per cent before the bulldozer worked over the tips and 104, 127 and 117 per cent afterwards. No further fill was placed over these tips for four months and during that time the ratio remained at about 100 per cent.

Because the tips had to be connected one at a time to the Bourdon gauges of the first apparatus, tests were made to find out how long each tip took to reach equilibrium. A pressure was built up at the tip by passing water into one connecting tube and measured with the Bourdon gauge connected to the other tube from the tip. At zero time the water supply was turned off and dissipation from the tip measured by the fall of pressure with time. The results from a typical test on tip no 9 at Usk are given by Fig 12. The pressure had only fallen to 25 per cent of its initial value after 3 hours. Because of this, the tips were left connected for at least a day each before routine readings were taken.

The results of a similar test at Daer, also given by Fig 12, show that equilibrium was reached after only 20 minutes.

FURTHER PORE PRESSURE MEASUREMENTS

Stability analysis of the Usk section, based on effective stresses, indicated that there would be an unacceptable factor of safety if the dam was built to full height with a pore pressure/overburden pressure ratio greater than 50 per cent. The dissipation of pore pressure during the winter shut-down was followed with keen interest.

A new gauge board with a separate Bourdon gauge for each tip, as described by Penman (1956), was constructed at BRS and fitted in the instrument house at Usk in January 1953. The existing tips 7, 8 and 9 were connected to it and five new tips (nos 10 to 14) were placed above them in the surface of the existing fill ready for new fill that might be placed in the Spring. A new connecting tube, made from polyethylene to the same size but in continuous lengths from tip to gauge board, was used. Figure 13 shows one of the tips in place.

It began to look as though sufficient dissipation would not occur before the next placing season began and there were contractual difficulties in reducing the rate of construction. It seemed unlikely that the stability of the dam would be at risk until near the end of the 1953 placing reason, but it might then have been necessary to delay construction for a year. At this stage further advice was sought from Prof Skempton at Imperial College. To check the pore pressures measured by the BRS piezometers 15 steel standpipe piezometers were driven on an adjacent cross-section of the dam into both downstream and upstream shoulders. Water levels in the standpipes rose a few metres above the surface of the fill, confirming the high values of pore pressure/overburden pressure ratio and showing that this condition was general in the fill of both upstream and downstream shoulders. Figure 14 shows some of the standpipes, painted white to mark the level of the water surface inside them.

The remedial measure adopted was to place drainage layers in the fill to reduce the drainage path length and thereby speed up pore pressure dissipation. The standpipes were cut off below fill level and 9 were connected to twin polyethylene tubes carried in trenches to a second instrument house built on top of the stone toe. To avoid passing the tubes through the puddle core, they were led from the upsteam standpipes to the right valley side and passed through the concrete cut-off wall under the clay core. A graded filter drainage layer consisting of river sand and gravel below and above a layer of crushed rock was spread over the fill to within 6 m of the puddle core before normal placing recommenced. Little (1958) has pointed out that the annual rainfall of 1.52 m at Usk was well distributed throughout the year. To minimise the amount of rainwater incorporated in the fill, scrapers were not used after 1952. Instead fill was won by face shovels and transported to site by tipper truck.

A second drainage layer was placed at about 2/3 full height of the dam. The success of the lower drainage layer can be judged by the improved dissipation that occurred (see Fig 7) as the 1953 fill was placed.

This first use of drainage blankets to control constructional pore pressures in a dam was repeated by Prof Bishop at Selset (Bishop et al, 1960) and the technique has since been used widely in embankment construction.

At Daer the BRS piezometers showed negligible pore pressures. As a check, following the work at Usk, holes 6 to 18 m deep were drilled into the fill when the bank was at about half height and standpipe piezometers installed in them. They remained dry and to test their operation, water was poured into them, but it flowed away leaving little or no residue (Little, 1973).

Four further tips were placed at Usk on the section of the standpipes at a level of about 293 m OD during the last week of July 1953. Two of them in the upstream shoulder were attached to connecting tubing that was placed when the standpipes were connected and taken through the concrete cut-off wall. It had been brought up with the puddle core and was laid in a trench across the fill when the tips were placed. The other two tips were in the downstream shoulder.

During subsequent readings it was found that one of the ex-standpipe tips required repeated deairing and contained gas that was found to be inflamable. Samples of the gas were collected for analysis and were found to have the composition given in Table 1. It was concluded that the gas originated from organic material and was not caused by decomposition of parts of the metal standpipe. Smaller samples of similar gas were recovered from tip 9 which contained no metal parts and was on a different section from the steel tubes.

Table 1. Gas analysis

	Ex-standpipe	Tip 9
Carbon dioxide (CO_2)	2.3	9.9
Unsaturated hydrocarbons ($C_n H_m$)	nil	nil
Oxygen (O_2)	1.1	1.1
Carbon monoxide (CO)	0.3	0.1
Hydrogen (H_2)	0.6	nil
Saturated hydrocarbons ($C_n H_{2n+2}$)	51.2	24.1
Nitrogen (N_2)	44.5	64.8
	100.0	100.0
'n' value in saturated hydrocarbons	1.0	1.0

DISCUSSION

The field observations made at these two dams aroused considerable interest and stimulated both analytical and instrumentation developments.

Pore pressures greater than the overburden could be due to lateral pressures induced by placing machinery or air pressures, as suggested by Cooling. In the early 1930s he had studied the properties of building stone (Cooling 1975) and used a suction plate apparatus to assess pore sizes. The suction plate itself was a fine pored stone which, after saturation, was connected to a variable head of mercury that produced a suction in the pore space of the stone. The pores of the stone were so small that the air/water meniscus at the surface of the stone prevented the ingress of air, but a sample put in contact with it would have some water sucked out of it. Croney (1952) has described a suction plate made from sintered glass with a pore size not greater than 1.5 microns (necessary to resist a suction of 1 atmosphere) used in a laboratory study of soil suction.

In an unsaturated fill, the pressure of the air in the voids can be expected to be higher than that of the water because of the shape of the water/air meniscus and the pressure of the water could be measured through the pores of a saturated porous plate with a pore size small enough to prevent the ingress of air at the maximum pressure differential between air and water in the fill. Work at BRS and Imperial College led to the fine pored piezometer tip described by Bishop et al (1960) and now generally used for hydraulic piezometers in unsaturated fills.

The filter stones used in the tips at Usk and Daer had relatively large pores, only designed to be small enough to prevent the entry of soil particles, and so measured pore air pressure. A comparison between the pore pressures measured by a coarse and fine pored tip was obtained during the construction of Chelmarsh dam . The

Fig. 14. Standpipes at Usk

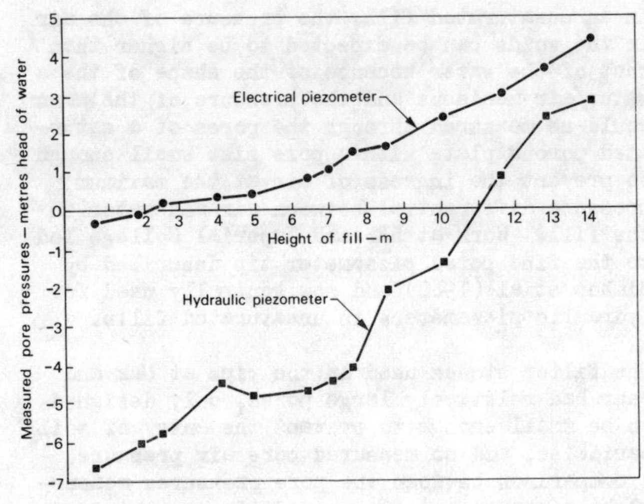

Fig. 15. Pore pressures measured by a fine and a coarse pored tip

two tips were placed close to each other in the fill and the pressures they measured are shown by Fig 15. The fill was placed dry of optimum and was quite strong when rolled, as reflected in the considerable suction developed in the pore water initially. At this stage the pore air was at about atmospheric pressure. As the total pressure increased due to bank construction, both pore water and pore air pressures increased. The rate of increase of pore water pressure was the greater and if the dam had been higher, the two pressures would have become the same when all the air had been compressed into solution and the fill became saturated.

Agriculturalists have an interest in water deficiency in the soil. The Rogers (1935) tensiometer was designed to measure pore water suction in the soil to indicate the degree of deficiency. It uses a mercury manometer and a saturated unglazed earthware pot that could support suctions up to 0.8 atmospheres with two copper connecting tubes for de-airing. Black et al (1958) described a tensiometer made entirely of glass to avoid air leakes, with a sintered glass filter that could measure a suction of 1 atmosphere.

This type of apparatus was not considered suitable for measurements in a dam during construction, but clearly more attention should have been paid to the need for a fine pored element.

Further developments relating to pore pressure were discussed during the 1960 BGS Conference on Pore Pressure and Suction in Soils (Proc Butterworths) and by Bishop et al (1964) and Vaughan (1974).

The doubly beneficial effects of even partial drainage of fill during construction has been assessed by Bishop (1957) and the consolidation of a layer of fill that is increasing in thickness has been analysed by Gibson (1958). Both papers use Usk as a field example.

Although there has been a move to place fill dry of optimum in the shoulders of a dam for the sake of stability, the water-tightness of a dam can be endangered by the extra stiffness of a core formed from dry fill. Cracking may be caused by differential settlements, as suggested by Low (1970) and a strong rigid core may transfer some of its weight to the shoulders (arching or silo action) so increasing the risk of hydraulic fracture. Penman (1976) pointed out that the total pressure in a core must exceed the pressure from the reservoir water at all heights to prevent hydraulic fracture and studies of the behaviour of cores by Penman and Charles (1973) have indicated that this occurs in cores placed wet of optimum. It can be argued that if the pore pressures in the core at the end of construction are equal or greater than the pressure from the reservoir water at all levels, there will be little risk of hydraulic fracture.

Tailings dams are usually built with fine material placed wet of optimum. Drainage blankets could usefully be incorporated in them during construction to control both construction pore pressures and seepage from the impounded waste. The efficiency of horizontal drainage blankets for accelerating the consolidation of clayey fills has been discussed by Gibson and Shefford (1968) and an assessment using three field cases has been given by Sills (1974).

ACKNOWLEDGEMENT
The valuable cooperation of Babtie Shaw and Morton and Binnie and Partners with the Geotechnics Division of the Building Research Station in the early research work described in this paper is gratefully acknowledged.

The paper is published by permission of the Director of the Building Research Establishment.

REFERENCES
1. BANKS, J A: Construction of Muirhead reservoir. Proc 2nd Int Conf SM&FE, 1948, vol 2, pp 24-31.
2. BANKS, J A: Problems in the design and construction of Knockendon dam. Paper 5852 Proc Instn Civ Engrs, 1952, part 1, vol 1, no 4, pp 423-443.
3. BARRON, R A: Consolidation of fine grained soil by drain wells. Trans Am Soc Civ Engrs, 1948, vol 113, pp 718-754.
4. BISHOP, A W: Some factors controlling the pore pressures set up during the construction of earth dams. Proc 4th Int Conf SM&FE, 1957, vol 2, pp 294-300.
5. BISHOP, A W and ELDON,G : Undrained triaxial tests on saturated sands and their significance in the general theory of shear strength. Geotechnique, 1950, vol 2, no 1, pp 13-32.
6. BISHOP, A W, KENNARD, M F and PENMAN, A D M Pore pressure observations at Selset dam. Pore Pressure and Suction in Soils, Butterworths, London, 1960, pp 91-102.
7. BISHOP, A W, KENNARD, M F and VAUGHAN, P R: Developments in the measurement and interpretation of pore pressure in earth dams. Trans 8th Int Congress Large Dams, Edinburgh, 1964, vol 2, pp 47-72.
8. BLACK, W P M, CRONEY, D and JACOBS, J C: Field studies of the movement of soil moisture. Road Research Technical Paper no 41, 1958, HM Stationery Office, London.
9. BRAHTZ, J H A, ZANGAR, C N and BRUGGEMAN, J R: Notes on analytical soil mechanics. Tech Memo no 592, 1939, Bureau of Reclamation, Denver, USA.
10. COOLING, L F: Correspondence on paper 5852 Proc Instn Civ Engrs, 1953, part 1, vol 2, no 2 pp 203-205.
11. COOLING, L F: The early history of soil mechanics at the Building Research Station. Geotechnique, 1975, vol 25, no4, pp 629-634.
12. COOLING, L F and GOLDER, H Q: The analysis of the failure of an earth dam during construction. Jnl Instn Civ Engrs, 1942, vol 19, no 1, pp 38-55.
13. COOLING, L F and SKEMPTON, A W: Some experiments on the consolidation of clay. Jnl Instn Civ Engrs, 1941, vol 16, no 7, pp 381-398.
14. CRONEY, D: The movement and distribution of water in soils. Geotechnique, 1952, vol 3, no 1, pp 1-16.
15. GIBSON, R E: The progress of consolidation in a clay layer increasing in thickness with time. Geotechnique, 1958, vol 8, no 4, pp 171-182.
16. GIBSON, R E and SHEFFORD, G C: The efficiency of horizontal drainage layers for accelerating consolidation of clay embankments. Geotechnique, 1968, vol 18, no 3, pp 327-335.
17. GOLDER, H Q: Correspondence on paper 5852 Proc Instn Civ Engrs, 1953, part 1, vol 2, pp 203-205.
18. HAMILTON, L W: The effect of internal hydrostatic pressure on the shearing strength of soils. Proc ASTM, 1939, vol 39, p 1100.
19. HILF, J W: Estimating constructional pore pressures in rolled earth dams. Proc 2nd Int Conf SM&FE, 1948, vol 3, pp 234-240.
20. KERR, H and LOCKETT, E B: Daer water supply scheme. Instn Civ Engrs, 1957, vol 7, pp 46-74.
21. LITTLE, A L: Discussion on paper no 6188. Proc Instn Civ Engrs, 1957, vol 7, pp 79-80.
22. LITTLE, A L: Compaction and pore water pressure measurements on some recent earth dams Trans 6th Int Congr Large Dams, New York, 1958, vol 3, pp 205-226.
23. LITTLE, A L: Experiences with instrumentation for embankment dam performance. Field Instrumentation in Geotechnical Engineering, Butterworths, London 1973, pp 229-239.
24. LOW, J: General Report - Recent development in the design and construction of earth and rockfill dams. Trans 10th Int Congre Large Dams, 1970, vol 5, pp 1-28.
25. PENMAN, A D M: Shear characteristics of a saturated silt measured in triaxial compression. Geotechnique, 1953, vol 3, no 8, pp 312-328.
26. PENMAN, A D M: A field piezometer apparatus. Geotechnique, 1956, vol 6, no 2, pp 57-65.
27. PENMAN, A D M: Leakage through fill cores. Discussion, Trans 12th Int Congr Large Dams, 1976, vol 5, pp 247-250.
28. PENMAN, A D M and CHARLES, J A: Effect of the position of the core on the behaviour of two rockfill dams. Trans 11th Int Congr Large Dams, 1973, vol 3. pp 315-339.
29. ROGERS, W S: A soil moisture meter. The Jnl Agricultural Soc, 1935, vol 25 pt 3, 326-43.
30. SHEPPARD, G A R and AYLEN, L B: The Usk scheme for the water supply of Swansea. Proc Instn Civ Engrs, London 1957, vol 7, pp 246-265.
31. SHEPPARD, G A R and LITTLE, A L: Stabilizing an earth dam foundation by means of sand drains. Trans 5th Congr on Large Dams, 1955, vol 1, pp 639-646.
32. SILLS, G C: An assessment using three field studies of the theoretical concept of the efficiency of drainage layers in an embankment. Geotechnique, 1974, vol 24, no 4, pp 467-474.
33. SKEMPTON, AW : Discussion on paper no 6210. Proc Instn Civ Engineers, 1957, vol 7, pp 267-9.
34. TERZAGHI, K and FRÖHLICH, O K: Theorie der Setzung von Tonschichten. F Denticke, Leipzig and Vienna (French transl. by M Adler: Theorie des Tassements des Couches Argilleuses. Dunod, Paris, 1939.
35. VAUGHAN, P R: The measurement of pore pressures with piezomters. Field Instrumentation in Geotechnical Engineering. Butterworths, London, 1974, pp 411-422.

S. H. PERRY, PhD, MICE, MIStructE, formerly
Sir Alexander Gibb and Partners, now Department of Civil Engineering, Imperial College of Science and Technology

Behaviour of fill from soft chalk and soft chalk/clay mixtures

A series of laboratory tests was undertaken to investigate the prolonged time taken by soft chalk and soft chalk/clay mixtures placed in embankments on the M4 motorway to stabilize sufficiently for further fill to be placed on top. It is shown that the presence of the clay both severely increases the setting time and reduces the ultimate bearing strength of the chalk. Also, that the air voids of uncontaminated soft Upper Chalk should be between 5% and 10% for maximum bearing strength immediately after compaction. It is argued that minimum compaction compatible with acceptable long-term settlement is the most effective method of forming soft chalk embankments with minimal risk of interruption due to instability of the chalk mass.

PRINCIPAL NOTATION

e	voids ratio
f	frequency
t	time
w	moisture content
G	specific gravity of solid particles
T	temperature
CBR	California bearing ratio
I_p	plasticity index
w_L	liquid limit
w_P	plastic limit
V_a	air voids

INTRODUCTION

It is well known that, under normal circumstances, uncontaminated chalk has a high bearing strength and can be used as subgrade for road construction with the minimum pavement thickness compatible with adequate protection from frost. However, soft Upper Chalk, with a natural moisture content of 28-30% (such as is found on the route of the M4 motorway in Berkshire), can present major earthworking problems (ref.1). In some instances, soft chalk embankments found to become unstable during earthworking have been removed as unsuitable material and replaced with other fill. Yet on the M3, Upper Chalk, which had to be specially imported from 23 miles away, was specified for the construction of bunds to retain gravel fill where the motorway passes through water filled gravel pits (ref.2). The difference between these examples lies in the method of winning the chalk. Nothing better than a lumpy cohesive paste can result when rubber tyred motor scrapers excavate a thin layer of chalk. The specification for the M3 bunds, however, which called for 60% by weight of the chalk to be retained on a 3 in. sieve, ensured that the method of excavation minimised crushing of the chalk.

The winning of soft chalk by motorscrapers and the subsequent placing and compaction (if any) as fill in embankments can result in severe crushing of the chalk and the consequential breakdown of the fissure structure. This tends to draw out the in-situ moisture which has been held in the fissures, causing positive pore pressures to build up and resulting in the formation of 'putty chalk'. In practice, layers of chalk are laid until the putty like instability of the mass of chalk prevents further laying. The material already placed has then to be allowed to stabilize before further fill can be placed on top.

Once putty chalk has been formed, pore pressure may be reduced by

1. reabsorption of water into fissures,
2. drainage,
3. evaporation, and
4. dilatancy.

Although almost immediately after recompaction some water appears to be reabsorbed through dilation of the chalk, what fissure structure remains after compaction will be clogged with powdered chalk, reducing the permeability to almost zero. Clogging of the fissure structure will be increased by the presence of fine silt or clay. Until either solution channels or a new fissure system open up through the mass of chalk, water loss through either drainage or evaporation will be slight, except in the extreme surface. Thus it may take several weeks for an embankment of soft chalk to stabilize sufficiently for earthworking to proceed.

Another factor which will assist stabilization of a chalk embankment is that when evaporation occurs calcium carbonate will be precipitated. Although necessarily a slow process, this precipitation will tend to cement the mass of chalk.

In contrast to the soft Upper Chalk, harder chalks (those with lower natural moisture contents - usually of the Middle or Lower Chalk beds) normally suffer little remoulding during earthworking. Even after compaction, such chalks tend to have a higher air voids than the in-situ material. This, together with the retension of much of their fissure structure,

Clay fills. Institution of Civil Engineers, London, 1978, 189-196

results in a relatively high permeability permitting pore pressures arising during recompaction to dissipate quickly.

Only recently, with several motorways in southern England driving right through the chalk beds, has the study of earthworking (as distinct from foundation problems) of chalk received much attention. Obviously no heavy plant was used for the construction of railway embankments in the nineteenth century, thus the problem of severe crushing, rapid rate of construction and positive pore pressures did not arise. However the fact that " - Recent experience during the construction of chalk embankments for motorways has shown that spongy-chalk embankments become quite stable when left for a few months, even during periods of wet weather,-" (ref.3) indicates that a material which, given sufficient time, would be suitable for the adequate support of a road, is fundamentally unsuitable when a rapid rate of construction is required to produce a stable embankment in a few days.

There is a need to differentiate between an unstable chalk embankment - one which is too spongy to permit further construction to proceed on it, or to be used as a haul route - and a stable one which might still, nevertheless, be subject to consolidation. When expediency demands, the latter can be tolerated while the former can not.

The route of the M4 motorway through Berkshire resulted in the formation level over several miles being close to the Upper Chalk/ sand - clay (of the Reading beds) interface, often just penetrating the chalk. This led to a considerable length of embankment (much of it over 20 ft. in height) being constructed of an intimate mixture of the chalk and clay, together with some sands and silts. One of the major problems encountered during earthmoving was the prolonged time taken by layers of soft chalk, and particularly soft chalk/fatty blue and red clay mixtures, to gain sufficient bearing strength for work to proceed.

In order to investigate these problems, two series of laboratory tests were devised. These attempted

1. to determine, for a fixed air voids, the relationship between bearing strength and the clay content of the chalk at various time intervals after compaction, and
2. to determine the relationship between the bearing strength and the degree of compaction of the chalk (as represented by the air voids), also at various time intervals after compaction.

The problem of earthworking and trafficking during construction on soils is essentially one of the bearing strength - or resistance to penetration - of the material. Thus use of the commonly available California bearing ratio test apparatus was considered appropriate. However, in order to similate the effect of the weight of the layers of pavement above, it is customary to place annular surcharge weights on the material in the mould during the penetration test. For these tests, since the problem was one of road construction, rather than the eventual bearing strength under the finished pavement, the CBR penetration test was done without surcharge weights. In this way, it was considered that the test conditions would approach more closely those of construction plant operating directly on the subgrade.

BEARING STRENGTH OF SOFT CHALK/CLAY MIXTURES

Materials

Chalk - fairly soft Upper Chalk from Frilsham (Berkshire), OS SU 534740. Samples taken from 20 ft. below ground level (where the chalk out-crops), were clean, flint free (but with flint bands every 4-8 ft in depth) and moderately block like when excavated. At this depth, it was assumed that the natural moisture content of the chalk was its saturation value.

specific gravity, $G = 2.58 - 2.60$

natural moisture content, $w_n = 28.2\%$

Clay - the Atterberg limits of 37 samples of clay and clay/sand mixtures found overlying closely the Upper Chalk in Berkshire between Theale and Hermitage were determined using standard techniques (ref.4). These are plotted on a modified Casagrande plasticity chart in Fig.1. Although the properties show wide variation, a frequently occuring plastic blue and red laminated clay from the Reading beds near Yattendon (Berkshire) was selected as typical. This had properties:-

$w_L = 69$

$I_p = 42$

$w_n = 22\%$

Experimental Moisture Content

At its natural moisture content of about 22%, the blue and red clay has a high bearing strength (with a CBR value of 10 to 50%) and when well compacted into an embankment would not be expected to present any problems during road construction. However, should the clay come into contact with water, for instance, from rain or wet chalk, then it will absorb water until its moisture content reaches equilibrium. The equilibrium moisture content varies with the soil pressure, decreasing as the pressure increases. Probable values for the equilibrium moisture content of cohesive soils under pavements in southern England have been presented by Black (ref.5). From Fig. 2, which has been derived from Black's Fig.9, a probable equilibrium moisture content of 32% is obtained for the clay used in the present tests, with a probable CBR value of about 3.0% (under surcharge weights).

Fig. 3 shows the relationship between the moisture content and CBR value for a typical remoulded sample of the clay, the CBR test being done with surcharge weights totalling 35 lb. This shows the clay to have a CBR value of 2.5% at 32% moisture content, in close agreement with the value obtained from Fig. 2. In general remoulded soils have a lower bearing strenth than in-situ or undisturbed samples. In this case, however, care was taken to recompact with large pieces of clay (nominally 1 inch size) which had not been broken down during the drying/wetting cycle of the moisture control.

No data is available to indicate the equilibrium moisture content of chalk/clay mixtures, with or without surcharge weights, but it seems

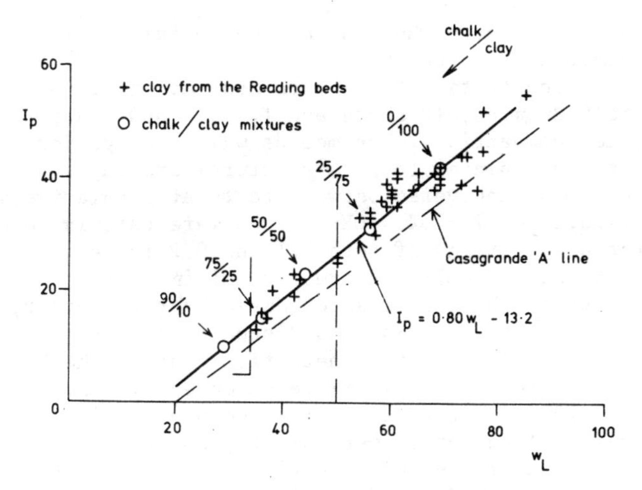

Fig. 1. Atterberg limits of clay from the Reading beds and of soft chalk/clay mixtures

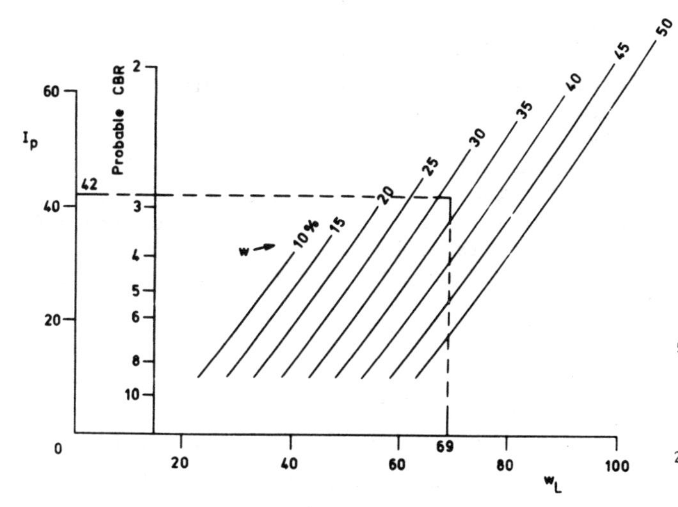

Fig. 2. Iso-moisture contours of plastic soils at probable equilibrium under pavement in southern England

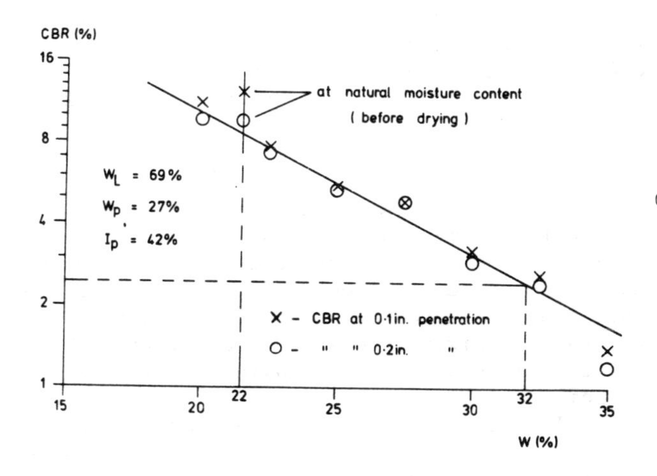

Fig. 3. Relationship between moisture content, w, and CBR for a typical plastic blue and red clay from the Reading beds

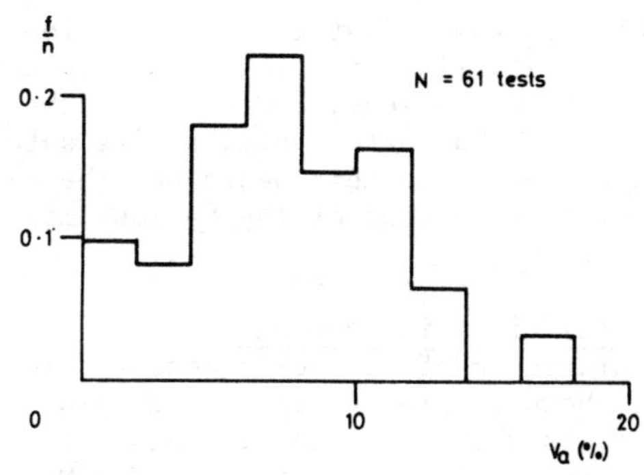

Fig. 4. Histogram of air voids content for various chalk/clay mixtures, recorded in the field

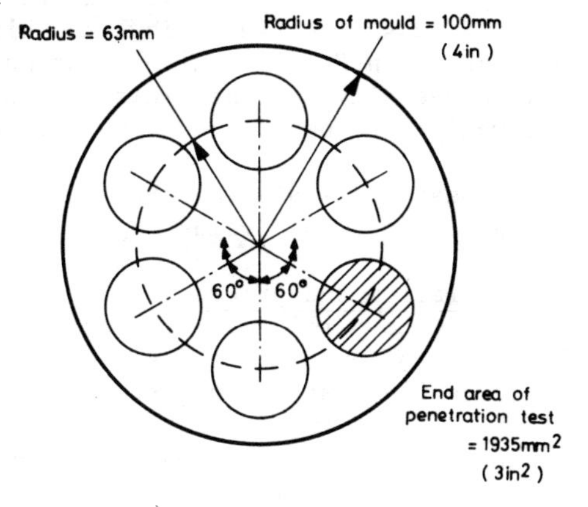

Fig. 5. Spacing of six penetration tests in a single mould

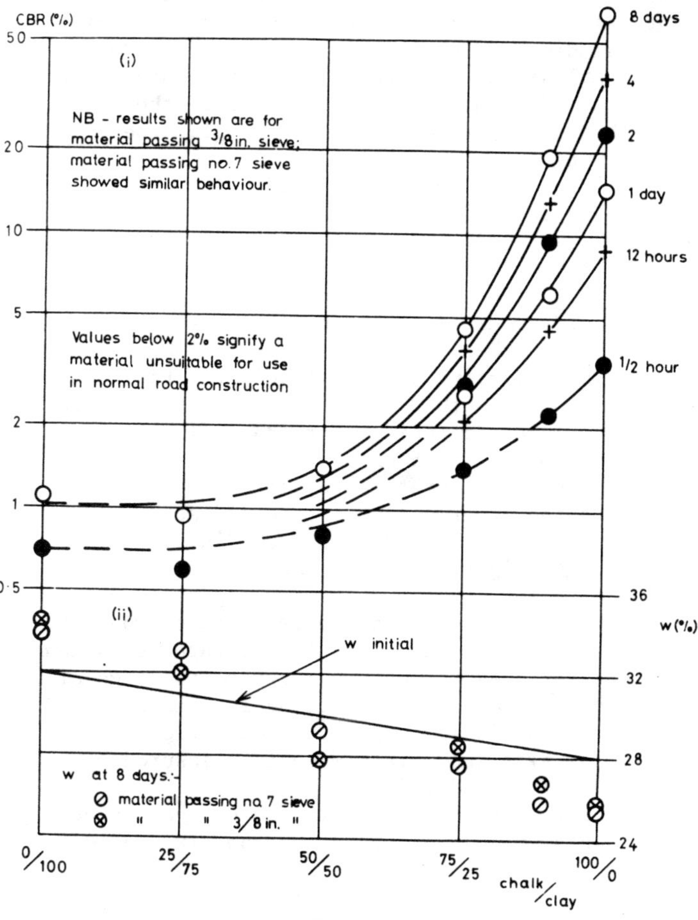

Fig. 6. (i) CBR of soft chalk/clay mixtures; (ii) change in moisture content of soft chalk/clay mixtures 8 days after compaction

reasonable to assume that the clay portion of such a mixture will absorb water to the same extent as the clay alone, and that the chalk will absorb or lose water until at its saturated moisture content. On this basis was the experimental moisture content of the various mixtures determined.

Experimental air voids content

A histogram of air voids measured in the field is shown in Fig. 4. The values plotted are for soils with moisture contents within 2% of those assumed for these tests. The M4 earthworks specification called for air voids in embankments of 10% or less, except in the 2 ft. immediately below formation level, where a maximum of 5% was specified. Thus the higher air voids shown in Fig. 4 (which occurred almost entirely in the clay and chalk/clay mixtures) would normally have been reduced by further compaction.

For these tests all the chalk/clay mixtures were compacted to 5% air voids.

Sample preparation and procedure

In order to permit easy mixing of the chalk and clay, and to obtain the desired moisture contents, remoulded specimens were used.

Inspection of embankments where instability of the chalk/clay fill had caused a suspension of earthworking showed that trafficking and/or compaction had resulted in a nearly homogeneous mixture, with no chalk fragments greater than ½ inch (12.7 mm) sieve size. Thus, after oven drying, the chalk and clay were broken down into two gradings - one of material passing a No. 7 (2.4 mm) sieve, the other of material passing a 3/8 inch (9.5 mm) sieve. Portions of the dry chalk and clay were then mixed to give the following by weight:

clay	100	75	50	25	10	0
chalk	0	25	50	75	90	100
w	32	31	30	29	28.4	28

for both coarse and fine sievings.

The standard method of measuring the CBR value of remoulded soils involves a single penetration in the centre of a 6 inch (150 mm) diameter mould, using surcharge weights. For these tests, moulds with 8 inch (200 mm) internal diameter were used. Using a standard plunger, of end area 3 in^2 (1935 mm^2), the larger moulds permitted six separate penetration tests to be made on the uppermost soil face of each mould. The spacing of the tests is shown in Fig. 5. Surcharge weights were not used. The disadvantage of doing all the tests for a particular mixture in one mould, since later tests would be affected to some extent by the disturbance caused by earlier ones and by proximity to the edge of the mould, was considered to be less important than the consistent degree of compaction, material size and moisture content ensured by the method adopted.

After thorough hand-mixing of the dry constituents, water was added and the mixture left sealed for five days. The moulds were then filled in the standard way (ref. 4), all the samples being compacted to 5% air voids. To minimise variation in the homogeneity and air voids content between different moulds, the same operator - who was skilled in this work - compacted all the specimens.

Penetration tests were done at approximately 30 mins, 12 hours and 1, 2, 4 and 8 days after compaction. The moulds which had perforated base plates to permit natural drainage, were kept uncovered between tests at a relative humidity of 70-75%. CBR values were calculated for penetrations of both 0.1 and 0.2 inches (2.5 and 5.0 mm). In order to minimise the effect of the disturbance caused by the plunger, each penetration test was discontinued after 0.25 inches (6.3 mm) penetration, rather than the 0.5 inches (12.7 mm) recommended by B.S. 1377:1967. For these tests, the penetration against load curves were smooth, and readings up to 0.25 inches were sufficient to enable the loads at both 0.1 and 0.2 inch penetrations to be determined without loss of accuracy. The moulds were weighed before and after each test in order to detect change in moisture content between tests.

Results

The Atterberg limits of all the chalk/clay mixtures containing clay were determined using standard techniques (ref. 4). In Fig. 1 plasticity index has been plotted against liquid limit on an extended Casagrande plasticity chart (ref. 6). The tendency for the plasticity of the chalk/clay mixtures to fall with decrease in clay content is to be anticipated. However, it is of interest to note that the values lie close to a line whose equation is

$$I_p = 0.80\, w_L - 13.2 \qquad \ldots\ldots (1)$$
$$= 0.80\, (w_L - 16.5).$$

This equation is similar to

$$I_p = 0.80\, (w_L - 16) \qquad \ldots\ldots (2)$$

obtained by Clare (ref. 7) for clays of the Chalk and Wealden beds, and to

$$I_p = 0.838\, w_L - 14.2 \qquad \ldots\ldots (3)$$

determined by Black (ref. 5) as characteristic of remoulded British clays in general. It appears that the relationship remains valid for mixtures of chalk and clay.

In Fig. 6 (i) the CBR attained at various times after compaction has been plotted against the clay content of the chalk. The steady increase in strength of mixtures with a high chalk content is to be expected, but the failure of mixtures with 50% or more clay to achieve any significant increase in strength during 8 days is perhaps surprising at first sight.

However, it can be seen from Fig. 6 (ii) that although kept at a humidity of 70-75%, during 8 days the moisture content of the pure chalk decreased by almost 10% while that of the pure clay increased by about 7%. Behaviour of the various mixtures was consistent, with the 50/50 chalk/clay mixture showing little change in moisture content.

The tendency for the unsurcharged clay to absorb water is to be expected - since the equilibrium values predicted by Black and assumed for these tests are based on the clay being confined beneath a 1½ ft depth of road construction. The reduction in moisture content of the chalk, initially in a saturated condition as when

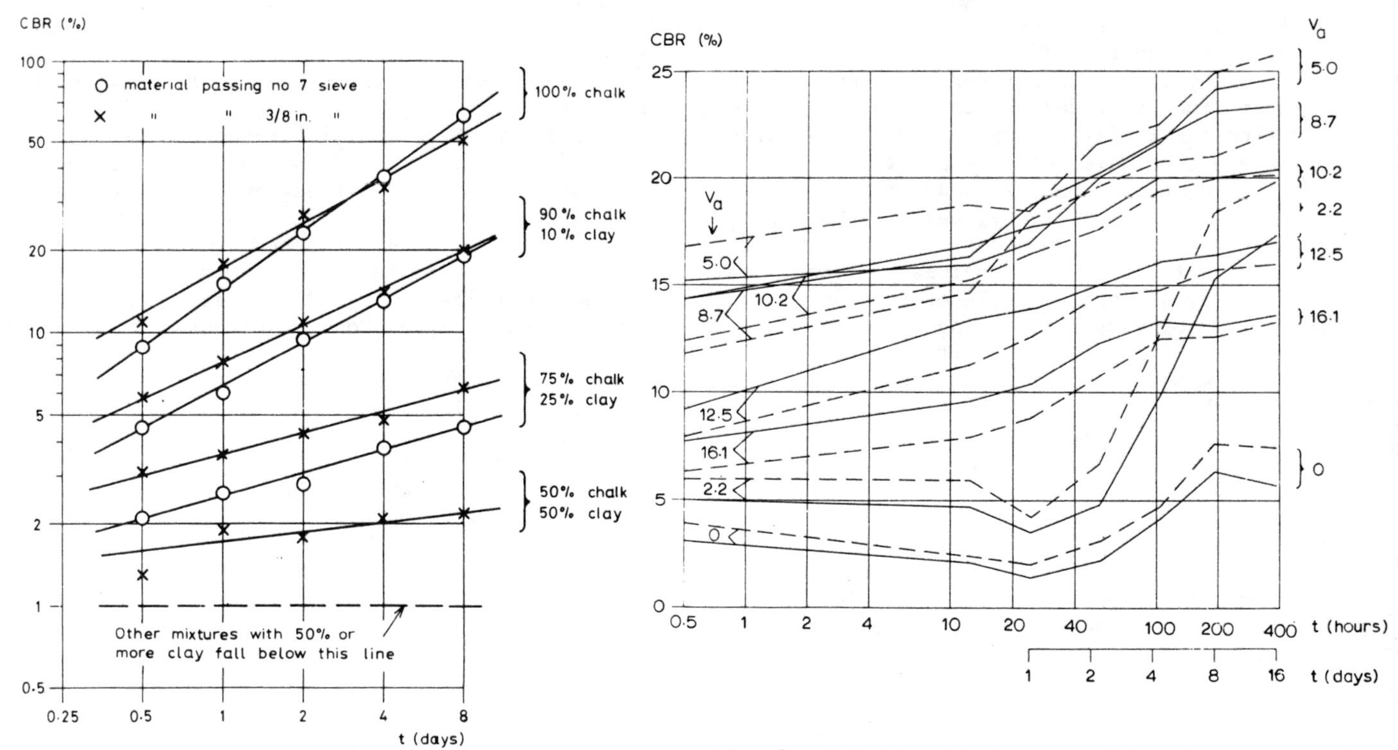

Fig. 7. Relationship between CBR and time for soft chalk/clay mixtures

Fig. 9. Relationship between CBR and time elapsing after compaction for soft chalk at various air voids from 0 to 15%

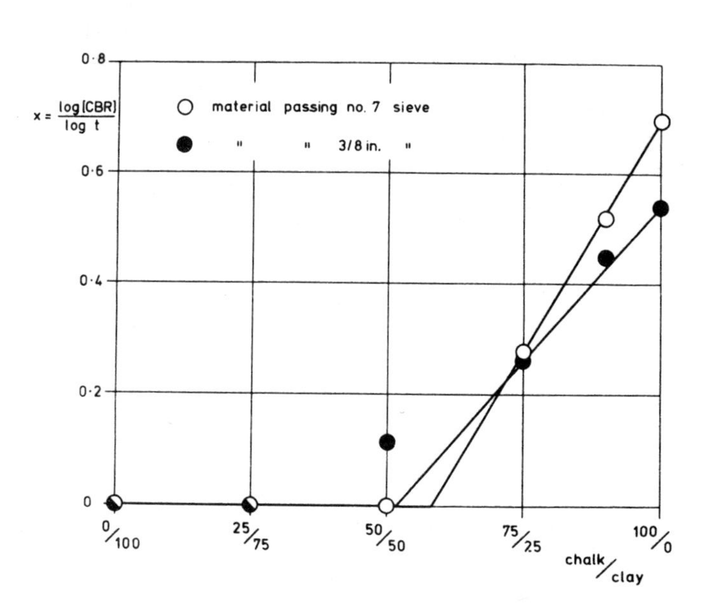

Fig. 8. Relationship between rate of increase in strength and clay content of soft chalk

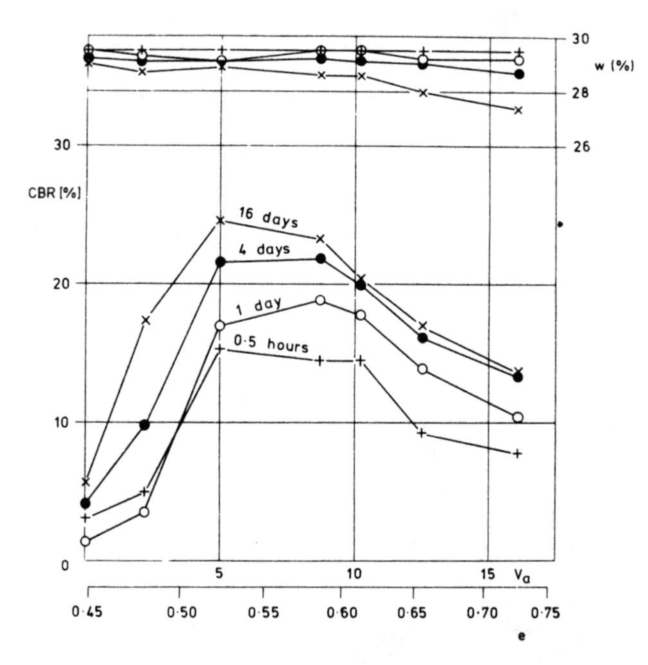

Fig. 10. Effect of air voids on the CBR of soft chalk at various intervals after compaction

freshly excavated, can be likened to the natural loss that would occur through drainage and evaporation if similar exposure at 70% humidity occurred during road-building.

On logarithmic axes, CBR has been plotted against time after compaction for both coarse and fine chalk/clay mixtures (Fig. 7). Results for $t = \frac{1}{2}$ hour have not been plotted in Fig. 7. Not only do these lie well away from the other results, but they are very sensitive to slight errors in the measurement of t. Furthermore, the CBR test is such that values of less than about unity are not reliable other than as an indication of very low bearing strength. Hence these also have not been plotted in Fig. 7.

For mixtures with 50% or more chalk, the relationship between CBR and t is seen to be linear during the period of the tests. It is in the form

$$(CBR) = at^{x} + C \qquad \ldots\ldots (4)$$

where a is a constant and C = CBR value at the time of compaction. The magnitude of x is a measure of the rate of increase in bearing strength with time. For the coarse 100% chalk sample, which showed the most rapid increase in strength, equation (4) approximates to

$$(CBR) = 14.1t^{0.67} + C \qquad \ldots\ldots (5)$$

In Fig. 8, x has been plotted against chalk/clay content. It can be seen that x is zero for mixtures with clay content greater than about 50%. These mixtures showed no increase in bearing strength during the period of the tests. While the results are insufficient to show clearly whether the relationship between x and clay content is linear, as indicated tentatively in Fig. 8, they do suggest a distinct difference in the rates of gain in bearing strength between the fine and coarse mixtures. The reason for this difference, which is particularly marked for the pure chalk, is not obvious.

BEARING STRENGTH OF CHALK AT VARIOUS AIR VOIDS

Materials

Fairly soft Upper Chalk from near to Pangbourne (Berkshire), OS SU 614738. Samples, taken from 8-12 ft below the interface between the chalk and the Reading sands and clays, were moderately block-like when excavated.

$$G = 2.58 - 2.60$$

$$w_n = 29.2\%$$

bulk density, $\gamma = 119.7\ lb/ft^3$ ($1917.4\ kg/m^3$)

dry density, $\gamma_d = 92.4\ lb/ft^3$ ($1480.1\ kg/m^3$)

Sample preparation and procedure

Freshly excavated chalk was broken down into passing 3/4 inch (20 mm) sieve size. Using standard techniques (ref. 4), large CBR moulds were filled to obtain samples with air voids of nominally 0, 2.5, (compacted in 5 layers), 5, 7.5, 10, 12.5 and 15% (compacted in 2 layers). The chalk was not crushed any more than was necessary to achieve these values.

Test procedure was similar to that already described for the chalk/clay tests except that an additional penetration test was performed on the undisturbed area in the centre of each mould Penetration tests were done at 30 mins., 12 hrs. and 1, 2, 4, 8 and 16 days after compaction of the moulds. Between tests, the moulds were kept covered at a humidity of 95-100%.

Results

The relationship between bearing strength, represented by the CBR value obtained without surcharge weights, and time is shown for the various air voids in Fig. 9. As is commonly the case, the bearing strengths calculated for 0.1 and 0.2 inch penetrations (both of which are shown in Fig. 9) differed by up to 15%. It is customary to take the higher of the two values (normally that at 0.1 penetration) for purposes of design. For these tests, at air voids of 7.5% or more, the 0.1 inch value was always the greater, but for the more highly compacted specimens, with air voids of 5% or less, the CBR value at 0.2 inch penetration was always the greater. Where the 0.1 inch value is the greater, failure is due principally to shear, whereas when the 0.2 inch value is greater, increase in bearing strength with depth of penetration is due to the soil behaving as if partly confined, with density approaching the bulk value.

Both the 0.0% and 2.5% samples showed a drop in bearing strength over the first 24 hours, but with the dissipation of positive pore pressures, the 2.5% sample in particular showed a rapid increase in strength between 1 and 8 days. The remaining samples, with 5% or more air voids, showed fairly steady gains in strength - of approximately linear form on CBR/log(time) axes - over the period of the tests.

The relationship between bearing strength (calculated at 0.1 inch penetration) and air voids, at various time intervals after compaction is shown in Fig. 10. This shows, immediately following compaction, the optimum air voids to between 5% and 10% for maximum bearing strength, although even at 15% the CBR value is sufficiently high to permit earthworking to proceed without difficulty. There is a sharp fall in the bearing strength for air voids of less than 5%. Although all the mixtures showed some increase in strength over the period of the tests, the optimum air voids for maximum strength remained at between 5% and 10%.

Also shown, in Fig. 10, are the changes in moisture content during the period of the tests. At the high relative humidity of 95-100%, water loss through evaporation was likely to be slight. The relatively small changes recorded in the overall moisture content suggest that water loss through drainage, also, was slight. As might be expected, specimens with high air voids showed the greater water loss.

No direct comparison can be made with the 100% chalk results from the chalk/clay tests, where remoulded chalk was used at a relative humidity of 70-75%. In practice, at the average ambient relative humidity between April and October in central-southern England of 73%, reduction in moisture content of the chalk through evaporation would be accompanied by an increase in the CBR value. This is not a feature of these tests where the high relative humidity of 95-100%, chosen to simulate the earthworking of chalk in inclement weather, prevented any significant decrease in moisture content during the period of the tests.

GENERAL DISCUSSION

Although the ability of clay fill to support construction plant has been discussin in general terms by Rodin (ref. 8), and some problems associated with the earthworking of soft chalk have been dealt with by Parsons (ref. 9), no work appears to have been done on the bearing strength of mixtures of soft chalk and clay.

In practice, considerable variation in the localised clay content of the chalk might be expected. However, close inspection of troublesome areas has shown that the disturbance caused by heavy rubber tyred motorscrapers struggling to pass through a spongy area and deposit their load has the effect of producing a near homogeneous mixture. Thus one of the main assumptions of the chalk/clay tests - that the materials were evenly mixed - would seem valid when investigating the problem of spongy areas.

Two other factors affect the relevance of the chalk/clay results, namely, the air voids content (5% for these tests) and the moisture content. There seems no reason why the former, based on the value both specified and achieved in practice, should not be typical of chalk/clay mixtures. The moisture content assumed for the tests represents a situation where free water is present to wet-up the clay during earthworking. Although a severe condition, it occurs frequently. Common sources of water are the chalk fissures, rain or springs.

The results show that for the particular materials and moisture contents used in the tests, mixtures containing more than about 25% clay do not achieve sufficient bearing strength in eight days to permit either the operation of normal earthmoving equipment, or the construction of a stable road. For each of these, materials with a CBR value of less than 2% are normally considered unsuitable. Even if left for much longer periods, it is unlikely that the mixtures with CBR values of less than 2% would show a significant increase in strength.

The clay used for the tests is a relatively poor road building material. However, the behaviour of a stiffer clay, with lower values of w_L and w_p, is difficult to predict. Certainly the lower equilibrium moisture content of such a clay would make it likely, provided excessive wetting did not occur during earthworking, that when mixed with chalk the inherent higher bearing strength of the clay would improve the strength of the mixture. Nonetheless, the clay would still tend to clog fissures and prevent the dissipation of positive pore pressures in the chalk. Thus it might well be that at lower moisture contents the bearing strengths of some of the mixtures would not only be lower than the chalk strength but lower than the clay strength also. This tendency is suggested by Fig. 6 (i) where, however, the apparent rise in strength for the 100% clay is within the limits of experimental error.

The chalk tests at various air voids indicate that for maximum bearing strength immediately after compaction, the air voids content should be between 5% and 10%. Fig. 10 shows that only zero air voids results in a CBR value less than that corresponding to minimum pavement thickness.

Opinions vary widely regarding the degree to which chalk should be compacted, although this is due in some measure to a failure to differentiate clearly between soft and hard chalk. Substantial compaction with heavy rollers is advocated by some authorities, including the TRRL book "Soil Mechanics for Road Engineers" (ref. 5) and Lewis and Croney (ref. 3); although the latter, together with Toms (ref. 10), recognise the danger of over-compaction. On the other hand, earthmoving contractors (for example Pierce, ref. 11) usually express the view that no compaction at all is best for chalk. Rodin (ref. 8) suggests that in order to minimise the constructional problems arising from over-compaction it is more important to work to a minimum air content than to a maximum.

Now the experimental results give no indication of settlement. From the results of tests Lewis and Croney (ref. 3) suggest that very little reduction in settlement would result from achieving a higher state of compaction than that equivalent to an air content of 10%. Investigations carried out by the TRRL have shown that where the dry density of chalk fill was approx. 1281.5 kg/m^3 (80 lb/ft^3) the settlement after one year was about 2.5%, but where the dry density was 1409.6 kg/m^3 (88 lb/ft^3) the settlement was negligible. Further, they concluded that on lightly compacted chalk embankments the recorded settlements after 5-7 years were from 0.4% to 1.1% of the height of the fill, the corresponding figure for better compacted fill amounting to 0.2%. In the exceptional case of chalk deposited at 12-15% air voids, settlement might be as much as 5% of the height of embankment.

To minimise settlement, it would seem best to compact so that the air voids content is only a little over 5%. Experience suggests, however, that to aim at 5% air voids would almost certainly result in some over-compaction. Thus as a general rule it is better to aim at air voids of 10-12% - almost certainly obtainable without special compaction - and accept some settlement, rather than run the risk of reducing the bearing strength by over-compaction with all the attendant delays and disruption to earthworking.

CONCLUSIONS

The conclusions may be conveniently incorporated in the form of proposals for good practice in the earthworking of soft chalk.

1. During the winning of the chalk, minimize mixing with other soils, especially clay.

2. Minimum compaction compatible with tolerable settlement is best for speedy earthworking. Normal settlement can be allowed for by slightly overfilling. Where a soft chalk embankment is of reasonable length, differential settlement should not pose a problem except at underbridges. For these, special precautions or design details such as run-on slabs or wedges of granular non-plastic fill would be required.

3. Where possible, earthmoving equipment with low bearing pressure - of which perhaps tractor towed boxes are ideal - should be used. This

will avoid the impact effect due to the bouncing of rubber-tyred scrapers, which tend to produce corrugations, and to crush the chalk.

Alternatively, the use of face shovels for excavation and dumpers for hauling and tipping the chalk would be beneficial in minimising crushing (ref. 12). This method has been cited by the earthmoving contractors for the M40 through the Chilterns as both the most economical and also the least likely to slurry the chalk (ref. 13).

4. If heavy rubber tyred motor-scrapers have to be used
   a) push load to enable a deeper cut and to minimize crushing during the winning of the chalk, and
   b) the chalk should be deposited outside the confines of suspected spongy areas and only a heavy duty tracked blade used to spread and compact the fill. The low bearing pressure of a tracked blade is well suited for compaction without undue crushing. Use of this procedure would generally prevent a highly pulverized wet impermeable skin being created, and could result in up to 1 m. of fill being placed every day instead of every 2 - 3 weeks.

5. Highly pulverized areas are most likely to develop on haul routes along chalk embankments. If only a single route exists, earthworking must cease should it become impassible, although the main area of deposition may be completely stable. This situation is most easily avoided by
   a) restricting the speed of the scrapers, and
   b) changing the position of the haul route sufficiently often to prevent positive pore pressures arising due to heavy trafficking.

6. A clay skin is sometimes laid along the haul route over a chalk embankment. This tends to reduce local pulverization of the chalk but is a practice which should be discouraged during the building of an embankment since clay is likely to be introduced into the fill from the haul route. This will have an adverse effect on the eventual bearing strength. Where a clay surfaced haul route is required for other construction traffic, it should be laid after the chalk embankment has been taken to full height (or to a height where it is going to be left for a period to consolidate). In the interests of minimum pavement construction depth, clay laid for a haul route is best removed and the final road laid directly onto chalk.

7. While it is good practice during construction to keep embankments graded to assist rainwater run-off, the spongy nature of chalk can make grading difficult. Also, since soft chalk is relatively impermeable, stabilization and consolidation of the mass of a chalk embankment is largely unaffected by surface saturation.

8. Chalk which, through a combination of clay contamination and wetting, has a CBR value of 2% or less is unlikely to show any significant gain in strength over at least 8 days, and normally should be classed as unsuitable material.

ACKNOWLEDGEMENTS

I am indepted to Mr. R.J. Coleman (Sir Alexander Gibb and Partners) and Dr. P. Vaughan (Imperial College) for encouragement and many helpful suggestions during the preparation of the paper. I wish to thank Sir Alexander Gibb and Partners for permission to publish the results used in the paper.

REFERENCES

1. ANON. M27 muck-shifter to claim £1/4 m. New Civil Engineer, 1973, 24th April.
2. ANON. M3 : Sunbury - Lightwater Section. Highways Design and Construction, 1973, 41, No. 1760, 29-32.
3. LEWIS, W.A. and CRONEY, D. The properties of chalk in relation to road foundations and pavements. Proceedings of the Symposium on Chalk in Earthworks and Foundations, Institution of Civil Engineers, London, 1966, 27-42.
4. B.S.1377 (1967). Methods of Testing Soils for Civil Engineering Purposes. British Standards Institute, London, 1967.
5. BLACK, W.P.M. Method of estimating the California Bearing Ratio of Cohesive Soils from Plasticity Data. Geotechnique, 1962, 12, No. 4, 271-282.
6. ROAD RESEARCH LABORATORY. Soil Mechanics for Road Engineers, H.M.S.O., London, 1962, 2nd Edition, 126-131.
7. CLARE, K.E. Laboratory studies relating to the clay fraction of cohesive soils. Proceedings of the Second International Conference on Soil Mechanics and Foundation Engineering, Rotterdam, 1948, 1, 151-158.
8. RODIN, S. Ability of Clay Fill to Support Construction Plant. Civil Engineering and Public Works Review, 1965, February, 197-202, 1965, March, 343-345.
9. PARSONS, A.W. Earthworks in soft chalk. A study of some of the factors affecting construction, R.R.L. Report L.R. 112, Road Research Laboratory, Crowthorne, 1967.
10. TOMS, A.H. Chalk in cuttings and embankments. Proceedings of the Symposium on Chalk in Earthworks and Foundations. Institution of Civil Engineers, London, 1966, 43-55.
11. PIERCE, MICKEY. Earthmoving and earthworks. Highways Design and Construction, 1973, November, 31-32.
12. PARSONS, A.W. Session B, Discussion, Proceedings of the Symposium on Chalk in Earthworks and Foundations, Institution of Civil Engineers, London, 1966, 103-104.
13. ANON. M40 earthworks contractor solves problems posed by Chiltern chalk, Highways Design and Construction, 1972, 40, No. 1756, 4-6.

G. H. THOMPSON, Cementation Ground Engineering Ltd,
and A. HERBERT, Menard Techniques Ltd

Compaction of clay fills in-situ by dynamic consolidation

Most techniques for compacting clay are based on the fill material being brought to site and placed and rolled in thin layers. There are, however, many sites now being developed in which the fill has already been placed for a number of years and was placed in an uncontrolled manner. The cost of excavating existing fill and re-compacting in thin layers can be prohibitive for deep deposits and compaction of the material in-situ from the surface to provide suitable bearing for structures or roads is most attractive. The methods of compacting the fill in-situ can be divided into two classes, non specialist techniques (e.g. pre compression or surcharging) and specialist techniques. In this paper, experience is drawn from various case histories of projects in Great Britain and France on which the dynamic consolidation process has been used to treat cohesive fills for a variety of purposes. A range of tests have been employed on the various projects and the test results together with settlement readings of completed structures provide useful data for evaluation of the effects of the treatment. The combined compaction of clay fill embankments together with the underlying material by dynamic consolidation from the top of the embankment has been used on the continent but appears to be little known in the U.K. The paper presents details of a motorway project in France where compaction was carried out on a clay embankment.

CONSOLIDATION OF CLAY FILLS

1. There are increasing economic restraints on construction of new developments requiring the use of less expensive methods of supporting structures on poor ground. At the same time sites which previously would not have been considered for development, often underlain by backfilled workings, are increasingly being employed for new building projects.

2. Except where the fill is strictly controlled for a specific purpose, the backfill material is usually a very heterogeneous mixture, often containing demolition debris and other waste materials. In bulk however, the fill will exhibit properties either essentially cohesive or cohesionless.

3. Various ground treatment processes for stabilising fills and rendering them suitable for building development have been used for many years, but it is only within the last eight years that ground compaction by the technique known as Dynamic Consolidation has become accepted since it was introduced in France by Louis Menard (ref. 1).

4. The Dynamic Consolidation process has been described elsewhere in detail and in this paper will only deal with the application of the process to specific projects. (ref. 2)

5. Initially Dynamic Consolidation was employed for compacting essentially cohesionless materials but increasingly clay fills have been successfully treated for both industrial and housing projects.

6. Of the 350 sites now treated by Dynamic Consolidation approximately 60% have been carried out on fill materials, the percentage being higher for contracts completed in the United Kingdom. For clay fills it is necessary to obtain, at the feasibility stage, the granulametric analysis, clay mineralogy, density, organic content and water table to estimate the degree of improvement that could be achieved. An equally important parameter is the stiffness of the clay lumps, for it has been found that in a firm clay fill only the top 8 to 10m can be compacted by usual 15tonne x 20m drop equipment, compared to 20m in a granular fill.

7. The large dredgers now in use for the placement of hydraulic fill are resulting in a larger proportion of clay and silt being placed within the fill. Occasionally trapped pockets of water are formed which often result in geysers appearing at the surface during treatment by Dynamic Consolidation.

8. The following five contracts illustrate the range of projects for which Dynamic Consolidation has proved successful in improving the clay fill characteristics.

I - Corby, Snatch Hill Experimental Area - 1973

9. Much of the area around Corby has been worked for ironstone and in the Snatch Hill area the seam of iron ore was at a depth of 30 metres. The overburden was first removed by drag line and replaced when the ironstone had been removed in a continuous hill and valley operation, the site ultimately being

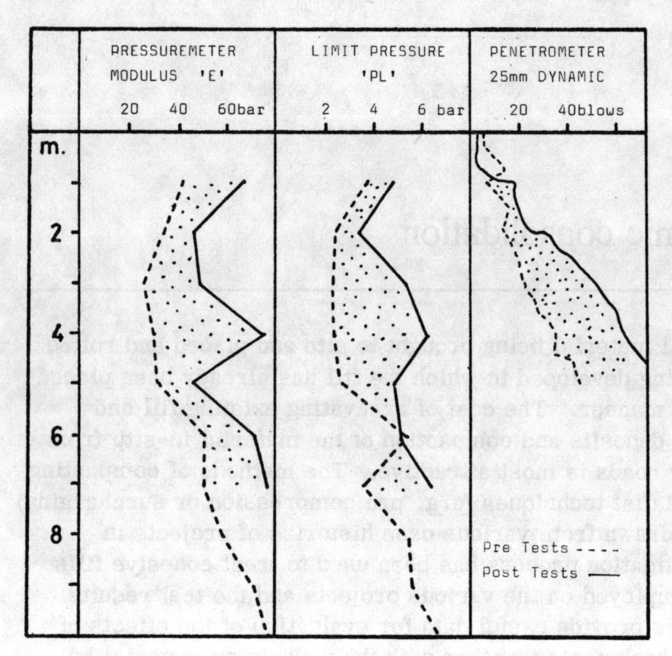

Fig. 1. Corby Snatchill average test results

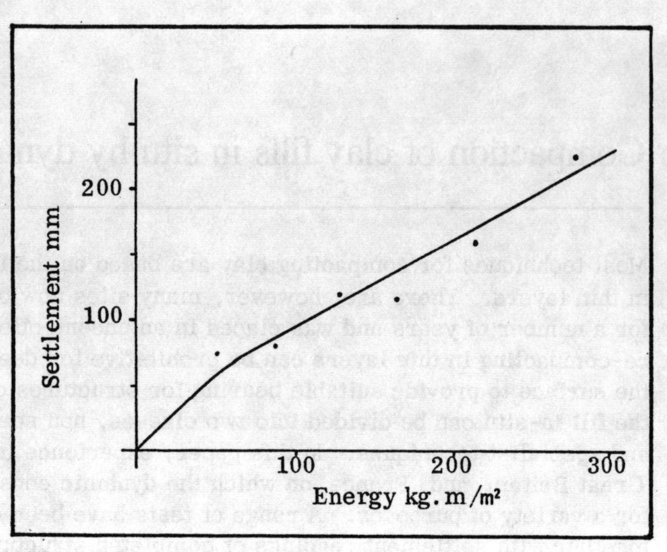

Fig. 3. Settlements at Corby induced by dynamic consolidation

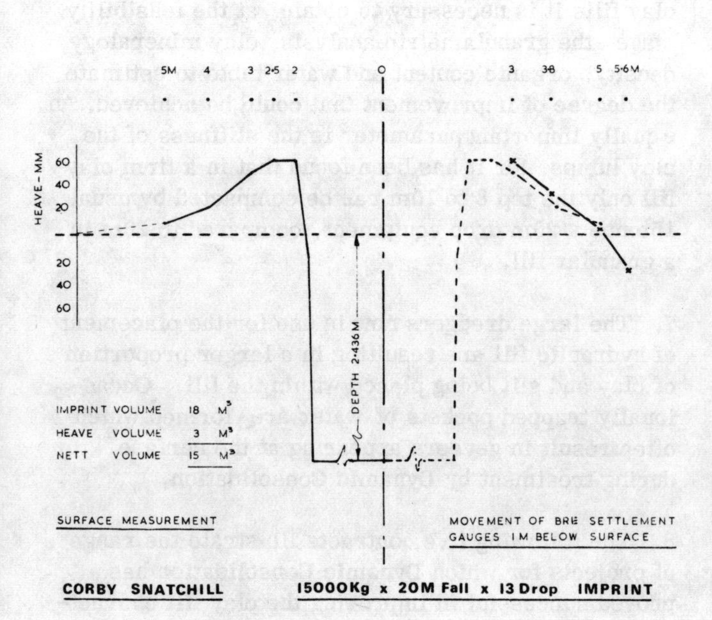

Fig. 2. Typical heave diagram

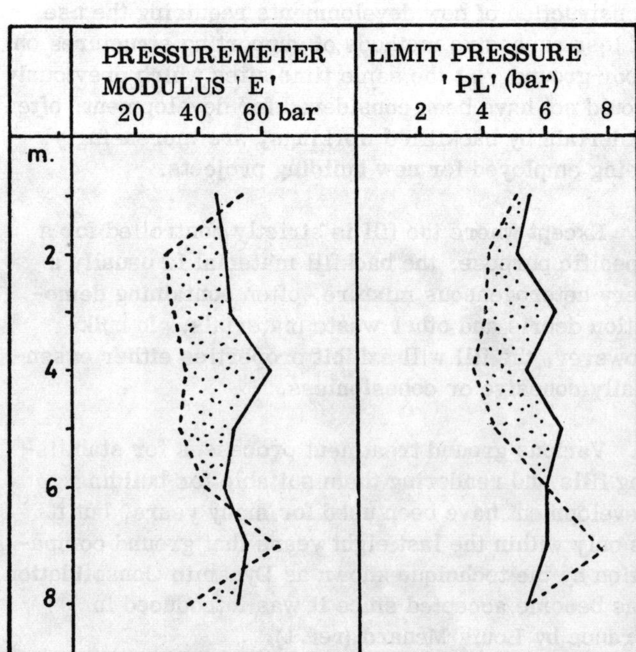

Fig. 4. Surrey Docks Zone B average test results

levelled by dozing.

10. Although some effort was made to replace the clay at the surface overlying replaced broken rock, inevitably some mixing took place and the upper 12m of the ground consists essentially of clay with sand, gravel and boulders. The clay fill had a moisture content generally ranging from 17% to 22% with a bulk density of 1900 Kg/m3 to 2200 Kg/m3 and cohesion between $40kN/m^2$ and $110kN/m^2$. After the backfilling had been completed for some five or six years, Corby Development Corporation decided to undertake trials in association with the Building Research Establishment to evaluate the affects of different methods of ground treatment to reduce settlements likely to affect proposed two storey housing developments. One of the methods chosen was the Dynamic Consolidation process to compact an area approximately 50m x 50m for a design bearing pressure of $100kN/m^2$.

11. In order to accelerate the settlement and consolidate a 10m raft of deep fill, compaction was carried out using a 15 ton tamping block dropped from a height of 20m on a 10m grid. A number of drops were performed at each point on the 10m grid, the number of drops being evaluated from what is referred to as a heave diagram (see Fig. 2).

12. Pressuremeter (ref. 3) and dynamic Penetrometer tests were undertaken across the site to evaluate the initial ground conditions and improvements obtained by compaction (see Fig. 1).

13. A method of control during ground treatment is to take levels on a grid of say 5m over the whole site area both before and after treatment. A total of some 110 positions were checked and the enforced settlement indicated by these levels suggests that the site surface settled some 25cm due to the ground treatment, approximately 2.5% of the depth treated (see Fig. 3).

14. In order to keep within the Development Corporation's budget figures a restriction was imposed on the equipment and the energy applied during treatment was 12.5 t. m/cu. m. Although the treatment has produced significant compaction of the ground subsequent experience with Dynamic Consolidation suggests that a higher energy level would now be appropriate.

15. Subsequently two storey houses have been constructed on the site using conventional foundations and have been occupied for some three years.

16. It is normally preferable to undertake Dynamic Consolidation treatment of clay fills of this type after they have been in place for at least 10 years in order to allow some settlement under self-weight of the deeper material. However on this site as the fill was only about five or six years old some continuing settlement of the deeper fill was anticipated and settlement checks before and after ground treatment generally coroborate this. At 10m depth settlement under Dynamic Consolidation was less than 10mm while under the area in which an earth spoil surcharge was applied, settlement was around 40mm. A paper by Dr. Charles of the Building Research Establishment gives more detail of the settlement readings which are still being taken for the structures on the area treated by Dynamic Consolidation together with details of other test areas employing other ground consolidation techniques.

II - Surrey Docks Development - 1978

17. As part of the overall plan to redevelop the London Docks, the Surrey Docks system is being infilled prior to housing and commercial development.

18. In order to provide overall site stability and safely support roads and services, the use of Dynamic Consolidation Ground Treatment was proposed and a trial compaction and testing programme undertaken on two typical areas.

19. The fill consisted essentially of soft to firm silty clay and clay silt with gravel, concrete and brick rubble, a little timber and occasional pockets of clayey sand to depths ranging between 6.6 and 10.9m. Beneath the fill, there was Thames ballast, a medium dense to dense gravel with sand.

20. A granular blanket approximately 500mm thick was first laid over the clay surface of the site and a grid of levels taken. Levels were taken at several stages in the contract to ascertain the amount of enforced settlement achieved by the treatment. After completion of the Dynamic Consolidation, the average enforced settlement was of the order of 0.5m.

21. Both 10 ton and 15 ton weights were used for tamping dropped from heights upto 20m with a 61RB Crane. The pattern of tamping was initially on a 10m open grid which produced craters over 2.5m deep after only two initial drops. The craters were backfilled with granular material and tamping repeated to provide the total energy input for that particular phase.

22. In subsequent phases the tamping was changed from an open grid to a continuous surface tamping to compact the upper layers. The total energy imput into the fill ranged between 18 tonne metres per m3 and 23 tonne metres per m3.

23. Pressuremeter Tests were taken prior to treatment and at various stages throughout the work to assess the improvement being achieved. Generally in clay soils, the full improvement will not be apparent until after the increased pore

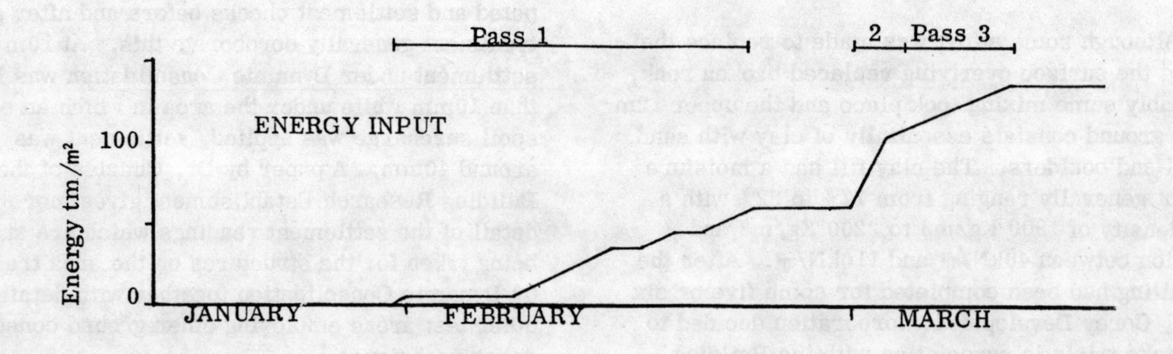

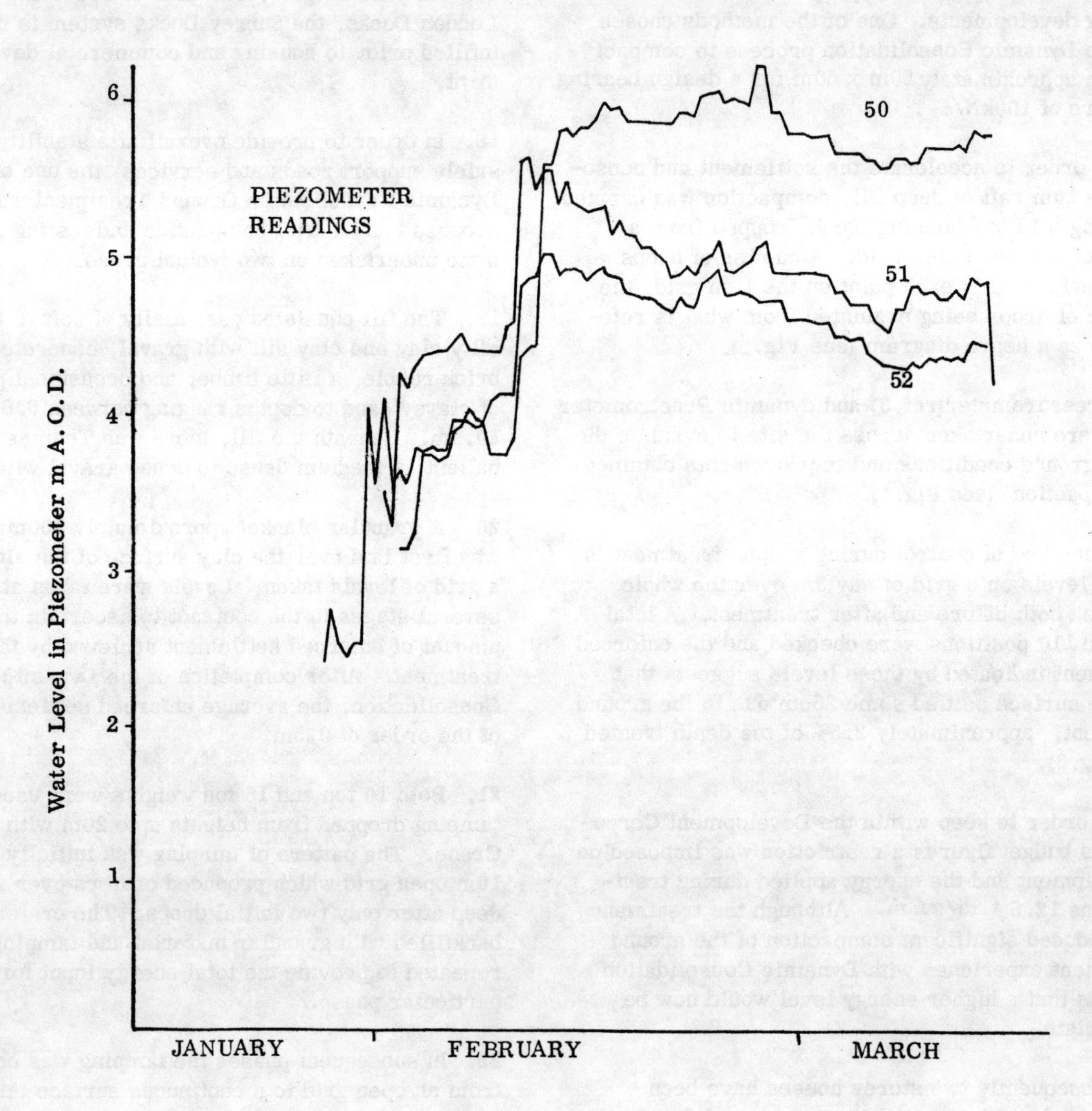

Fig. 5. Surrey Docks Zone C piezometer readings

pressures generated by the treatment have fully dissipated. However, the Pressuremeter Tests taken within two weeks after completion of the work show some overall improvement but the main impr-ovement is in the more uniform ground conditions obtained (see Fig. 4).

24. Both stand pipe and closed cells piezometers were installed around the edge of the compaction zone. These indicated immediate rises in hydro-static water level immediately following the initial phases of treatment, in some cases with a water head well above ground level. In subsequent phases of the treatment however, the piezometer response was not so marked but raised pore pressures were still evident one month after completion of the work. Fluctuations in the water level in the piezometers was observed to be related in some instances to rainfall which caused short term fluctuations during periods when no energy was being applied to the ground (see Fig. 5).

25. Based on the results of the Pressuremeter Tests the amount of settlement to be expected of the treated ground due to self-weight, according to the method proposed by Menard (ref. 4), is negligable and the calculated settlements for a typical slab 3 x 3m loaded up to 42kN/m² are in the range 5mm to 16mm.

III - Leucate-le-Barcares, France 1972

26. Autoroute 9 in the Narbonne district was opened to traffic in the spring of 1971 (ref. 5). Part of the road was constructed in cut and fill both longitudi-nally and in section. The clay fill proved to be of poor quality and was supplemented by imported fill but compaction proved difficult in the winter months. The depth of fill ranged from an average of 5 metres to a maximum 15 metres and soon after the road was opened large longitudinal cracks appeared in the wearing course with differential settlements in excess of 10cm appeared. A report by the regional laboratory at Toulouse showed the dry density of the fill to be too low in relation to the optimum Proctor value and the following remedial measures were considered :

1) The top 3 metres to be rebuilt with more suitable fill.
2) Vibro replacement stone columns
3) Dynamic Consolidation.

The recommended method of Dynamic Consolidation was carried out using a 9.5 Tonne weight with a first pass grid of 10 metres.

27. The application of this technique to a road in use presented a number of special problems from the presence of underground services, embankment slope stability and structures such as bridges etc. 3m deep cut-off trenches were dug 10m from piles in the bridge abutments to reduce the vibrations from the effects of the tamping operations by a factor of 20.

28. To prevent service ducts in the embankment from collapsing, the interiors were supported by an inflated membrane to 30kN/m² pressure.

29. Due to the unusual technique adopted, the traditional controls could not be employed, thus the specification consisted of post treatment soil density greater than or equal to 95% of the optimum Proctor value from the surface to 5m depth and 90% from 5 to 10m.

30. Fill material. The fill consisted of a clayey marl with the following properties :

- Atterberg Limits
  LL = 38%, PL = 17%, PI = 21%
- $CaCO_3$ content = 42%
- in-situ water content 17 to 18% and degree of saturation averaged 89%
- Proctor optimum : 15.2% water content, Density 17.8 kN/m^3

These results indicated that the fill material was extremely difficult to place and relative to the optimum Proctor value is in a 'weak' state with densities between 14 and 16 kN/m3.
Test results on the fill before Dynamic Consolidation

- Pressure limit, 150 to 200 kN/m²
- Undrained triaxial tests, ϕ' = 25° and c' = 15 kN/m²

31. Improvements achieved. Density measure-ments from gamma-backscattering show that Dyn-amic Consolidation has homogenised the fill material, eliminating the numerous voids, and improving the average specific density as follows:

		Pressure Limit
- pre tamping	γh = 18.7kN/m^3	200kN/m²
- after 1st pass	γh = 19.5kN/m^3	
- after 3rd pass	γh = 19.7kN/m^3	500kN/m²

The second and third passes were performed when the dissipation of excess pore water pressure reached 25 and 50% respectively. The fourth pass, a month after the 3rd pass was made at 65% dissi-pation, and the enforced settlement was estimated at 5% of the depth of the fill, thus the density after the final pass was estimated as 20.5kN/m3.

32. The corresponding dry densities are :

- pre γd = 15.8 kN/m^3
- post 1st pass γd = 16.5 kN/m^3
- post 3rd pass γd = 16.6 kN/m^3
- post 4th pass γd = 17.3 kN/m^3 (estimated)

As the optimum Proctor density is 17.8kN/m^3 the measured densities were $\geq$ 95% OPN. By comparing Pressuremeter results this condition was satisfied when PL ⩾ 600kN/m² and this criteria was used as the acceptable minimum.

33. The enforced settlements varied between 0.2 and 1.4m equivalent to a densification of 5 to 13%. Lateral displacement of the side slopes averaged 0.6m at the embankment crest.

34. Since its re-opening to traffic, the embankment has settled a maximum of 20mm, and less than 10mm over the majority of its length.

IV - Ambes, France - 1974

35. The site lies near Bordeaux, 3Kms upstream from where the Dordogne and the Garonne rivers meet. A 55m diameter tank, with a design bearing pressure of 145kN/m² was to be built on site.

36. Calculations made using the pre consolidation results indicated that the total edge settlements would be in the order of 1.28m, a figure confirmed by observation of nearby tanks.

37. The site consisted generally of 18m of recent soft river-clay deposits underlain by approximately 5m of clean sand and gravel which overlayed stiff marl.

38. Soil parameters. Laboratory tests on the Alluvial clay revealed the following properties :

Particle Size Distribution	By Weight
Smaller than 50 microns	98%
Smaller than 20 microns	74%
Smaller than 2 microns	32%

Atterberg Limits (Average Values)

Depth :	0 - 10m	10-18m
Liquid limit	46%	61%
Plastic limit	21%	22%
Moisture content	48%	64%
Casagrande Classification	CI	CH
Average degree of Saturation 'Sr'	98%	
Organic Matter Content	1.5%	
Total $CaCO_2$	6.8%	

Clay Classification

In order to correlate and classify the Dynamic Oedometer Laboratory Tests, mineralogical analyses were carried out which revealed the following proportion of the constituents :

Kaolinite	24%
Illite	25%
Smectite	39%
Various	12%

The Ion exchange capacity method (for 42milli. equi/100gr)

Kaolinite	33%
Illite	36%
Smectite	31%

Exchangeable Cations (filtrated with NH_4^+)

Ca^{++}	36.9
Mg^{++}	6.64
K^+	0.62
Na^+	1.90

39. Mechanical Laboratory Tests. Undrained triaxial compression tests were made to determine the undrained shear strength of the soft clays. Between depths of 0 to 10m, the Cu value averaged 12kN/m² (V. Soft), and a gradual increase, to approximately 27kN/m² is noted at greater depth.

Table 1. Co-efficient of consolidation

	Average Results			
Depth	0 - 4m	4 - 10m	10-18m	0 - 20m
C_v Values cm2/sec.	$2.5x10^{-4}$	$3x10^{-3}$	$2.5x10^{-4}$	$2x10^{-3}$
m²/year	0.79	9.5	.79	6.3
	(V. Slow)	(Rapid)	(V. Slow)	(Rapid)

40. Compression Index. Values ranged between 0.19 to 0.90 with an average value of 0.42

41. In-situ tests. Several penetrometer tests were carried out in order to determine the thickness of the various layers. Pressuremeter tests were also carried out to determine the Limit Pressure and modulus values of the material.

42. Pore pressure. In order to accelerate the dissipation of excess pore pressure set up during Dynamic Consolidation, 3.5m deep drainage trenches were constructed and backfilled with free draining material. Piezometers were placed in Zone 'A' at depth of 3, 5, 7, 9 and 11 metres. The measured porewater pressure at ground level prior to tamping was 0.2Kg/cm2. The maximum pore pressure increase measured during tamping was 0.45Kg/cm2 at 11m, 0.6Kg/cm2 at 9m, 0.4Kg/cm2 at 7m and 0.3Kg/cm2 at 3m depth. Dissipation took 3 to 4 weeks which is fairly common in this type of soil.

43. Pressuremeter results. Typical Pre and Post Pressuremeter results are shown in Fig. 7. These show that the geometric mean of the Pl has been doubled down to a depth of 18m. Dynamic Consolidation of the Alluvium caused a large rise in pore pressure which dissipated slowly resulting in a gradual but definite improvement in the soil characteristics with time.

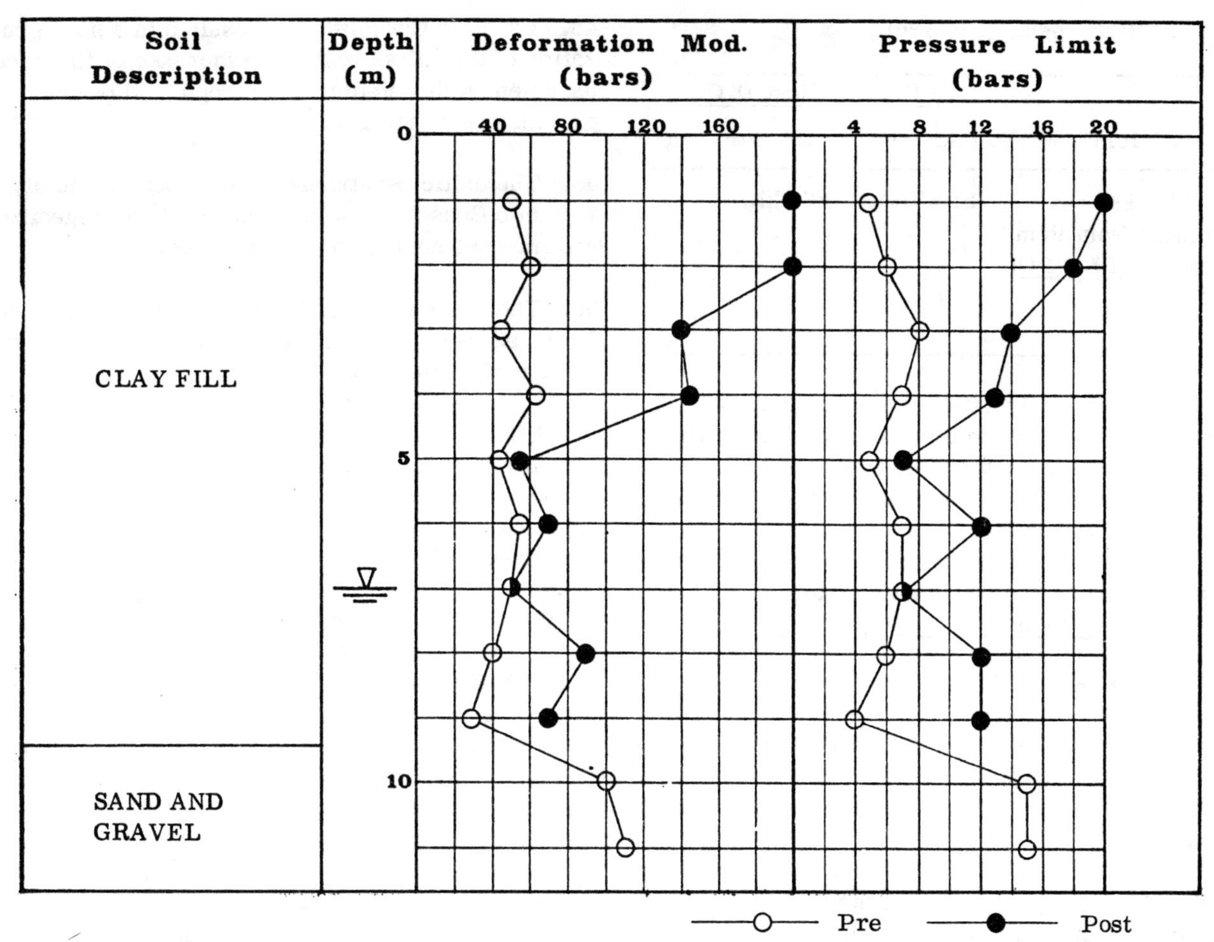

Note : 1 Bar = Kg/cm^2 = $100kN/m^2$

Fig. 6. Asnieres average pressuremeter test results

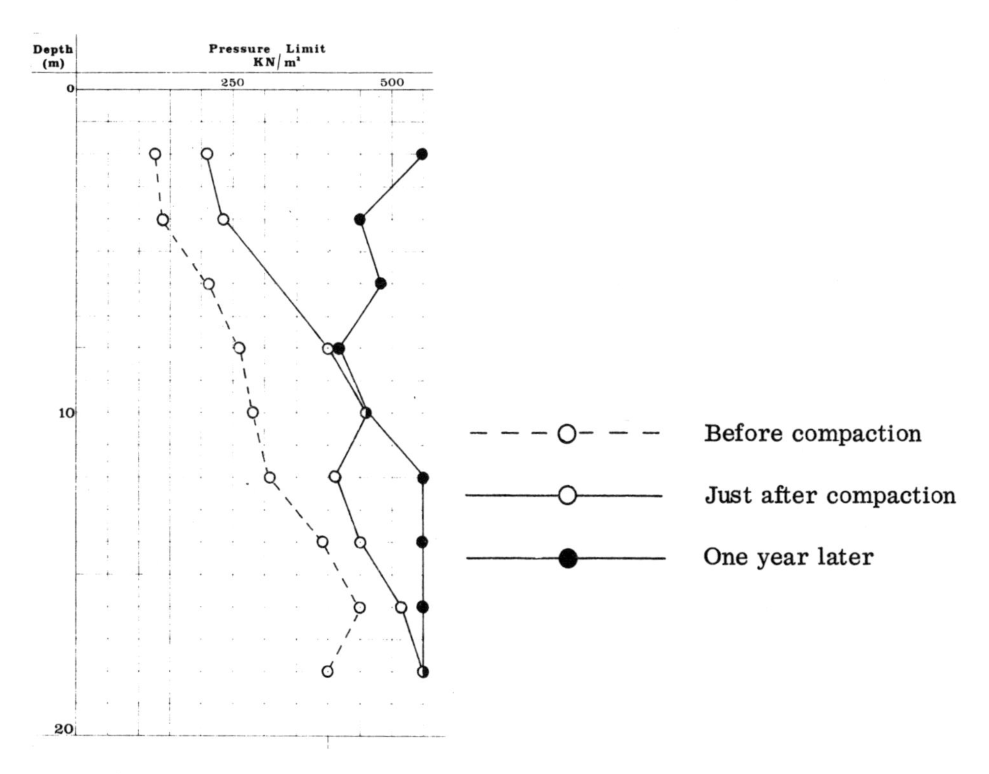

Fig. 7. Ambes average pressuremeter limit pressure results showing thixotropic recovery

Table 2. Vane tests (in situ)

	Pre D. C.	Post D. C.
0 - 10m	11kN/m²	40kN/m²
10 - 18m (correlated from P/m test results (PL - PO) / 7	24 kN/m²	70 kN/m²

44. Conclusion. Calculations based on the post treatment results indicate that the total edge settlement for the 55m diameter Tank loaded to 145kN/m² would be less than 1/3 of the value calculated using pre-treatment results.

V - Saint Martin Printing Plant, Asnieres, France - 1975

45. The soil comprised a clay fill 8m deep with the water table at 6m.

46. The construction included a one-storey office building of 60 x 12m contiguous to a 60 x 38m workshop sheltering heavy vibrating machinery (80 to 120 tonnes).

47. The heterogeneous nature of the fill excluded the possibility of using superficial foundation inspite of the fairly high mean value (6 bars) of the limit pressures measured. The founding of each machine on 12m piles also had to be envisaged.

48. Dynamic Consolidation resulted in a homogenization of the fill as well as an increase of its strength parameters threefold upto 4m depth and twofold at greater depth (see Fig. 6.)

49. The static penetrometer tests before and after Dynamic Consolidation confirmed this homogenization and improvement of the mean values.

50. The buildings were founded on strip or isolated footings loaded at 2 bars and the machinery has given no sign of settlement.

REFERENCES

1. MENARD L. 'An economical method of foundation on filled sites reclaimed from the sea'. Sols-Soils 1971, No. 24.

2. 'Ground Treatment by deep compaction' I.C.E. 1976.

3. BAGUELIN F. et al. 'The pressuremeter and foundation engineering'. Trans Tech Pub. 1978

4. MENARD L. 'Dynamic Consolidation of foundation soils', ITBTB Conference, Paris 1973

5. DEBAIRE, MIEUSSENS and COLOMB, 'Compactage d'un remblai routier par consolidation dynamique' Bulletin Liaison Laboratory Ponts et Chaussees No. 66 July 1973, pp 69-77.

6. MENARD L. 'The use of Dynamic Consolidation to solve foundation problems for the construction of quays, terminals, storage tanks and artificial islands on soft soils', Int. Harbour Conference, Antwerp 1978.

P. R. VAUGHAN, BSc, PhD, MICE, and D. W. HIGHT, MSc, MICE, Imperial College of Science and Technology, V. G. SODHA, BSc, MSc, MPhil, and H. J. WALBANCKE, BSc, PhD, MICE, Binnie and Partners

Factors controlling the stability of clay fills in Britain

The design of clay fills, such as those used for embankment dams, is governed primarily by stability, and hence by strength and the development of pore pressures during construction and in the long term. The paper will review laboratory and field data for the drained and undrained strength, and consolidation and pore pressure characteristics of fills derived from a range of the clays encountered in Britain. The paper will concentrate on problems associated with construction and with the ultimate stability of grassed slopes. Stability during rapid drawdown will not be considered.

INTRODUCTION

1. Clay fills up to some 40 m in height have been constructed in Britain, using material derived from glacial tills and from overconsolidated sedimentary clays. High fills have been used mainly for embankment dams. Road embankments are generally much lower, but some problems are common to all embankments irrespective of height. Factors controlling both the long and short term stability of clay fills, constructed in British conditions, are considered in this paper. While, frequently, clay fills must be placed on clay foundations the properties of which control the embankment design, the safe slopes which can be formed in the clay fill itself are always of importance. Stability during rapid drawdown of the upstream slope of an embankment dam will not be considered here.

SOURCES OF CLAY FILLS IN BRITAIN

2. Overconsolidated sedimentary clays occur in the Lias and in younger strata; the principal ones are indicated on Fig. 1. These sedimentary clays may often be bonded and exhibit the properties of soft mudstone when in situ: for example, the lower parts of the Oxford Clay behave in this manner (refs 1 and 2). However, degradation during excavation and placing by modern plant is sufficient for clay fills to be formed from these materials (ref. 3).

3. The older clay marls of the Trias usually contain clay minerals in stable, well bonded aggregates, and exhibit the properties of siltstones and silts unless the cementing materials are removed by a high degree of weathering (ref. 4) or they are subject to severe mechanical degradation. Fills derived from these marls do not generally show full breakdown to clay-like behaviour and they will not be considered here.

4. The mudstones of the Carboniferous only yield clays when cementing agents are removed or when they are very strongly degraded. This degradation does not occur when they are placed as fill (refs 5 and 6). Strata older than the Carboniferous, which contained clay when originally deposited, have usually been sufficiently altered subsequently for clay minerals not to be liberated by the removal of cementing agents or by mechanical degradation alone.

5. Tills are widespread within the limit of glaciation indicated on Fig. 1. Ice movements have been mainly southwards and eastwards and the gradings of these tills reflect the strata over which the ice has passed (ref. 7, Fig. 4).

6. A distinction, which is relevant to the subject of this paper, can be made between tills derived from the Carboniferous and younger strata, and those derived wholly from rocks older than the Carboniferous. The former contain a clay matrix, with the clay content increasing to the south and east where materials from the marls and sedimentary clays have been included in the till. The latter do not contain clay within the matrix even where abrasion has produced a predominance of fine material and a low permeability. The presence and proportion of clay within the grading will influence shear characteristics, sensitivity of strength to water content change, compressibility and permeability, probably in that order of importance.

7. Clay fills may be derived from materials outside the broad classification attempted above, such as tills derived from weathered basalts in Northern Ireland, and deeply weathered igneous rocks in the West Country. However, the comments made in this paper will be mainly directed towards fills derived from tills containing a clay matrix and from sedimentary clays which are of Jurassic age or younger. Typical index properties of these materials are indicated on Fig. 2.

EMBANKMENT SLOPE STABILITY

8. Embankment slope stability will be considered within the classical framework of limit equilibrium analysis. The two limiting cases are long term failure with pore pressures in equilibrium under operating conditions, and undrained failure during construction. The consolidation and swelling characteristics of the

Clay fills. Institution of Civil Engineers, London, 1978, 205-217

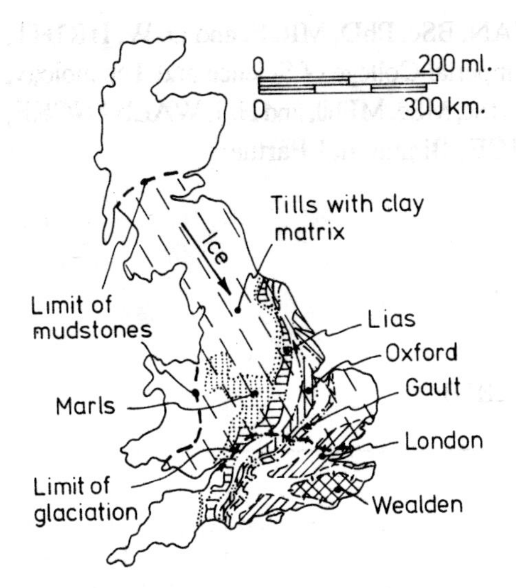

Fig. 1. Occurrence of clays in Britain

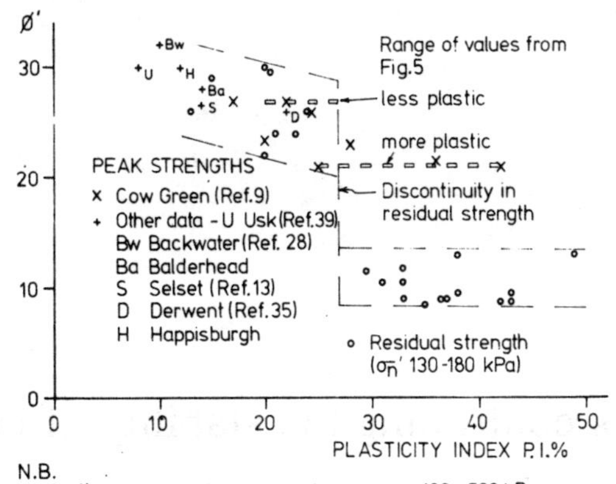

Fig. 4. Variation in drained strength with plasticity

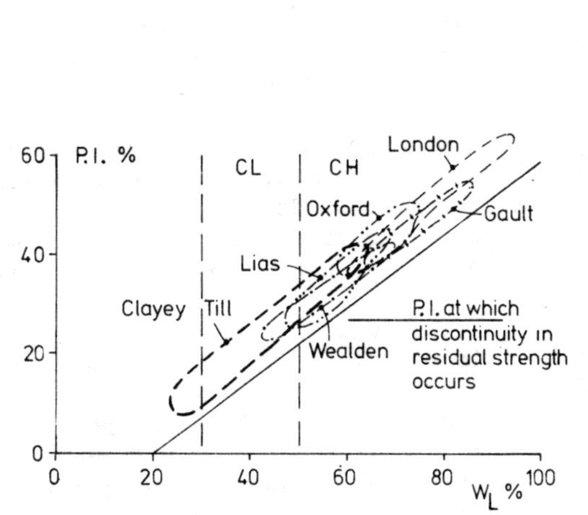

Fig. 2. Plasticity of typical British clays

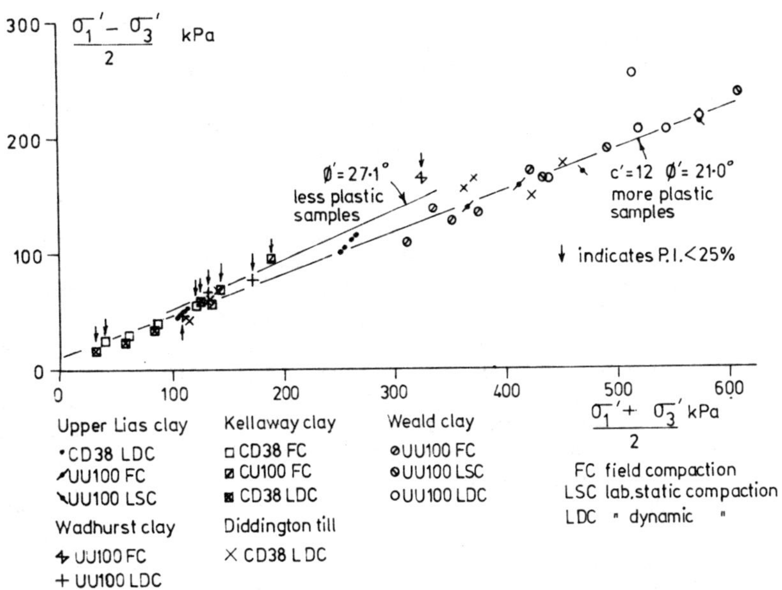

Fig. 5. Peak strength of some compacted clays in terms of effective stress

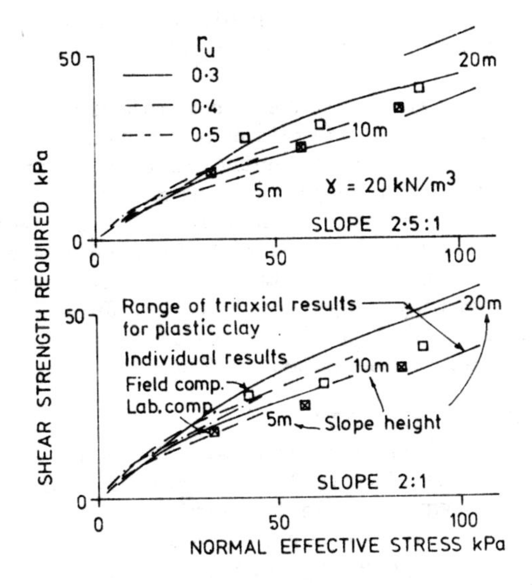

Fig. 3. Resistance envelopes of clay embankment slopes

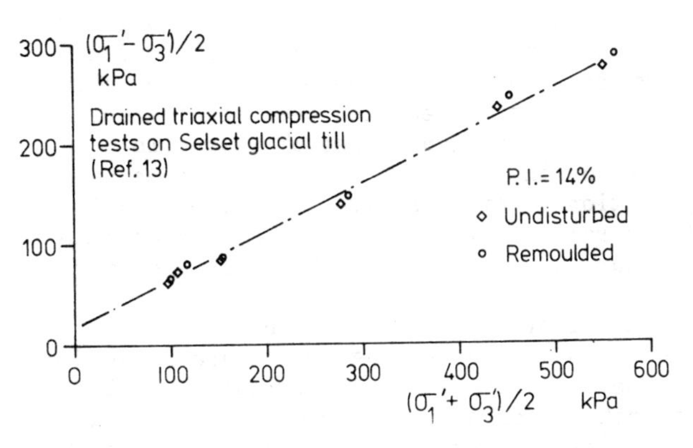

Fig. 6. Peak drained strength of intact and remoulded sandy clay till

fill determine the rate at which pore pressures equilibrate, the delay before long term conditions develop, the extent to which undrained conditions can exist during construction and the design of drainage works if these are used to accelerate consolidation during construction. The design parameters required for long term stability studies are the operational shear strength in terms of effective stress and the equilibrium pore pressures. The design parameters required for analyses of end-of-construction conditions are the undrained strength of the fill, or the undrained pore pressure response characteristics which can be used together with the effective stress strength envelope. Additionally, if consolidation must be considered, the undrained pore pressure characteristics and the coefficient of consolidation must be used. The relevant design parameters and assumptions will be considered in turn.

9. It may be noted that placement by modern rubber tyred plant requires a minimum undrained strength to ensure trafficability. This minimum undrained strength is typically about 30 kPa and rather higher if severe rutting is to be avoided. Simple calculation shows that with such a strength an embankment, even with steep side slopes, must be built to a height in excess of 10 m before stability problems are likely. Thus undrained stability problems during construction are a feature only of relatively high embankments.

DRAINED STRENGTH OF CLAY FILLS

10. Drained strength controls ultimate stability with equilibrium pore pressures operating. It is often convenient to compute the resistance envelope required for stability (ref. 8) rather than a factor of safety. This procedure has been followed for two typical slopes and is discussed subsequently; the results are shown in Fig. 3. It is demonstrated that strength is critical at low normal effective stress, i.e. approximately up to 60 kPa for a 20 m high slope and 30 kPa for a 10 m high slope. Thus, if the strength envelope is defined conventionally in terms of c' and ϕ', it is the value of c' which is more critical.

11. Vaughan and Walbancke (ref. 9) correlated peak drained strength of tills with plasticity index. This correlation is shown on Fig. 4, together with other data from sedimentary clays. There is a gradual and systematic decrease in ϕ' with increasing plasticity.

12. Vaughan and Walbancke (ref. 9) and Vaughan, Davachi, El Ghamrawy, Hamza and Hight (ref. 10) show that there is a discontinuity in the relationship between residual drained strength (as determined in the ring shear apparatus) and plasticity index; for typical British clays this discontinuity occurs at a plasticity index of about 27%, refer Fig. 4, and coincides with the change in classification from CL to CH which is defined on Fig. 2. The discontinuity apparently arises when the proportion of platey, low friction clay minerals to coarser sub-angular particles in the soil rises to a critical value at which it is possible for a highly orientated failure zone to form within the clay mineral fraction of the soil, without being continuously disrupted by the coarse particles. These two contrasting types of behaviour in drained shear will be subsequently associated with 'sandy clays' and 'plastic clays'. In the sandy clays the normally consolidated soil is not brittle and the peak and residual drained strengths coincide.

13. Progressive failure and the extent to which pre-existing shear surfaces can reduce bulk strength are only of potential significance in plastic clays of low residual strength. Shear surfaces do not exist in clay fills unless they are formed during placement*, or a shear failure has occurred. Progressive failure involves loss of strength by post peak displacement along an incipient failure surface; it results in overall failure taking place at average shear stresses less than the peak strength observed when the material is sheared under uniform stresses. Its influence depends on the rate at which strength drops with post peak displacement, as well as on the magnitude of the drop from peak to residual. It also depends on the scale of the slide, since the magnitude of the displacement on the incipient failure surface which can be accommodated prior to overall failure depends on this scale (ref. 11). The rate at which strength drops with post peak displacement in compacted plastic clays appears to be low, as discussed subsequently in connection with undrained strength. Thus progressive failure is likely to have little influence in small scale, superficial slides, such as are likely to be critical in embankment fills. Its influence may be greater in deep seated slides of major embankments which usually involve failure of the foundation.

14. The influence of the structure induced by compaction on peak strength in terms of effective stress appears to be slight (ref. 12). Typical results for a range of compacted clays are shown on Fig. 5. Differences in structure induced by laboratory dynamic, laboratory static and field compaction have no systematic influence on strength. It is of interest that a number of different plastic clays give very similar strengths. Clays of lower plasticity give higher strengths, although by an amount which is less at low effective stresses.

15. The effect of structure on the strength of a sandy clay till is illustrated on Fig. 6, after Skempton and Brown (ref. 13). There is little difference between the strength of the

---

*If shear failure occurs beneath the wheels of plant placing a plastic clay fill, shear surfaces may be formed in profusion. Such behaviour was observed during placing of the core of the Empingham Dam. The reduction in bulk drained strength due to the presence of such surfaces may be of significance.

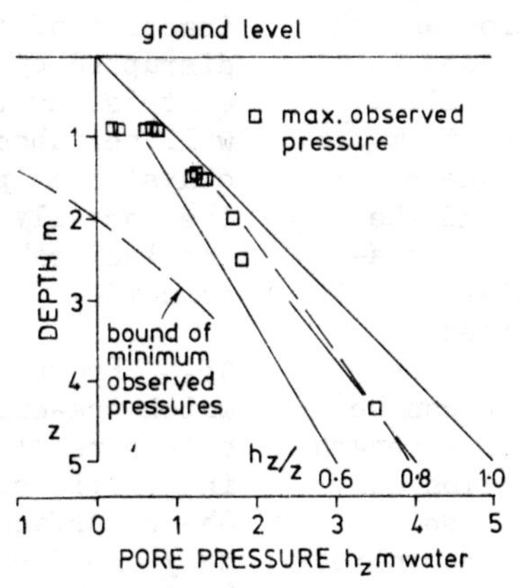

Fig. 7. Seasonal pore pressures observed in some grassed clay-fill slopes

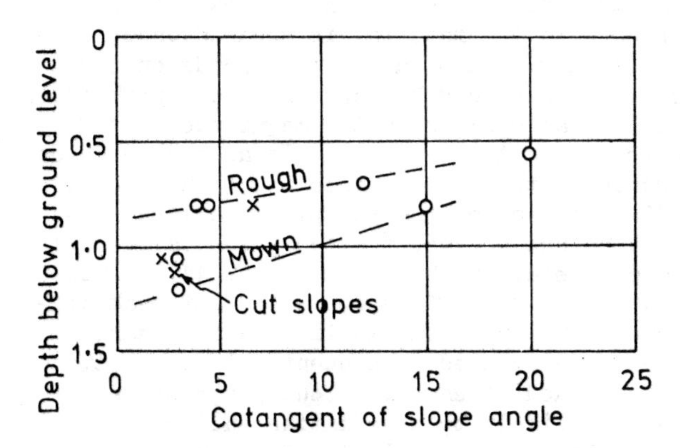

Fig. 8. Approximate depth of zero average pore pressure in some grassed clay-fill slopes

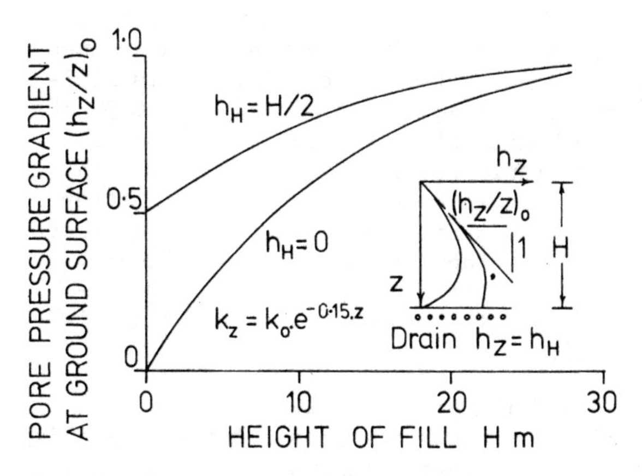

Fig. 9. Perched water table effects due to an increase in permeability with depth - one-dimensional flow with underdrainage

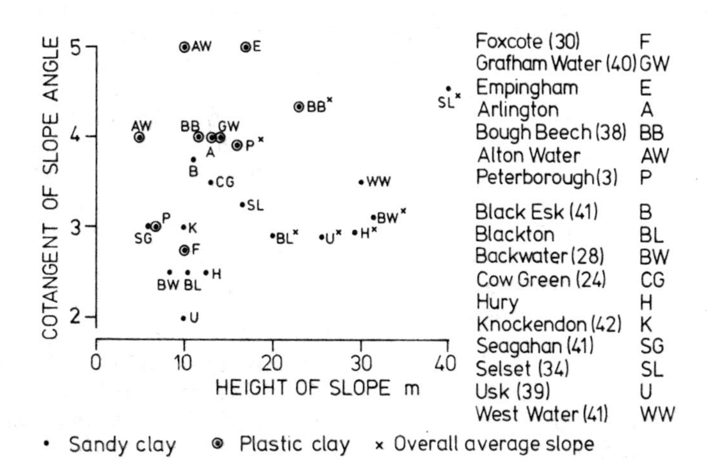

Fig. 10. Typical downstream slopes of British embankment dams with clay fills

intact clay and the same clay after full remoulding.

16. To the authors' knowledge, there has been no systematic study of the strength of compacted clay fills at the low effective stresses which govern slope stability. It may be conjectured that the cohesion intercept relevant to ultimate stability increases with the fill density which exists after pore pressure equilibration. There is no evidence that changes in compacted density and compaction water content within the range normally specified have much influence on the cohesion intercept. The limit set by the undrained strength which allows placement must also imply a significant cohesion intercept. However, loose, uncompacted fill may well have a much reduced cohesion intercept after swelling and softening at low stress, and reasonable compaction of embankment slopes is clearly desirable. Chandler and Pachakis (ref. 14) give details of a 12 m high embankment with average side slopes of 2.2:1 which failed 80 years after its construction. The fill was formed from the Upper Lias clay, probably from tunnel spoil, and, because of its age, it would have been dumped in place in a loose state. Back-analysis of the failure in terms of effective stress shows the average cohesion intercept of the fill to be negligible.

ULTIMATE PORE PRESSURES IN CLAY FILLS

17. A field study of the factors controlling ultimate equilibrated pore pressures in clay fill embankments was made by Walbancke (ref. 15). The equilibrium pore pressures are mainly controlled by the slope surface boundary conditions. After consolidation and swelling, density will increase and permeability decrease with depth and increasing effective stress (refs 9 and 12). This gradient leads to 'perched water tables', and, in a high embankment, pore pressures within the fill may be relatively unaffected by underdrainage at the embankment foundation.

18. British climatic conditions give an excess of infiltration over evaporation in slope surfaces. The pore pressure at the slope surface fluctuates with climatic conditions and cyclic swelling and consolidation occurs. Rapid fluctuations may take place near the ground surface but at a depth of about 2 m the fluctuations become seasonal (and measurable with adequate accuracy by conventional standpipe piezometers with comparatively long response times). Seasonal fluctuations penetrate to about 5 m, the magnitude of the fluctuations decreasing with depth. Typical pore pressures observed beneath grass-covered clay fill slopes are shown on Fig. 7.

19. The factors relevant to slope design are the maximum values attained during seasonal fluctuations, and, for high embankments, the average surface value, which controls pore pressures below the zone of fluctuation. Impounding water against the upstream slope of a clay embankment has little influence on the equilibrium pore pressures within the downstream slope. There is some evidence that the highest pressures at shallow depths occur when wet weather follows a prolonged dry period, possibly due to flooding of deep shrinkage cracks. An approximate estimate of average surface values, from Walbancke (ref. 15) is shown on Fig. 8. The effect of slope vegetation can be significant. Black, Croney & Jacobs (ref. 16) found that pore pressures were lower below a grass cover than below a bare earth surface. Shrubs and trees which increase evapo-transpiration may decrease pore pressures further, as well as providing root reinforcement to the soil (ref. 17).

20. The development of a 'perched water table' when the embankment is underdrained can be examined with sufficient accuracy by considering one-dimensional flow from the surface boundary to the underdrain. The variation of permeability with depth (ref. 15) can be represented conveniently by:

$$k_z = k_o e^{-az} \qquad (1)$$

where, k_z is the coefficient of permeability at depth z ,

k_o is the permeability at the ground surface, and

a is a constant, typically equal to $0.15\ m^{-1}$.

Assuming zero pore pressure at ground level:

$$h_z = z - (H - h_H)(1 - e^{az})/(1 - e^{aH}) \qquad (2)$$

where, $h_z \gamma_w$ = pore pressure at depth z,

H = depth of underdrain, and

$h_H \gamma_w$ = underdrain pressure.

The pore pressure gradient at the ground surface is:

$$(\delta h_z/\delta z)_{z=0} = 1 + a(H - h_H)/(1 - e^{aH}) \qquad (3)$$

Equation 3 is plotted in Fig. 9. Perched water table effects may be of significance even in quite low fills. The introduction of horizontal drainage layers in a slope will substantially reduce these effects.

21. A gravel layer has been placed beneath the topsoil on the clay fill slopes of many embankment dams. Observations of superficial pore pressures in such slopes (ref. 15) show that the gravel layer substantially reduces the depth and magnitude of the seasonal pressure variations, and the average surface pore pressure can be taken as zero, acting at or just below the base of the gravel layer. The effect of such a layer, typically, is to increase minimum effective stresses near the slope surface (and so to improve superficial stability), and to increase very slightly the pore pressures at depth where they are controlled by average surface boundary conditions. Granular road bases may have similar effects on pore pressures in road embankments.

SAFE LONG TERM SLOPE ANGLES

22. A condition close to limiting instability is acceptable for clay fill slopes for relatively low and wide road embankments, where shallow failures do not normally affect road usage. It is of interest to make predictions of these slopes from the information given in the previous two sections. Maximum pore pressure ratios ($r_u = u/\gamma z$) may be of the order of 0.5 near the ground surface dropping to values of the order of 0.3 - 0.4 at depth (see Fig. 7). The strength envelopes required for stability of slopes of 2.5:1 and 2.0:1, with heights increasing from 5 m to 20 m, and with pore pressure ratios decreasing from 0.5 to 0.3 with increasing slope height, are shown on Fig. 3. The strength data from Fig. 5 for the more plastic clay fills is also shown. Comparison of the envelopes with this data indicates that slopes of 2.5:1 up to 10 m high are likely to be stable, provided that progressive failure has a negligible effect. Slopes of greater height might develop problems and these could involve deep failures and considerable maintenance. Progressive failure may also have a greater influence in higher slopes. Fig. 3 indicates that slopes of 2.0:1 in compacted plastic clay are likely to show long term instability, particularly where they are higher than about 5 m. This conclusion seems to be in general agreement with field experience.

23. Slopes of compacted sandy clay fill should remain stable when somewhat steeper and higher. 2:1 slopes up to about 10 m high should be acceptable. Slopes steeper than this may show superficial instability of topsoil, particularly if they are used by grazing animals (ref. 9).

24. The risk of even superficial instability cannot be tolerated in embankment dams, and slopes flatter than those used in roadworks are generally adopted. Some typical slopes (where these are controlled by fill rather than by foundation properties) are shown on Fig. 10.

STRENGTH AND PORE PRESSURES DURING UNDRAINED CONSTRUCTION

25. As stated previously, placement by modern plant requires a minimum undrained strength such that undrained stability problems are unlikely in relatively low embankments. However, the problem of undrained stability during construction has been of major concern to designers of embankment dams since the failure of Muirhead Dam (ref. 18).

26. The stress path followed by the conventional undrained triaxial compression test (including the initial application of the confining stress) is a reasonable approximation to the stress paths followed by elements in the fill during embankment construction, and this test may be used legitimately to study undrained strength and pore pressure generation. If total stress methods of formulation and analysis are used, as is permissible when pore pressure is only a function of total stress, then a curved failure envelope will be obtained. A linear approximation, defined by c and ϕ, over the appropriate stress range can be used in analysis. Marked differences between the undrained shear behaviour of relatively stiff sandy clays and that of the more plastic clays have been noted (refs 9, 10, 19 and 20). These two types of behaviour will be summarised and discussed.

Plastic clays

27. The undrained behaviour of fills of plastic clay (refs 12 and 19) is illustrated in Fig. 11, which shows the results of undrained compression tests on specimens of Upper Lias clay, as compacted in the field in the Empingham dam, and in the laboratory by dynamic compaction. The laboratory compacted specimens were prepared from the field specimens after these had been tested. Tests on 'undisturbed' specimens of the clay, taken from about the same level in the borrow pit as the fill, are also shown. The intact Lias clay was brecciated and did not contain fissures. Similar results for compacted Weald clay are given in ref. 12. A summary of these results is presented in Table 1. It can be seen that undrained behaviour cannot be studied reliably by testing specimens compacted dynamically in the laboratory. Specimens produced by static laboratory compaction show behaviour nearer to that of field compacted specimens, but do not reproduce it (refs 12 and 21).

28. The undrained behaviour of the plastic clay fills shows the following features:-

Table 1. Undrained strengths from triaxial compression tests on compacted Weald clay (P.I. 30 - 43% av. 34%) - influence of type of compaction and shear rate

Relationship	FS/LDS	LSS/LDS	FQ/LDQ	FS/FQ	LDS/LDQ
Ratio	0.66	0.87	0.82	0.79	0.95
No. of pairs of tests	7	3	6	5	4

FS: Field compacted, slow
LSS: Lab. static compacted, slow
LDS: Lab. dynamic compacted, slow
FQ: Field compacted, quick
LDQ: Lab. dynamic compacted, quick

slow: $\delta\epsilon_v/\delta t$ = 0.05%/hr.
quick: $\delta\epsilon_v/\delta t$ = 36%/hr.

All specimens tested at a confining stress of 414 kPa.

Range of undrained strengths: 80 - 250 kPa

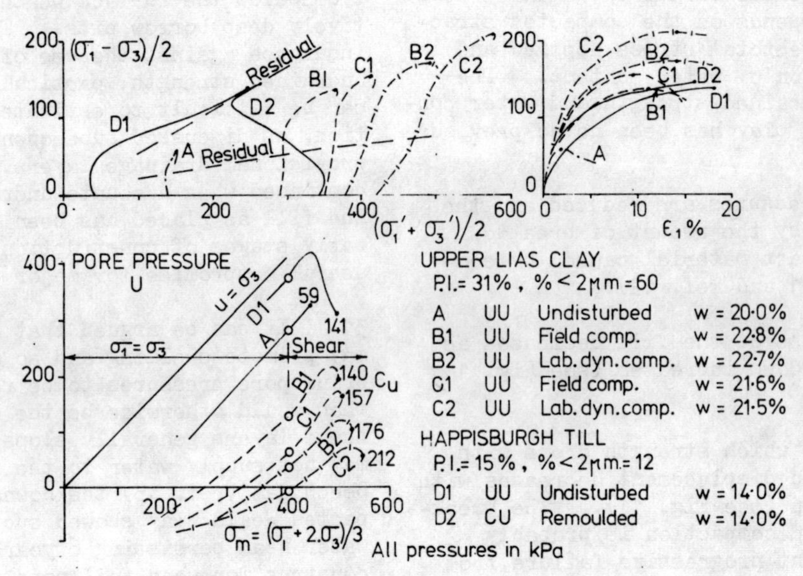

Fig. 11. Typical undrained shear behaviour of compacted clays - triaxial compression tests

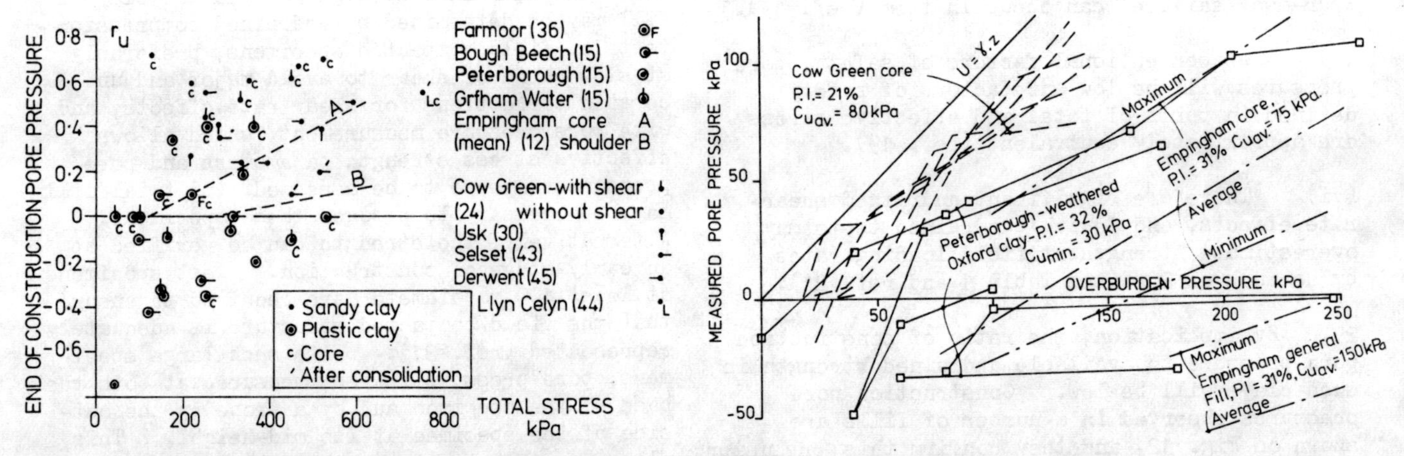

Fig. 12. Typical construction pore pressures in British embankment dams

Fig. 13. Generation of undrained construction pore pressures in low fills

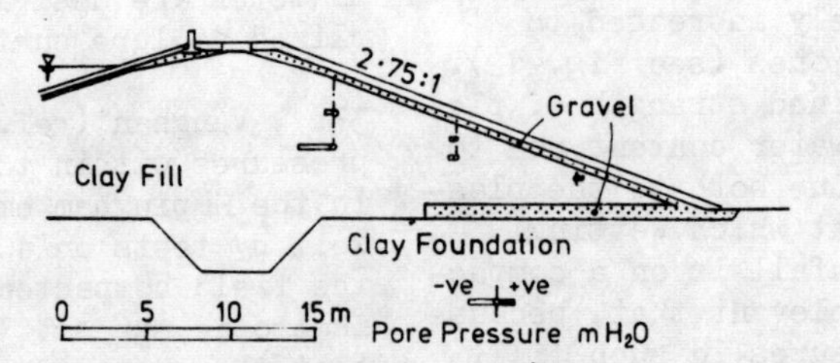

Fig. 14. Negative pore pressures in Foxcote Dam 20 years after construction and impounding

(i) The average strength is a function of the average effective stress in the specimen. This effective stress depends on the compacted structure and the average total stress applied and is not a direct function of water content. The independence of undrained strength and water content for the London clay has been noted previously (ref. 22).

(ii) The pore pressures are reduced and the strength increased by the amount of breakdown of the original intact material caused by compaction (see Fig. 11 and ref. 12).

(iii) The stiffness of the fill decreases as the amount of breakdown increases (see Fig. 11 and ref. 12).

(iv) The rate at which strength drops with post-peak strain and displacement decreases with increasing breakdown (see Fig. 11). The breakdown caused by field compaction is probably sufficient to prevent progressive failure from having a significant influence on the average strength operating at failure, although, even after full remoulding, these clays have a low residual undrained strength (ref. 22) and thus 'run-away failure' can occur in them (ref. 19).

(v) At conventional factors of safety, pore pressures will be low and factors of safety defined in terms of total and effective stress are approximately equivalent (ref. 19).

(vi) There are significant undrained shear-rate effects, and 'quick' undrained tests may overestimate strength at field loading rates by as much as 20% (see Table 1 and ref. 21).

29. By implication, the ratio of construction pore pressure to available undrained strength in such clays will be low. Construction pore pressures observed in a number of fills are shown on Fig. 12, and they confirm this conclusion. There appears to be no record of significant construction pore pressures occurring in embankment dam fills of this type when the clay is obtained from relatively deep borrow pits, although such pore pressures have been recorded where water contents have been deliberately increased to promote flexibility in core zones (see Fig. 12). The sensitivity of the undrained strength of plastic clay fills to increased water content due to rainfall is relatively low, due both to the plasticity and to the slow rate at which wetting penetrates (provided the rainfall is on a compacted surface). It is also relevant that, because of the low general pore pressures in such fill, excess pore pressures generated in a wet, weak layer inadvertently included in the fill will dissipate into the surrounding drier fill throughout construction, and the weak layer will gain undrained strength.

30. The use of horizontal drainage layers to allow excess pore pressures to dissipate and strength to be gained during construction has been a feature of embankment dams constructed from clays in Britain. There is little evidence that these have served much purpose in fills up to 35 m high which have been formed from sedimentary over-consolidated plastic clays excavated from below the surface weathering zone in relatively deep borrow pits. They may be a useful insurance against the use of fill of inadequate undrained strength, particularly as this strength can be difficult to estimate prior to construction, as discussed subsequently. The use of provisional drainage layers, which can be discontinued when adequate undrained strength in the fill as placed has been proved during the early stages of construction (ref. 23), is a useful compromise for major embankments.

31. It can be argued that such drainage layers may promote swelling and so allow stable equilibrium pore pressures to be reached much sooner than would otherwise be the case. However, these layers generally slope outwards and they may not supply water to the fill. At Bough Beech dam (ref. 15) the downstream fill of compacted Weald clay showed suctions up to 5 m of water head persisting 6 years after the end of construction when full pore pressure equilibration between the relatively closely spaced drainage layers would be predicted.

32. The undrained strength of plastic clay fill may be determined by undrained compression tests on field compacted specimens, preferably at slow rates of shear to avoid major and uncertain corrections for shear rate effects, and with pore pressure measurement, which allows effective stress strength parameters and pore pressure response to be assessed. A trial fill may be necessary to produce these specimens. Alternatively, the strength can be verified at an early stage of construction. Test specimens at least 100 mm diameter are required to ensure that the field compacted structure is adequately represented (ref. 21). With such large specimens, pore pressure is best measured at both the base of the specimen and by a probe in the outside of the specimen at its mid-height. This allows shear rates to be used which are faster than those required for full pore pressure equalisation. The use of lubricated end plattens is uncertain for slow tests, and conventional test specimens of a length twice the diameter are desirable to accommodate the inclined failure surfaces which normally form.

33. Vaughan (ref. 12) showed that pore pressures within the compacted Upper Lias clay in the Empingham embankment were predicted quite well by tests on 100 mm diameter specimens of the field compacted clay, loaded undrained under isotropic stress. This follows from the fact that the shear stress has only a small influence on the pore pressures generated in this material. Thus such tests offer an alternative and complementary method of estimating undrained behaviour. Field observations of pore pressure may also be used to deduce undrained strength, since, if the influence of shear stress on pore pressure is neglected, it may be estimated from the average effective stress in the embankment and the strength parameters in terms of effective stress. The application of this method to the results from Empingham gives undrained strengths within ± 15% of those measured by field control tests.

34. Where it is necessary to estimate undrained strengths from preliminary tests on specimens compacted in the laboratory, the ratios given in Table 1 may be of value. The tests from which this data was obtained were performed in pairs, a laboratory dynamically compacted specimen being made at the same water content from each field or laboratory static compacted specimen after it had been tested. The strength of field compacted clay tested slowly, which is believed to approximate to that available in the field, may be less than two thirds of the strength determined by standard quick triaxial tests on specimens compacted dynamically in the laboratory from the same clay at the same water content. The strengths of these test specimens were high and the same ratios may not operate at lower strengths. The specimens for all the tests were allowed to equilibrate under the confining stress before shearing. Quick tests on unequilibrated specimens may show lower strengths because of heterogeneity.

Sandy clays

35. The undrained behaviour of typical sandy clays has been discussed in some detail in refs 9, 10, 19, 20 and 24. Some aspects of this behaviour are illustrated on Fig. 11, which shows undrained compression tests on undisturbed and remoulded specimens of Happisburgh Till (a clayey sand from Norfolk). The behaviour may be summarised as follows:-

(i) Residual drained strengths are high, and the decrease in pore pressure caused by dilatant behaviour allows the clay to be non-brittle in undrained shear (at least under triaxial loading).

(ii) The residual strength in terms of effective stress is mobilised when the ultimate undrained strength is fully developed. However, this is only at very large strains, and unacceptable deformations are likely to occur before ultimate stability is reached (ref. 19).

(iii) Progressive failure cannot occur (ref.19).

(iv) The structure induced by compaction has negligible influence on the ultimate undrained strength, which is, to at least an adequate engineering approximation, a function of water content (ref. 10).

(v) The behaviour at small strains is a function of compacted structure (see Fig. 11).

(vi) The same factor of safety used in total and effective stress analyses gives very different answers, the effective stress analysis being much more conservative (ref. 19).

(vii) The effect of rate of undrained shear seems to be small (ref. 24).

36. It follows from the above that the reliable ultimate undrained strength of clay fills of this type can be estimated from undrained compression tests on compacted or remoulded test specimens without undue difficulty. However, this strength must be used in total stress analyses with caution, as unacceptable deformations may develop at apparently satisfactory factors of safety (refs 19 and 24).

37. The estimation and interpretation of undrained construction pore pressures presents greater difficulties. It follows from the type of shear behaviour illustrated on Fig. 11 that pore pressures during undrained shear are controlled as much by shear stress as by average total stress. This behaviour has been observed in the field (ref. 19, Fig. 6.18 and ref. 24, Fig. 15). The development of high construction pore pressures at an early stage of construction, at least when water contents are around optimum or above, are a feature of such fills. Fig. 13 shows pore pressures generated at an early stage of construction of the core of the Cow Green dam (ref. 24), which was placed at a controlled water content to give an average undrained strength of 80 kPa. The effective stresses indicated are much lower than would allow the available undrained strength to be mobilised without a major decrease in pore pressure due to shear.

38. The explanation of these high early construction pore pressures in terms of initial placement conditions is not clear. Recent tests on the Happisburgh Till have shown that compaction and remoulding of initially saturated clay increases pore pressure and reduces effective stress. The effect of traffic and compaction may also play a part. The generation of pore pressure by trafficking is reported by Cocksedge and Hight (ref. 25). Such high pore pressure may also be connected with the resilient behaviour of fill surfaces reported by these authors. Certainly the low effective stresses indicated on Fig. 11 suggest that, under wheel loading, the strength in terms of effective stress must be fully mobilised. If this is so, the mobilisation of sufficient undrained strength to resist failure requires large plastic strains, and the occurrence of large, recoverable plastic strains under rolling wheel loads can be expected. These high pore pressures, which seem to be, at least in part, a function of placement conditions, prevent the prediction of construction pore pressure by means of laboratory tests, and they also prevent the prediction of ultimate undrained behaviour from interim field measurements of pore pressure.

39. Because of low plasticity, the undrained strength of the sandy clays is sensitive to increases in water content due to rainfall. A 50% reduction in strength of the sandy clay fill at Cow Green (average P.I. = 21%) required an increase in water content of 2%. By contrast, a 50% reduction in strength in the plastic clay fill at Empingham (average P.I. = 31%) required an increase in water content of 4%. The high pore pressures generated during the early construction of sandy clay fills, largely irrespective of the undrained strength at which they are placed, has a further implication. If a layer of wetter, weaker fill is inadvertently included, there will be little difference

between the pore pressures in this layer and in the fill surrounding it until the later stages of construction, and so there will be little local consolidation and gain in strength of the layer. It will thus persist as a weak layer, unlike the situation in plastic clay fills described previously. These two factors suggest that if undrained strength is relied upon for the stability of sandy clay fills, without the provision of internal drainage to ensure consolidation during construction, then particular care must be taken to prevent the inclusion of wet layers.

40. The subdivision into the two types of undrained behaviour according to plasticity seems to coincide with the discontinuity in residual strength described previously. However, there is little data available from which the transition between the two types of behaviour can be studied, and care should be taken in ascribing one or the other to a fill on the basis of plasticity alone. Intermediate behaviour may also occur, possibly where sedimentary clays of medium plasticity are used as fill.

CONSOLIDATION AND SWELLING CHARACTERISTICS

41. As shown on Figs 12 and 13, fills of sandy clay, even of modest height, frequently develop high pore pressures during construction, whereas negative pore pressures are typical of fills of more plastic clay, even when these are of considerable height. Equilibration of pore pressures involves consolidation in the former, and swelling in the latter. The simple theory of consolidation is adequate to describe these processes, despite the complications of a two phase compressible pore fluid. Values for the coefficients of consolidation and swelling for various compacted clays, deduced from laboratory measurements and from analysis of field piezometers, are summarised on Table 2.

42. Three conclusions can be drawn. Firstly values for sandy clay fills are only slightly higher than those for the more plastic clays. Secondly, values are generally low, implying slow pore pressure equilibration unless drainage paths are short. Thirdly, field values are only slightly higher than those deduced from conventional laboratory tests on small specimens, generally formed by dynamic compaction in the laboratory.

43. A further factor influencing pore pressure equilibration is the manner in which c_v changes with effective stress. Typically, in situ saturated sandy clays show c_v increasing with effective stress and consolidation, whereas in situ plastic clays show the reverse (refs 26 and 27). Intermediate behaviour was observed at the Cow Green dam (ref. 24), where the foundation of locally derived till had the unusually high average P.I. of 25%. Here, the laboratory tests showed c_v increasing with stress, while the field measurements showed the opposite trend. Sandy clay and plastic clay fills also show these opposing trends (refs 26 and 28) although in compacted sandy clays the tendency for c_v to increase with effective stress is less marked than for the in situ soil, and may be overestimated by laboratory tests. There is no information from which swelling trends in the plastic

Table 2. Coefficients of consolidation c_v, and swelling, c_s, for compacted clays from laboratory measurements and analysis of field performance

Embankment Site	Laboratory c_v (m^2/yr)	Field c_v (m^2/yr)	PI (%)	Ref.
Sandy clay embankments				
Usk	8.9	12.3	8	33
Selset	0.9-3.1	1.4-3.6	14	34
Derwent[1]	1.3	1.0-1.7	19	35
Derwent[2]	-	1.6-3.7	22	
Balderhead	2.3-15.8	9.3-11.9	14	26
M6 Kendal	1.4-3.8	1.9-5.8	7	
Backwater	1.5-13.0	3.6-13.9	10	28
Llyn Celyn[1]	3-6	-	14	36
Llyn Brianne[1]	15-20	-	10	
Plastic clay embankments				
Grafham Water	0.4-1.5	0.8-1.3	38	37
Farmoor	0.1-0.4	-	40	36
Arlington	1-2	-	25	
Bough Beech	0.9	-	-	38
*Peterborough	0.3-0.6	-	5-35	21
*Peterborough[1]	-	0.8-1.8	5-35	15

Embankment Site	Laboratory c_s (m^2/yr)	Field c_s (m^2/yr)	PI (%)	Ref.
Plastic clay embankments				
Grafham Water	-	0.5	38	26
Bough Beech[1]	-	6.7-6.9	-	15
Bough Beech[3]	-	1.6-1.7	-	
*Peterborough	0.4-1.3	-	5-35	21
*Peterborough[4]	-	0.7-2.4	5-35	15

* fill also contained sandy clay
1 core
2 fill
3 shoulder
4 upstream shoulder and downstream toe

clays can be deduced. The direction of change of c_v or c_s with stress is important, as it differentiates between an accelerating and decelerating process.

44. An example of the slow rate of pore pressure equilibration, in this case swelling, is provided by the Peterborough dam (refs 3, 12 and 29). The fill was a mixture of clayey sand and stiff clays of varying plasticity. It was placed dry and showed negative pore pressures at the end of construction. Seven years after the reservoir was impounded, the positive pore pressures due to seepage had only penetrated 5 m from the upstream face. A further example is provided by the small embankment dam at Foxcote, Buckinghamshire, (ref. 30) a section of which is shown in Fig. 14. The rolled fill is from a relatively plastic till derived from local clays. Suctions still persist at depth in the downstream slope some 20 years after construction and impounding of the reservoir.

45. The implications of this slow swelling on the long term stability of the slopes of even low embankments of clay fill which has negative pore pressures at the end of construction, are considerable. Full equilibration of pore pressure to long term values may take decades, or even centuries. A decrease in permeability with depth implies more rapid equilibration of pore pressure near the surface than is predicted by simple theory (ref. 31) and this is also the zone influenced by seasonal fluctuations. Pore pressures may become close to equilibrium in this zone in 5 - 10 years, and failures may be induced by periods of wet weather (ref. 32). Subsequently, seasonal fluctuations will be imposed on a slow and slight upwards trend in average pore pressure due to swelling at depth. In the long, relatively steep slopes which are typical of road embankments, in which some variation in average strength and swelling rate can be expected from place to place, delayed failures are likely to start occurring after periods of wet weather some years after construction,the number of failures depending on the severity of each wet period. A prolonged wet period induces failures in the more critical parts of the slope, and it will reduce the frequency of subsequent failures elsewhere until a comparably wet period occurs. The general picture which can be predicted from the superposition of climate-induced fluctuations on a gradually increasing average pore pressure is of a high frequency of local failure a few years after construction, followed by a period of some decades during which failures occur with decreasing frequency. The general time scale will be longer in higher slopes.

46. The prediction and magnitude of the settlement of clay fills is essentially beyond the scope of this paper, but the following points may be noted. Post construction settlements are of most interest. In sandy clay fills placed wet enough for high excess pore pressures to be generated, consolidation settlements will occur. Such clays usually have low compressibilities, which reduce the magnitude of these settlements. Unless they are placed very wet, fills of plastic clay are likely to swell after construction. If the fill has been placed dry and poorly compacted, the increasing pore pressures may induce collapse settlement. In well compacted fills, heave can occur. Swelling involves a decrease in average effective stress and, because the average shear stress remains constant, an increase in stress-ratio. The strains and settlements induced by these stress changes require further study, but it is likely that shear strains will compensate for volumetric swelling. Swelling tests on specimens compacted in the laboratory will also overestimate heave, as the negative pore pressures in such specimens will be greater than in the field compacted soil.

CONCLUSIONS

47. The following conclusions may be made:-

(i) The shear behaviour of clay fills may be subdivided generally into two types according to grading and plasticity.

(ii) Little information is available concerning the drained strength of such fills at the low effective stresses which govern slope failures, but this strength is mainly governed by the cohesion intercept. It seems to be similar for most plastic clay fills and is not sensitive to compacted structure. Progressive failure, which can only occur in the more plastic fills, probably has little effect on superficial long term failures. Long term pore pressures can be high due to infiltration and perched water table effects. The strength and pore pressure data available suggests that slopes of 2.5:1 should be just stable in the more plastic fills, and steeper slopes are permissible in the sandy clays.

(iii) For the same undrained strength, construction pore pressures are much higher in sandy clays than in the plastic clays. Undrained strengths and construction pore pressures in plastic clays can be estimated reasonably from laboratory tests, but results are sensitive to compacted structure, and field compacted specimens should be used. Undrained strengths can be inferred from field pore pressure measurements. The undrained strengths of the sandy clays are not sensitive to compacted structure and may be determined more easily. However, their mobilisation requires large strains and potentially unacceptable deformations. Ultimate undrained behaviour cannot be inferred from field pore pressure measurements.

(iv) Pore pressure equilibration in clay fills is slow, and swelling of more plastic clay fills will lead to a history of delayed slope failure.

ACKNOWLEDGEMENTS

48. Acknowledgement is made to the Anglian Water Authority, the Southern Water Authority, the Bucks Water Board, the East Surrey Water Co., the C.E.G.B., Rofe Kennard & Lapworth, Binnie & Partners and Watson-Hawksley for the use of field data. Field and laboratory work at Imperial College has been supported by the Water

Resources Board and the Science Research Council.

REFERENCES

1. PARRY, R. H. G. Some properties of heavily overconsolidated Oxford Clay at a site near Bedford. Géotechnique, 1972, 22, No. 3, 485-507.
2. BURLAND, J. B., LONGWORTH, T. I. and MOORE, J. F. A. A study of ground movement and progressive failure caused by a deep excavation in Oxford Clay. Géotechnique, 1977, 27, No. 4, 557-591.
3. VAUGHAN, P. R. Discussion. 8th International Congress on Large Dams, Edinburgh, 1964, 5, 301-304.
4. CHANDLER, R. J. The effect of weathering on the shear strength properties of Keuper Marl. Géotechnique, 1969, 19, No. 3, 321-334.
5. KENNARD, M. F., KNILL, J. L. and VAUGHAN, P. R. The geotechnical properties and behaviour of Carboniferous shale at the Balderhead Dam. Quarterly Journal of Engineering Geology, 1967, 1, No. 1, 3-24.
6. VAUGHAN, P. R. Discussion. Proceedings of the Institution of Civil Engineers, Part 2, 1973, 55, September, 697-699.
7. DERBYSHIRE, E. The distribution of glacial soils in Great Britain. Proceedings of the Symposium 'The Engineering Behaviour of Glacial Materials', Birmingham, 1975, 6-17.
8. JANBU, N. General report: slopes and excavations. Proceedings of the Ninth International Conference on Soil Mechanics and Foundation Engineering, Tokyo, 1977, 2, 549-566.
9. VAUGHAN, P. R. and WALBANCKE, H. J. The stability of cut and fill slopes in boulder clay. Proceedings of the Symposium 'The Engineering Behaviour of Glacial Materials', Birmingham, 1975, 209-219.
10. VAUGHAN, P. R., DAVACHI, M. M., EL GHAMRAWY, M. K., HAMZA, M. M. and HIGHT, D. W. Stability analysis of large gravity structures. Proceedings of the Symposium 'Behaviour of Offshore Structures', Trondheim, 1976, 1, 467-487.
11. BISHOP, A. W. The influence of progressive failure on the choice of stability analysis. Géotechnique, 1971, 21, No. 2, 168-172.
12. VAUGHAN, P. R. Panel discussion: Session 3. Proceedings of the Ninth International Conference on Soil Mechanics and Foundation Engineering, Tokyo, 1977, 3.
13. SKEMPTON, A. W. and BROWN, J. D. A landslide in boulder clay at Selset, Yorkshire. Géotechnique, 1961, 11, No. 4, 280-293.
14. CHANDLER, R. J. and PACHAKIS, M. D. Long-term failure of a bank on a solifluction sheet. Proceedings of the Eight International Conference on Soil Mechanics and Foundation Engineering, Moscow, 1973, 4, 45-51.
15. WALBANCKE, H. J. Pore pressures in clay embankments and cuttings. Ph.D. Thesis, University of London, 1976.
16. BLACK, W. P. M., CRONEY, D. and JACOBS, J. C. Field studies of the movement of soil moisture. Road Research Laboratory Technical Paper, No. 41, H.M.S.O., 1958.
17. GRAY, D. H. Reinforcement and stabilisation of soil by vegetation. Journal of the Geotechnical Division of the American Society of Civil Engineers, 1974, 100, GT6, 695-698.
18. BANKS, J. A. Construction of Muirhead Reservoir, Scotland. Proceedings of the Second International Conference on Soil Mechanics and Foundation Engineering, Rotterdam, 1948, 2, 24-31.
19. VAUGHAN, P. R. Undrained failure of clay embankments. Proceedings of the Roscoe Memorial Symposium 'Stress-Strain Behaviour of Soils', Cambridge, 1971, 683-691.
20. VAUGHAN, P. R. and HAMZA, M. M. Clay embankments and foundations: monitoring stability by measuring deformations. Proceedings of Speciality Session No. 8, 'Deformation of Earth-Rockfill Dams'. Ninth International Conference on Soil Mechanics and Foundation Engineering, Tokyo, 1977, 37-48.
21. SODHA, V. G. The stability of embankment dam fills of plastic clay. M. Phil. Thesis, University of London, 1974.
22. BISHOP, A. W. Shear strength parameters for undisturbed and remoulded soil specimens. Proceedings of the Roscoe Memorial Symposium, 'Stress-Strain Behaviour of Soils', Cambridge, 1971, 3-139.
23. VAUGHAN, P. R., WERNEK, M. L. G., and HAMZA, M. M. Discussion. Proceedings of the Symposium 'Field Instrumentation in Geotechnical Engineering', London, 1973, 617-623.
24. VAUGHAN, P. R., LOVENBURY, H. T. and HORSWILL, P. The design, construction and performance of Cow Green embankment dam. Géotechnique, 1975, 25, No. 3, 555-580.
25. COCKSEDGE, J. E. and HIGHT, D. W. Some geotechnical aspects of road design and construction in tills. Proceedings of the Symposium 'The Engineering Behaviour of Glacial Materials', Birmingham, 1975, 209-219.
26. BISHOP, A. W. and AL DHAHIR, Z. A. Some comparisons between laboratory tests, in situ tests and full scale performance with special reference to permeability and coefficient of consolidation. Proceedings of the Conference 'In situ Investigations in Soils and Rocks', London, 1969, 251-264.
27. ROWE, P. W. Twelth Rankine Lecture: the relevance of soil fabric to site investigation practice. Géotechnique, 1972, 22, No. 2, 195-300.
28. WILKINSON, W. B., BARDEN, L and ROCKE, G. An assessment of in situ and laboratory tests in predicting the pore pressure in an earth dam. Proceedings of the Conference 'In situ Investigations in Soils and Rocks', London, 1969, 277-284.
29. WALBANCKE, H. J. Discussion. Proceedings of the Symposium 'Field Instrumentation in Geotechnical Engineering', London, 1973, 552-555.
30. LITTLE, A. L. and VAIL, A. J. Some developments in the measurement of pore pressure. Proceedings of the Conference 'Pore Pressure and Suction in Soils', London, 1960, 75-80.
31. SCHIFFMAN, R. L. and GIBSON, R. E. Consolidation of non homogeneous clay layers. Journal of the Soil Mechanics and Foundation Division of the American Society of Civil Engineers, 1964, 90, SM5, 1-30.

32. VAUGHAN, P. R. and WALBANCKE, H. J. Pore pressure changes and the delayed failure of cutting slopes in overconsolidated clay. Géotechnique, 1973, 23, No. 4, 531-539.
33. SKEMPTON, A. W. Discussion. Proceedings of the Institution of Civil Engineers, 1957, 7, 267-269.
34. BISHOP, A. W. and VAUGHAN, P. R. Selset Reservoir: design and performance of the embankment. Proceedings of the Institution of Civil Engineers, 1962, 21, February, 305-346.
35. ROWE, P. W. Derwent Dam - embankment stability and displacements. Proceedings of the Institution of Civil Engineers, 1970, 45, March, 423-452
36. BEAVAN, G. C. G., COLBACK, P. S. B. and HODGSON, R. L. P. Construction pore pressures in clay cores of dams. Proceedings of the Ninth International Conference on Soil Mechanics and Foundation Engineering, Tokyo, 1977, 1, 391-394.
37. AL DHAHIR, Z. A. Correlation between field and laboratory measurements on earth dams. Ph.D. Thesis, University of London, 1967.
38. HALLAS, P. S. and TITFORD, A. R. The design and construction of Bough Beech Reservoir. Proceedings of the Institution of Water Engineers, 1971, 25, 293-314.
39. SHEPPARD, G. A. R. and AYLEN, L. B. The Usk scheme for the water supply of Swansea. Proceedings of the Institution of Civil Engineers, 1957, 7, 246-274.
40. HAMMOND, T. G. and WINDER, A. J. H. Problems affecting th design and construction of the Great Ouse Water Supply Scheme. Journal of the Institution of Water Engineers, 1967, 21, 15-66.
41. LUCKS, A. S. The measurement of construction pore pressures in earth dams. Institution of Civil Engineers Medal and Premium, 1966. Unpublished paper.
42. BANKS, J. A. Problems in the design and construction of Knockendon Dam. Proceedings of the Institution of Civil Engineers, Part 1, 1952, 1, 423-443.
43. BISHOP, A. W., KENNARD, M. F. and PENMAN, A. D. M. Pore pressure observations at Selset Dam. Proceedings of the Conference 'Pore Pressure and Suction in Soils', London, 1960, 91-102.
44. CRANN, H. H. The design and construction of Llyn Celyn. Journal of the Institution of Water Engineers, 1968, 22, 13-43.
45. BUCHANAN, N. Derwent Dam - construction. Proceedings of the Institution of Civil Engineers, 1970, 45, March, 401-422

Technical editors: D.W. HIGHT, MSc, MICE,
Imperial College of Science and Technology, and
D.M. FARRAR, MSc, MInstP, Transport and Road
Research Laboratory

Discussion: Engineering properties and performance of clay fills

DR A.D.M. PENMAN, *Building Research Station*
A dam retaining 90 m depth of water has to support a horizontal thrust of more than 4000 t/m length of dam. It was to be expected, therefore, that a dam would be pushed downstream by a measurable amount when the reservoir was first filled.

The horizontal plate gauge,[1] which has given extensive information about the movements of discrete points within the fill of embankment dams, has made it possible to measure the movements of several points on the downstream face of the clay cores of two dams during reservoir impounding. The positions of the gauges on the major section of the two dams are shown in Fig.1, in which C indicates clay core, R indicates the rockfill shoulders and P the positions of stable reference pillars to which core movements could be related with an accuracy of ± 2 mm. Measurements were made several times as the reservoirs filled; the horizontal movements are given in Table 1. Dam A was 70 m high and measurements were made where the top and third from the top gauges went into the downstream face of the core. Dam B was 90 m high and measurements were made on the downstream face of the core at the crest and at each of the three horizontal plate gauges.

The movements were surprisingly small and indicate that the horizontal thrust due to pressure from the clay cores against the upstream and downstream shoulders must have been, in the case of dam A, greater than the thrust from the full reservoir and, for dam B, comparable with the water thrust. The clay for the core of dam A was placed well wet of Proctor optimum and it had values of c_u = 35-100 kN/m^2. The core of dam B was placed at a lower water content and could be expected to be slightly stronger.

The beneficial effects of a substantial pressure from the clay core can be appreciated by considering the interface between the base of the core and the dam foundation on the floor and sides of the valley. The interaction pressure across this interface must be greater than the pressure of the reservoir water to maintain a watertight contact. This principle applies throughout the height of the core: at any level the clay pressure must exceed the pressure of the reservoir water to prevent hydraulic fracture.

Unfortunately differential settlements between the core and shoulders usually reduce the total pressure in the core because some of its weight is carried by the shoulders by 'silo' or 'arching' action. At dam B earth pressure cells placed in the core during construction at positions shown in Fig.2 measured vertical pressures (σ_v) of only 0.6 and 0.7 of the overburden pressure (σ_o) near the centre of the core at the upper and lower positions. In the 90 m high dam described by Coumoulos and Koryalos (pp 73-78), earth pressure cells at the centre of the core and at about third height measured a vertical pressure of only 0.5 σ_o (see their fig.6). This core was placed at about Proctor optimum water content, whereas the core of dam B was placed at 1.5% above Proctor and that of dam A at an even higher value.

It has been argued that differential settlements between core and shoulders could be reduced by placing the clay core dry of optimum.

Table 1. Horizontal movements in mm
- downstream, + upstream

Dam A

Depth of water, % of full depth	Top gauge	Gauge third from top
14	0	0
55	-5	+4
61	+4	+16
80	-4	+10
99	0	+1

Dam B

Depth of water, % of full depth	Crest	Top gauge	Middle gauge	Bottom gauge
0	0	0	0	0
34	+4	-	-	-
59	0	+2	0	+5
75	-2	-3	-6	-7
78	-2	-11	-6	-3
100	-26	-51	-29	-9

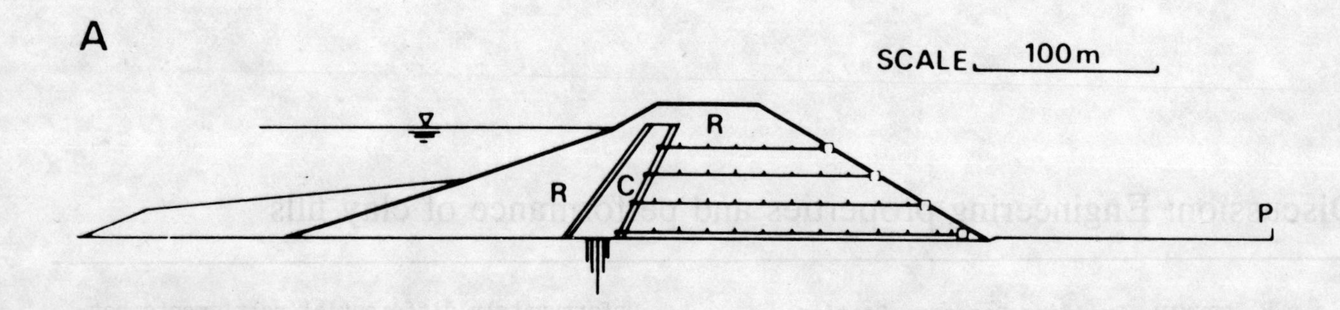

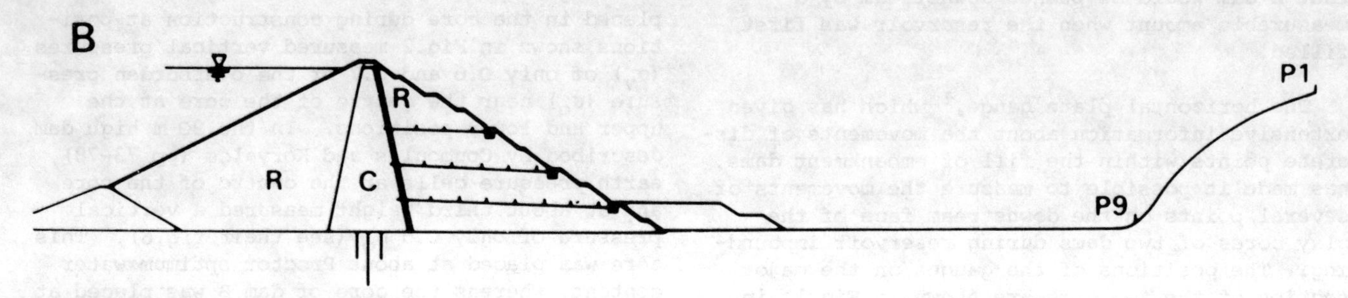

Fig. 1

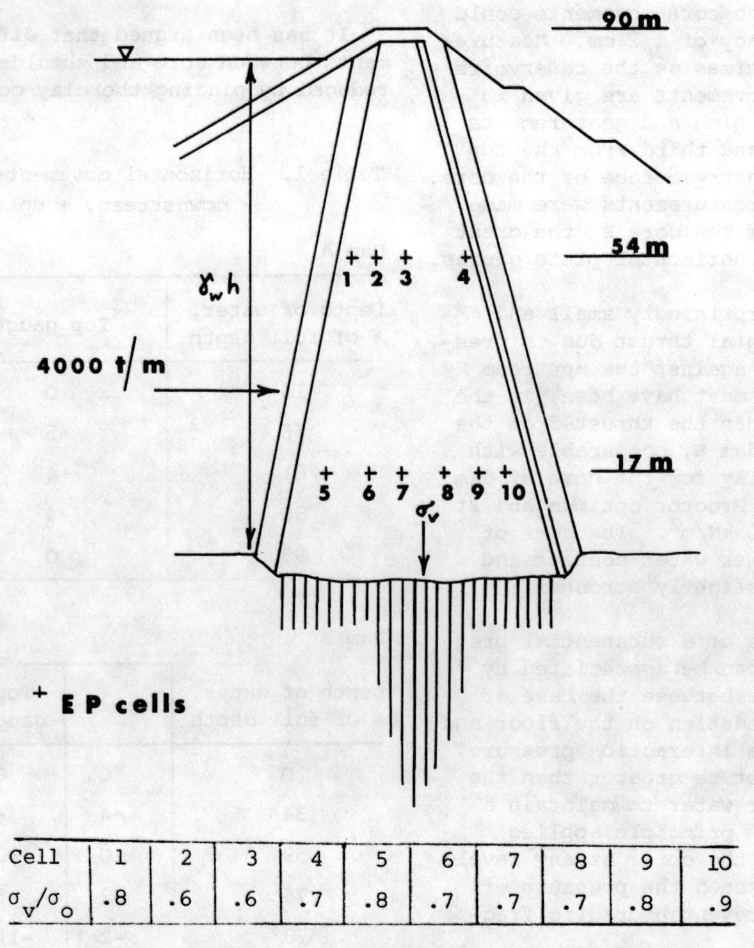

Cell	1	2	3	4	5	6	7	8	9	10
σ_v/σ_o	.8	.6	.6	.7	.8	.7	.7	.7	.8	.9

σ_v must exceed $\gamma_w h$ to avoid hydraulic fracture at the core/foundation interface

Fig. 2

(This would also reduce construction pore pressures in the core and it has been argued that an improvement in overall stability would result.) However, the above field measurements and those made at other dams indicate an opposite behaviour, i.e. the wetter cores exhibited less reduction of vertical stress. This can be attributed to the lower undrained shear strength (c_u) of the wetter cores which limits the amount of weight that can be transferred to the shoulders.

Clay cores placed wet of optimum usually develop construction pore pressures almost equal to the total stress in the core. Values of the ratio pore pressure/overburden (r_u) often lie in the range 1-0.95 in the early stages of construction, but as the height of the core increases, these values fall to perhaps 0.5 or lower. This is sometimes regarded as a good sign, from the point of view of stability, but in fact it is usually simply an indication of the reduction of the ratio σ_v/σ_o.

In order to maintain a sufficiently high value of σ_v a core should be placed wet of optimum and, to ensure that the pressure from the clay is greater than the pressure from the reservoir water, the zero pore pressure line should be at least at dam crest level at the end of construction and on first filling. When the reservoir is full the phreatic line will then join the reservoir surface to the zero pore pressure line of the core and the clay will suffer neither swelling nor consolidation.

Placement control by water content as a given percentage above Proctor optimum is not easy and it is refreshing to see the move towards control by measured c_u value as described by Kennard et al. (pp 143-147). They advocate the use of low values of 50-100 kN/m^2. Dennehy also speaks of c_u values (pp 87-94) but in relation to the ability of placement plant to be supported adequately by the placed fill. He shows that values of 40-60 kN/m^2 are the lower limits for some types of scraper, so it appears that cores should be placed just dry enough to carry the placing machines.

Coumoulos and Koryalos (pp 73-78) compare the observed settlements of the core with values predicted from oedometer tests made on undisturbed samples taken from the core. They attribute the apparent overestimate obtained from oedometer tests to the dissipation of pore pressure in the laboratory tests which was faster than that which occurred in the dam.

It appears from their figs 10 and 11, however, that core settlement has been related to overburden pressure, whereas their fig.6 shows that the total vertical pressure at about one third height on the centre line of the core was only half the overburden. If measured settlement were related to measured total vertical pressure it would show a closer agreement with the oedometer results.

A simple comparison with oedometer results is impossible. Allowance must be made for pore pressures as well as the actual total stresses in the core. The inclinometer tube, presumably placed close to the downstream face of the core, showed that there was no appreciable horizontal movement during construction. This indicates that a perfect balance was always maintained between the opposing horizontal pressures from the sand and gravel shoulders and the clay core. This is a remarkable result in itself and it would be valuable to know the movements that were measured by the inclinometer during first reservoir filling.

DR D.G. COUMOULOS, *Castor Ltd, Athens*
Settlement data during the construction of a large embankment dam, with a maximum height of 165 m and central core (Fig.3) are shown in Fig. 4, plotted against elevations from two locations along the axis of the dam and for different dates. The clay core material of this dam has similar characteristics to those of the dam described on pages 73-78.

The curves in Fig.4 are similar to those in fig.9 on page 76. They indicate that during the initial stages of construction, maximum settlements occur at approximately mid-height of the embankment, but as construction proceeds the point of maximum vertical settlement moves towards the one third point of the embankment height.

Compressibility characteristics of the clay core material derived from Fig.4 are shown in Fig. 5 in terms of percent vertical strain against overburden pressure. These results do not agree well with those obtained from oedometer tests on laboratory compacted specimens (Fig.5). On the contrary, Figs 10 and 11 on page 76 show that there is good agreement between compressibility observed in the field and that obtained from laboratory oedometer tests on undisturbed specimens from the actual fill.

This supports the view that the influence of structures imposes limitations on the use of laboratory compacted specimens at the preconstruction stage.

Dr Penman suggests in his discussion that if the computation of the overburden pressure on the horizontal axis of Figs 10 and 11 on page 76 were based on the results from the earth pressure cells (fig.6 on page 74 which shows that measured vertical pressure is half the overburden, with which Dr Penman agrees), and the results of the piezometers, then there would be much better agreement between compressibility observed in the field and that obtained from oedometer tests on either laboratory compacted specimens or undisturbed specimens obtained from the actual fill (figs 10 and 11) on page 76 for the latter case).

I agree with Dr Penman's comments. However, Mr Koryalos and I were reluctant to use data from the dam instruments because, although there were many instruments installed in the dam, we felt there were uncertainties in the data

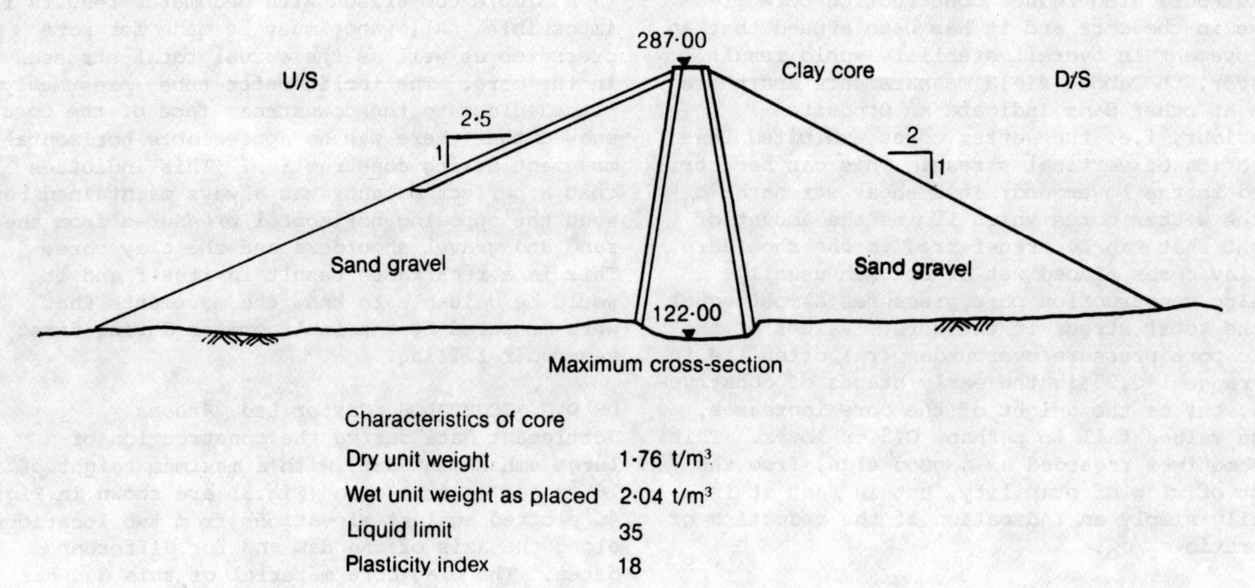

Fig. 3. Maximum cross-section of Kremasta Dam (*courtesy Public Power Corporation of Greece*)

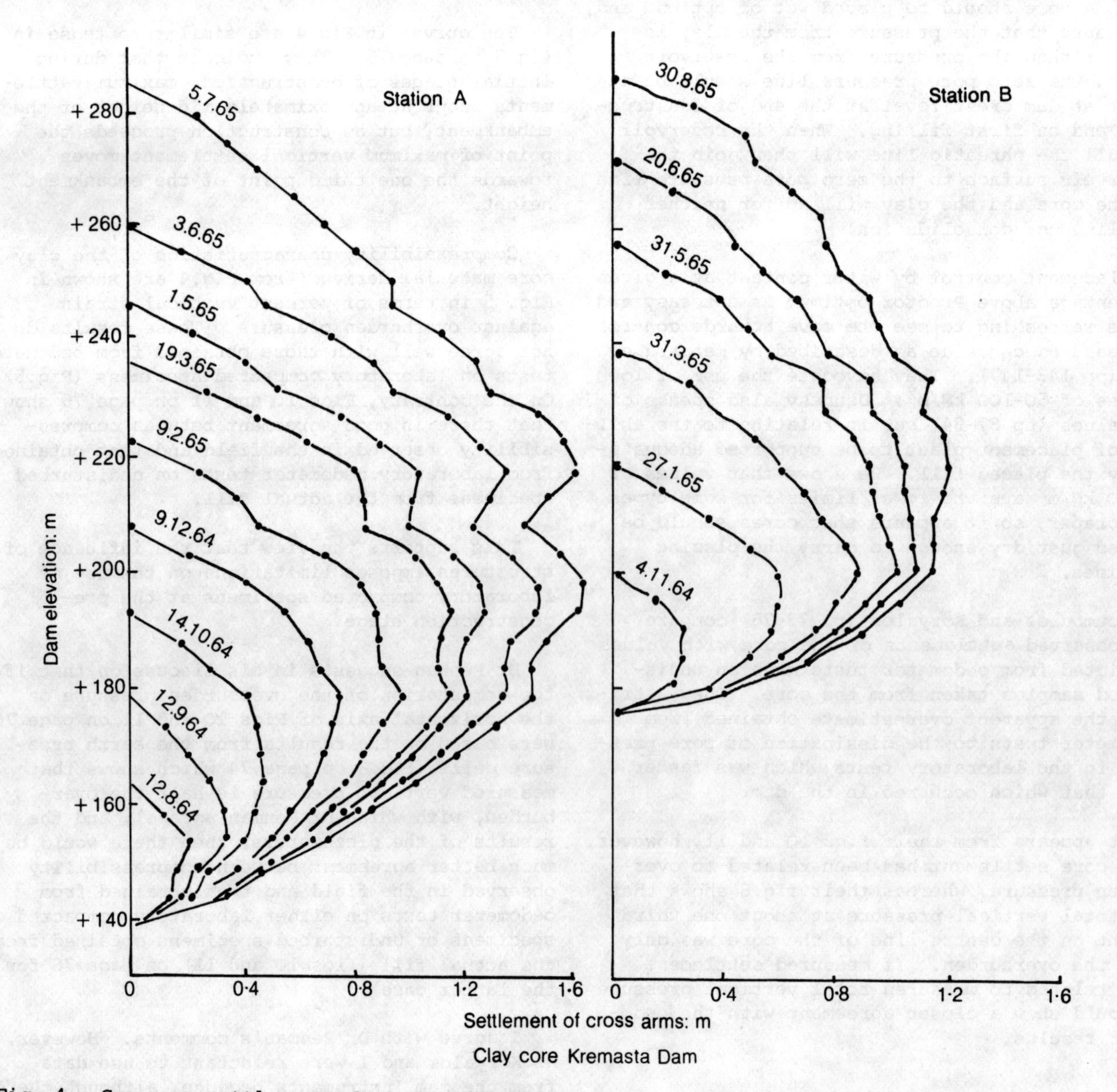

Fig. 4. Crossarm settlements during construction at two locations along the axis of the dam

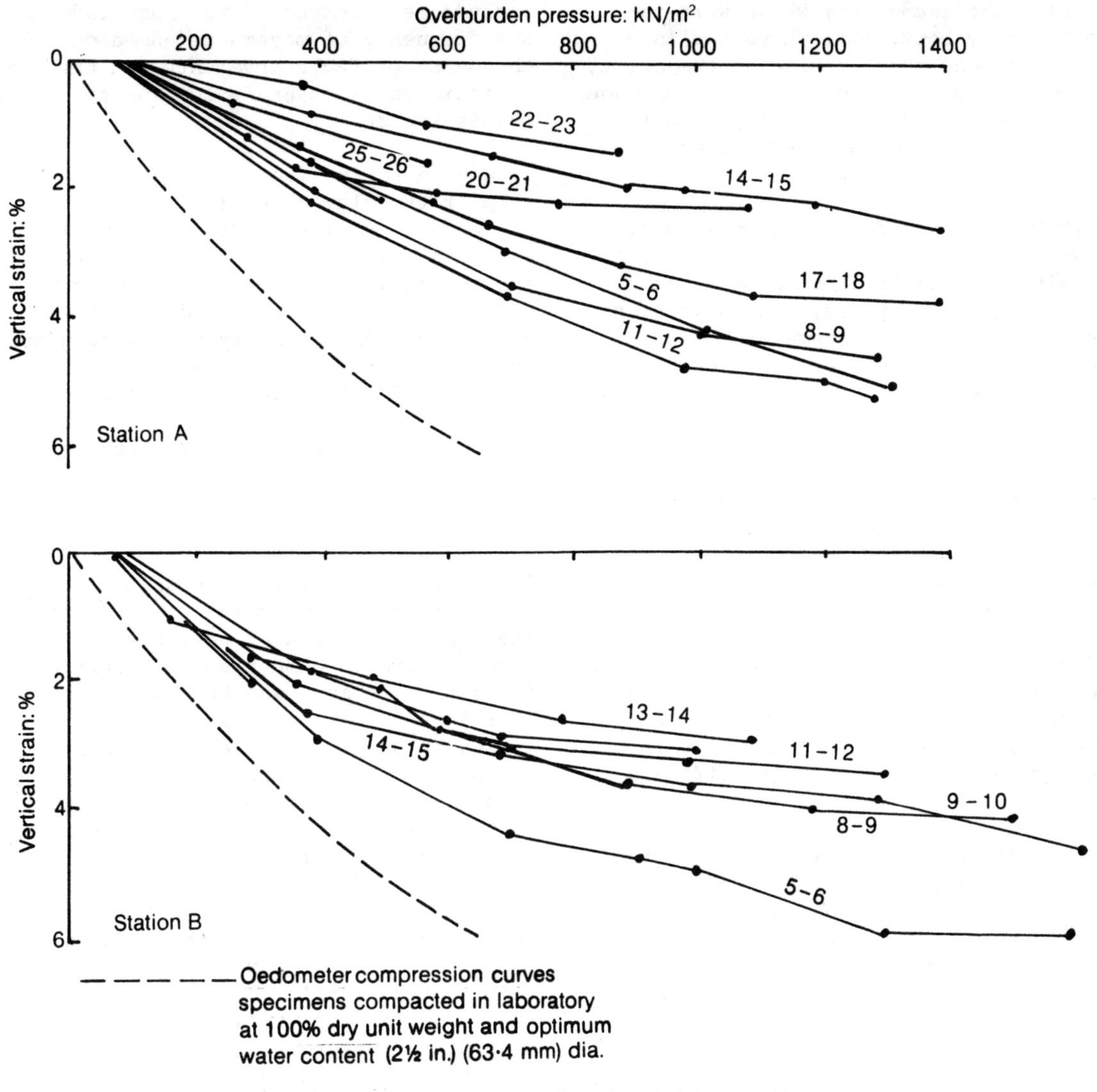

Fig. 5. Consolidation of the clay during construction

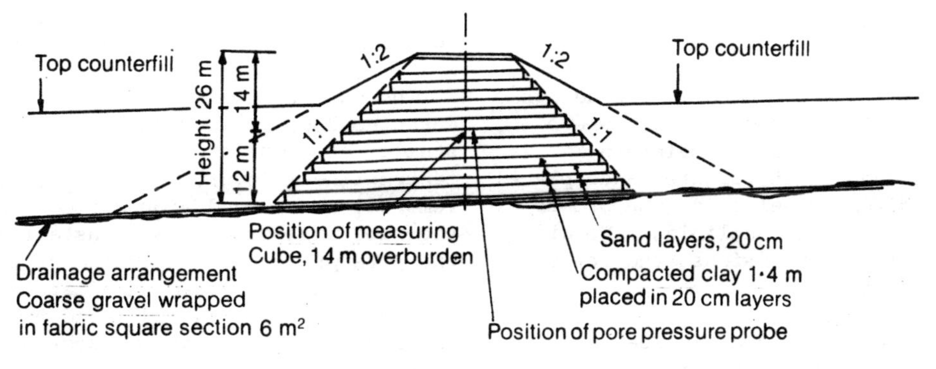

Fig. 6. Cross-section of compacted fill with cube for the measurement of horizontal and vertical stresses

obtained from these instruments (see paragraphs 9-13 on page 75).

At Dr Penman's suggestion, we have examined inclinometer data during filling of the reservoir from three inclinometers, which were installed in the clay core. All three inclinometers were at the downstream side of the crest, the middle one being that shown in fig.8 on page 76. Based on the inclinometer data, no horizontal displacements were recorded during filling of the reservoir.

MR R.E. HARPSTER, *Woodward-Clyde Consultants, San Francisco*
Based on design and construction experience of the California aqueduct system, case histories of changes in the physical parameters of clay material used in embankments are discussed. Changes in parameters cited were small, but real. It is important for designers to be aware of these.

At Cedar Springs Dam (see pages 119-125), due to handling there was an increase in the percentage of clay sized particles and of the liquid limit measured. No design changes were needed but the contractor did file a claim.

Due to the method of excavation at the borrow pit for Del Valle Dam there was a change in the particle size distribution of material. The method of excavation was corrected, and this enabled the gradation and density specification to be met.

In the San Jolquin Valley small changes in plasticity and compressibility of material in clay embankments were observed. X-ray analysis showed that there had been cation exchange which occurred when the clay had been moisture conditioned with water of different quality from that occurring naturally in the borrow areas.

With regard to the design of clay cores and clay sublinings for dams and canals crossing active faults, since the Baldwin Hills Dam failure a more conservative approach has been adopted in dam design and construction. The crossing of active faults should be avoided if at all possible, and, in cases where the assessed risks are high, totally avoided.

I wish to discuss the configuration and dimensions of clay zones. Other standard considerations must also be included in the design analysis. However, in the special case where the clay core or sublining is to maintain its integrity, even though fault rupture takes place, priority must be given to selecting the proper dimensions for clay core or sublining. To aid selection of the proper dimensions, the following studies are essential.

Regional and specific site geological studies are necessary to demonstrate the presence or absence of faults in the foundation.

Specific studies must be made to define the limits of the fault, e.g. the width of zones of disturbed rock, the planes where shear has occurred, the length and direction of the slip planes and the pattern of faulting.

Also studies to reconstruct the history of fault movements must be made. These studies include measurement of the magnitude, direction and frequency of movements and a correlation of the sense of fault movement with the regional tectonic environment reflecting the present stress regime.

Consideration must be given to the relationship of the clay zone with respect to its boundaries - specifically drainage and filter zones, and transition core protection zones - and to the interrelationship of clay zones with the cutoff, with grouted zones and with the freeboard elevation. All boundary zones, foundations and crest restrictions must be designed with the same consideration for the magnitude and direction of anticipated slip as is given to the design of the core. This is necessary to ensure the integrity of the clay zone should rupture occur.

One method used to arrive at a figure for a design slip requires the possible magnitude of slips to be expressed in probabilistic terms. The final selection of design slip can then be obtained from a comparison of estimated risks for various slips, and the corresponding probability of occurrence. A limit will be established by selecting an acceptable level of risks. In the USA the responsibility for the choice of an acceptable level of risk lies with the owner of the dam pending approval by the appropriate review and regulatory agencies.

DR H. ØSTLID, *Norwegian Road Research Laboratory*
The recently completed E6 motorway north-east of Oslo has a fill of maximum height of 26 m and is constructed of silty clay or similar materials. The Norwegian Road Research Laboratories decided that pore pressure and stresses should be measured during and after construction.

A special steel cube was designed and earth pressure cells were fitted on all six faces, facilitating the measurements of stresses in the three principal directions. This cube was also fitted with inclinometers in order to measure any rotation of the cube itself. Both measuring systems were of the vibrating wire type and the actual recording was done with a frequency meter.

The cube was made of 35 mm steel plates and measured 236 mm along each edge. It was installed at about mid-height in the fill, on the centre line (Fig.6). Soft clay, with all stones removed, was placed around the cube. Close to the cube a piezometer was installed to measure the pore pressure. Recording of stresses and pore pressures started immediately after installation on 29 August, 1977, and will continue for as long as the cube and piezometer function.

A typical cross-section of the fill is shown in Fig.6. The construction is of the sandwich type with 20 cm thick horizontal sand layers spaced at vertical intervals of 1.4 m. These sand layers are used to keep pore pressures down

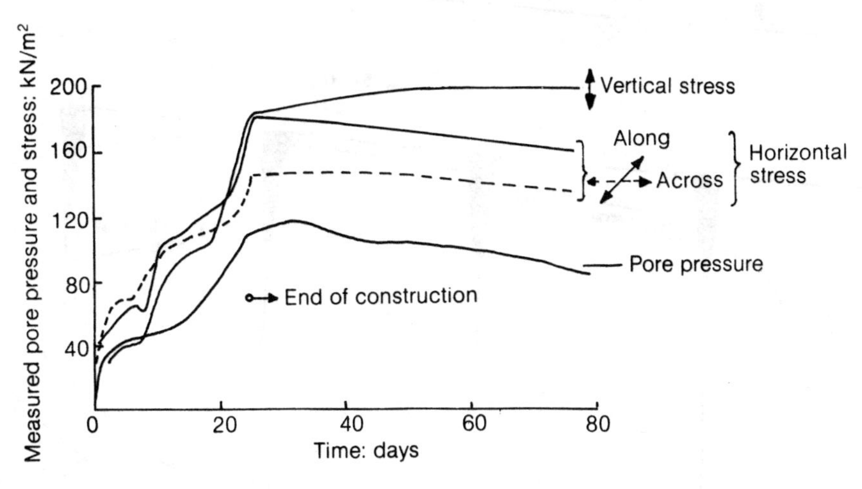

Fig. 7. Average vertical and horizontal stresses and pore pressure measured during the early stages

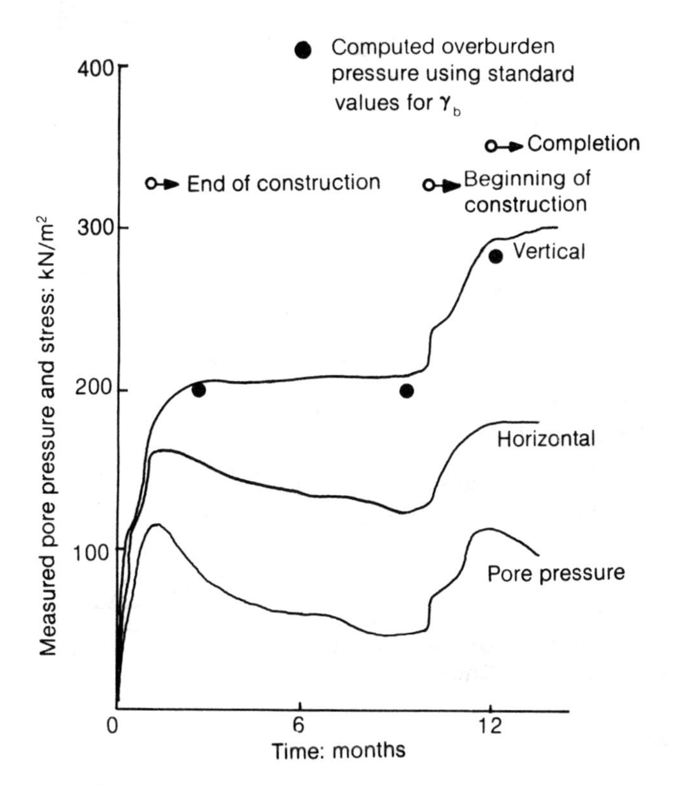

Fig. 8. Average vertical and horizontal stresses and pore pressures recorded to date

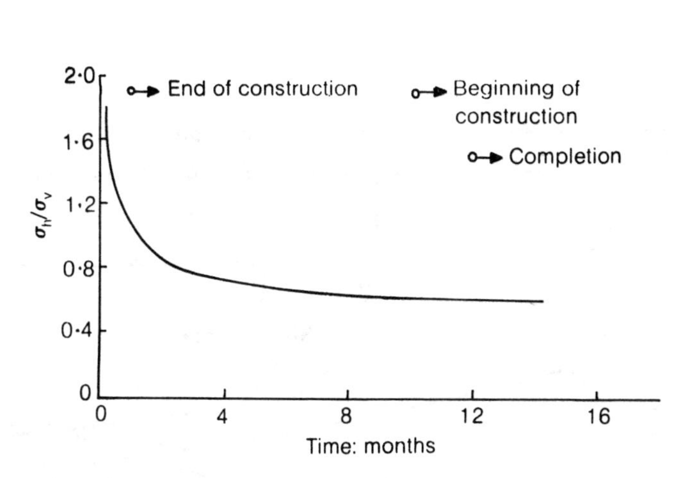

Fig. 9. Development of σ_h/σ_v with time

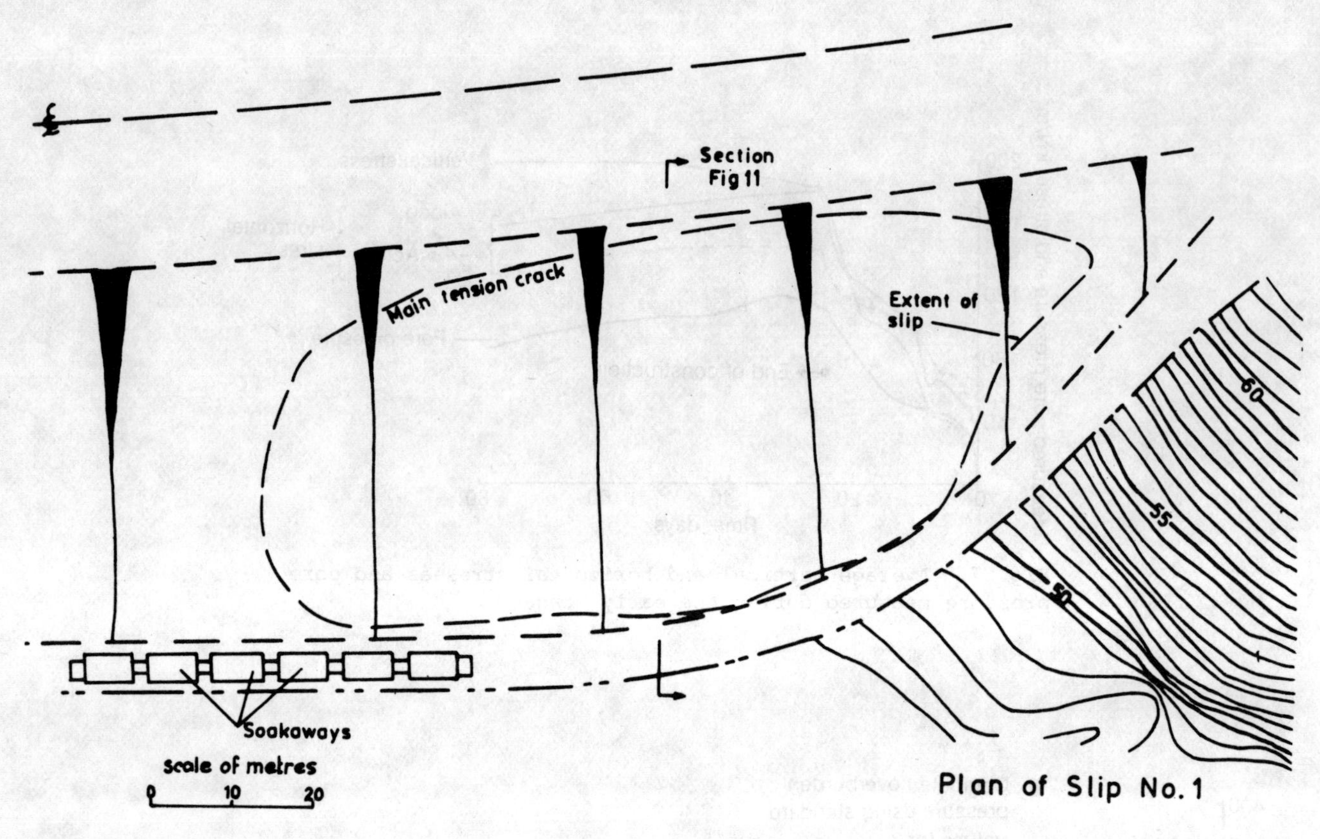
Section
Fig 11
Main tension crack
Extent of slip
60
55
50
Soakaways
Scale of metres
0 10 20
Plan of Slip No. 1

Fig. 10

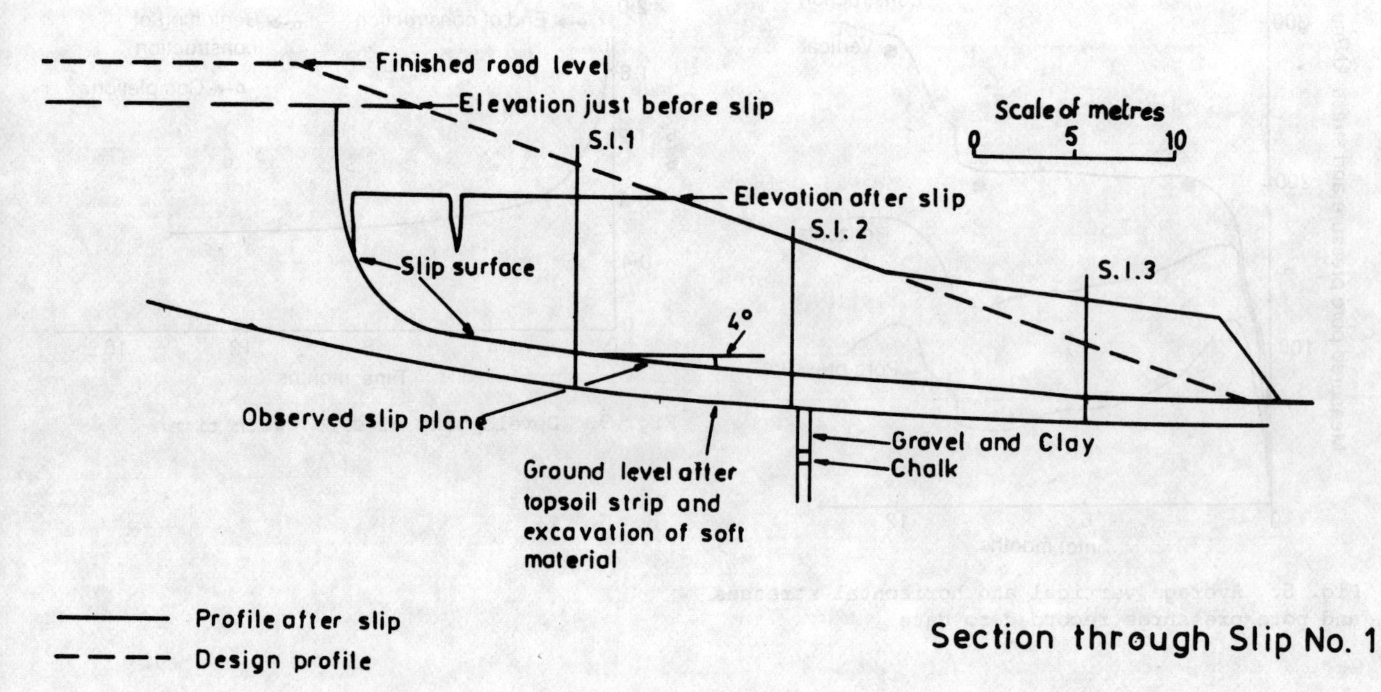
Finished road level
Elevation just before slip
Scale of metres
0 5 10
S.I.1
Elevation after slip
S.I.2
Slip surface
S.I.3
4°
Observed slip plane
Gravel and Clay
Chalk
Ground level after topsoil strip and excavation of soft material
Profile after slip
Design profile
Section through Slip No. 1

Fig. 11

during construction and also to increase the rate of settlement.

The fill material was mainly a normally consolidated silty clay with a maximum water content of 30%. The liquid and plastic limits were 40% and 20% respectively and the optimum water content was 18-20%. The fill was placed in 200 mm thick layers by light bulldozers.

The vertical stress increased roughly linearly with increase of fill height and similar behaviour was observed in the horizontal stress development (Fig.7).

When construction was stopped before the winter, the increases in the horizontal and vertical stresses stopped. The steady increase in pore pressure also stopped. During the winter little work was done on the fill and the stresses were nearly stationary: the vertical stress increased slightly and the horizontal stresses and the pore pressure decreased a little. The rates of dissipation of horizontal stresses and the pore pressure were roughly equal.

The horizontal stresses were not quite equal in the two directions. The stresses along the centre line were about 15-20% higher than the stresses across the fill. Lateral deformations may have caused this effect. The higher stresses along the centre line may also be caused by an increase in stresses due to consolidation; the valley has a very sharp V form.

The dissipation of pore pressure may be caused by the sand layers, lateral deformation or natural redistribution of pressure.

Figure 8 shows the development of stresses. When work was resumed in summer 1978 all the stresses reacted immediately and the recorded values seem to be reasonable. The vertical stress increases more than the horizontal stress and the general stress development changes sharply when the filling operation stops.

Pore pressure follows the general trend of the other stresses and drops after the end of construction.

If the average values of σ_h/σ_v are computed (see Fig.9), this factor is as high as 1.8 in the early construction stage. It drops to about 0.6 when construction is finished and this value is of the order that may be expected in a normally consolidated clay.

MR J. WRIGHTMAN, *South East Road Construction Unit*

I should like to describe two slope failures which occurred during the construction of road embankments of London Clay. The likely cause of failure is the building in of smooth planes formed by a smooth-wheeled roller coupled with a slight wetting of the fill surface and an outward slope of 4° or more formed to shed water during breaks in construction.

Figure 10 shows a plan of the first slip and Figs 11 and 12 show sections through the first and second slips respectively.

At the site of the first slip, bank construction began in late April following removal of topsoil and poorer subsoil, and the construction of five precast concrete drainage soakaways at the bank toe. Failure occurred 28 days later when the bank slope height was 14 m and the maximum height of fill 9 m, following a weekend when no fill was placed. The slip was about 100 m long and involved about 25 000 m^3 of soil. It occurred suddenly with no advance warning cracks, and the only eye witness said, 'There was a kind of bang'. Initially there was a vertical tension crack with a 2 m drop in fill level and the toe pushed out within the bank fill exposing large stiff lumps of slickensided clay. No water issued from the toe.

Investigations started within hours of the failure and three slip indicators were installed. The rapid installation of slip indicators while the slip was still moving permitted early and positive identification of the slip surface so that it could be examined and sampled from trial pits later, and back analyses made for planning remedies.

The second slip was similar to the first, except that the bank had been partly built the previous summer. It was left over the winter with a 6° slope at the fill surface and construction continued with Norwich or Red Crag after some attempt to dry out the London Clay.

For the first slip, back analysis in terms of total stress of a section after failure indicated that the undrained shear strength c_u along the known slip surface should be 11-16 kN/m^2 for a factor of safety of unity. c_u values for the bank material ranged from 72 kN/m^2 to 168 kN/m^2 with an average of 120 kN/m^2 and so the factor of safety should have been quite high.

Examination and sampling were carried out from trial pits at the slip surface which was clearly defined by the deformed slip indicator (Fig.13). This revealed a thin band of softened clay up to 6 mm thick which appeared to be continuous. Samples from this plane were difficult to obtain and test and the water content ranged from 30% to 44% with an average of 35.7%. The average water content of the bank fill was 26%. c_u was determined on samples which were remoulded at a range of moisture contents, and from this relationship the c_u value related to a moisture content of 35.7% was 36 kN/m^2. This still did not explain the failure.

Effective strength parameters were determined in triaxial compression tests, and peak and residual shear strengths were found from shear box tests. A likely value of pore pressure/overburden ratio (r_u) was assumed based on laboratory measurements and nearby instruments in a similar bank. Analyses using peak shear strength parameters gave a factor of safety of 1.4 and using residual shear strength parameters a factor of safety of 0.9 was obtained. Even when an upper bound value of r_u was used in the analysis with peak strengths the factor of

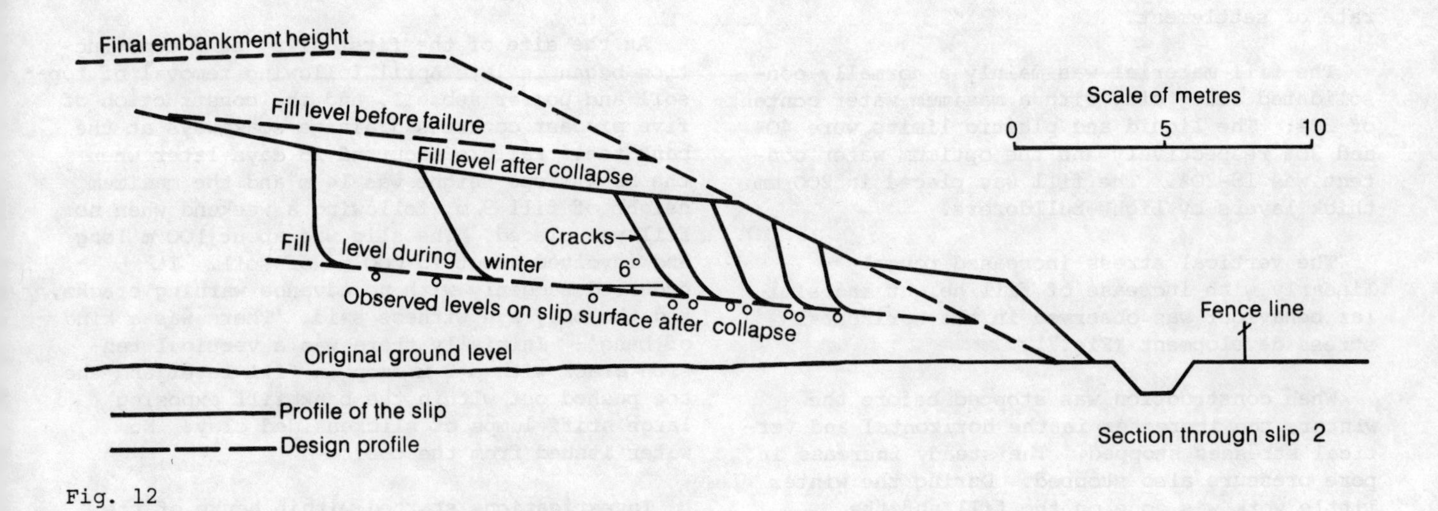

Fig. 12

Fig. 13

Fig. 14

Fig. 15

safety was greater than unity. The failure was consistent with the fill having a shear strength between peak and residual values.

Further undrained tests were done on remoulded samples at a range of moisture contents, but with a plane preformed in the test specimens to simulate the planes formed by a smooth-wheeled roller. The c_u value related to a moisture content of 35.7% was 8.5 kN/m^2 and this gave a factor of safety of 0.85.

The slip cannot be explained in terms of either total or effective strength parameters but only in terms of a plane of weakness being built into the soil. A typical example is shown in Fig.14 and an example of such a plane formed by a smooth-wheeler roller is shown in Fig.15. The analyses indicate, as one might expect, that such a plane has only a partial effect. The analyses carried out for the second slip lead to similar conclusions. The angle to the horizontal of successive construction layers has a marked effect on factors of safety, and this is taken into account in earthworks specifications for cohesive soils.

This work was carried out in the Department of Transport and is submitted by permission of Mr R. Bridle, Chief Highway Engineer. Mr T. Lumbe, lately of the Department of Transport, Mr P. Kearns of Suffolk County Council and Mr D. Farrar of the Transport and Road Research Laboratory collaborated in the work.

MR E.A. SNEDKER, *Midland Road Construction Unit* Recent years have been particularly dry but the exceptionally dry summer of 1976 was followed by a period of unusually intense rainfall in early 1977. This rainfall resulted in many slides in highway and other slopes around the UK. Of particular interest was the number of embankment fills showing signs of distress. I was involved in the examination of a number of these highway fill areas. The degree of failure varied from slumping, affecting only the fill side slope, to movement also affecting the pavement in the form of subsidence, cracking or, in the worst cases, a combination of both. Movements were all in the embankment fill and did not involve the foundation material.

Some of the fills involved had been constructed for 2-8 years, others had been constructed late in 1976. Water was associated with each problem. What proved to be of interest was the form of construction and construction history. The fills were mixtures in one form or another of permeable and less permeable materials. Some of the typical situations which were found to exist are shown in Figs 16-21.

Figures 16 and 17 typify some of the problems in the Midland boulder clays which contain frequent lenses of sand and silt. The problem shown in Fig.16 occurred before paving was started; the situation in Fig.17 caused failure of the pavement surface on a 10 m fill. In both cases water percolated through the reconstituted haphazard arrangement of more permeable silts and sands from a water source to induce the problems.

Figure 18 shows a situation which occurred in an 11 m fill. Earlier problems with the fill had required rebuilding of the pavement on a slip road. The rubble from the pavement excavation with rock and sub-base had been used to make good the road foundation. Water storage in the foundation was also exaggerated by tipping of snow adjacent to the area. Water was again allowed to flow through more permeable lenses to the slope face and failure of the slope and pavement was again induced.

The configuration shown in Fig.19 occurred on an unopened stretch of road on a 6 m fill and was first noticed because of the settlement of the carriageways. The movement occurred in winter and spring 1977 during the very wet period. A stand-pipe later inserted in the central reserve showed a rapid response to short periods of rainfall, indicating that considerable free water would have been stored within the fill in the period of heavy rain.

Trial pits dug at the site shown in Fig.20 indicated the presence of water under the pavement (concrete) which eventually subsided and required remedial work. It was significant that a number of the areas involved were at the transition between general embankment fill and special granular fill placed adjacent to structures. Deficiencies in the general fill had, perhaps, been made up with granular material.

A further problem occurred in a 10 m fill on a heavily trafficked section of motorway where a more detailed investigation was made. The initial failure was reported following the period of 11 days' continuous rain in February-March 1977. An independent assessment had diagnosed a serious situation within the fill and had recommended construction of a berm. My involvement with the area began when the remedial berm collapsed. A number of boreholes were placed in the fill and showed the zoning of materials to be typically as in Fig.21. Remedial work to the slope exposed many water seepages, some carrying considerable quantities of water. Most of these occurred in the upper clay shale, presumably confined by the heavy clay layer 2-3 m below the top of the fill, but some were evident at a depth of 6-7 m, within the colliery waste.

Investigations during even short storms over the 1977-78 winter indicated that considerable backing up was occurring in the central reserve french drain. The outfall in fact passed down the problem slope and the fully silted misshaped pipe was eventually exposed and pinpointed as a problem. Excavation of the central reserve drain also indicated broken and misaligned pipe surrounded by substantial drain stone creating water storage at the centre of the fill.

Insertion of slip gauges, piezometers and stand-pipes was included in the investigation. What became significant was that in spite of the rectification of drainage problems which had been found, piezometers and stand-pipes in some

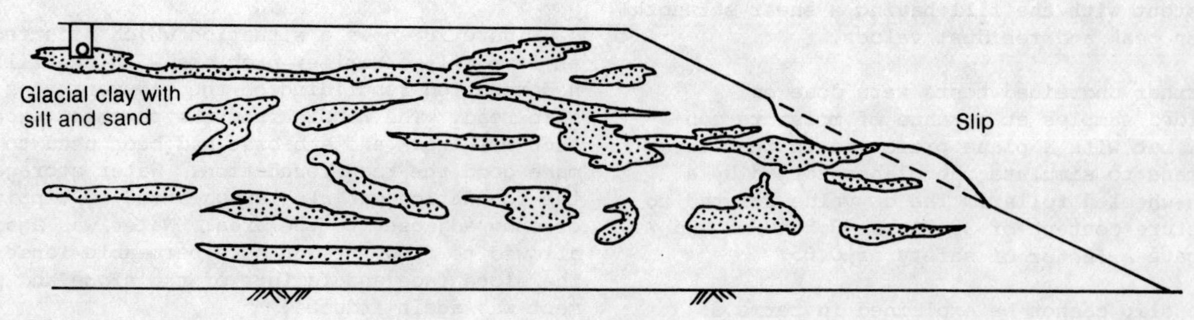

Fig. 16

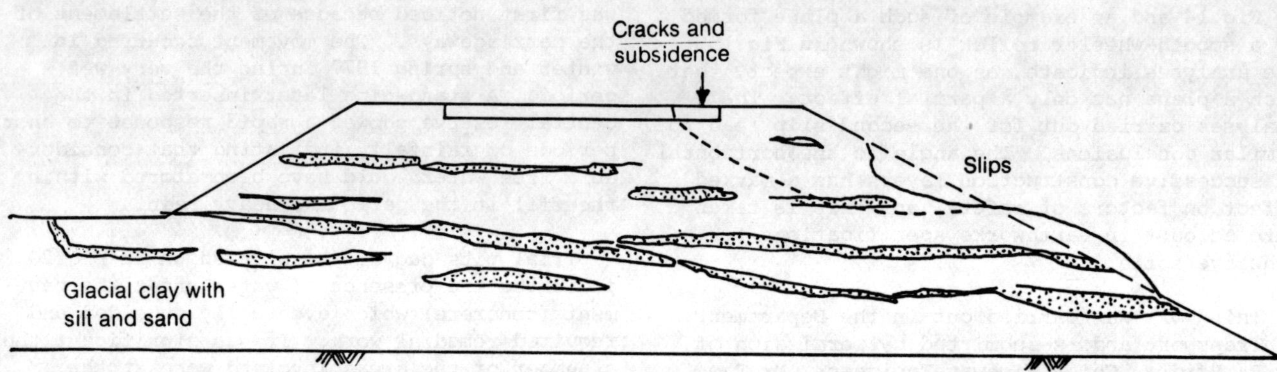

Fig. 17

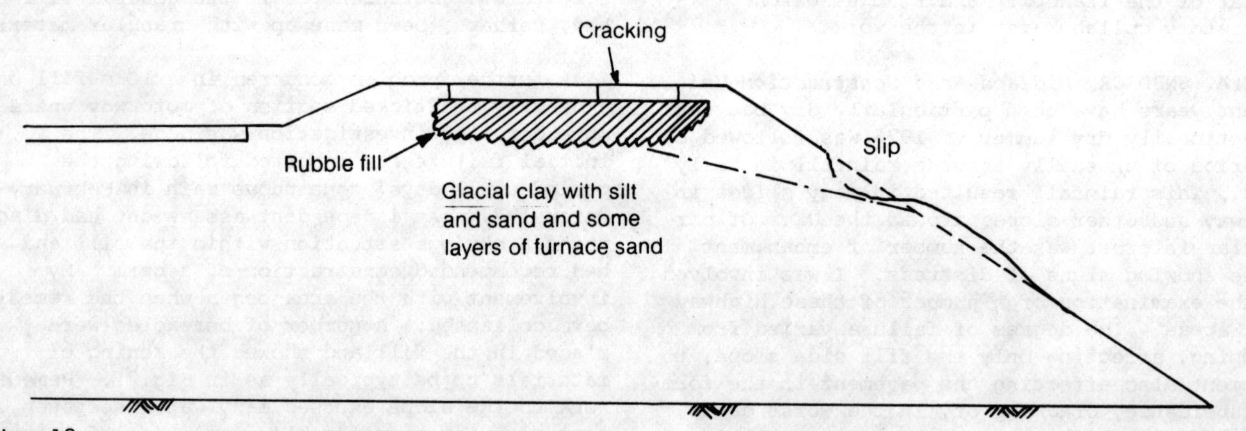

Fig. 18

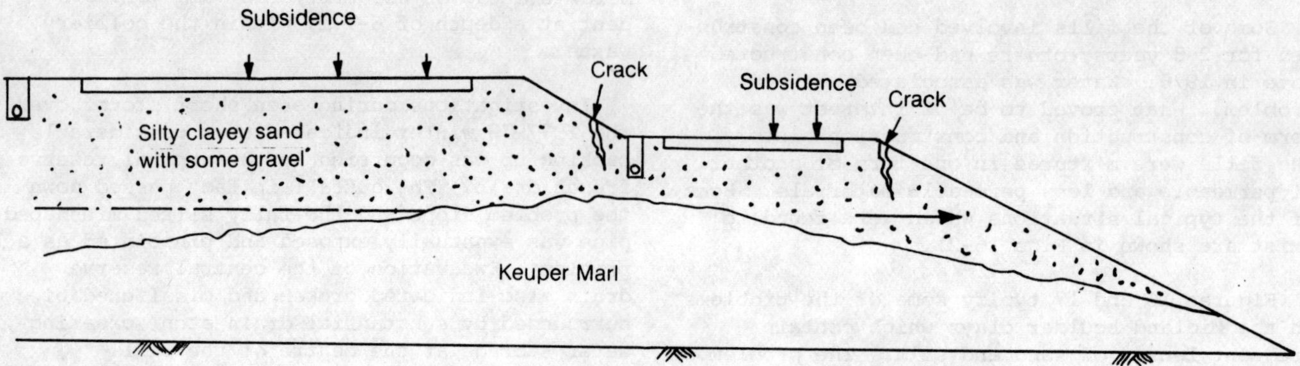

Fig. 19

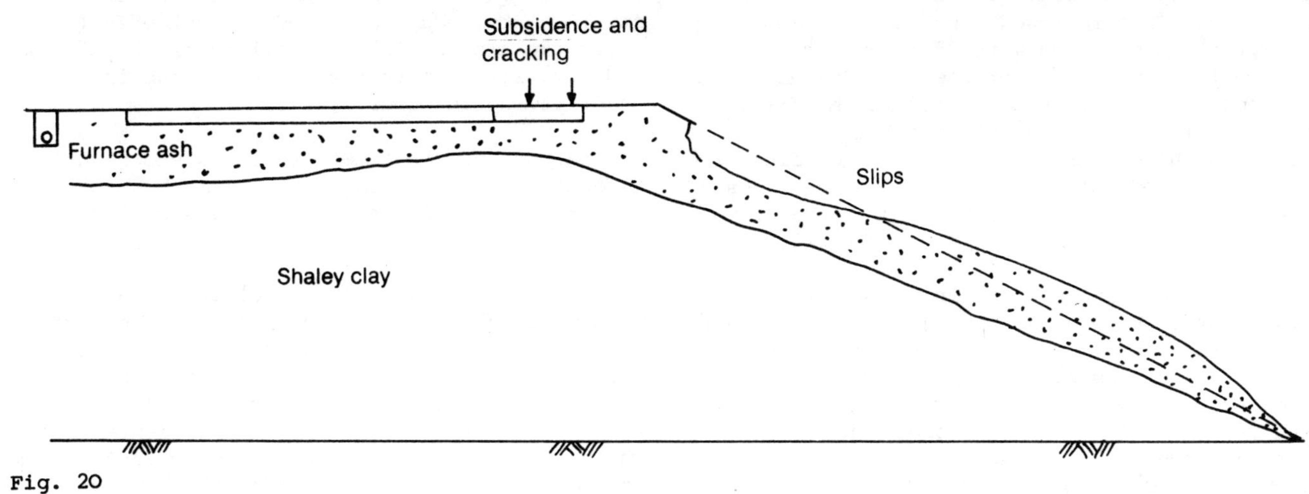

Fig. 20

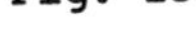

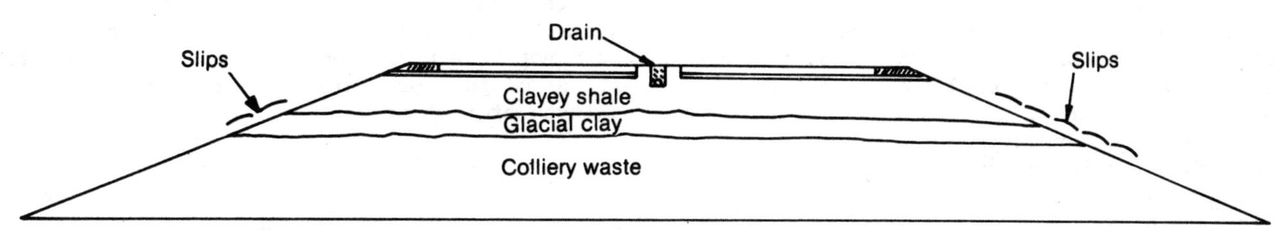

Fig. 21

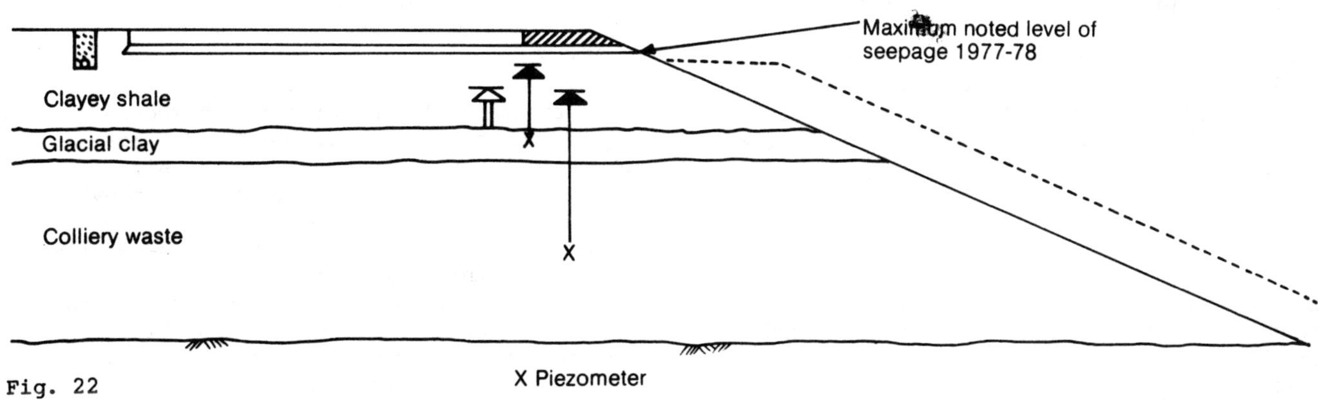

Fig. 22

places still showed substantial water heads, particularly under the northern verge (the most troublesome area). In March 1978, one year after the troubles first occurred, readings of piezometers and stand-pipes - on the northern side in particular - indicated heads of 1-1.5 m of water above the boulder clay level, as shown in Fig.22. This followed a month of 31 mm of precipitation compared with 98 mm during the same period a year earlier when the drainage problems were still incorporated in the fill. The water level build-up at that time must, therefore, have been substantial. Recordings were maintained until July 1978 when the tubes were lost due to reconstruction works, but levels continued to fall off, as did the rainfall.

There is nothing new in finding water in embankments. Porous and open-jointed drains are used in the top of embankments and water is also carried in sub-bases. Where materials permit percolation of water through the fill, water will find its way to the batter slopes. Where such areas are found by routine inspection they can be cured rapidly by drainage. With the current lack of maintenance funds it is doubtful if such inspection, let alone the remedial work, is likely to be done until the area is showing obvious or catastrophic signs of distress.

It is clear that had the fill been impermeable in all the situations described, the problems would not have occurred. Also had there been no water entering the fills as constructed, no trouble would have arisen. Engineers are faced with using indigerous materials and the solution to such problems appears to lie in preventing the entry of large quanitities of storm water into the fill. This suggests the need for more positive drainage systems.

MR L. THREADGOLD, *Exploration Associates (Warwick) Ltd*
In the Eastern Road Construction Unit area an embankment (see Fig.23) was constructed of London Clay in 1973. The material had a liquid limit of 71-89% (mean 79%) and plastic limits of 22-30% (mean 27%). The embankment has shown signs of instability: relatively minor slips occurred shortly after completion and there was a larger slip during 1978.

Boreholes were drilled from the road verge using shell and auger techniques and standard U102 samples were taken at 1 m intervals. Standpipe piezometers were installed in two of the boreholes near the slip; the highest recorded levels are shown on Fig.23(b). Triaxial compression tests undertaken on samples from the boreholes revealed the shear strength/depth relationship shown in Fig.23(a). Shear strengths reach a peak of 180 kN/m^2 at about 2.4 m depth and reduce to 60 kN/m^2 at 5.4 m depth.

Simple oedometer tests were undertaken on samples from 3.4 m and 6.4 m. The initial loading and saturation stages of the test on the sample from 3.4 m showed that a load in excess of 200 kN/m^2 was necessary before the tendency to swell was arrested; the sample from 6 m showed no such tendency above a load approximately equivalent to overburden.

It is postulated that the strength and consolidation data, although limited, serve to indicate that the pore pressures in the embankment are not yet in a state of equilibrium and that potential for swelling with associated softening still exists. Such a time-dependent effect would explain the delayed failure of the slopes and suggests that further movement and instability could occur.

Analysis of the latest slip has shown that effective stress parameters of $c' = 5$ kN/m^2 and $\phi = 18°$ are consistent with the instability which has occurred.

Two possible explanations of the behaviour of this material have been considered. The first is that the lower sample is near to the pre-existing ground level and has had the opportunity of reacting to the proximity of water from the natural ground immediately beneath the embankment and is therefore close to equilibrium. No drainage layer was placed at this interface. The second explanation is that the material at the lower level was obtained from near surface soils from an adjacent cutting 6-7 m deep. The clay within the central zone of the embankment was from the base of the same cutting. The very short period between excavation and placing would not have been adequate for equilibrium conditions of pore pressure and stress to be achieved at that stage.

Both effects could be in operation at the same time but the latter explanation coincides with observations made at other sites and is believed to be the predominant effect.

A second case concerns a road embankment near Grantham which was constructed in 1961-62 to a height of up to 7 m, using mainly Lias Clay. The embankment slopes are generally in the range 1:1.9 to 1:3; liquid limits of the clay range from 50% to 65% and the plastic limits are 25-35%.

At the time of construction a number of deviations from normal practice were made, particularly in relation to pavement design and construction. Positive drainage was not provided from the median strip between the carriageways, kerbing and associated gulleys were generally omitted and the limited drains in the verge were taken to outfalls which discharge near to the top of the embankment slopes. Instabilities have been occurring particularly since 1976, notably at three locations.

Although no specific evidence of water being transmitted from the central area to the outside of the embankment such as that obtained by Ingold and Clayton (pp 133-136) was obtained, a similar mechanism must have been in operation. Cracking in the road pavement also allowed ingress into the subgrade and embankment soils.

Drainage from the median drain, carriageway and from local discharges on to the slope has clearly created groundwater conditions which

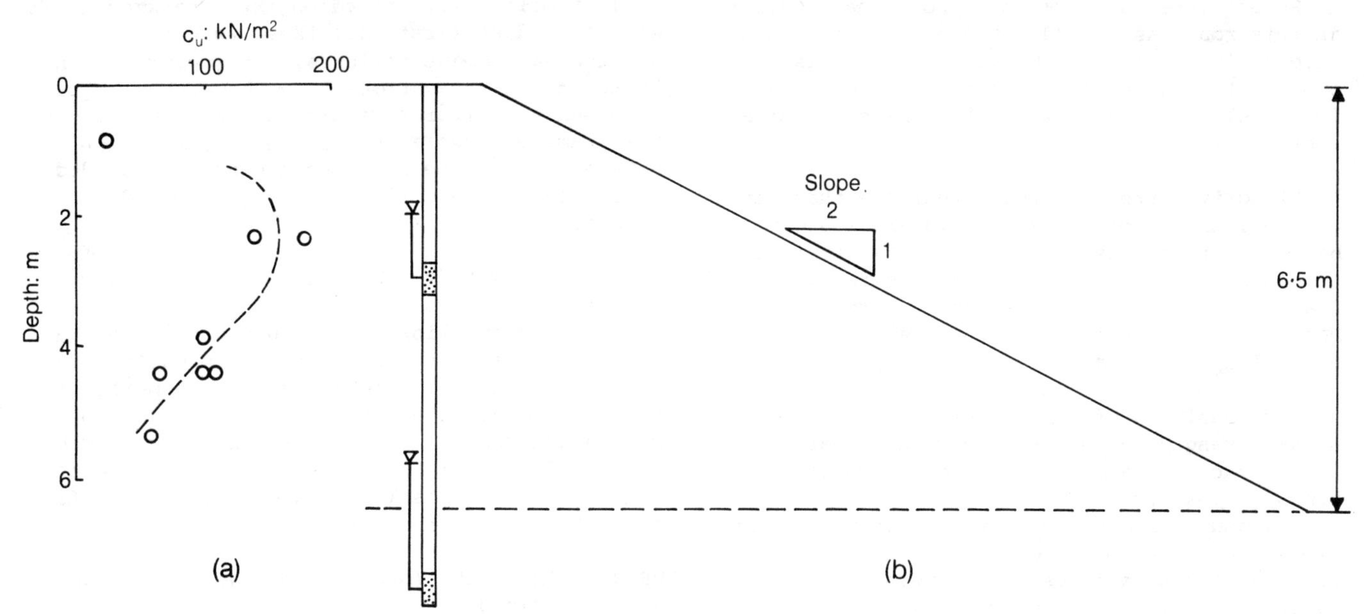

Fig. 23. Shear strength/depth profile and cross-section through an embankment in London Clay

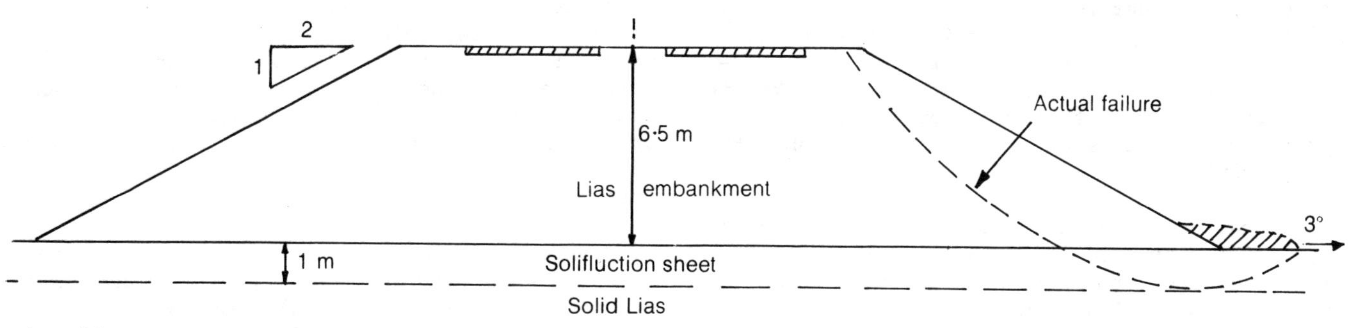

Fig. 24

reduce stability by reducing the effective stresses within the slopes. It also provides a source of water to allow the more rapid dissipation of negative pore pressures and hence brings forward the time when the slope behaves in the manner appropriate to equilibrium effective stress conditions.

Recent remedial work to improve the pavement on this road has included the provision of positive drainage to all embankments and it is anticipated that this will reduce the pore pressures in the slope and at least delay the onset of any instability.

Effective stress analyses indicate that, in general, slopes of between 1:2.5 and 1:3.0 will be necessary to ensure the long-term stability of slopes in the Lias Clay assuming fully drained conditions and peak shear strengths. Hence further instability could well occur in clay slopes which are steeper than this.

Back analyses of slips have indicated average shear strength parameters along the shear surfaces of $c' = 0$ and $\phi' = 18°$, although combinations of positive c' values and reduced ϕ' values could also be consistent with the observations. Where slips have occurred, therefore, reduced effective stress parameters have to be taken into account in designing remedial measures.

Both these case histories suggest that in fills constructed at relatively steep side slopes in clay of high plasticity, delayed failure does occur. Instability may not have occurred at other locations with similar geometry because of local variations in material type, groundwater conditions or geological conditions in the materials beneath the embankments. However, it may be that when equilibrium pore pressure conditions are approached, instability will result.

In paragraph 18 of the paper by Vaughan et al. (p. 209), conventional stand-pipe piezometers are considered appropriate to the measurement of pore pressure below about 2 m below slope surface. As negative pore pressures within slopes cannot be measured by such instruments, what does this mean?

DR P.R. VAUGHAN, *Imperial College of Science and Technology*
Observations by Walbancke,[2] summarized in fig.7 on page 208, indicate that pore pressures become positive as ultimate equilibrium is approached, and stand-pipe piezometers are then effective. Failure is unlikely to occur while negative pore pressures persist. The comment in paragraph 18 on page 209 refers to response time. Pore pressures at shallow depths fluctuate with changing weather conditions; these fluctuations cannot be measured accurately by piezometers with the long response times typical of stand-pipes.

MR J. HILL, *Lincolnshire County Council*
Further to Mr Threadgold's remarks, Fig.24 shows an embankment typical of several constructed up to 7 m high in the building of the 10 km long Grantham Bypass A1 trunk road in 1961-62. The drainage system, during the years since construction, has remained substantially the same. During the wet winter of 1977-78, several areas of instability were noted on the embankment. It was assumed at first that these were typical shallow embankment failures, but subsequent investigation and excavations have shown them to be deep seated failures within the embankment. In the example shown in Fig.24, a failure plane extending into the existing ground and bounded by the base of a solifluction sheet was discovered.

This shows that embankments built of overconsolidated clays may take several years to reach their equilibrium moisture content and at that stage may not themselves be in equilibrium. It is clear that the embankments were built at too steep a slope angle, but this was common with embankments built at that time. The problem now for the maintenance engineer is how to stabilize such embankments and prevent significant failures in the future.

DR R.J. CHANDLER, *Imperial College of Science and Technology*
The phenomenon of water accumulation within embankments is a general one, not necessarily associated particularly with local areas of granular fill. When an embankment is formed of well compacted, saturated clay the pore pressures within the fill will reflect the stress changes from the in situ stresses in the borrow area to those in the same element of soil within the embankment. Often there will be a reduction in stress between the borrow area and the fill, and the pore pressures in the fill may then be low, or even negative, particularly at shallow depths. The effective stresses and the corresponding soil strength will thus be high.

However, in most parts of Britain, precipitation over the year exceeds evaporation and transpiration from vegetation. Thus on surfaces where infiltration is possible, such as the central reservations and verges of motorway banks (and, to a lesser extent, side slopes), there will be net infiltration over the year as a whole and it becomes possible for a rise in the low pore pressures in the bank to occur (although perhaps very slowly), with consequent loss of strength of the fill, particularly near the side slopes where infiltration is most likely. This process will be accelerated where drains are faulty and if it is possible for water to accumulate in the sub-base beneath the road pavement.

Thus there is likely to be long-term swelling of clay embankments in a manner similar to the long-term swelling of cutting slopes, leading in both instances to the possibility of delayed slope failures. Examples of long-term slope failures of cohesive embankments where there was a relatively high water table are given in references 3 and 4.

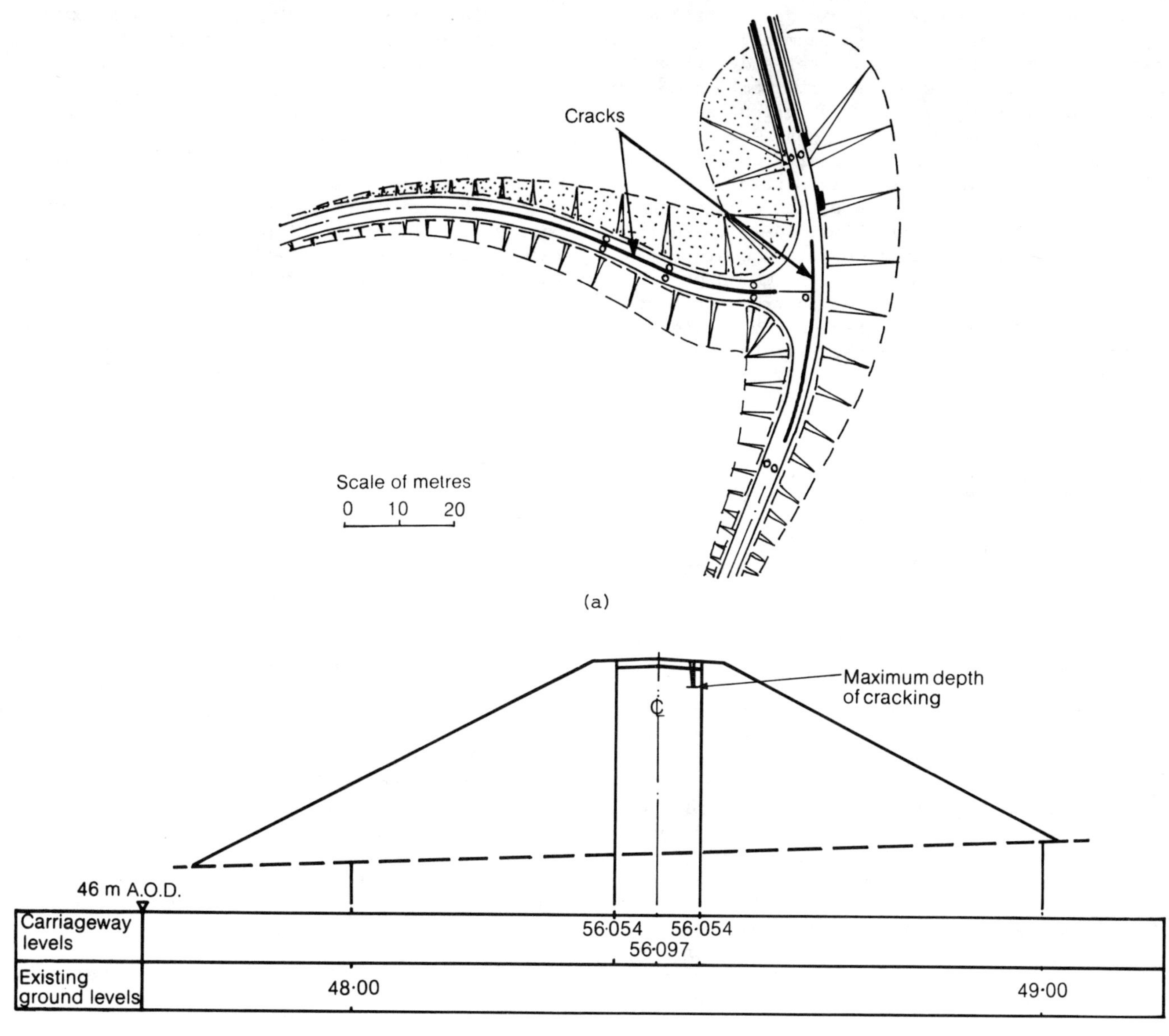

Fig. 25

Fig. 26

Fig. 27

MR R.J.G. EDWARDS, *Eastern Road Construction Unit*
A pavement failure on an accommodation bridge approach embankment has been examined. The location and layout of the embankment and fill are shown in Fig.25. The banks vary from 1 m to 7 m in height and were constructed during late summer 1976 of chalky boulder clay - a stiff overconsolidated slightly fissured glacial silty clay containing discrete chalk particles up to 20 mm in diameter. The properties of the general fill as laid are: plastic limit 14-17%, plasticity index 25-32%, natural moisture content 14-17%, vane shear strength >120 to 600 kN/m^2 (latter figure based on hand penetrometer test), CBR value >20, 7-30% retained on 425 μm sieve (mainly chalk particles); the fill is described as very stiff to hard grey or grey brown silty clay with chalk fragments, some flinty gravel and occasional sandy partings.

The pavement failed by being split longitudinally (Figs 26 and 27). The properties of the fill adjacent to the cracks are: plastic limit 14-16%, plasticity index 25-30%, natural moisture content 18-22%, vane shear strength 16-100 kN/m^2; and the fill there is described as soft to firm grey brown and grey silty clay with chalk fragments and some flint gravel, sand and silt lenses and partings.

The following tentative conclusions are drawn. It is postulated that the cause of failure is that the fill was placed dry of optimum moisture content (BS 4.5 kg rammer) in large scraper operations late in the contract and was placed quickly using material from the bottom of earthwork cuttings. The density achieved using a heavy smooth-wheeled roller was such that the air voids were more than 5%. The fill was thus placed in a largely unremoulded condition with relatively high air voids. This condition has been exacerbated by the availability of water from surface run-off, infiltration and absorption.

The embankment is narrow with respect to its height and thus the subsequent elastic heave and swelling, caused by destressing and water absorption respectively, have caused the bank to deform by a lateral as well as a vertical movement.

It is suggested that the use for embankments of overconsolidated stiff fissured clays at moisture contents less than the optimum (BS 4.5 kg rammer) may well lead to pavement failure problems, particularly for non-motorway embankments. In extreme cases there may also be embankment stability problems.

MR J.B. MILLER, *HM Inspectorate of Mines and Quarries*
My main interest is in tips at mines and quarries where considerable quantities of clay are tipped annually. At present the fill is placed mainly in thin layers using wheeled scrapers and/or dump trucks, and bulldozers are used to spread it. In many cases the fill is used to contain tailings. Compaction is achieved, either by the use of towed rollers or by routeing the dumping plant. Unlike many structural uses of fill, construction of tips goes on throughout the year despite weather conditions, resulting in considerable problems in achieving uniform compaction and permeability and avoiding high pore water pressures.

There are three broad types of tip failure. The first type is the flow slide or liquefaction failure that occurs when the failed material moves at speeds up to 20 km/h and is associated with high pore water pressure. The second type occurs when the major slip moves off the slope face up to a distance equal to the slope height. This appears to be due to failure of the foundation, or sometimes failure within the lower level of the fill, and to be associated with high pore pressures. The third type is the shallow-seated surface slide from slopes similar to the cases shown at this Conference. Observation suggests that most of these slips involve poorly compacted material which has spilt over the edge as banks are raised in thin layers. This poorly compacted material tends to slip down side slopes especially as a result of weathering, ultimately forming a slickensided surface which impedes horizontal drainage from the layered construction. Eventually the centre line side of the slicked surface is broken and water enters the joint. Later, failure occurs in the form of a visible surface slip, conditions in the slip plane being wet and showing sliding striations. The area dries out fairly quickly in suitable weather. Piezometers placed in the fill immediately after a slip indicate random perched water tables in the layered construction.

I suggest that many shallow-seated surface slips from embankments are due to the sealing effect of poorly compacted soil at the edge, combined with agricultural operations aimed at grassing slopes.

MR I.P. LIESZKOWSZKY, *Dominion Soil Investigation Inc., Toronto*
Total stress analysis (TSA) can be used to assess the short-term stability of fills and is valid for the limited condition of the load applied rapidly with a shear occurring under undrained conditions.

Effective stress analysis (ESA) can be applied to a wide range of stresses, rates of loading and strain, provided the effective shear strength parameters - ϕ' and c' - have been properly assessed from appropriate tests and pore pressures are known or can be predicted. However, there are inherent difficulties in predicting all the parameters necessary for ESA.

Although shear strength parameters ϕ' and c' can be determined in advance reasonably well, it is difficult to predict the construction pore pressures, particularly for varved clay deposits. Experience with Fort Creek Dam (pp 157-164) substantiates this. Piezometers P_1 and P_5 (see fig.6 on page 162) installed in the foundation material near the toe of the dam and under the stabilizing berms showed much higher pore pressures than predicted. (Pore pressures approach those recorded in P_7 and P_3 under the centre line of the dam.) The rate of pore pressure dissipation could also not be predicted: pore pressures in P_1 and P_5 were still increasing when P_3 and P_7 already recorded a drop. The piezometers installed in the compacted clay core (E_1, E_6 and E_8) also responded in an unpredictable manner (fig.7 on page 162).

This difficulty in predicting the pore pressures imposes a serious limitation on the use of ESA for the analysis of the short-term stability of dams and embankments. However, TSA requires only one parameter - the undrained shear strength c_u - which is much easier to predict. At Fort Creek Dam, there was good agreement between the results from tests on laboratory compacted specimens and on samples obtained from the actual fill (fig.4 on page 160). The relationship between moisture content and unconfined compression strength was, within narrow limits, the same for both materials. The difference in the structure of the fill and the laboratory compacted material appeared to have played only a minor role, if any. However, this may not be true for clays of different geological origin.

The TSA yielded a factor of safety of 1.2 in the end of construction case. The ESA carried out during construction, using the measured pore pressures, yielded safety factors as low as 1.03 at a stage when the dam had not yet reached its final height and the placing of the fill was temporarily halted. Unfortunately, there are insufficient data available for it to be possible to determine the true factor of safety, but I believe that in this case the TSA had overestimated the factor of safety.

The advantages of TSA are that it is simple, quick, the parameters are predictable and desired end results can be easily obtained by specifying the shear strength or a range of compaction moisture contents that will assure the necessary values of c_u. The disadvantages are that it is inaccurate and applicable only to limited cases of rate of loading and deformation.

ESA is probably more accurate, is valid for a wide range of stress, strain, rate of loading and deformation, and makes it possible to define shear strength parameters more accurately. However, it cannot predict pore pressures with sufficient accuracy and therefore cannot be used for a short-term case unless the pore pressures are actually measured.

MR I.L. WHYTE, *University of Manchester Institute of Technology*
Slides in steep embankment slopes are to be expected if seepage and/or positive pore water pressures occur near the surface. For example, a slope of 1:2 in Lias or London Clay fill embankments corresponds to a slope angle of about 27°. The effective angle of friction for first time slides in these soils could be about $\phi' =$ 20-23°, and thus a slope angle of 27° can be stable only if pore water suction exists. In the extreme case of seepage from the bank face, the maximum stable slope angle is reduced to about $\phi'/2$, i.e. 10-12°, and thus, as fully drained conditions develop with time, steep slopes will fail, particularly if seepage conditions are allowed to develop from leaking drains and so on. Slopes with angles similar to, or exceeding, the friction angle should be designed either on the understanding that local failure can occur in high pore pressure/seepage regions, with maintenance costs in the future, or positive drainage must be included to protect the slope face from high pore pressures and seepage.

MR K.W. COLE, *Ove Arup and Partners*
I should like to describe a case history in Africa.

The road was constructed along the upper levels of a long hill ridge bounded by two deeply incised rivers. The road alignment was such that for much of its length it was below the crest of the ridge and the embankments were less than 5 m on the uphill side; on the downhill side the toe to crest height reached 20 m.

On the uphill side the ground was shaped to provide a 5 m minimum width catchment area for run-off water, and outlet pipes passing through the embankment were provided at intervals. At places where the road alignment changed from one side of the ridge to the other, generally at a slight dip in the ridge forming a shallow saddle, the road was raised on a small embankment with adjacent wide catchment areas. At two of these positions small slips had occurred on the downhill face of embankments less than 3 m high (see Fig.28).

At other places along the road large areas of the downhill face of the road had slipped, the slips generally being of shallow nature, extending only to depths of 1-2 m below the original slope surface (see Fig.29).

The soil is highly weathered residual lateritic clay derived under conditions of intense tropical weathering from the underlying basalt. It was placed in 200 mm thick layers and compacted by smooth-wheeled rollers. After each layer within a fill area had been placed and compacted, it was checked for acceptable density by at least two sand replacement tests. As these tests took two or more days to complete, during which time no further layers were allowed to be placed, the surface of each layer was subject to desiccation under the generally fine weather conditions.

As may be seen from Fig.30, it was possible to to expose the individual layers in an excavation into the slope. At the same location water was seen to be seeping from the slipped zone, the level of seepage being the same as the level of the catchment area on the opposite (uphill) side of the embankment, where some ponding was evident after heavy overnight rain. It was therefore concluded that run-off water, temporarily ponded in catchment areas and seeping through the more permeable zones at the upper surfaces of layers, was the cause of the embankment slope instability. The placing and compacting of the clay at about optimum moisture content forming overlying layers, was evidently not able to

Fig. 28

Fig. 29

Fig. 30

cause remoulding and thus recompact the desiccated surfaces.

The primary cause of the problem could be held to be over rigid adherence to a performance specification (that used was similar to the 4th edition of the road and bridgeworks specification[5]) and it is likely that, had the specification been a method specification, the time interval between placing succeeding layers would have not been sufficient for the degree of desiccation encountered to have occurred.

With the soil type used, however, more rapid construction of the fill may have given rise to other problems, such as the generation of pore pressures sufficiently large to cause local instability.

With reference to the adoption of effective stress methods of calculating embankment stability, I would like to say that this method is acceptable for forward prediction provided the stress paths for the soils in their compacted state are well understood. With long experience and detailed knowledge it is possible to select an average stress path which will satisfactorily predict the factor of safety at the next stage of loading. It is probably easier and more reliable to estimate the average increase in undrained shear strength around a potential slip surface from laboratory tests which determine the change in this strength with change in effective consolidating stress, and make the calculation in undrained (total stress) terms. As field measurements of changes in piezometric level can be compared with settlement measurements, they are more likely to be reliable when used to calculate changes in effective stress than when the same measurements are used to determine the absolute value of effective stress.

MR D.J. AYRES, *British Railways Board*
The suggestion that clay embankments should have a limiting safe slope of $\phi'/2$ depends on the assumption that the water table would become coincident with the surface. In observing clay railway embankments over 100 years old it seems that the length of clay embankment which has not obviously slipped is greater than that which has. My department has installed several thousand simple polyethylene tubes to detect and monitor slip movement since developing the method in 1955, and as the tubes were in effect unsealed stand-pipes the water levels were also checked. The patterns of water table noticed did not show a water level coincident with the surface and this was probably due to the protection afforded by vegetation and to the permeability of the fill as a whole.

The strain which occurs in the embankment slip is noticed by its effect on rail level and alignment to fine accuracy. The time lag between precipitation and movement and between the onset of dry weather and reduction of movement indicates an effective stress response.

British Rail's standard method of dealing with such slips is by a combination of grouting techniques (clacquage, compaction and penetration) as this gives the most efficient and economic results. The expulsion of water in a substantial quantity from the surface of the slope has been observed both from the clay areas and from the ballast pockets of ash and stone at the top of the slope placed to support the track to level. This may indicate inadequate original construction techniques with air spaces between large clay masses. However, inspection of trenches dug into the slope has not revealed these voids.

Slip planes could extend or be supplemented by deeper and wider planes in an existing slip so that early grouting was desirable. One such slip, in a fill composed probably of Oxford or Kelleways Beds clays, where the tension crack was at the shoulder, had altered over four months to include the centre 2 m way of the two tracks. Slip surfaces appeared to approximate more to a circular shape in embankments than in cuttings and in a significant number of cases went below the natural ground surface.

DR P. SMART, *Glasgow University*
Figure 31 shows a Cambridge q:p plot with failure assumed along q = Mp and the stress path of the most critical point in the structure. In practice the stress paths for several points should be considered.

There are two approaches

(a) Find the point on the stress path which is closest to the failure line, drop the perpendicular from this point on to the failure line and let a and b be as shown in Fig.31. Then

$$\text{factor of safety} = a/b$$

(b) Draw the tangent from the origin to the stress path, let the slope of this tangent be θ and the slope of the failure line θ_f (see Fig.32). Then

$$\text{factor of safety} = \theta/\theta_f.$$

MR N.A. SKERMER, *Golder Associates, Vancouver*
A few years ago my firm was asked to investigate the collapse of a number of large corrugated metal plate culverts. These structures had been intended for highway stream crossings in Northern Canada within the Arctic circle. In cross-section the culverts were horizontal ellipses with a span of about 11 m and a height of about 8 m. Such structures have a fairly flat roof arch, and they are unable to tolerate much movement at the haunches without roof failures occurring. The fill around the sides of such structures needs to be very well compacted. In this instance a rippable mudstone had been compacted by grid rollers in very cold winter conditions. The fill was said to have been hard and compact at the time of placement. It might well have been hard, because it was probably frozen solid. The following spring when the ground thawed a number of the culverts collapsed. The thickness of fill above the culverts was about 10 m on average.

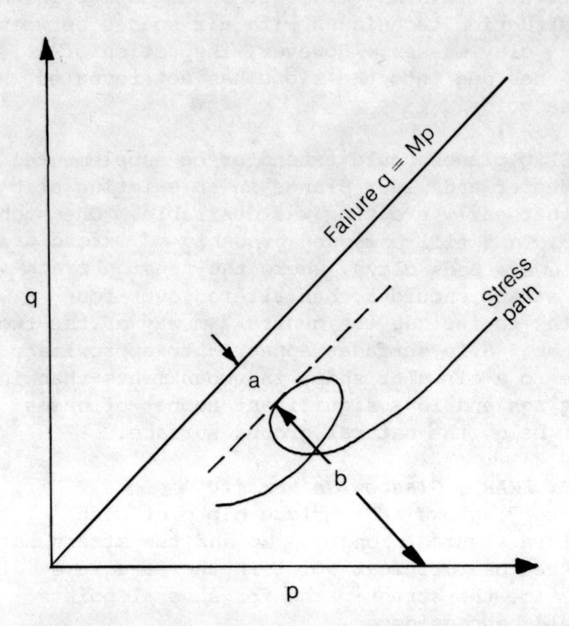

Fig. 31

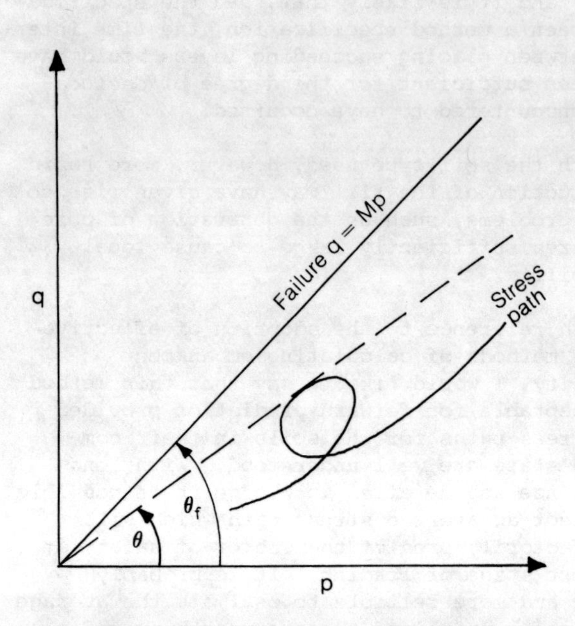

Fig. 32

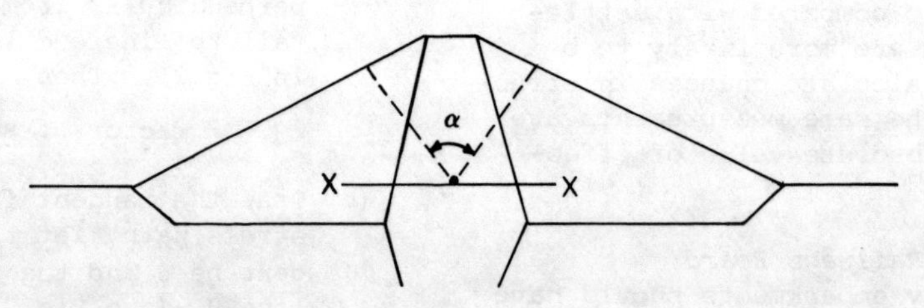

Fig. 33

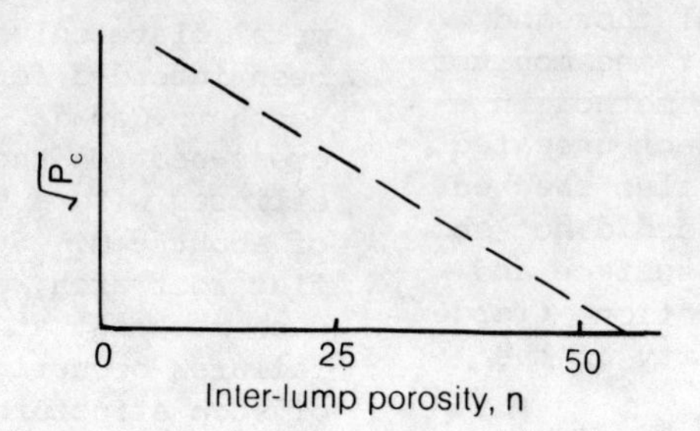

Fig. 34

I carried out field density measurements in the less disturbed fill, and these indicated that in fact the fill was very loose. A few laboratory shear strength tests were made on the fill in a loose condition. Simple earth pressure calculations showed that the fill alongside the culverts must fail by bulging under the superimposed roof loads.

Subsequently the culverts were rebuilt, but the fill was placed in the late summer under cool, but not freezing conditions. The same fill was compacted with vibratory rollers and high densities were achieved. The rebuilt culverts have performed satisfactorily since then.

As a general rule and for specification purposes, fill placement is allowed under freezing conditions down to -7°C, and in some cases in small areas being continuously and intensely worked this limit is extended to -12°C. Field density measurements and close visual inspection can show whether or not fill is freezing, and aid judgement on whether or not compaction is acceptable.

DR F.W. SHERRELL, *Frederick Sherrell*
When using mudstone fill, I suggest that consideration should be given to the chemistry of the material, especially if the mudstone contains iron sulphide (pyrite). Oxidation of pyrite, in the presence of moisture, can lead to the formation of ferrous sulphate. If there is a source of free lime in the fill, or perhaps in the road sub-base material, the lime may react with the ferrous sulphate to form gypsum. The conversion from pyrite to gypsum involves a volume increase eight times that of the original pyrite and can result in appreciable expansive forces; it has been suggested that the forces required to resist expansion are as high as 500 kN/m^2. It is prudent during the site investigation stage to include chemical analyses, to determine the total sulphur content of clays or mudstones in which the presence of pyrite might be suspected; a content of about 1% may be sufficient to cause a swelling problem.

MR E.H. STEGER, *Pell Frischman and Partners*
It must be realized that the specifications for, and indeed the properties of clay cores and road embankments can rarely be equated. In the case of a clay core for an earth dam one generally deals with the nearest approximation to a uniform homogeneous material an engineer is ever likely to achieve. This is because the clay is usually carefully selected and put through a puddle mill to become thoroughly remoulded. Thus, to specify placement moisture contents and density to within ± 2% is realistic, and feasible. In the case of long embankments for motorways, conditions are quite different in that, generally, for economy and expedience the fill material is obtained from borrow pits along or near the proposed route. This means that one is likely to have to deal with greatly varying types of material, including lumps of clay of almost any size, remoulded clay, shales, silts and sands. To place this to precise specifications is unrealistic and substantial settlements are likely to occur.

DR T.K. CHAPLIN, *University of Birmingham*
I wish to refer first to the vertical stress σ_v in the lower parts of dam cores. Fig.33 shows that only a small proportion of the weight of the fill placed near the top of the dam contributes to σ_v at level XX, because of the small included angle α. This effect is very noticeable with granular cores of dams.[6]

During rolling operations at Trimpley Reservoir, adjacent to the River Severn, a layer of fill, approximately 2.5 m long, moved forward in front of the roller. This might have resulted from very low values of E near the surface or to sliding on a previous lift. Presumably this would aggravate the orientation of particles produced by the roller surface.

With regard to the relationship between the crushing strength of lumps in an aggregate and the inter-lump porosity under pressure, for brittle lumps the tensile strength σ_t approximates to

$$\sigma_t = \frac{2}{3}\frac{P_c}{d^2}$$

where d is the sieve size and P_c is the load required to crush the lump. I have found that the square root of P_c and the inter-lump porosity n is as shown in Fig.34.

REFERENCES

1. PENMAN, A.D.M. et al. Observed and predicted deformations in a large embankment dam during construction. Proc. Instn Civ. Engrs, 1971, vol. 49, May, 1-21.
2. WALBANCKE, H.J. Pore pressures in clay embankments and cuttings. PhD thesis, University of London, 1976.
3. CHANDLER, R.J. and PAHAKIS, M. Long-term failure of a bank on a solifluction sheet. Proc. 8th Int. Conf. Soil Mech., Moscow, 1973, vol.2.2, 45-51.
4. CHANDLER, R.J. et al. Four long-term failures of embankments founded on areas of landslip. Q.J. Eng. Geol., 1973, vol.6 (3/4), 405-422.
5. MINISTRY OF TRANSPORT. Specification for road and bridgeworks, 4th edn. HMSO, London, 1969.
6. CHAPLIN, T.K. Discussion on Selset reservoir, by J. Kennard et al. Proc. Instn Civ. Engrs, 1962, vol.23, Dec., 737-740.

Technical editor: E.W.H. CURRER, FInstHE,
Transport and Road Research Laboratory

Discussion: Road subgrades

DR R.H. JONES, *University of Nottingham*
The suction index approach provides a useful framework for discussing pavement design and has also been used as a basis for establishing criteria for muck suitability. Although my main interest is in the ancillary uses, I should like to discuss the method in the context of pavement design.

Figure 1 on page 38 of the paper by Black and Lister appears to be an extension of fig.9 presented by Black in reference 1, which was basically for saturated clays. In reference 1 a correction was proposed for unsaturated materials, i.e. those having a lower ρ_d. This correction could lower the predicted CBR by 20% at v_a = 5%. To what degree of saturation does fig.1 on page 38 apply? Are corrections envisaged for other values?

I should like clarification on the accuracy of the method and the desirability of calibrating. The suction index method is based on a number of approximations and simplifications, some of which are open to question.[2] Fig.2 on page 38 which is on a logarithmic scale, indicates that a measured strength might well be only half the estimated strength. Figs 8 and 9 on page 45 suggest that for this case the deflexions would increase by 50% and the critical life of the road would be reduced to less than one third of that expected. In these circumstances is not some check or calibration testing essential rather than merely desirable?

The estimation of the moisture content appropriate for check testing should be as close as possible to the predicted equilibrium for the worst case. Its prediction will involve the use of a suitable value of α in the equation $U = S + \alpha P$. It should be recognized that α is not a constant for a given soil but varies between 0 and 1 for every soil. Values of α can be computed from the results of suction and oedometer tests for the limiting case U = 0, and from the slope of the shrinkage limit curve for the limiting case of P = 0. For intermediate conditions approximate values of α can be obtained by interpolation on a plot of ρ_d versus w (see Fig.1 or reference 3). A typical variation of α with water content is shown in Fig.2.

A number of direct measurements within the range 2.2 < pF < 2.7 (i.e. suctions of 1.75-4 m of water) resulted in the relationship shown in Fig.3[4,5] that is applicable to the working range beneath most British pavements. The relationship to the α values appropriate to the extreme values of pF, that is for saturated or dry soils, is shown clearly. The variability of α was explicit in some of the earliest work by the Transport and Road Research Laboratory in this field and is implicit in the paper by Black and Lister (pp 37-48). However, this point is not always appreciated and considerable confusion can result.

MR W. BLACK, *Transport and Road Research Laboratory*
The CBR, consistency index, plasticity index correlation in the paper by Lister and myself (pp 37-48) is substantially the same as that in reference 1 with some minor alterations to CBR values of less than 2% as the result of reanalysis of the original data. The relationships were treated as applicable to fully saturated soils even though it was realized that this would not be correct at the higher suctions. It is assumed that the air voids correction, being experimentally determined, corrects for this assumption. This implies that the degree of saturation must be known if a correct assessment of strength is to be made at high suctions and that strengths estimated on the basis of full saturation may be in error to some extent. Hence calibration is essential if accurate prediction is the aim.

With regard to the accuracy of fig.2 on page 38, it must be remembered that none of the data used was produced primarily to compare strengths derived from plasticity data with measured strengths. In most cases the plasticity data were merely typical values for the soils undergoing strength testing rather than good average values carefully related to the actual samples used in the strength tests. This being so it is felt that the scatter in fig.2 on page 38 should not be used to question the accuracy of the life of roads quoted later in the paper. This is particularly so because it is stated there that the life predictions were only indications based on the limited data at present available.

It is recognized that α for any soil can vary from zero to one; α = 1 for a heavy clay and α = 0 for sand at the usual equilibrium suctions beneath roads. Although it is not explicit in our paper, it is implicit that a range of α values has been used which is related to the soil type. The way in which α is used in the paper is as follows. At low suctions when the soil has settled under low overburden $\alpha = 1$ is assumed for all soils. This is the same as saying that the soils are fully saturated and the principle of effective stress applies precisely. This is not completely true, but since the analysis involved the product αp it is the combined effect of these two variables which is important. It is assumed that p is constant with change of water content, whereas in practice p increases as the soil wets up due to the increase in the bulk density of the soil. It is also true that the lateral pressure increases as the soil wets up because the lateral expansion is constrained when the soil tries to swell. If p is the correct effective soil pressure at high suction, then at low suction p ought to be increased. The result is that the value of α taken in the calculation may have been too high at low suction and the value of p too low, but it is unlikely that the product would be greatly in error. Preliminary analysis of the results of a full-scale experiment at the Transport and Road Research Laboratory designed to measure the effect of water table level on soil strength suggests that the method used in our paper correctly predicts the strength of a silty clay when the water table rises to formation level.

At equilibrium suction beneath the pavement the effect of α was allowed for as in reference 1. Investigation at that time showed that the equilibrium water table beneath covered areas over a heavy clay subgrade, was within 50 cm of formation level and α was equal to unity. At the other extreme, for low plasticity soils, the equilibrium water table depth was probably nearer 150 cm and α was one fifth of the heavy clay value. The total road thickness differed in the two cases by about 0.5 m. As a result it was concluded reasonable to assume that, taking account of the different α values and different water table depths, a constant equilibrium suction of 180 cm $(u + \alpha p)$ of water could be applied to all pavements and all soils, thus obviating the need for estimates of α for each case. The further justification for this approach is that in practice the approximations give estimates close to the measured values.

It would be a mistake to overclaim for the suction index method. It is appreciated that even when calibrated it may provide only broad indications of likely strength changes for the very large changes in moisture content considered in fig.2 on page 38. Over the narrower range of soil consistencies encountered in road subgrades there is ample evidence that the method provides adequate estimates of strength. No other method is available to the engineer from which he can readily infer likely changes in soil strength. It is clearly advantageous for design purposes to be able to give an indication of the relationship between strength measured at the time of a survey and the ultimate strength likely to be achieved under a pavement.

MR C.E.J. WOOD, *Department of Transport*
Rather than referring to 'non-positive' and 'positive' drainage systems in connection with clay fills, it might be better to speak of 'combined' and 'separate' systems when referring to surface water and subsoil water. Dual carriageway roads in the UK invariably have a filter drain in the central reserve when built on embankments and this often receives water non-positively where one carriageway, due to superelevation, is higher than the other. A positive system can be installed with a shallow surface channel at the central edge of the carriageway with outlets connected to a central drain. This may be a filter drain or a separate drain beneath the filter drain. The drain invert is at least 600 mm below the top of the embankment and the sides of the trench are normally unsealed even where a concrete bed has been used to seal the floor of the trench.

During heavier storms, which have a greater intensity in the drier parts of the country, a head of water may be present in the trench and water may enter the fill. Nevertheless, this system works well with few failures. On chalk, however, overnight settlement has occurred on embankments in the South East as the result of inundation.

The use of channels at pavement edges can enable water to enter the construction through open longitudinal joints.

MR E.J. ARROWSMITH, *North Western Road Construction Unit*
I would not favour more authority being given to the Site Engineer to vary the thickness of the pavement layers during construction. I consider that a more realistic assessment of the CBR value and closer site control is a better procedure to follow in order to avoid foundation failures.

In the North West the practice is to design on a CBR value of 2-3% for clay subgrades whatever the CBR values measured at the soil survey stage. A subgrade layer 400 mm thick of coarse granular material with not more than 10% passing the 75 μm sieve is used as standard, thickened up to 600 mm at cut/fill lines or similar areas of poor drainage. A sub-base layer 150 mm thick is placed on this subgrade layer and, after trimming, is tested by plate bearing tests or deflexion beam. A modulus of subgrade reaction of 68 MN/m^2 per metre for a 760 mm dia. plate is considered acceptable in the plate bearing tests. If this value is not reached the cause is investigated by digging trial holes. Typical causes of failure are found to be saturated sub-base, sub-base mixed with clay, inadequate compaction and insufficient sub-base thickness. On rare occasions it may be necessary to replace the subgrade layer and deepen the construction or to stabilize with rock fill. This test is not treated as a contractual acceptance test. It should be carried out under the supervision of an experienced engineer.

Special measures are sometimes necessary to protect subgrades which are particularly susceptible to water. In the North West a clause has been used in the Specification to warn the

Contractor of this possibility in his choice of sub-base material and programming of the Works. Unfortunately the warning clause has not been too successful and it is now considered advisable to require the use of a bound sub-base to keep out water in these situations. Attempts to seal the sub-base with tar spraying have been only partially successful as the process prevents drying out once water has got in.

No foundation failures have occurred in the North West on motorways since these procedures have been adopted.

MR N.W. LISTER, *Transport and Road Research Laboratory*
I agree with Mr Arrowsmith that during construction as much site testing as possible should be carried out, providing it can be correctly interpreted. On some contracts in Scotland three sub-base thickness options were permitted. These could be varied by the Engineer depending on subgrade conditions during construction and this approach worked well.

On the question of sub-base testing by either plate bearing or deflexion beam, I think either test would be an excellent idea in the hands of an engineer who has an adequate amount of the correct experience on which to make a judgement. However, Mr Arrowsmith does not use these tests as acceptance tests and I would not like to have to write a specification round either test for this purpose. At the Transport and Road Research Laboratory we have monitored, by deflexion beam, the stiffness changes in roads from the time the sub-base is placed to completion of the road and thereafter. Some roads can stiffen up simply because moisture conditions change for the better; in other cases, for example in deep cut, wetting up occurs as might be expected and deflexions increase as the road gets weaker. It would be impossible to guarantee by plate bearing or any other strength test on the sub-base that a pavement will perform satisfactorily under traffic. However, I agree that with this type of testing in the hands of an engineer with experience in the interpretation of the results a great deal can be done during the construction phase to improve the subsequent performance of the road.

Tests on the sub-base will reveal the classic weak spot where some action will be required but I would not advocate the use of strength tests during construction to determine whether or not the pavement structure as a whole is under-designed.

MR H. GRACE, *Scott Wilson Kirkpatrick and Partners*
During construction of the South Cheshire section of the M6 motorway progress with the earthworks was maintained during both wet and dry weather by judicious use of the sands and lean clays present in the area using whichever material could be worked in the prevailing weather conditions and with the fill thickness appreciably larger than the cuts.

The earthworks were completed with a 4 ft layer of sand in both cut and fill and the whole formation was proof rolled with a loaded scraper. Any soft spots were dug out and recompacted and this ensured a subgrade layer of uniform bearing capacity.

On top of the sand layer was placed a 6 in. layer of sand cement having two main functions: to protect the earthworks so that they would be unaffected by weather and to provide a platform on which plant could run during construction of the pavement.

Some doubts were expressed concerning the ability of the sand cement layer to carry the construction traffic. I considered that, based on my experience of the relatively thin pavements founded on sand and used by heavy bombers during the Second World War, a 6 in. sand cement layer would be adequate to carry construction traffic. In the event the sand cement layer performed well, requiring only minor maintenance in one area, and the completed motorway has performed well with only minimum maintenance since being opened to traffic in 1962.

MR D.J. AYRES, *British Railways Board*
In developing a pavement design system for railway track I have noticed a response of rail movement to precipitation where the track passes over weathered overconsolidated clays on natural ground and on fill composed of overconsolidated clays (Fig.4).

The movement is the same as for embankment slips mentioned on page 239 but in this case the ground alongside heaves instead of settling. Water-table observations showed a direct relation between a high water table, loss of track level and heave of adjacent ground. It is possible on railway embankments to trench or pipe through the heave of impermeable clay to release the trapped water and achieve track stability. On level ground or in shallow cuttings where the subgrade is weathered overconsolidated clay, methods to divert water from the loaded areas such as by grouting, plastic film or bituminous impregnation have been used to stabilize the system.[6] It is obviously an effective stress failure and the strains which occurred may have orientated the clay particles to a residual state. In such a circumstance the design approach of Black and Lister (pp 37-48) based on soil suction is highly commendable and has had my enthusiastic support for many years. However, at failure, the soil approximated to the 'continuously disturbed' condition as represented by the unique line on the pF - moisture content curves of Croney and Coleman.[7]

Why did Mr Black and Mr Lister not choose this unique line as the basis for pavement design?

MR W. BLACK, *Transport and Road Research Laboratory*
If the shear stresses in the soil generated by traffic were large enough to change the soil suction so that it would want to draw in water to get back into equilibrium with the water table after many repetitions of load, the unique line

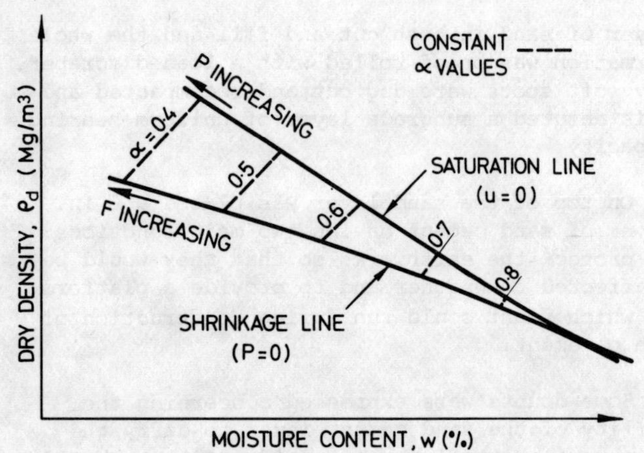

Fig. 1. α relationships (after Coleman and Croney[7])

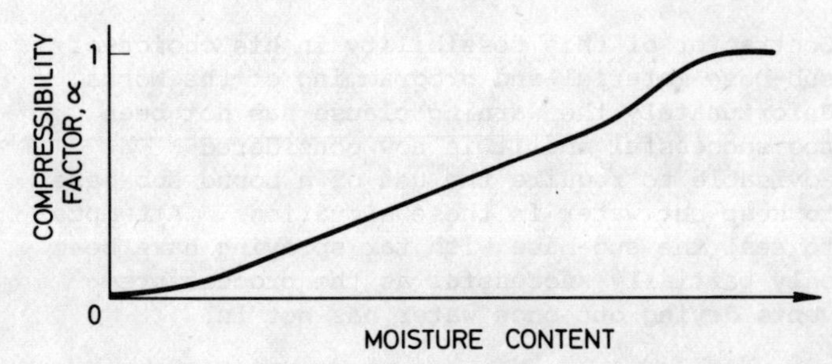

Fig. 2. Variation of α with moisture content

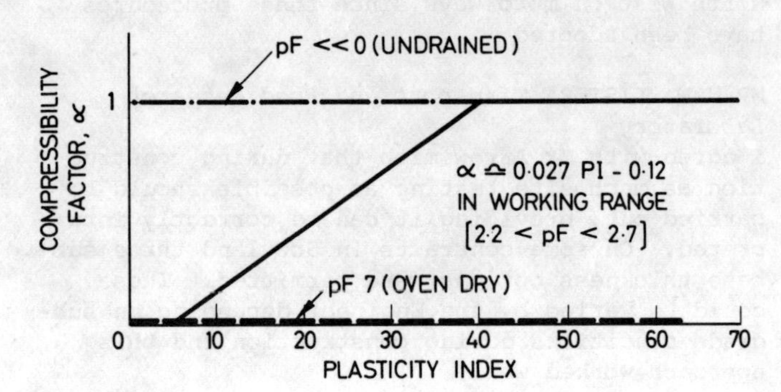

Fig. 3 (right). Relationship of α to pF and plasticity index

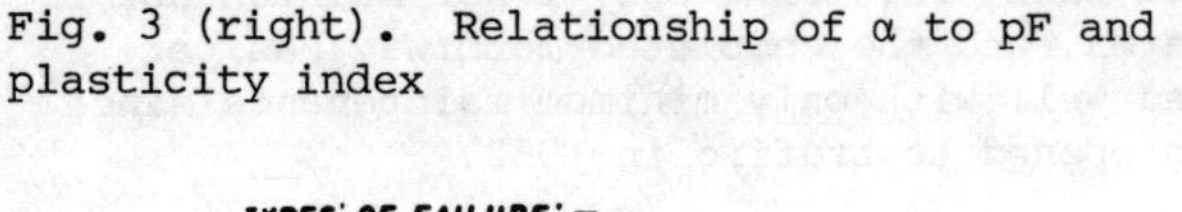

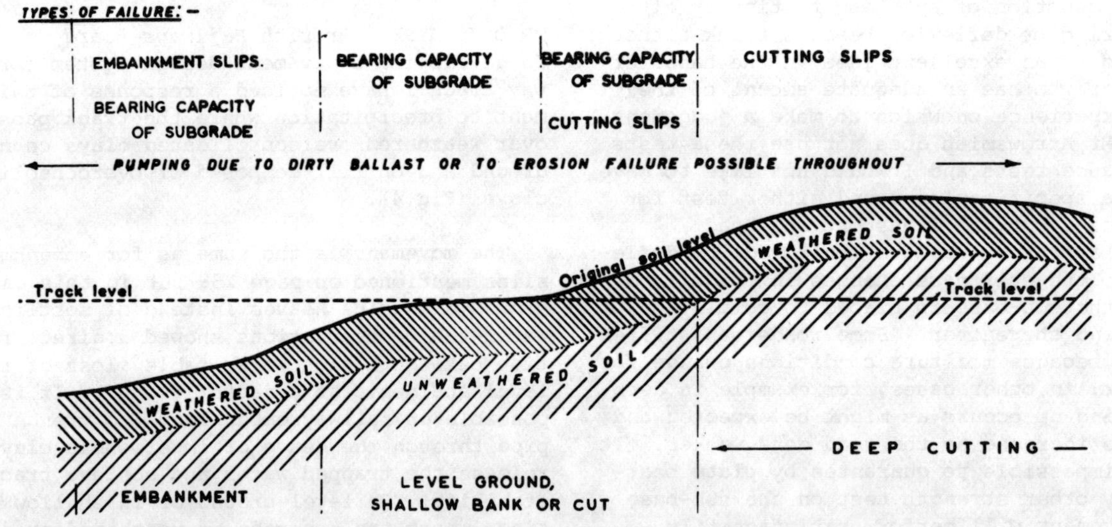

Fig. 4. Track movements associated with soil conditions in clay areas

would be reached; we would then have to design for this condition and would need to know the number of loadings necessary to reach the ultimate stable value. It is possible that in railways the stresses are high enough to induce this mechanism and Mr Ayres should design for this situation.

In the case of road foundations, so far investigations do not indicate a measurable increase in moisture content with time in clay soils that can be attributed reliably to this mechanism. The hysteresis referred to in the paper (pp 37-48) is not associated with this mechanism but refers to the difference in moisture content between the wetting and drying of overconsolidated clays without shear distortion; such distortion does not cause a constant drift towards the unique line.

REFERENCES

1. BLACK, W.P.M. A method of estimating the California bearing ratio of cohesive soils from plasticity data. Géotechnique, 1962, vol.12, Dec., 271-282.
2. FORDE, H.C. Wet fill for highway embankments. PhD thesis, University of Birmingham, 1975.
3. COLEMAN, J.D. and CRONEY, D. The estimation of the vertical moisture distribution with depth in unsaturated cohesive soils. 1952, RN/1709/JDC.DC. (Unpublished)
4. RUSSAM, K. Sub soil drainage and the structural design of roads. Road Research Laboratory, Crowthorne, 1967, Report LR 110.
5. CRONEY, D. The design and performance of road pavements. HMSO, London, 1977.
6. AYRES, D.J. The treatment of unstable slopes and railway track formations. Trans. Soc. Engrs, 1961, vol.52, no.4, 131-135.
7. CRONEY, D. and COLEMAN, J.D. Pore pressure and suction in soil. *In* Pore pressure and suction in soils. Butterworths, London, 1961, 31-37.

Technical editors: M.J. DUMBLETON, BSc, PhD, FGS, Transport and Road Research Laboratory, and D. BURFORD, DIC, Building Research Establishment

Discussion: Construction, placement and methods of treatment of clay fills

MR D.W. QUINION, *Tarmac Construction Unit*
The papers indicate the great deal of information which is becoming available on the appropriate properties of clay soils used in earthmoving. The purpose of this information must be to influence economically the work on site. This is where the work is done and the aim must be to produce the most economical solution, bearing in mind construction and maintenance costs. The best value for money is required, and the construction process is all-important.

It has become increasingly evident that it is possible to regularize, by analysis, the better construction methods and temporary works procedures to improve the general level of performance. It is not always possible to see in advance how this will be achieved; nevertheless it is happening.

The earth-mover uses expensive plant resources and intends that the use of these resources shall be profitable. Delays and hesitancy cost money. Both Contractor and Engineer must be prepared to predict the possibilities posed by changes in soil and weather conditions so that production is maintained to the benefit of all the interested parties.

DR R.H. JONES, *University of Nottingham*
My remarks concern the difficulty, before construction, of estimating correctly the quantities of fill which will ultimately be judged suitable or unsuitable. I have had several years' first-hand experience as a member of a major contractor's team involved in estimating, contract planning and the preparation of claims, followed by the opportunity to undertake a limited amount of research.

For roadworks, the estimate of quantities of suitable and unsuitable fill involves the choice of a limit (based on factor x PL, minimum shear strength, moisture condition value (MCV) and so on), estimating the quantities at the billing and contract planning stage on the basis of the site investigation report, and testing the material immediately before placement and deciding its classification.

Parsons (pp 169-175) has made progress with the MCV approach; in particular he is investigating the sensitivity of a soil to moisture content changes so that appropriate limits can be set. However, even when the limit is chosen correctly and is based on a method which is feasible for site operation, the actual utilization of the muck may still differ considerably from expectations due to changes which occur between the site investigation and placement. These usually result in claims which may cause the job to cost more than anticipated.

Mechanisms by which the material properties, particularly the moisture content, may change during this period have been detailed by Jones[1] and Al-Shaikh-Ali (pp 15-23). Intermixing seems to be the largest single cause of poor quantification. It would be interesting to know whether or not this view is held generally. Certainly, two recent case studies from the Midlands pointed to this source, supporting the impression which has been built up over the years. If intermixing is so important, then soils should be mixed in the laboratory and subjected to testing.

Other potential hazards, which apply even when intermixing can be avoided, are high surface run-off, which may be concentrated into limited areas, and the interception of saturated throughflow.

All the mechanisms involve interaction between material properties, weather conditions and environmental (particularly geomorphological) factors. Material behaviour can be judged from laboratory tests and the probability of given weather can be estimated from previous records. However, relatively little attention has been given to environmental factors, despite the obvious benefits of early recognition of problem areas so that appropriate ground investigation can be devised.

At the University of Nottingham we are evaluating the possible use of geomorphological mapping to identify problem areas. Such mapping has been used for a variety of geotechnical purposes overseas, but under UK conditions its use to date has been mainly in instability mapping. However, recently it has been used to assist in the location of boreholes and trial pits. At this stage, geomorphological units are being correlated successfully with weathering profile and geotechnical properties.[2] Ultimately, it is hoped to interpret this work in terms of muck suitability.

Clay fills. Institution of Civil Engineers, London, 1979, 247-282

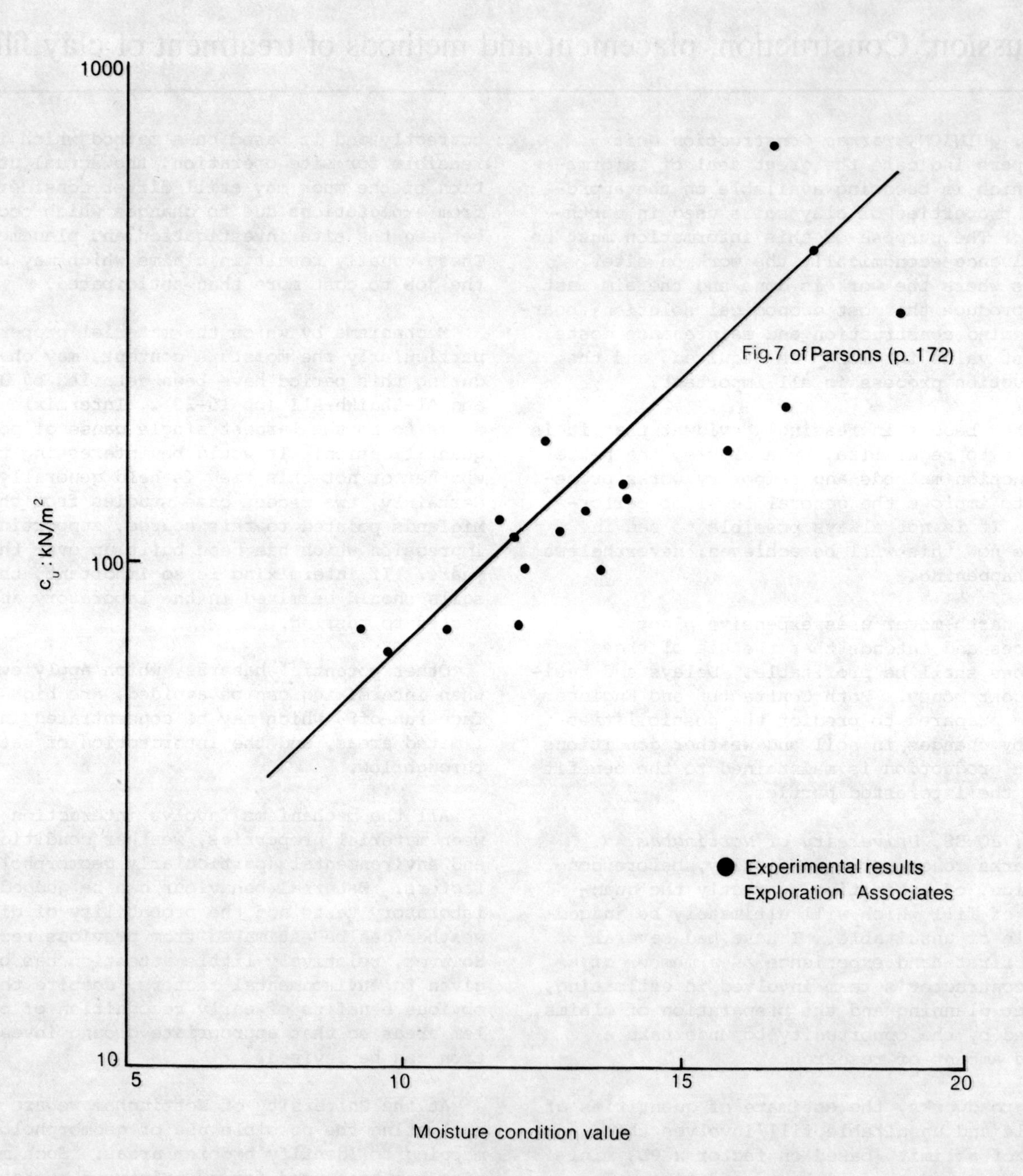

Fig. 1. Relationship between undrained shear strength c_u determined in the triaxial apparatus on remoulded samples and moisture condition value for London Clay

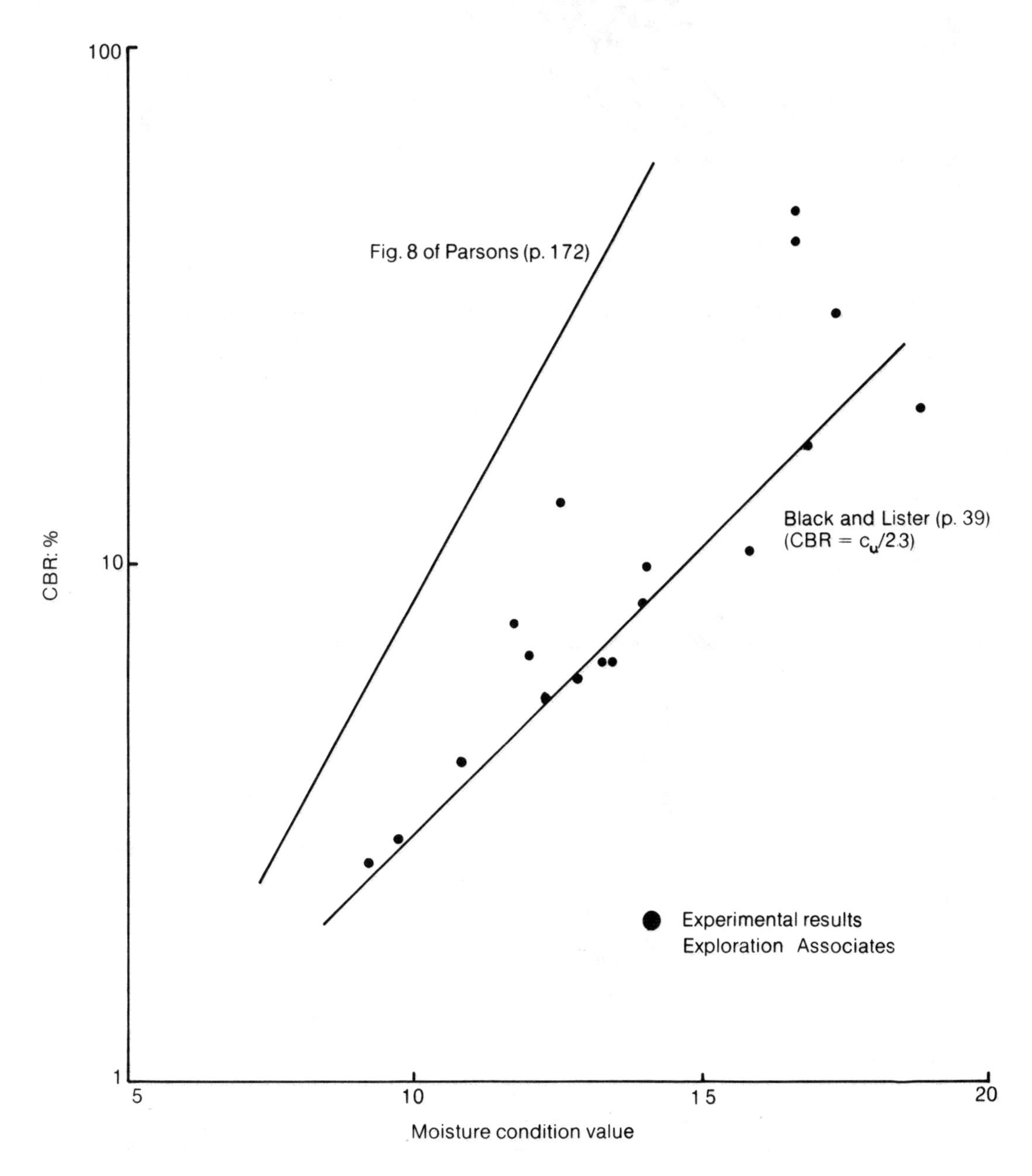

Fig. 2. Relationship between CBR and moisture condition value

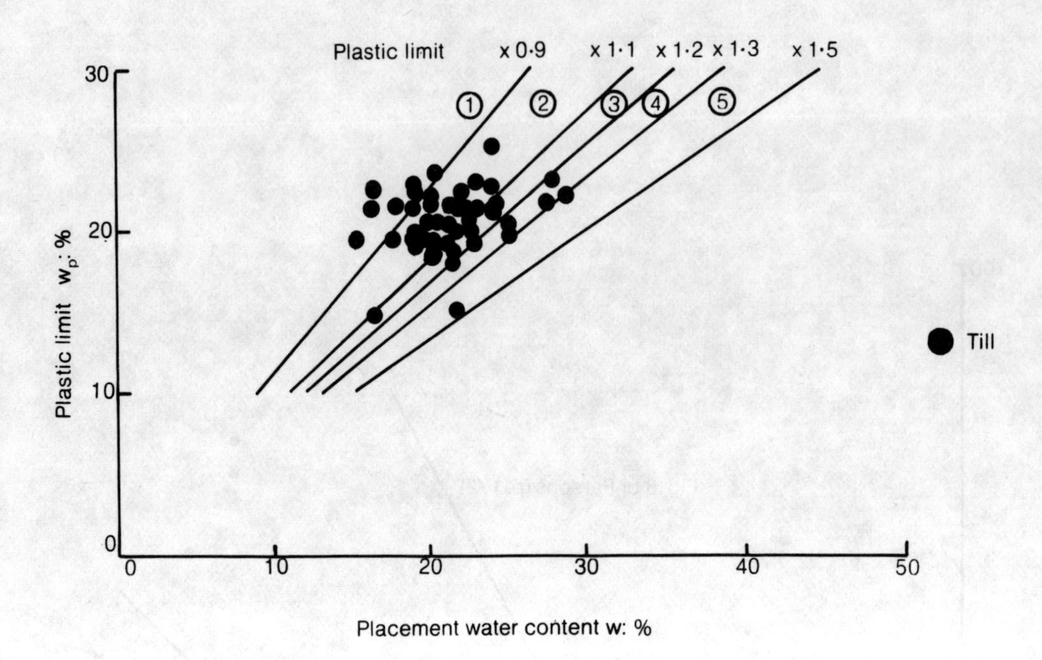

Fig. 3. Relationship between plastic limit and water content for a glacial till

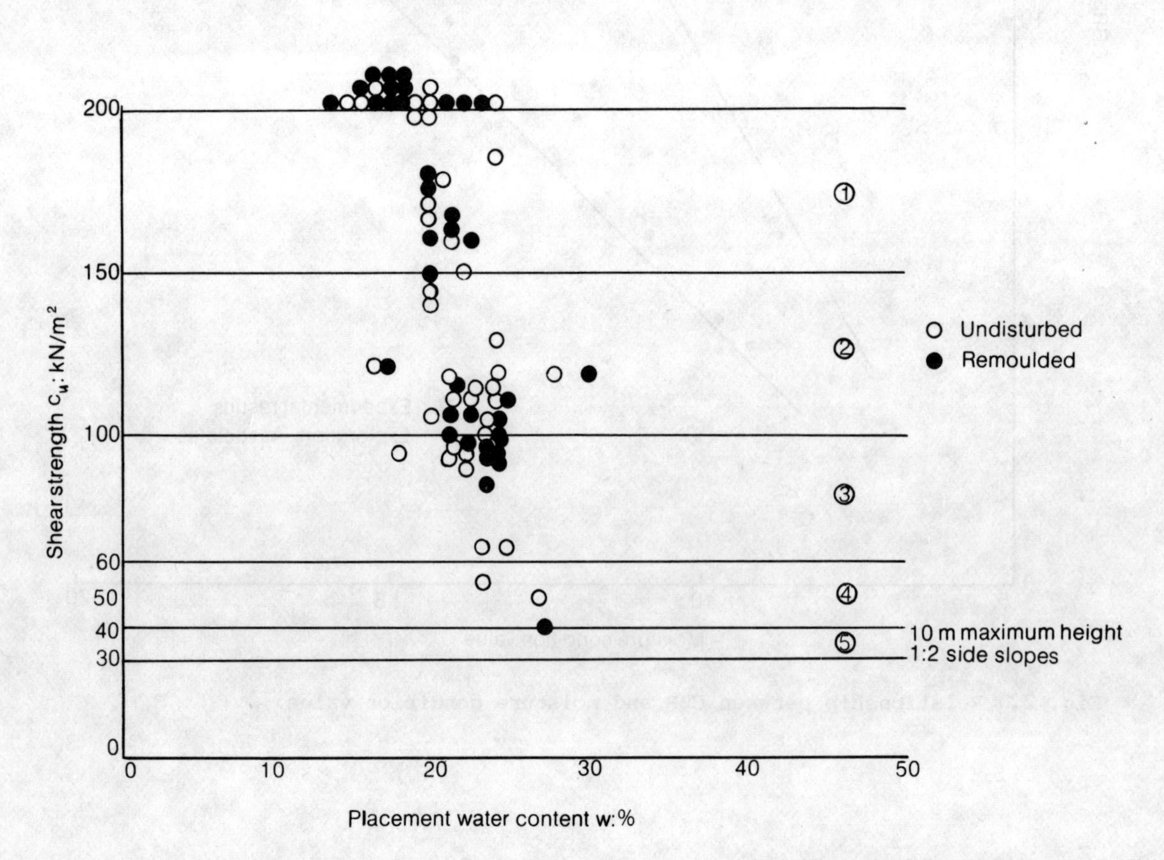

Fig. 4. Relationship between undrained shear strength and water content for a glacial till

As with many geotechnical problems, the need is to study case histories. This is particularly difficult in this context because

(a) of the long time-scale through the various stages of inception, investigation (including mapping), design, construction and performance

(b) of the commercial sensitivity of the information in relation to claims (the sampling and test results which eventually become available may be anything but random)

(c) site records are often poorly kept so that it is difficult or even impossible to relate the ultimate decision on the muck classification to particular test results in the ground investigation report.

I would therefore appeal for more careful recording of data on site so that the behaviour of the materials can be evaluated scientifically after the dust of the claims has settled. There may be a case for setting up a small working party to consider this.

MR L. THREADGOLD, *Exploration Associates (Warwick) Ltd*
I wish to provide some data which have recently been obtained from tests to identify suitability criteria for material which it is proposed to use in embankments.

In order to identify suitability criteria for the London Clay a programme of testing which included natural water content, plastic and liquid limits, undrained shear strength, CBR and MCV testing was undertaken.

Figure 1 shows the correlation between undrained shear strength (using sets of three 38 mm dia. specimens) and MCV, obtained for samples of London Clay remoulded at a range of water contents. The results show good agreement with the correlation line shown by Parsons in his fig.7 (p.172), although it is possible to infer a slightly shallower angle. The deviations from this line are within the order of scatter which he obtained.

The correlations obtained between CBR and MCV (Fig.2) indicate a flatter slope than the theoretical line indicated by Parsons in his fig. 8 (p.172). Fig.2 also shows a line derived from the correlation between shear strength and CBR quoted by Black and Lister (p.39) (CBR = $c_u/23$) and using the correlation line with MCV from fig.7 of Parsons. There is a good correlation between the data and the line.

These tests tend to indicate that the MCV is a useful parameter for identifying the behaviour of material appropriate for use in earthworks.

As indicated by Arrowsmith (pp 25-36), the principle of identifying suitability criteria for clay soil using a natural water content/plastic limit relationship has been established by the Ministry of Transport.[3] At the site investigation stage for roadworks, it has therefore been necessary to attempt correlations between such relationships and shear strengths so that the implications of the use of soils having properties to either side of the hitherto commonly accepted criterion of 1.2 x plastic limit could be identified.

This led to the development of Figs 3 and 4 which relate plastic limit and undrained shear strength to water content and were derived on the basis of the work of Rodin[4] and Snedker.[5] It was intended that soils falling within zones 1-5 on each plot should have broadly similar behaviour in relation to both types of plant and slope stability.

The zones chosen for types of plant most suitable for various shear strengths are broadly in agreement with those identified by Arrowsmith (pp 25-35), Dennehy (pp 87-94) and Parsons (pp 169-175): zone 5 corresponds to small tracked plant, zone 4 to tracked and medium/light and towed plant, zone 3 to medium/heavy plant and zone 2 to heavy motorized scrapers. Soils in zone 1 should allow efficient movement of all plant, but above this zone in Fig.4 compaction criteria begin to predominate. The plotted examples show results from an investigation of a proposed cutting in glacial till.

The correlation between water content/plastic limit criteria and undrained shear strength appears reasonably good in this instance. The shear strengths were measured in the triaxial apparatus on both undisturbed samples (38 mm in diameter) and remoulded samples (38 mm and 102 mm in diameter). No particular influence of size of sample on results could be detected. The glacial till at this site had an average of 19% of material retained on a 425 μm sieve.

The criterion for the suitability of a 10 m high embankment under undrained total stress conditions (Fig.4) is at the lower end of the range of strengths and material on which normal plant can operate and does not therefore represent a critical factor.

The long-term stability will depend on effective stress parameters which cannot be related directly to initial shear strength. Such considerations should be taken into account at the design stage in conjunction with the design of cutting slopes.

It is clear from these data and others from Arrowsmith (pp 25-36) and Al-Shaikh-Ali (pp 15-23) that the choice of a given upper limit of water content reflects the availability of fill materials of a given quality on the site rather than any absolute performance criterion. It is important that this subjectivity of limits is fully appreciated by those who specify the limits for use in earthworks contracts and that the limits are not regarded as absolute.

In my opinion the use of water content/plastic limit relationships introduces an unnecessary complication into the establishment of suitability criteria because shear strength

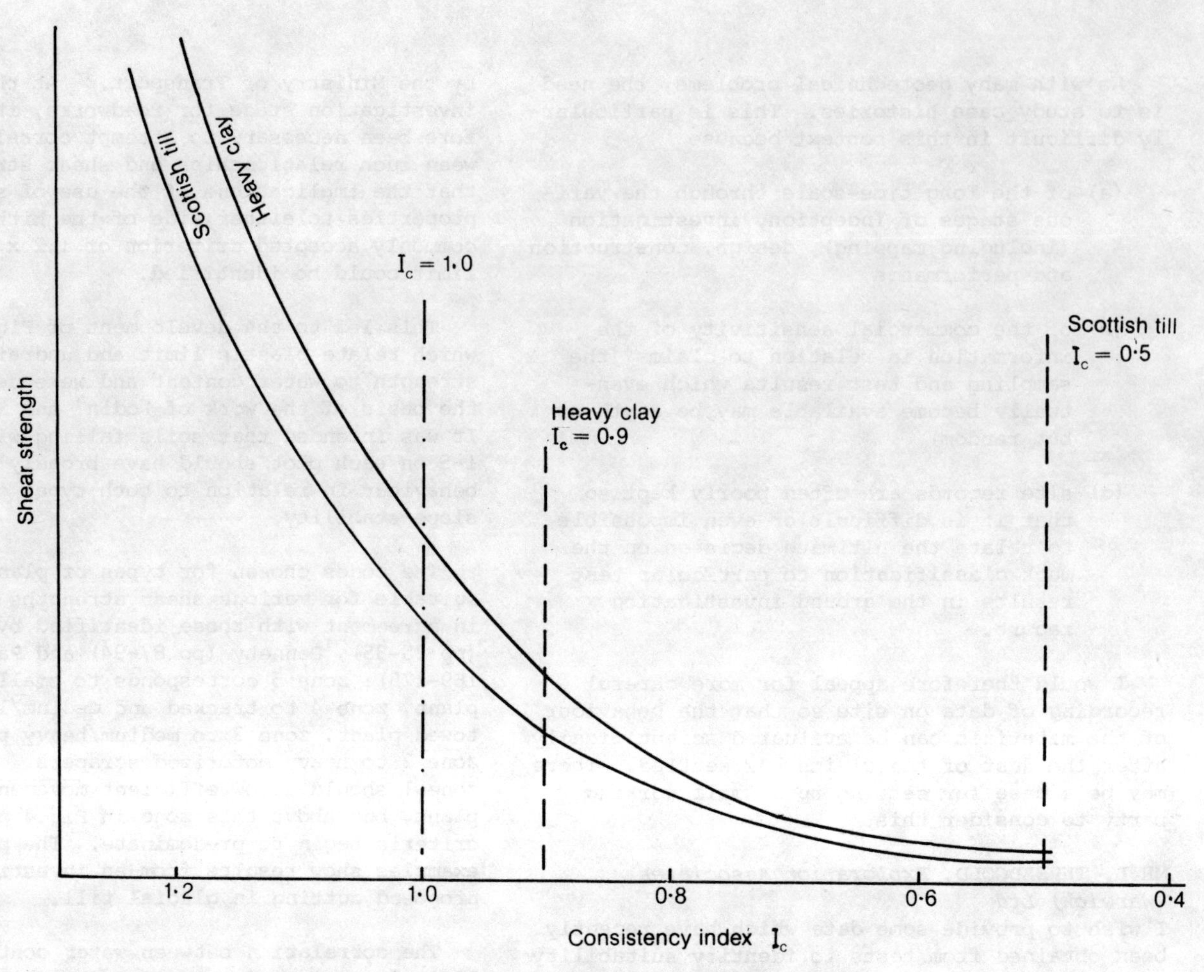

Fig. 5. Effect of change of moisture content on shear strength

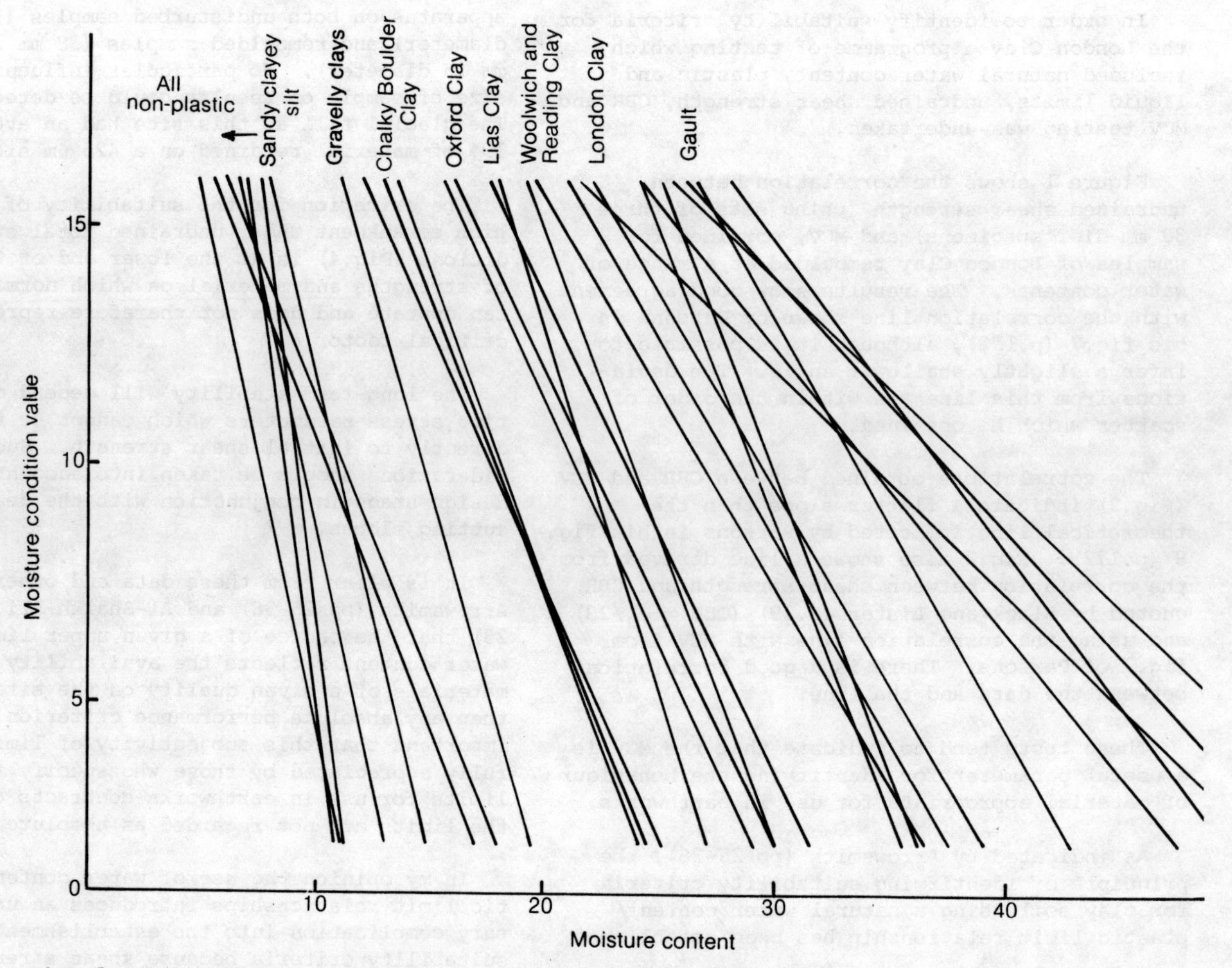

Fig. 6. Calibration of moisture condition value against moisture content for a range of B soils

provides the primary criterion for assessment of fill suitability in the short term in relation to both plant movement and stability.

Other methods such as the MCV test, which can be closely correlated with shear strength, CBR and compaction performance and is capable of rapid operation, are to be encouraged.

MR E.A. SNEDKER, *Midland Road Construction Unit*
The MCV test has obviously been well applied on wet soils. Although less frequent, particularly in the UK, problems of suitability and trafficability do occur in the drier range of moisture content on materials such as clays, sand and gravel mixtures and PFA. The analogy of trafficking a dry beach illustrates one type of problem. It becomes clear that at certain places in a fill the material is far less confined than it is during the MCV test. It seems therefore that as one approaches drier materials the MCV would begin to give erroneously higher strengths than might be valid due to the confinement of the test material in the mould.

MR A.W. PARSONS, *Transport and Road Research Laboratory*
There is no doubt that trafficability problems can be experienced with dry sands and gravels. With such materials earth-moving plant has been observed in difficulties even though very high MCVs have been measured with the material.

The major intended role of the moisture condition test is the assessment of the quality of fill material and in that role the relation between confined shear strength and MCV must be a major consideration. This relation will remain valid under the conditions described by Mr Snedker.

DR M.C. FORDE, *University of Edinburgh*
I welcome the concept of the moisture condition test for application to the low plasticity stony fills or boulder clays of the north of England and Scotland because

(a) it is portable
(b) it is simple to operate
(c) it gives an immediate result both on site and in the laboratory
(d) it tests the soil actually to be compacted except where material is greater than 20 mm in diameter.

Mr Parsons (pp 169-175) has demonstrated the use of the test on a high plasticity clay and an intermediate clay and has indicated that it may be possible to propose a certain MCV (e.g. 8) as the limit of suitability. This approach, if viable, would be of substantial value to both Contractors and Engineers.

I should like to consider the implications of a variation in moisture content of 5% for the heavy clay used by Parsons and for a Scottish till, assuming that the soils are initially at their plastic limits. Table 1 shows the plasticity data for the soils and the consistency indices at the plastic limit (w_p) and at a moisture content of w_p + 5%. The consistency indices of the soils are 1.0 at w_p. The addition of 5% of water to the heavy clay reduces its consistency index only slightly - to 0.9 - but an equal addition of water to the Scottish till reduces its consistency index to 0.5. The effect of these changes on the shear strength is shown in Fig.5, which is based on work by Forde[6] and Skempton and Northey.[7] From the hypothetical trend lines (based on experimental trends) it can be seen that when both the soils are at the plastic limit (consistency index 1.0) their shear strengths are comparable. However, at w_p + 5% the consistency index of the heavy clay is much the higher and its shear strength is therefore much the greater. This indicates the sensitivity of low plasticity soils to variations in moisture content.

Has anyone experience of using the MCV test on low plasticity clays and stony clays having regard to the following aspects of the MCV test?

(a) MCV is based on the log of the number of blows.
(b) Figure 7 of Parsons (p.172) is effectively a log-log plot, with shear strength variations of 100% at a given MCV.
(c) The low plasticity clay in Parsons' table 1 (p.173) has the lowest correlation coefficient for the log-log plots.
(d) For stony clays the larger stones are discarded.

MR PARSONS
There is at least one moisture condition apparatus in use in Scotland and, as far as I know, no difficulty is being experienced with Scottish tills that contain plastic fines. There will be a value for the proportion of material exceeding 20 mm in size (and therefore excluded from the test) beyond which the test will no longer reflect the properties of the soil as a whole. If this should become a serious problem there is no reason why the principle of the moisture condition test should not be applied using a larger-scale apparatus compatible with the size of the larger soil particles.

MR R.J.G. EDWARDS, *Eastern Road Construction Unit*
The MCV calibration lines for a range of soils in the east of England have been obtained. The grading of these soils varies from non-plastic sandy gravels and clayey gravels through to highly plastic silty clays, covering all soil groups from the Lower Lias of the Jurassic Age through Cretaceous, Tertiary, Pleistocene to Recent. Typical calibration lines for MCV covering this range of materials are shown in Fig.6. The shear strength of these materials was obtained using a laboratory vane test on the actual MCV sample after each individual MCV test. This value for representative type samples is shown in Fig.7 for MCVs of 7, 12 and 16. It is apparent from these tests that at low values of MCV (4-8) the test is an indirect measurement of shear strength and is consistent over a wide range of material types. At higher values of MCV a relationship clearly exists but is not as consistent; the scatter of results is also

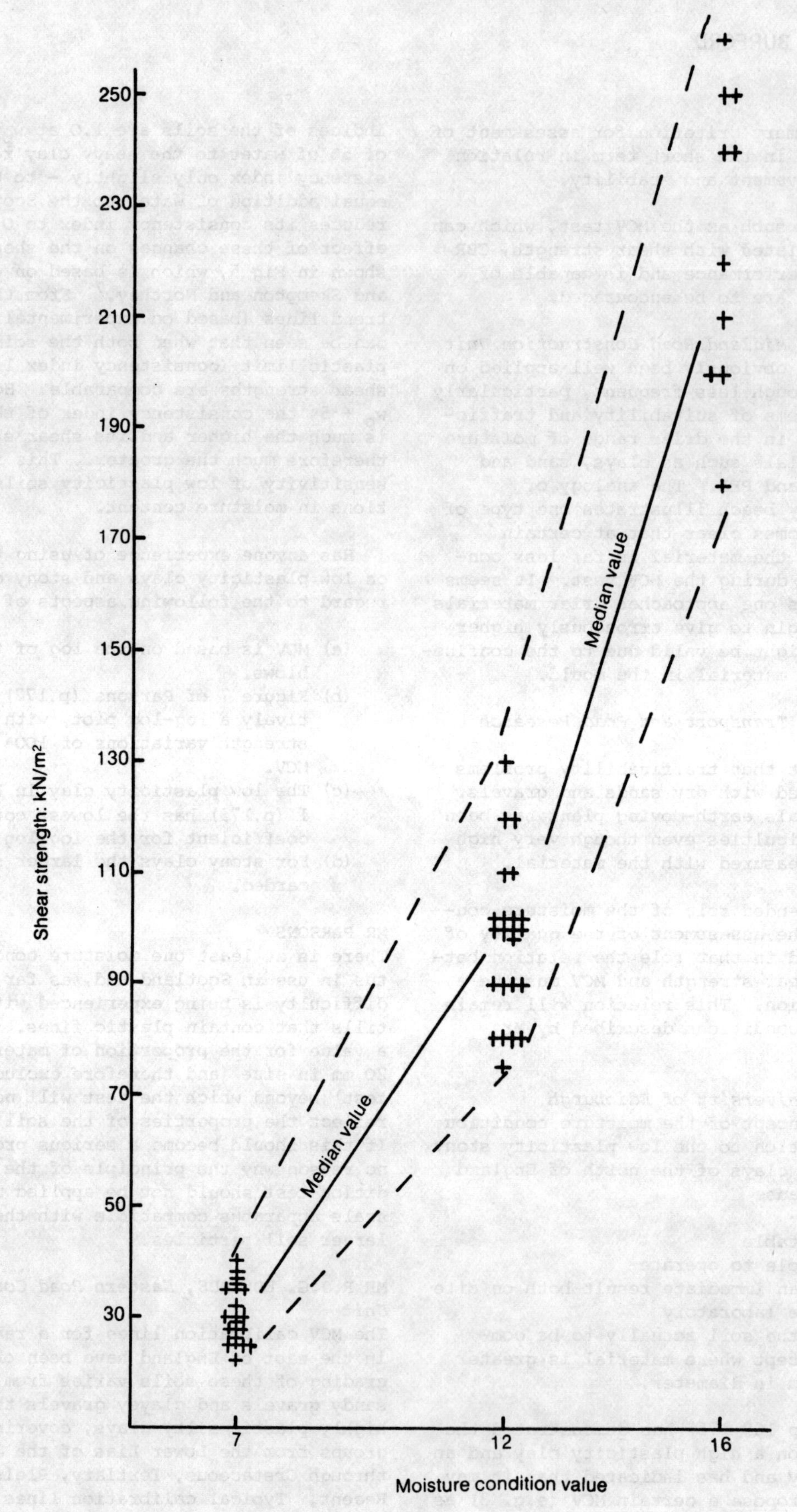

Fig. 7. Relation between laboratory vane shear strength and moisture condition value

Table 1. Effect of change of moisture content on the consistency index

Soil	Liquid limit	Plastic limit	Plasticity index	Consistency index at plastic limit	Consistency index at plastic limit + 5%
Heavy clay	76%	27%	49%	1.0%	0.90%
Scottish till	28%	18%	10%	1.0%	0.50%

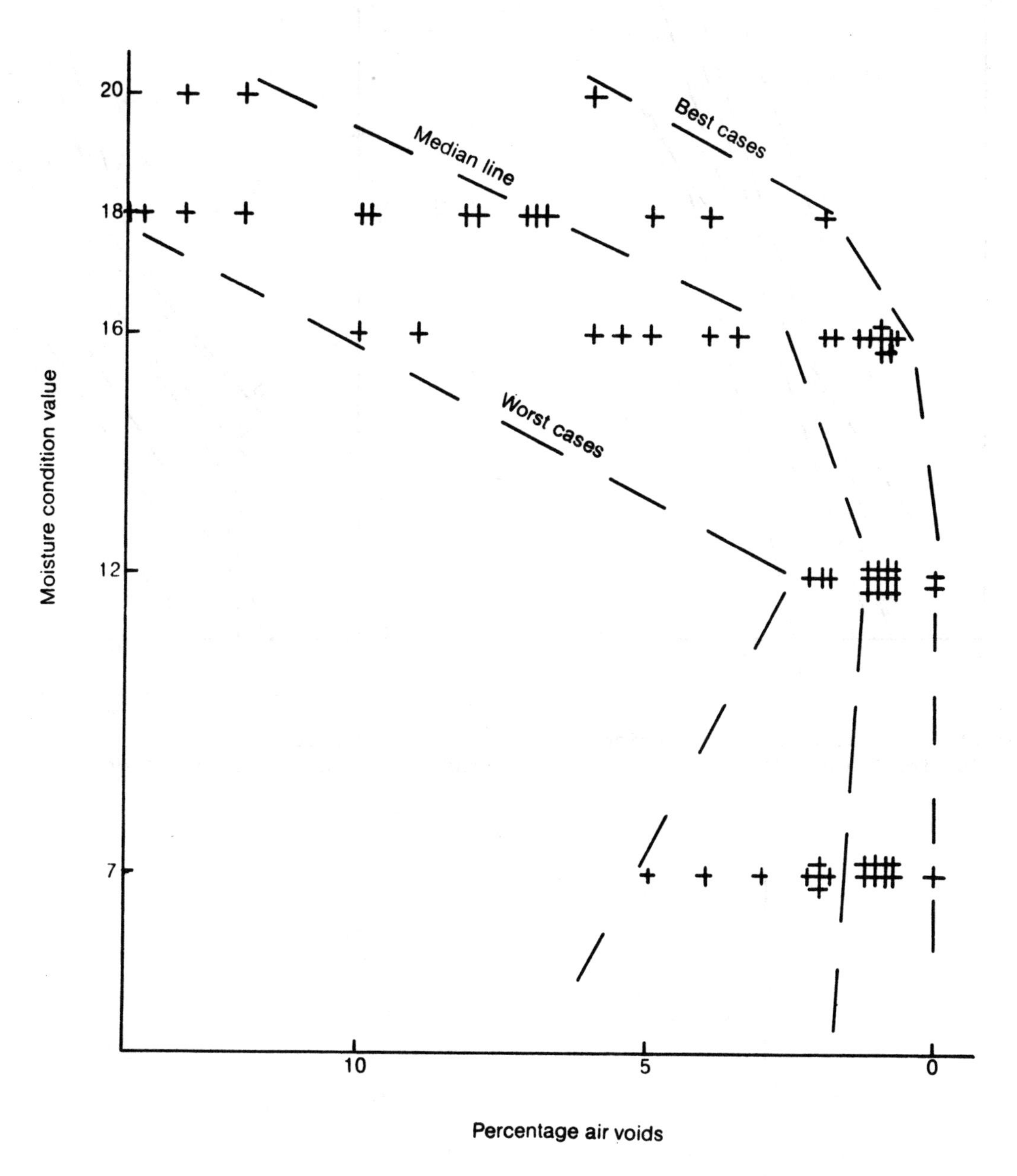

Fig. 8. Air voids in compacted soils related to moisture condition value

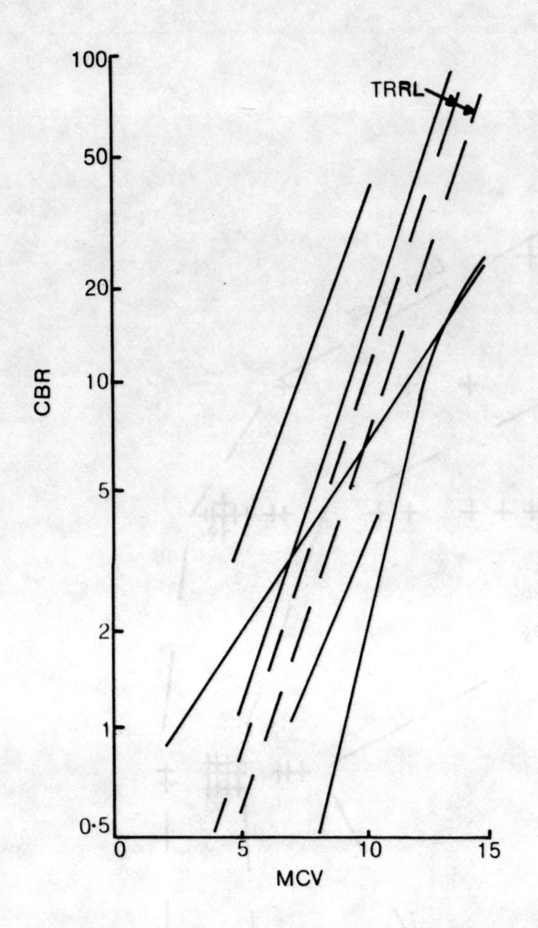

Fig. 9. Relationship between CBR and moisture condition value for Boulder Clay

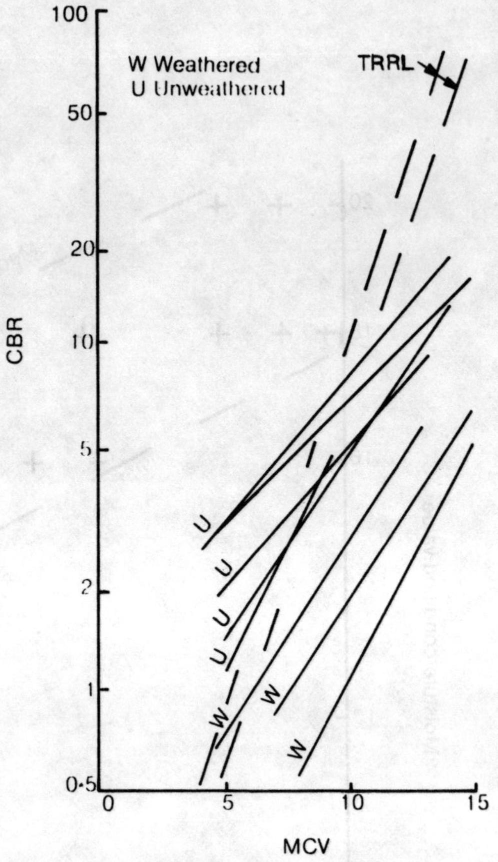

W Weathered
U Unweathered
TRRL
CBR
MCV

Fig. 10. Relationship between CBR and moisture condition value for London Clay

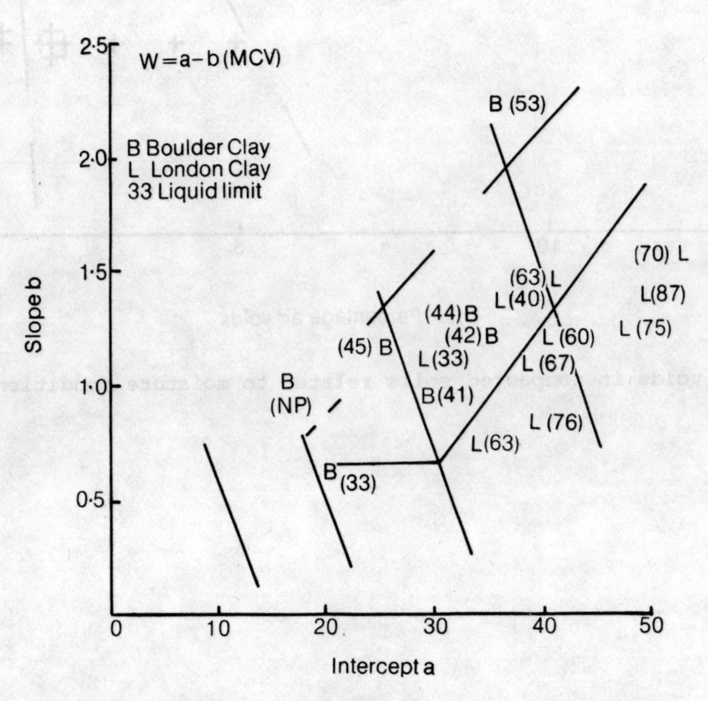

Fig. 11. Results plotted on Parsons' classification chart (fig.6 on page 172)

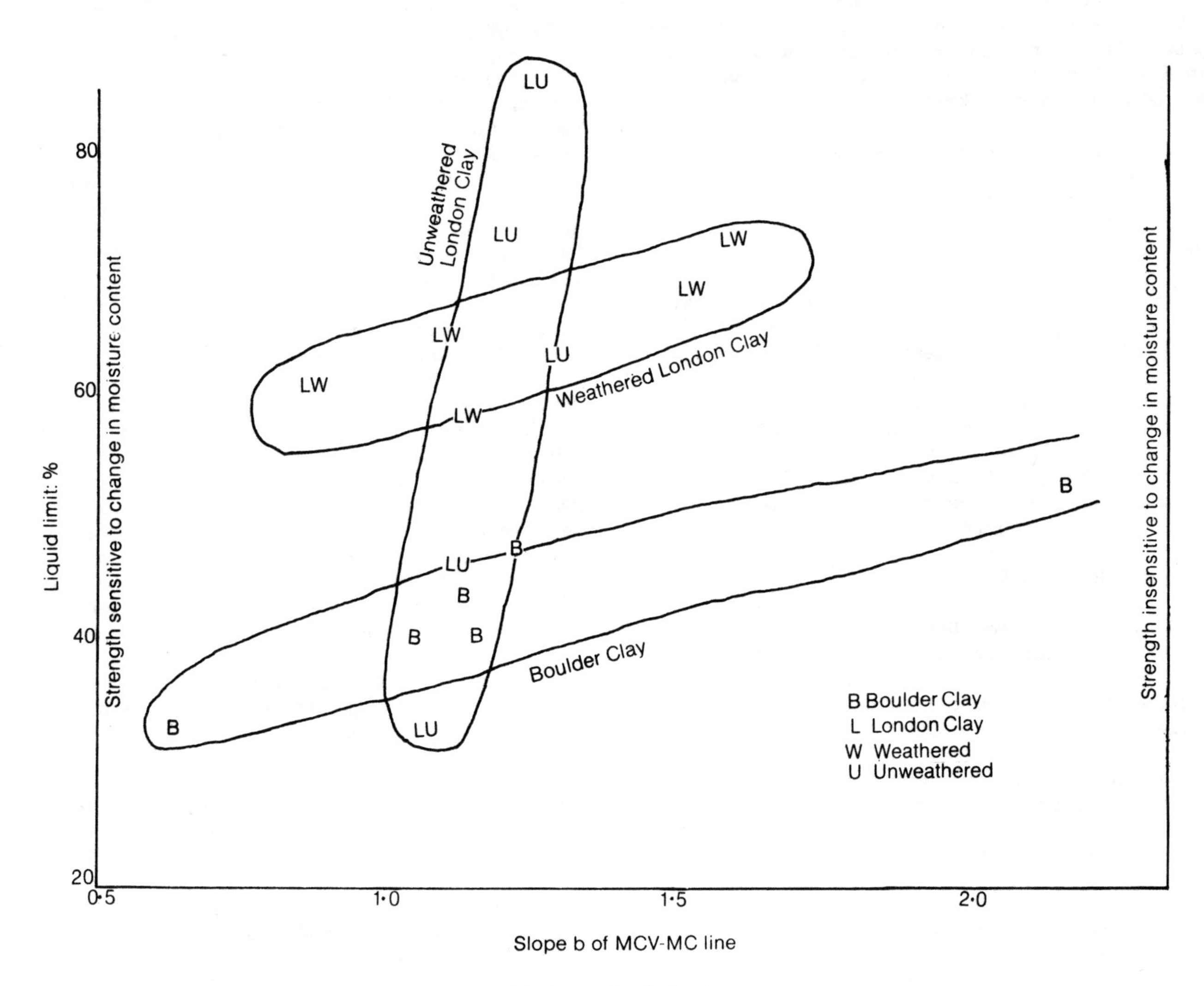

Fig. 12. Relationship between liquid limit and slobe b

greater.

At MCVs of 16 and above it is apparent that the density achieved in the test moves away from the zero air voids line (Fig.8). It is therefore suggested (and field experience confirms this supposition) that at moisture contents equivalent to an MCV of 16 and above materials over the range of types examined cannot be compacted to less than 5% air voids. Experience with respect to standard compaction plant (table 6/2 of reference 8) suggests that the value of 5% air voids at MCV = 16 can be achieved only with carefully selected plant ideally suited to the purpose. At lower moisture contents or with less than ideal plant and site control, air voids of 10% plus are more likely to be achieved.

It would appear from experimental data that the moisture condition value can be directly related to the ratio of moisture content/plastic limit. Thus for MCV = 7 the following relationships would be expected.

(a) For silty clayey gravels of low plasticity, vane shear strength = 25-30 kN/m^2 at moisture content = PL x 1.0, equivalent laboratory CBR = 3-4.
(b) For medium plastic gravelly or sandy silty clays, vane shear strength = 30 kN/m^2 at moisture content = PL x 1.2, equivalent laboratory CBR = 2-3.
(c) For silty clays of high plasticity, vane shear strength 30-40 kN/m^2 at moisture content = PL x 1.5, equivalent laboratory CBR = 1-1.5.

MR S.B. WEBB, *Wimpey Laboratories Ltd*
The MCV test has promise of greater speed and less operator dependence than have existing methods of assessing the suitability of soil as fill.

Wimpey Laboratories have carried out a series of MCV tests mainly on London Clay and Boulder Clay. Many of the tests were simply a measurement of the MCV at natural moisture content. Breaking up samples of stiff to very stiff clay to pass a 20 mm sieve is time consuming, and a test can easily take over an hour.

To calibrate for the relationship between moisture content and MCV, tests were made at natural moisture content and then at several wetter and drier moisture contents. In some cases it was found that the MCV at natural moisture content tended to lie above the line given by other determinations and it is thought that the most likely cause of this is the difference between undisturbed and remoulded shear strength. Tests on stiff to very stiff clays which have been subjected to varying degrees of remoulding would be valuable in assessing the importance of this factor; it appears to be less noticeable for softer samples. A few tests carried out with with material consisting of lumps of various sizes show that undisturbed lumps of 15-20 mm do give a higher MCV than samples with well remoulded fines.

Although the MCV apparatus includes a disc to reduce extrusion of soil between the rammer and the mould, this is only effective to a rather limited extent with soft soils. Tests where soil has got above this disc tend to give false, high MCV values. It is suggested that it should be a requirement of the test to note the presence of any soil above this disc, or perhaps to weigh it. In some of our tests there was 20-30 g of soil.

Most of our samples of Boulder Clay (Fig.9) gave relationships between CBR and MCV fairly similar to those shown in Parsons' fig.8 (p.172), although they occupied a rather wider band. This may have been due to differences in plasticity characteristics, although the range was not great, or to differences in weathering. For London Clay (Fig.10), however, the slope of the lines was substantially flatter. The overall band in Fig.10 is rather broad, but all the lower lines relate to weathered clay and the upper ones to deeper, less weathered clay. Similar relationships were found between the log of triaxial shear strength and MCV.

The Boulder Clay samples follow the classification in Parsons' fig.6 (p.172) well, but lines further to the left of Parsons' fig.6 would more nearly match our results for London Clay (Fig.11). If the slope of the moisture content versus the MCV line is plotted against liquid limit (Fig.12) again there is a marked difference not only between Boulder Clay and London Clay, but also between weathered and unweathered London Clay. This suggests that both Boulder Clay and weathered London Clay can be much more sensitive to softening by change in moisture content than is deeper, unweathered London Clay.

Have other users of the MCV apparatus recorded soil working its way up between the rammer and the mould?

Is it right to assume that the scatter of the CBR values on which Parsons' fig.8 is based is similar to that for the shear strengths shown in his fig.7?

How many of the soils tested by Parsons were overconsolidated and undisturbed when taken for test, as distinct from those which were remoulded either by earth-moving plant or in the laboratory? Is there any minimum of remoulding needed for these soils to give results similar to those of well worked soils?

MR PARSONS
Some extrusion of soil around the sides of the mould occurs during the determination of MCV at high moisture content levels and is generally acceptable. The relative diameters of the rammer of the moisture condition apparatus and the lightweight disc that is placed on top of the soil are such that any extruded soil should not interfere with the free fall of the rammer. Neither should the rate of extrusion be so great that the criterion for the change in penetration of the rammer at the MCV would be affected. However, if attempts were made to estimate the bulk density of the material after completion of the test, then the extruded soil could have a

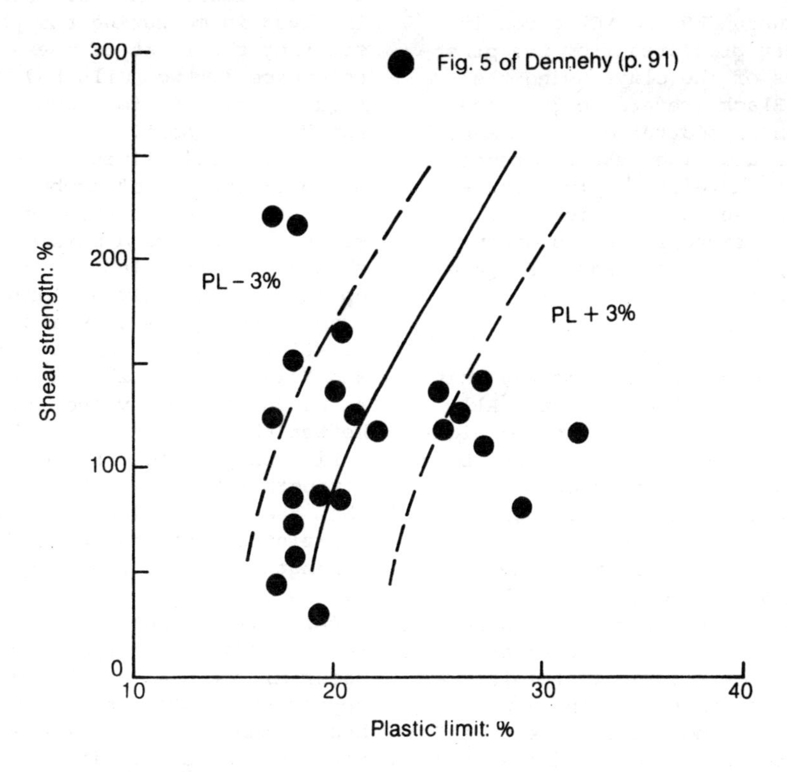

Fig. 13. Influence of the plastic limit on remoulded soil strength

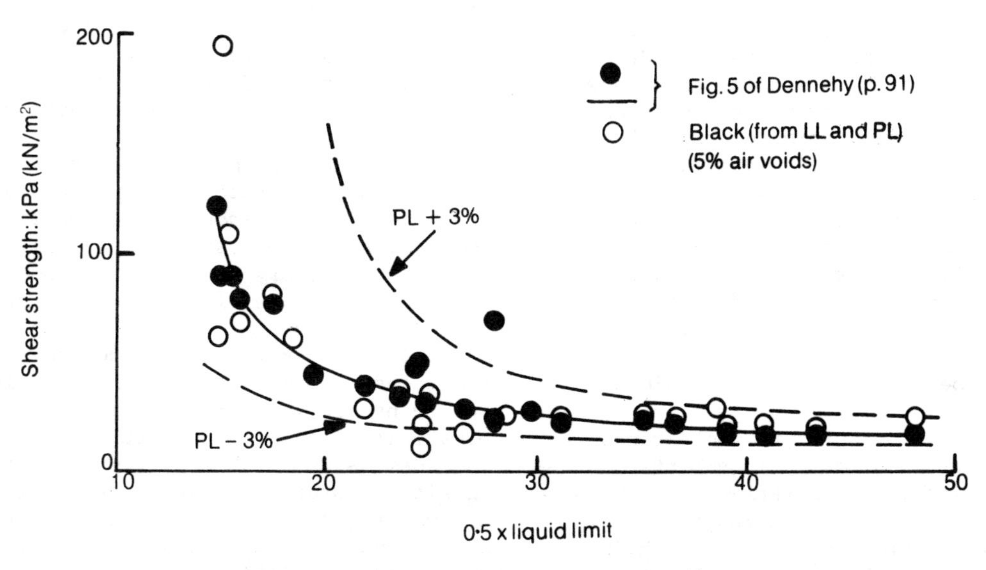

Fig. 14. Influence of the liquid limit on remoulded soil strength

serious effect on the result. The apparatus was not designed with the determination of density in mind.

The relations between CBR and MCV given in fig.8 of my paper were predicted from the plasticity characteristics of the clays using the method described by Black (reference 8 of the paper). The question of scatter of experimental results, as given for undrained shear strength in my fig.7, does not therefore apply. Variations in practice from such predictions could arise from a number of factors, including the degree of remoulding of the soil and the state of compaction used in the CBR test.

Differences in MCV at varying degrees of remoulding can be expected in the same way that differences in shear strength can occur. All results shown in my fig.7 were obtained on highly remoulded samples of soil, i.e. they had been subjected to a laboratory mixing process. The differences in MCV that occur when various degrees of remoulding are applied to soils in the laboratory indicate the ability of the moisture condition test to provide a meaningful strength scale. The main problem arises when one is trying to simulate, in the laboratory, the conditions on site over a range of moisture contents. If the degree of remoulding applied to the soil varies as the moisture content is increased, as can often be the case, there is a tendency for the calibration line relating moisture content to MCV to become a curve.

MR W. BLACK, *Transport and Road Research Laboratory*
Dennehy (pp 87-94) makes a good case against the use of a moisture content related to the plastic limit for control of the suitability of fill in earthworks if soil strength is the intended criterion of suitability (Fig.13). Also shown in Fig.13 are the mean relation between plastic limit and shear strength estimated using the suction index method, and the likely limits of strength for the usual variability in plastic limit of ± 3%. The analysis confirms that poor correlation between plastic limit and strength would be expected. At any given plastic limit the estimated shear strength could vary by a factor of not less than four. The scatter of results shown by Dennehy would be expected and they lie mainly between the predicted limits.

Dennehy shows that a moisture content related to the liquid limit provides a much better correlation with strength, particularly at the higher liquid limits. The predicted strengths are shown in Fig.14 and closely follow Dennehy's results. Fig.14 also shows the likely limits of strength for the plastic limit varying by the usual ± 3%. Dennehy's results do not show anything like the predicted scatter in strength at low liquid limits, but this can be explained by the fact that his measurement did not have a wide scatter of plastic limits for any given liquid limit. On the Casagrande plasticity chart[9] all except two of his soils fall within 1% of the mean line through them. This degree of correlation is unusual, as can be seen from fig.2 of Vaughan et al. (p.206). It is implicit in Dennehy's correlation that either at a given liquid limit strength differences are small or all differences in plasticity index at a given liquid limit are due to difficulties involved in measuring the plastic limit. Considering the first of these possibilities, in reference 10 two soils had virtually the same liquid limit: 49 for Mombasa Road soil and 50 for Naivasha soil. At a moisture content of half the liquid limit the ratio of strengths was 10 to 1; the CBR of Mombasa Road soil was 3.5 and that of the Naivasha soil was 35. Considering the second possibility, Casagrande showed that the plasticity index could vary by up to 25 at any given liquid limit. In the above example there was only a modest difference in plasticity index; the values for Mombasa Road and Naivasha were 27 and 22 respectively. Hence although control by the liquid limit is much better than with the plastic limit, sampling variability could provide poor control of strength for the lighter soils. Fig.15 shows that the suction index method provides good correlation between strength estimates and equal strength lines in fig.7 of Dennehy.

As soil strength is likely to become the preferred parameter for assessing the suitability of material for use in embankment fills, it would appear better to achieve this by direct strength measurement. In cohesive soils of low stone content and low sensitivity in situ measurements using a penetrometer would be feasible. Low sensitivity is important because the penetrometer would probably be used primarily to test undisturbed soil, whereas the soil strength required is the remoulded value. For more general use, the MCV test of Parsons (pp 169-175) provides a rapid reproducible means of arriving quickly at a remoulded strength value, the actual strength which he correlates with his MCV being a function of his method of sample preparation and method of strength measurement.

MR A.C. POWELL, *Lancashire County Council*
The suitability criterion for cohesive soils in the *Notes for guidance on the specification for road and bridge works*[3] has been criticized but there is obviously a sound basis behind a method which has been used in motorway construction for more than ten years. Until 1965 suitability was defined by natural moisture contents of less than plastic limit plus 2%, but problems arose on a contract in east Lancashire where the high plasticity clays in the cuttings were being assessed as unsuitable by virtue of their high in situ moisture contents. This was illogical because the clays were of stiff to very stiff consistency and were relatively insensitive to wet weather conditions.

The first approach to the problem was on the lines of a consistency index concept, but this was not followed up because of problems with sliding of the silty clays within the cup apparatus then in use for the liquid limit test. An empirical approach, using a factor times plastic limit, appeared to be the best alternative and good correlation was soon found with local clays: each soil type appeared to have a unique factor, generally increasing from 1.1 for

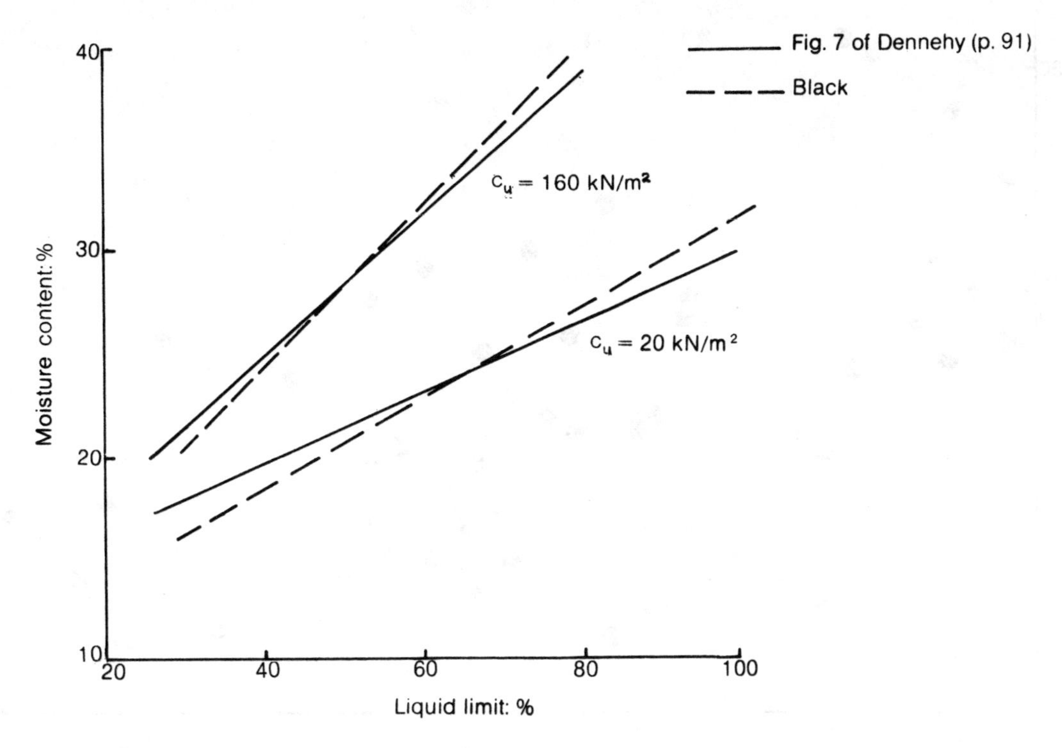

Fig. 15. Remoulded undrained shear strength contours

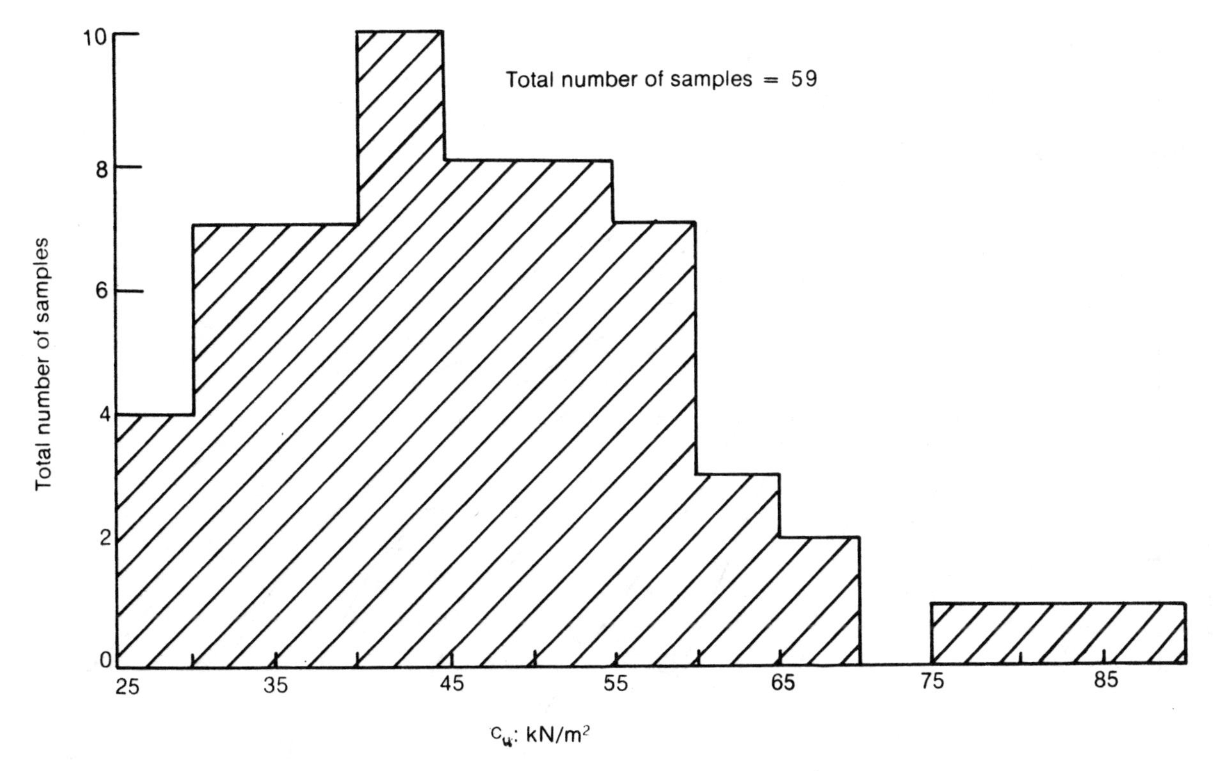

Fig. 16. Variation of undrained shear strength at moisture contents equal to 1.2 × plastic limit

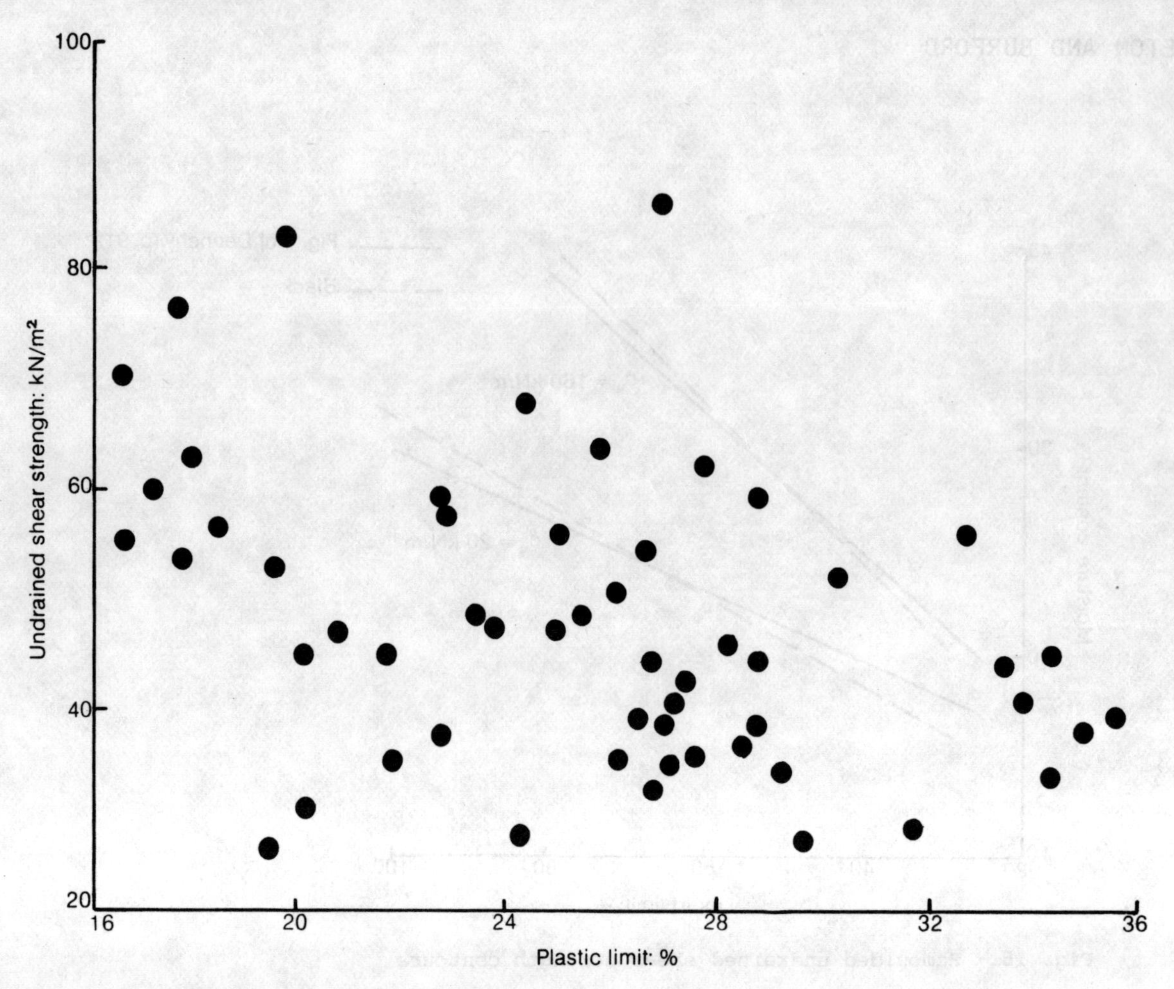

Fig. 17. Relationship between undrained shear strength and plastic limit at moisture contents equal to 1.2 × plastic limit

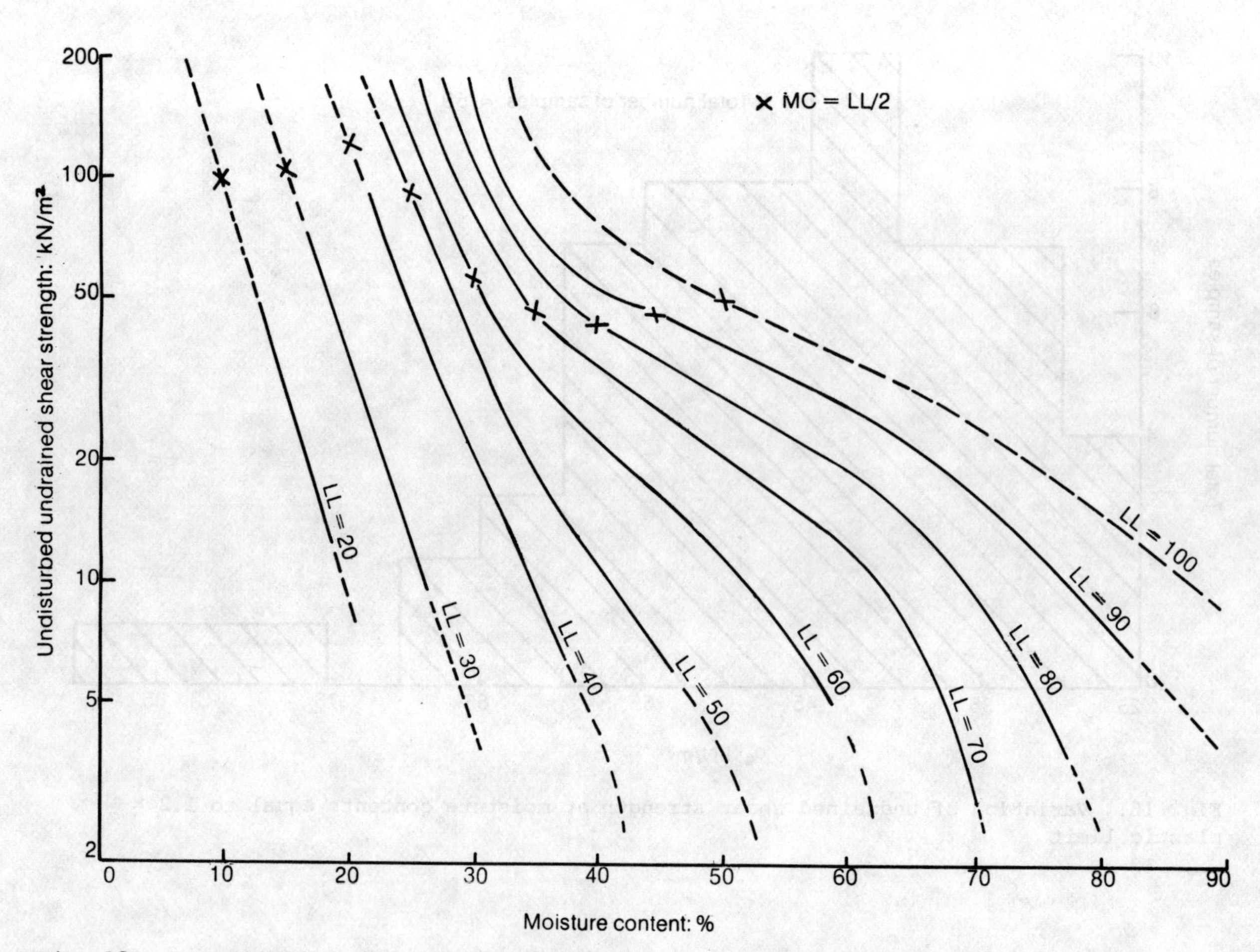

Fig. 18. Strength plotted against moisture content for soils of given liquid limit

low plasticity clays to 1.3 for high plasticity clays. This method was approved and soon found satisfactory in use.

Figure 2 of Vaughan et al. (p.206) shows a narrow band of results grouped just above the A line. The best fit through the origin follows the straight line PI = 0.6 LL or PI = 0.4 LL.

Referring to fig.1 of Black and Lister (p.38) and taking the generally accepted critical value of 50 kN/m^2 for shear strength or 2% CBR, it is possible to select consistency indices for varying plasticities. A consistency index of 0.8 is required for a high plasticity clay; substituting LL = 2½ PL in the consistency index equation, the in situ moisture content has a critical value of 1.3 PL. By similar reasoning, a low plasticity clay with a consistency index of 0.9 has a critical value of 1.15 PL. Both these results agree well with the empirical values.

The establishment of these relationships gives a clearer insight into the reasons for the wide scatter of results obtained when the moisture content/plastic limit ratio is plotted against strength. Clays with higher than normal PI values, plotting well above the A line, will be relatively insensitive to change in moisture content. Clays which plot well below the A line will generally be silty or organic and likely to be unsuitable by virtue of their high natural moisture content. The sandier clays, which are permeable and very sensitive to changes in moisture content, present classification problems. However, there will always be a practical problem here, irrespective of the method used to assess suitability at the design stage, because changes in moisture content rapidly affect the inherent strength and trafficability of these clays.

There are clearly disadvantages inherent in the plastic limit test, and the poor reproducibility indicated by Sherwood[11] could be partly to blame for the scatter of results in the strength relationships. With the increased accuracy now obtainable from the cone-type liquid limit test, there may be sound reasons for a change to a suitability criterion based on this test. Although it is still open to criticism on the basis that the test is carried out on the material passing a 425 μm sieve, it is now possible to carry out this test on the full range of cohesive soils. Using the same equations as before, the likely limit for suitability will be about 0.5 times liquid limit.

The liquid limit test may well be successful in establishing a new suitability criterion but the application is still likely to be limited to typical British clays. Local clays may not always be amenable to a standard treatment and a strong plea is made for a flexible approach to the whole issue of suitability criteria.

MR PARSONS
Arrowsmith (pp 25-36) and Dennehy (pp 87-94) have illustrated the poor correlation of the ratio of moisture content to plastic limit with undrained shear strength. During studies involving the moisture condition test my team has made measurements of undrained shear strength over a range of moisture contents on a variety of clay soils. The variation of undrained shear strength at a moisture content of 1.2 x PL is given by the histogram in Fig.16. Undrained shear strengths ranged from below 30 kN/m^2 to above 85 kN/m^2, with most values in the range 35-55 kN/m^2.

When the same values of undrained shear strength are plotted in relation to the plastic limit (Fig.17) there is no trend of increase in strength with increase in plastic limit, as would be expected if allowance were made for the usual increase in plasticity index with plastic limit. The conclusion reached is that considerable experimental error occurs with the measurement of plastic limit and that this is the overriding factor in the lack of agreement with strength measurements.

MR WEBB
With regard to the suggestion that the suitability criterion for soil should be based on the liquid limit, taking a maximum moisture content of perhaps half the liquid limit, Fig.18 shows the relationship between undisturbed undrained shear strength and moisture content for soils of given liquid limit, as measured in 2000-3000 quick triaxial tests. The dotted lines are for areas where the points are either sparse or excessively intermingled. For the drier and lower plasticity soils, cell pressures will have had an effect on the strengths and this may account for the curve near the top of some lines, as many of the deeper, and therefore drier, samples were tested at substantial pressures. It can be seen that strengths at half the liquid limit range between about 100 kN/m^2 for low plasticity clay (based on sparse information) to about 50 kN/m^2 for high plasticity clays. These strengths will be reduced on remoulding during excavation and compaction, but as clays of high sensitivity are the exception in the UK, an average strength of perhaps 40 kN/m^2 can be expected.

DR FORDE
I consider that the approach adopted by Arrowsmith (pp 25-36) of assuming stones larger than 5 mm in diameter to be spherical and surrounded by a water layer 0.23 mm thick while material between 5 mm and 425 μm has a natural moisture content of 9%, is applicable more to granular than to cohesive material. The approach which I have adopted, which is also oversimplified and may not be any more accurate, is as follows.

The coarse fraction, retained on the 425 μm sieve, is non-plastic and will have a comparatively low moisture absorption compared with the overall moisture content. If it is assumed that the moisture within a sample is distributed such that the fraction retained on the 425 μm sieve contains the maximum moisture absorption value, then the remainder of the moisture is concentrated in the fraction passing 425 μm. Table 2 lists the natural moisture contents, the average moisture absorption of the granular

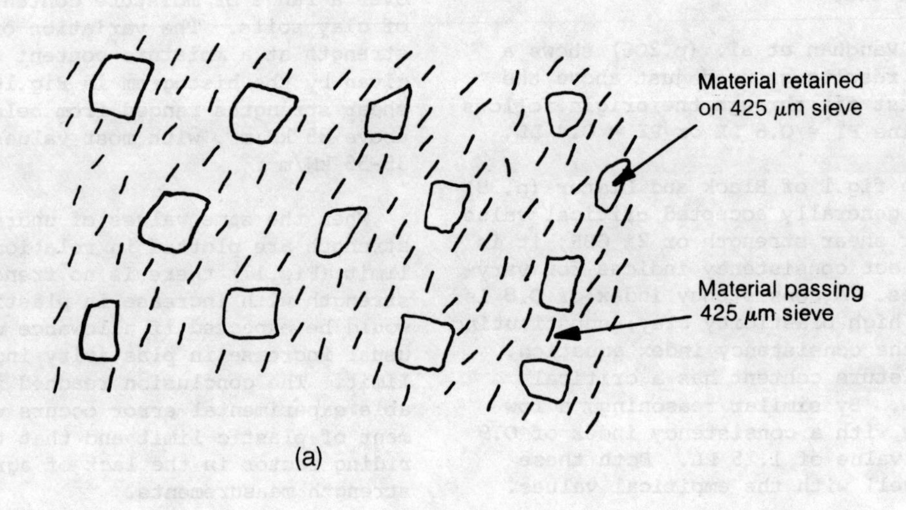

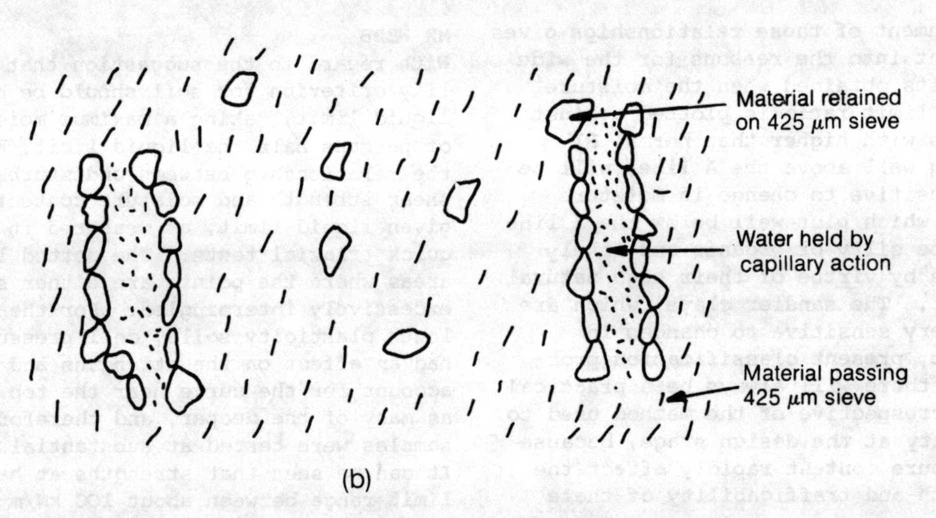

Fig. 19. Moisture distribution in clays containing coarser material

Table 2. Estimates of the moisture content of the matrix (passing 425 μm sieve) of samples of till

Sample number	1	2	3
Natural moisture content	24.51%	33.90%	20.51%
Moisture absorption of granular particles	3.94%	8.13%	6.05%
Percentage retained on 425 μm sieve	48	59	45
Liquid limit	45%	35%	42%
Plastic limit	21%	21%	20%
First trial corrected moisture content	43.49%	69.05%	32.34%
Final corrected moisture content	34.00%	51.49%	26.43%

particles, the percentage passing the 425 μm sieve and the liquid and plastic limits of three samples of till from the Isle of Skye. Using the above assumption, the first trial corrected moisture content of the matrix given in Table 2 was obtained. Physically the assumption is that each coarse particle (retained on the 425 μm sieve) is surrounded by particles passing the 425 μm sieve (Fig.19(a)). For the soils with a substantial fraction retained on the 425 μm sieve this assumption may not be wholly justified; a proportion of the water will be held by the capillary action in the coarse fraction (Fig.19(b)). In view of these considerations a final compromise corrected moisture content of the matrix has been calculated by taking the mean of the natural moisture content and the first trial moisture content.

MR E.J. ARROWSMITH, *North Western Road Construction Unit*
I have not done any further work on this subject since the Lancaster Bypass was constructed in 1958 as subsequent schemes in Lancashire were further south where the boulder clay was less stony and so did not present a problem. However, it is current practice in the North West to require the measurement of the moisture content of cohesive soils to be carried out on samples containing no stones larger than 5 mm. (This is the size which is large enough to be picked out by hand.)

On the M6 in Cumbria a large compaction mould of 0.0423 m^3 capacity, employing an 84.5 kg dropping hammer, was used for comparing the densities of less than 75 mm material with the same material after varying proportions of the stones between 75 mm and 19 mm had been removed. It was concluded that for up to 40% of stone between 75 mm and 19 mm there was only a slight increase in bulk dry density (24 kg/m^3). On one sample the moisture content of the passing 5 mm material was 12% compared with 9.2% for that passing 19 mm and this is typical of this boulder clay.

DR P. SMART, *Glasgow University*
The following have been useful when predicting moisture holding characteristics (e.g. optimum moisture content and liquid limit) from composition.

(a) Allow for interaction, e.g.
$m = m_c c^2 + m_m 2cs + m_s s^2$
for a two-phase soil comprising a fraction c of clay and a fraction s of sand, where m is the moisture content and m_c, m_m and m_s are constants. This extends at once to multi-phase soils.
(b) Use volumetric water content
$\theta = \frac{\text{vol. water}}{\text{vol. solids}}$
(c) Separate samples into families according to both geology and grading (and possibly other factors) using separate analyses for each family.

MR G. ROCKE, *Babtie, Shaw and Morton*
Kielder Dam, situated in the North Tyne valley, Northumberland, is 140 m above sea level in an area of relatively high rainfall and humidity. The valley is covered with glacial drift, most of which is glacial till, often described as boulder clay. The till lies in convenient depths and quantities near the dam to provide sources of fill material for the earth embankment which is 52 m high and requires 4 x 10^6 m^3 fill of all types. In general the glacial till is highly weathered for a depth of 2-3 m from ground surface level and below this it is very stiff. Although large boulders occur infrequently, the till contains numerous medium-sized boulders and cobbles which were recognized at the site investigation stage as a problem to specification, construction and material testing. In the case of the specification for the dam it was the intention to provide a framework within which the Contractor could achieve flexibility in his choice of plant and an end product meeting the design requirements. The specification called for compaction to be to a high standard based on the results of undrained strengths and air contents obtained from trial compaction strips on site. From time to time, in the light of test results from actual construction operations, target undrained shear strength values were to be assessed by the Engineer in collaboration with the Contractor. However, at no time was material in the core of the dam to have a strength of less than 60 kN/m^2 or greater than 140 kN/m^2 or 80% of the samples from a given area to have a strength of less than 70 kN/m^2 or greater than 130 kN/m^2. In the case of the shoulders of the dam the material was required to have a strength greater than 100 kN/m^2 and an air content of no greater than 5%. The specification included as a general guide to the Contractor a table relating plant type, layer thickness and number of passes to satisfy these requirements, although it was made clear that the table did not prejudge the results of the site trial compaction strips.

The method of measurement of the undrained strengths was not specified because it was realized that conventional site laboratory methods would not prove entirely satisfactory and in situ methods would have to be developed.

Earth-filling operations in operation during 1976 and 1977 were linked largely with the infilling of limited areas such as the core trench excavation and areas adjacent to the outlet culvert. 1979 will be the first season when large areas of dam will be available for earthmoving operations. To date three difficulties have arisen in the compaction of the glacial till. The first was expected and involved the removal of boulders by machine and by hand. An attempt was made to accommodate a larger proportion of these in thicker layers of compacted fill in the dam but new problems arose in controlling the thicker layers and meanwhile the compacted thickness has been reduced in the core. The second difficulty concerned adjusting the moisture content of the material on the dam before compaction to meet the strength specification for the core material. Different items of plant and methods of application were tried out during 1978, the most successful to date being a high-powered water spray-gun mounted on tanker lorries. Further trials are necessary to avoid the tendency for water to land unevenly. The third difficulty arising from the size of the plant required on such a large earth-moving operation, is how to break down substantial lumps of very stiff glacial till to enable the compaction plant to operate efficiently. It has been found that limiting the layer thickness is essential to control this problem and to enable heavy sheep's-foot machines to break the lumps down to manageable sizes.

When testing fill material, it was decided to measure the specified undrained shear strengths initially by means of 100 mm dia. samples taken from the compacted fill in a mainly random fashion supported by fewer tests where areas of poorly compacted fill were suspected. The samples were then confined in a triaxial machine and sheared under an all round pressure of 100 kN/m^2. This control method has proved inefficient and expensive in man-hours. As many as six samples are attempted before one is found suitable for testing or can be patched well enough for testing. Stones and such like ruin many samples taken from the fill after a great expenditure of effort. The concept of in situ testing has been and still is the main objective at Kielder Dam. In 1978 preliminary designs and trials of a trailer equipped to carry out plate bearing tests were made. The initial results have been promising and the method will be used in 1979 on a greater scale. Co-operation between the Engineer and the Contractor lies at the root of all such efforts and this as much as the physical product is crucial. Trials have also been made of a large-diameter in situ density test and which will be developed further in 1979.

In the case of moisture content measurements on large samples in the site laboratory speed, accuracy, economy and reliability have been obtained using infra-red lamps suspended above material spread out in shallow trays.

Earth-filling operations at Kielder Dam are still at an early stage, but their scale, the climatic conditions and the bouldery and stiff nature of the materials have presented problems, some well known but other a product of modern plant and stringent economic conditions.

DR A.D.M. PENMAN, *Building Research Station*
Years ago clay fill for dam cores had to be placed wet because the compacting machinery was the heels of many men's feet. The low energy required a low shear strength if air was to be driven out. As labour became more expensive attempts were made to simulate the human heel by projections welded to the underside of a crane bucket. Unfortunately the soft clay adhered to the bucket and various types of pneumatic punner were tried without particular success.

So that machines could be used, much drier material was placed to form rolled clay cores. These were not regarded as being watertight as the puddled clay cores and were usually made much wider, both to reduce the hydraulic gradient across them and to give more room for placing machinery. The lower construction pore pressures were regarded as an advantage, particularly as there had been cases of shear failure during construction, and even drier cores were used in the interests of stability.

A few examples of damage attributed to hydraulic fracture have now drawn attention to the need for a soft, flexible core. Fortunately developments in machine design now allow their use on much softer fills. A two-engined scraper can work on fill with c_u = 25 kN/m^2 and a towed scraper can work when c_u = 35 kN/m^2. This means that the low strengths in the range 50-100 kN/m^2 described by, for example, Kennard et al. (pp 143-147) can be used without difficulty.

An example of the flexible behaviour of a soft clay in a modern 70 m high dam is given by the movements measured during core placement. The core was being placed at a water content several per cent above Proctor optimum: it had c_u values in the range 35-100 kN/m^2. A 2 m long strain gauge was placed horizontally across the centre line of the core as part of the instrumentation of the dam. When a further 1 m of fill had been placed over the gauge it was found to have extended 97.5 mm; its range was only 150 mm. At that time the surface of the downstream filter was nearly 2 m low (Fig.20).

In order to reduce the 97.5 mm extension of the gauge, a D4 bulldozer was driven along the core just upstream of the gauge. After one pass the extension was 95.8 mm, after two passes it was 94.9 mm and after three passes 93.1 mm. This indicated that a great number of passes would be required to return the gauge to its installed condition. During the next hour its extension increased to 98.0 mm, so it was dug out and replaced further along the core where the upstream fill was higher.

MR G. PAVLAKIS, *Public Power Corporation, Greece*
In summer 1978 laminations in the clay core were encountered during the construction of the Pournari Dam in Greece.

When the fill in the clay core was about 10 m high it was discovered by accident that when the material was excavated by means of a bulldozer or a backhoe one could find smooth horizontal surfaces having the appearance of a slickenside at almost every boundary between layers (Fig.21) and sometimes many such surfaces within one layer. The area over which these laminations were continuous varied, the maximum observed being some 10 m square. By means of undisturbed sampling they were found to extend all the way down to the bedrock foundation. Visual examination of the walls of test trenches had not revealed this phenomenon.

The clay used was from alluvial deposits having a CL classification with a PI of about 17%. The optimum water content (the average for May and June 1978) was 17%. The average fill water content was very close to optimum. The layers were about 20 cm thick before compaction, which was effected by means of a twin drum self-propelled tamping foot roller. The cause of the laminations within layers was attributed to bearing capacity failure due to heavy traffic. The cyclic loading of the soil - because almost

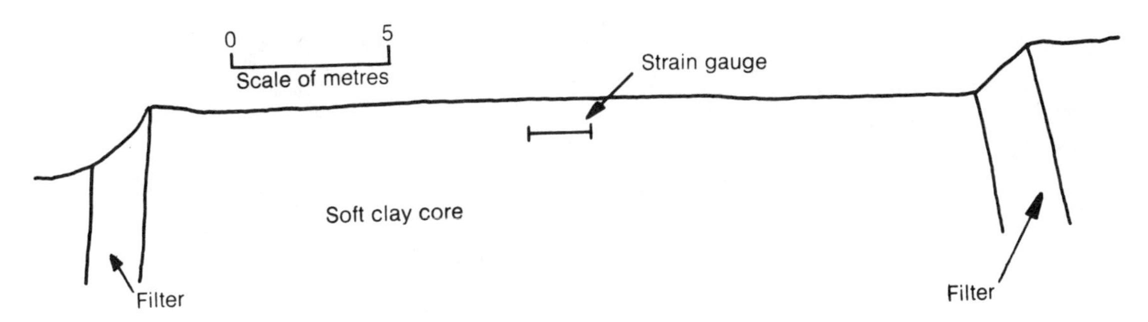

Fig. 20. Position of strain gauge in clay core

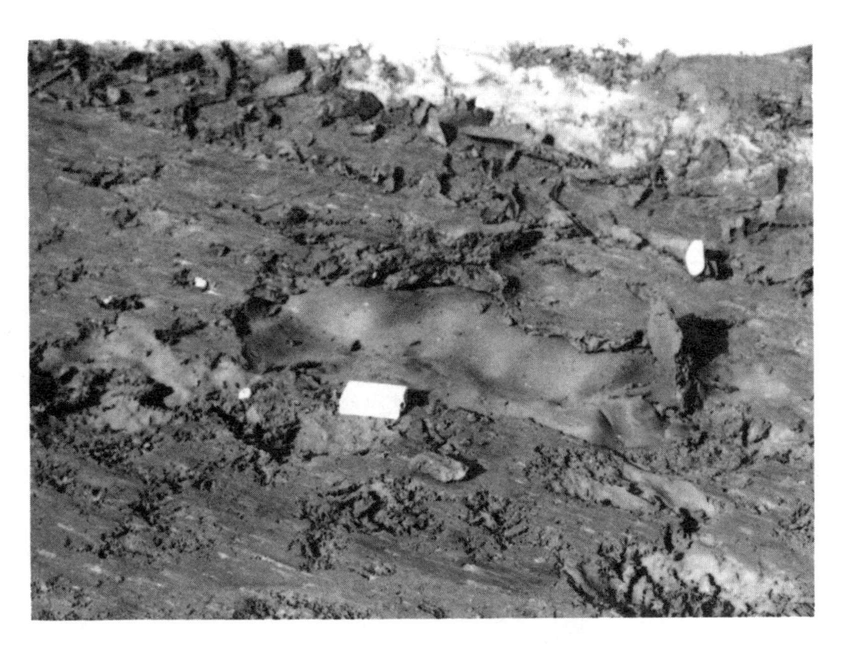

Fig. 21. Smooth surface between layers of fill

Fig. 22. Feet of compactor

Fig. 23. Scarifying attachment to bulldozer

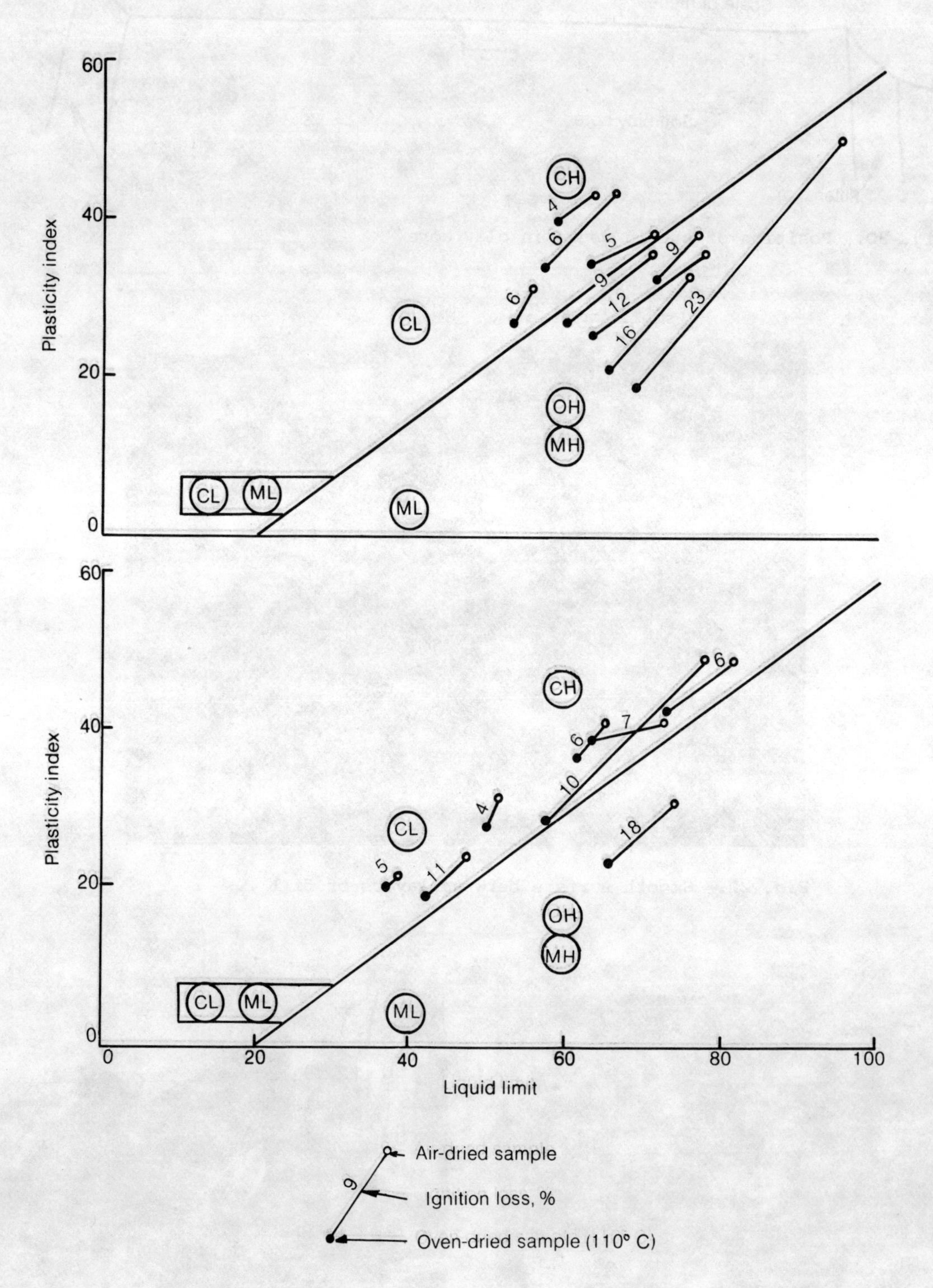

Fig. 24. Effect of loss on ignition of Atterberg limits of clay samples containing lignite

without exception the trucks followed the same path - was also a contributing factor.

The laminations between layers could be accounted for by any or a combination of the following mechanisms

(a) high water content gradient at the boundary between layers because of water sprinkling before placement of a new layer when the top of the compacted layer was dried out
(b) the type of feet of the compactor (they were not long enough and had a relatively large area, Fig.22) in combination with clogging
(c) for water contents on the wet side of the optimum the compaction process produces smooth areas separated by small walls of soil; when the new layer is being compacted the lateral pressure may shear off these walls resulting in a continuous wavy but smooth surface
(d) acceleration or deceleration of heavy equipment may shear the underlying soil.

The solution applied to the problem was double scarification by a scarifying attachment, consisting of two staggered rows of teeth, at the back of a bulldozer (Fig.23). The first scarification was applied after compaction to remove the inprints of the tamping feet as well as any superficial shear surfaces. The second scarification was applied after placement of the new layer to remove the ruts of the clay hauling equipment as well as any shear surfaces caused by the heavy traffic. Some additional measures were to disperse the routes of the clay hauling equipment and to build a temporary road on the clay, using gravel, for the upstream-downstream communication.

After these measures the laminations were drastically reduced and the density of the clay did not appear to have been affected.

MR PARSONS
The horizontal polished surfaces in that clay fill could be associated with the phenomenon described by Chaplin (p.241) where a surface layer of soil is transported bodily along the surface of the fill ahead of a heavy roller.

The clogging of tamping and sheep's-foot rollers has also been observed in the UK. During compaction tests at the Transport and Road Research Laboratory homogeneous soils have been compacted over a wide range of moisture contents and such clogging has not been experienced, but on sites where more heterogeneous soil conditions are encountered - in particular, mixtures of wet and dry soil - clogging appears to be a major problem.

MR F.D.T. CHARTRES, *Rofe, Kennard and Lapworth*
A similar phenomenon was observed in the construction of the Ardingly Dam in Sussex during the extremely high temperatures of summer 1976. When the sheep's-foot roller was operating entirely within the clay of the abutment cut-off, the roller remained clear. While clay was being placed on the core by tipping from articulated dump trucks, the compactor moved onto the dusty surface of the adjacent shoulder and within a short time the sheep's-foot roller became a smooth roller and was then of little use in core compaction. With a shear strength specification, it was possible to dispense with the sheep's-foot compactor and use a D6 bulldozer. The articulated dump trucks moving along the length of the core gave a practical macro-test of the strength of the clay.

DR D.G. COUMOULOS, *Castor Ltd, Athens*
I would suggest that criteria have to be established for the acceptance or rejection of organic clays or clays containing coal for their use in various types of fill. These criteria may, for example, be

(a) the maximum permissible amount of organic content or coal content - in connection with this criterion a simple method for determining organic and/or coal content has to be accepted and become the standard method of testing (e.g. the loss on ignition method, which can be applied easily in a field laboratory); Fig.24 shows the effect of loss on ignition on the Atterberg limits determined on air-dried and oven-dried samples
(b) a certain value of maximum dry unit weight as obtained by a standard compaction method, below which materials will have to be rejected - as can be seen from Fig.25 the organic clay OH yielded a low maximum dry unit weight and a compaction curve (curve C) which is considerably flatter than the compaction curves A and B of the clays CL and CH respectively; naturally, a flat compaction curve means that material cannot be compacted.

In paragraph 4 of the paper by Koryalos and myself (p.73) it is explained that the quality control method applied during placement and compaction of the clay core was aiming at spotting and correcting doubtful material. When there were no doubtful areas, sampling aimed at verifying acceptability. This type of quality control needs experienced field personnel. With this method no defects are accepted. The Contractor has to meet the specification requirements over the whole area where compaction operations take place. This type of sampling is called 'purposive'. It has the advantages that fewer tests are needed and fewer personnel are needed to man the field laboratory. Consequently, field quality control becomes considerably less expensive when compared with the extensive testing which is required for the random sampling procedures.

MR G.H. CHILD, *Soil Mechanics Ltd*
I speak from the point of view of a site investigator and hence deal with prediction rather than construction control. If shear strength is the criterion to apply to determine suitability of a clay for fill, the strength measured should take into account the soil structure. Therefore, I consider that vane tests or 38 mm dia. compression tests are too small and may

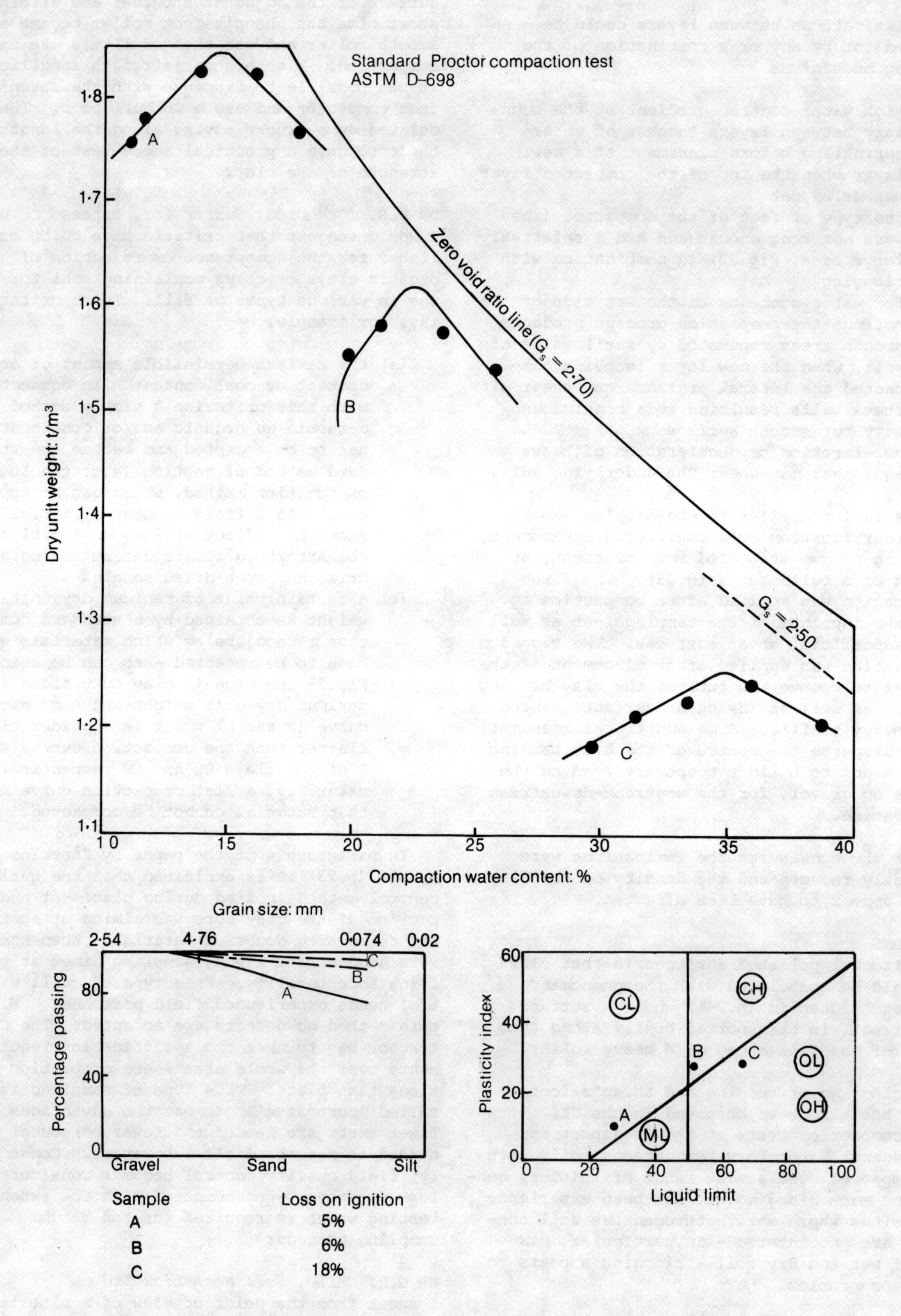

Fig. 25. Compaction characteristics of clay samples containing lignite

give a considerable scatter of results. This problem has long been recognized in in situ structured clays, e.g. London Clay, where the intact strength may be as much as five times the mass strength. If shear strength is the right criterion, larger samples (e.g. 100 mm in diameter) should be used to obtain more representative values (see Kennard et al., pp 143-147).

However, if some function of the plastic limit is to be the criterion and the present plastic limit test is considered unsatisfactory, why not replace the present test with a core test on the lines of the liquid limit core test?

MR D.W. COX, *Polytechnic of Central London*
A procedure which I have employed, together with F. Graham and Partners, at a number of sites which were backfilled and then built on, has been to use rapid plate bearing tests conducted with a heavy hydraulic excavator as kentledge. The procedure has been to excavate backfilled sites and recompact in a sequence similar to strip mining. The fill is thoroughly mixed into a uniform mass by the excavation process. The uniformity and shear strength are tested by frequent rapid plate tests. Over 1000 such tests have been performed on various sites at an average cost similar to that of a laboratory triaxial test. The completed fill is tested by plate tests at footing level and below, with plates extended down a test pit using stools.

The settlement characteristics are determined by a 1 m square reinforced concrete base with a rubbish skip above as kentledge. These take between a few days and a month to settle depending on fill type.

Many of the problems of testing fill relate to the large size of a representative sample. At the Polytechnic of Central London there is a 1 m cube container which can be pressurized by inflatable pressure bags on all six sides. There are access ducts through the bags to the fill, for instrumentation. The apparatus is available at agreed times to anyone wishing to compress large samples or calibrate pressure cells in a particular soil. A 2 m cube is under construction. The apparatus can be transported.

DR P.R. VAUGHAN, *Imperial College of Science and Technology*
The control of fill placing by the use of undrained strength measurement is advocated by Kennard et al. (pp 143-147). There is little doubt that this method offers advantages over that of controlling water content by relating it to various index properties. The specific need to determine water content is avoided and, while the undrained strength is not a unique parameter and it depends on method and rate of testing, it does relate directly to short-term stability and to trafficking problems.

The method is of particular value for stony, sandy clay fills, where remoulded test specimens with the larger stones removed can be used, as discussed by Kennard et al. for the Cow Green Dam. Corrections to the water content for stone content are avoided.

It is more common to use undisturbed test specimens taken directly from the fill. These can be used also for density and air void determinations. Kennard et al. refer to the large scatter in results obtained with the method, particularly where water is added to the fill in embankment dam cores. This method of shear strength control was used for the Empingham Dam, referred to by Kennard et al., in which brecciated Upper Lias clay was used for both general fill and for the core. This clay is very heterogeneous, with a wide local variation in water content.

Figure 26(a) shows the distribution of water content in the general fill (line A) as observed in the control tests, and the distribution in the core (line B). Distributions of the water content in the borrow pit for the general fill (line C) and for the core (line D) are also shown. Lines C and D involve some estimation, as it is not known precisely what parts of the borrow pit were used in which parts of the dam, but the water contents cannot be in error by more than 0.5%. The distributions of the undrained strengths measured in the control tests on the two types of fill (using quick triaxial tests on undisturbed 100 mm dia. samples with a confining stress of 250 kPa) are shown in Fig.26(b).

There is no evidence that the general fill changed in average water content during placing (comparing lines A and C). Placing was done from September to November 1972, April to August 1973 (until September for the core) and from April to June 1974. However, about 10% of the fill was placed wetter than the clay in the borrow pit by up to 2%. The water content of the core was increased by nearly 3% - by disc harrowing, rotavating and watering on the dam before compaction. During this the scatter in water content was increased considerably (compare lines B and D).

The core was specified originally to have undrained strengths of 60-90 kPa. Only 45% of the test results fell within this range, although the average strength achieved (74.9 kPa) was exactly in the middle of the specification. Even had there been no increase in the spread of water content due to processing, only about 60% of the results would have fallen within the specified range. This shows the importance of specifying realistic strength limits related to borrow pit conditions, and of using average results, as stressed by Kennard et al. The relationship between in situ strength in the borrow pit (as measured by quick undrained triaxial tests on 100 mm dia. undisturbed samples obtained using standard open-drive sample tubes) and the strength of the fill as placed is shown in Fig.26(b). The strengths were increased by about 60% by placement.

It is probable that a reduction in scatter of strengths measured would be achieved if larger samples were tested. This is not practical using laboratory equipment, but might be achieved by using 100 mm dia. or 150 mm dia. plate tests, which would also increase the

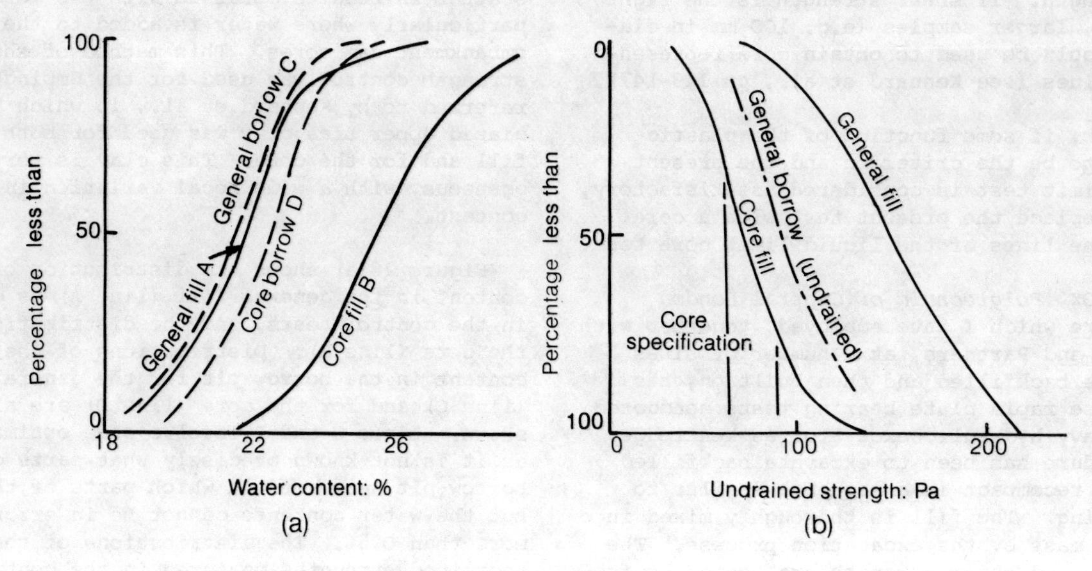

Fig. 26. Empingham Dam: control tests on fill

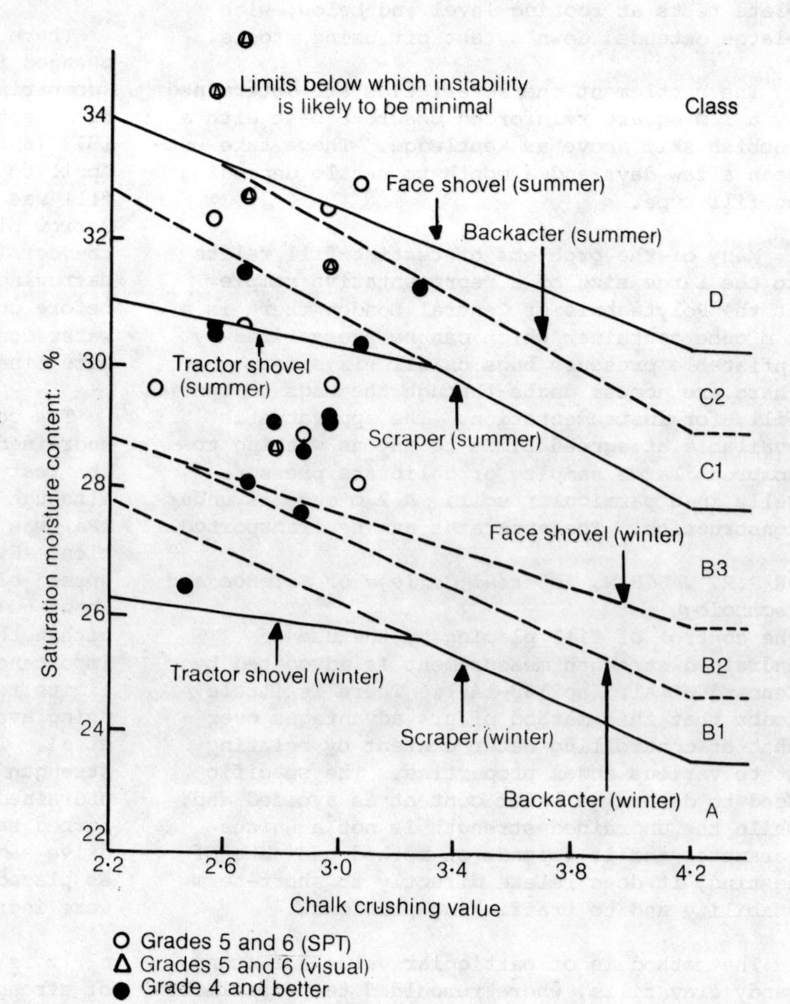

Fig. 27. Relationship between saturation moisture content and chalk crushing value obtained from tests on Upper Chalk, subdivided on the basis of grade of chalk

speed of testing.

The relationship between undrained strength measured in the quick undrained triaxial test and the strength relevant to short-term stability and to trafficking is discussed on pages 283-295. It is sufficient to note here that when loading rates are sufficiently slow for local pore pressure redistribution to occur, then the strength is likely to be significantly lower and the variation in strength is likely to be much less.

DR L.M LAKE, *Mott, Hay and Anderson*
With regard to index property data for general correlation and earthworks prediction use, there is no doubt that reliability is difficult with low and marginally plastic soils, but as the plasticity increases, extremely good correlations are possible, especially using the liquid limit. Plotting water content against undrained shear strength on a semi-log plot, I have obtained good linear correlations for soils with given liquid limits.

Regarding the use of a multiple x plastic limit for earthworks prediction and control purposes, I have always regarded the appropriate figure not as a precise correlation factor but rather as an indicator of the likely lower bound for general earthworks suitability and Fig.16 illustrates this interpretation. The 59 results showed strengths of up to 85 kN/m^2 but only two or three were below 30 kN/m^2 - the strength commonly regarded as a lower bound for trafficability by general construction plant; this also corresponded with the commonly employed factor of 1.2 x PL.

I agree that such relationships become more problematic with leaner and semi-plastic clays, especially when partially saturated.

My firm has recently carried out a dewatering, excavation and compaction trial in Essex, involving Claygate and Bagshot Beds material comprising very loamy clays and clayey fine sands and silts. Observed trafficability with rubber-tyred and tracked plant and a smooth-wheeled tandem roller was found to be accurately predicted by the moisture condition test results obtained by staff of the Transport and Road Research Laboratory. Trafficable limits lower than 1.0 x PL were found to apply. Dewatering before excavation and stockpiling for several weeks appeared to have no beneficial effect on performance. Lime treatment was effective only with the more clayey material.

Concerning more contractual matters, is it reasonable to expect contractors to stock only plant that can accommodate the worst possible conditions and materials? Such plant will generally be among the most expensive to buy and and run and there are many contributory factors to plant selection for particular projects, all of which have to be considered. Is perhaps the problem under discussion self-generated as a result of imposed impracticable specification requirements?

MR E.H. STEGER, *Pell Frischmann and Partners*
It may be considered that some of the compaction and field testing techniques mentioned were mooted many years ago. The frequent mention of shear strength test, with or without moisture content relation, reminds me of Golder's proposed pavement design method[12] which was based on the shear strength of the formation material.

It is known, from the point of view of bearing capacity and settlement, how bored piles behave but it is not known why they behave so; engineers still argue about the values for shaft friction, settlement calculations, effective stress analyses and so on. However, piles can be placed successfully and failures, if any, are the result of poor workmanship.

In embankments or clay cores, the material can be placed and the machinery, including Irishmen's heels, exists to achieve a certain amount of compaction. Field control and prediction of behaviour are not so well established or reliable. It should be realized, therefore, that the stage has been reached in this branch of engineering where construction techniques have, generally, outpaced theoretical and design knowledge. One reason for this is pure economic pressure, which resulted in the manufacture of heavy equipment designed for specific tasks. The other is, to my mind, that natural deposits are involved which cannot, and never will, be made to fit simply into one or two theories and then into a computer program.

MR R.S. DENGATE, *Cementation Construction Ltd*
There is unfortunately no precise answer as to which type of plant is the best to use in motorway construction for excavating, hauling and placing clay. In addition to the properties of the clay, one has to consider

(a) the topography of the site and the haul roads
(b) the length of haul
(c) any obstruction along the haul road
(d) the location of the project, which affects the cost of road rental and the cost of tips and borrow areas
(e) the effect on the programme of hauling through construction areas, which perhaps influences decisions to tip in one location and to borrow in others
(f) the balance of plant needed to cater for local areas that do not conform to the general pattern
(g) the availability of plant, often during a very short period, together with all the necessary back-up plant and maintenance facilities.

MR QUINION
The selection of plant is based essentially on the experience of field staff who operate this equipment under different conditions and on a variety of sites. They are not necessarily able to identify the most suitable plant. The selection must take into account rates of excavation, ability to compact and the length and nature of the haul route. There are limited earth-moving fleets of different types available. As the

weather and/or the soils change, the performance of the plant is affected. Ideally the Engineer and the Contractor should get together before work starts and identify the options available, combining areas of excavation or sources of borrow and the areas of fill or tip. These can be priced in an attempt to predict the economic options available as the job proceeds.

The Contractor welcomes as complete information as possible, which he reflects in his mass haul diagrams and tender planning. He does not always have soil data at the depths or locations he most desires. He must particularly predict the period of the year in which the work will begin, and the likely weather based on historical records.

Haul roads and distances are particularly important and influence unit costs. Cost records are not easy to achieve in a useful form, much as contractors would like to have them. Earth-moving plant is very expensive and the capital invested in the construction of a major contract is an appreciable percentage of the contract sum. Utilization and maintenance costs are significant factors which the Contractor must take into account.

MR M.M.H. AL-SHAIKH-ALI, *North Western Road Construction Unit*
In addition to the factors already discussed it is pertinent to mention that effective management and technically experienced personnel from the Contractor's side have a great deal to contribute, in the way of handling clay fills and the like, and in satisfactorily completing the work in the prescribed time.

In my own experience there have been several occasions when the Contractor has dispensed with the appointment of a materials engineer on a major motorway contract. For instance, two contractors started construction of twin motorway contracts at the same time, in the same locality, subjected to the same weather and ground conditions and having similar in situ materials. The experienced contractor had an experienced soils engineer on site and the other one had not. The experienced contractor completed the work in time and the other one was bogged down for several months. As a result, substantial amounts of money were lost, in addition to effor effort and valuable time of both the Engineer and the Contractor being wasted. The inexperienced contractor put the blame on the materials, the weather and the specifications.

I firmly believe that the inclusion of technically experienced personnel on the Contractor's side at all levels, and effective management at the top, has paramount importance in handling clay fills and the like and satisfactorily achieving a good standard of performance. The attitude of contractors to the appointment of an experienced materials engineer on major contracts should be positive. This appointment, the appointment of technically experienced personnel, and a requirement for experienced and able subcontractors, should be written into the specifications for all major contracts.

These views are personal and do not necessarily conform to those of the Department of Transport.

DR JONES
Although there is a bias towards deterioration of the muck between ground investigation and construction, improvements are by no means unknown. However, even these can lead to claims, e.g. for loss of profit on imported fill. The point is that a reliable estimate is required for average weather conditions and also for the likely extreme conditions.

With regard to the repeatability of the plastic limit test, detailed revisions were made to the test as specified in BS 1377 in 1975.[13] It would be interesting to know the improvement to the repeatability which followed from this revision.

The factor x PL approach appears wrong in principle. If a criterion based on limits is to be retained then it should be in terms of consistency index, perhaps redefined entirely in terms of the cone penetrometer test.[14] However, at this stage the MCV approach appears more versatile and deserving of the major share of research and development effort.

The importance of the earth-mover's experience in assessing the likely behaviour of the muck has been emphasized. Without detracting from the value of this approach, it should be noted how often the final judgement accords with that based on a scientific assessment of the data. Furthermore, the scientific assessment can be made at an earlier stage.

MR J.P. DENNEHY, *Norwest Holst Soil Engineering Ltd*
Despite the presentation of data which shows the wide inaccuracies involved in relating w_p to undrained shear strength (pp 25-36, 87-94, 169-175, 260), there appears to be some reluctance to dispense with the parameter as a control. Additionally proposed controls such as the consistency index (e.g. p.253) involve the use of w_p. However, it has been proved[15] that wherever w_p is involved in a correlation with shear strength no accurate relationship will result, and this applies to soils covering the main spectrum of British clays as shown in table 2 on page 90. A further factor affecting the relationships of w_p with shear strength is saturation. At the limits of soil suitability clays placed in embankment will be near saturation. It would appear incongruous to use w_p to relate to the strength of these soils because, by the nature of the test, the plastic limit is the moisture content of an unsaturated soil. (The shrinkage limit test may give the theoretical saturation limit.) However, by definition in the w_L (liquid limit) test the soil is saturated. In terms of assessing soils on the basis of wetness index (Dohaney and Forde, pp 95-100) the lack of definition of an optimum moisture content (see, e.g., Kennard et al, p.143) would counteract the other accurately related components of w and w_L

in its formula.

A further problem with the use of correlation indices such as liquidity index, consistency index, wetness index and MCV is that their reduced scale renders them insensitive to small changes in the moisture content component and this must inhibit their use as prediction parameters.

It is gratifying to see that the predominant influence of w_L on strength is largely confirmed by Black (p.260). It may therefore be possible to use fig.7 on p.91 instead of fig.1 of Black and Lister (p.38) as a more accurate adjunct to the suction index method. It is also interesting that the low remoulded shear strengths of sensitive soils quoted by Kuno et al. (p.155) can be predicted using an extrapolation of fig.7 on p.91 and the data in table 1 on p.150, values of 10-20 kN/m^2 being predicted compared with measured strengths after compaction of about 20 kN/m^2, the latter value possibly being higher due to restricted remoulding. From the data of Kuno et al. a further line could be added to fig.6 on p.91, below the line for montmorillonite, for allophane and halloysite.

With regard to the moisture condition test, the relationship of MCV with strength in fig.7 of Parsons (p.172) is broadly based, and data presented by Edwards (pp 253-258) show that the spread of c_u values in different soils may increase with increasing MCV. This may partly be explained by the less sensitive nature of the MCV scale in clays, when compared with, say, moisture content. Particularly, at drier moisture contents a small change in moisture content which might be unnoticeable in terms of MCV can have a large effect on shear strength (see, e.g., fig.2 on p.91). At drier moisture contents there may also be a tendency for air to be trapped between the original 20 mm lumps because of the limited facilities for air escape in the apparatus. In terms of the MCV as a trafficability control, there may be some doubts as to its field modelling capacity because there is lateral restraint and the soil is forced to compress one-dimensionally, rather than remould, under the action of the snugly fitting rammer. In the field, at trafficking limits, there is a tendency towards slight lateral restraint, exuding, soil displacement and reworking.

Calibrations of remoulded undrained shear strength or MCV with moisture content for principal cutting soil types are essential for assessing quantities of suitable and unsuitable materials at the site investigation stage, as well as for use as field controls. However, such calibrations are indicative only of the general behaviour of the stratum and significant quantities of index testing, which reveal the depositional and lithological variations within a stratum, are considered essential. The effect of such variations in conjunction with moisture content changes may be assessed, for example, by fig.7 on p.91. The problems of intermixing at the construction stage can also be assessed by index testing mixtures of various proportions of adjacent strata, whereas modelling by shear strength or MCV testing may be more difficult. The critical intermixing ratio of the soils can be determined with reference to the test results obtained (e.g. fig.4 on p.19, and fig.7 on p.35 in conjunction with fig.7 on p.91). (Judgement is required as to the mix ratio most likely to pertain on site.) Intermixing produces a greater numerical drop in w_L than it does in w_p and this, in conjunction with the higher overall moisture content that results, may explain why two suitable soils when mixed become unsuitable.

Apart from the problems of climatic variations and inclusions, differences between construction properties and site investigation data may result from delayed testing at the investigation stage. This can be overcome on the larger contracts by duplicating all disturbed sampling, and testing one set of samples immediately, either on site or by dispatch to the laboratory. A greater intensity of moisture content testing is also advocated as much can be gained from it, at little cost, both in absolute values and in profiling strata. At the construction stage moisture contents and related indices may be an efficient control if a microwave oven is employed, especially in view of the observation by Webb (p.258) that MCV determinations on heavy clays may take an hour. In the latter case, some drying of the soil will occur during the present process of sample preparation.

MR J.B. BODEN, *Transport and Road Research Laboratory*

If the Engineer wishes to make the maximum economic use of on-site soils then, in certain cases, he may have to take construction processes into account in the formulation of his design and specification for the works. For example, a chalk which is suitable before excavation in the cut area can be rendered unsuitable by the time it is placed in the fill area as a result of the method of handling (see pp 137-142).

Consider a chalk with a natural moisture content of 26%, a saturation moisture content of 31% and a crushing value of 3.7. If the crushing of the compacted chalk fill can be kept below about 70% passing a 20 mm sieve size then instability is unlikely to occur (see fig.1 of Ingoldby, p.138). Excavation by face-shovel and transport by dump truck would probably result in a compacted fill with about 65% or less passing the 20 mm sieve size (see fig.6 of Ingoldby, p.140). However, if the chalk were excavated and transported by scraper then a grading with about 85% passing the 20 mm sieve size could be produced and instability in the fill area would be highly likely. Thus the same chalk could end up as a stable or unstable fill depending on the method used to handle it.

Ingoldby's classification is based on practical observations made at eleven earth-works sites over about one and a half earth-moving seasons. In many respects Ingoldby has merely placed the experience and expertise of, say, eleven plant foremen within a rational framework for use by others. As such, the classification should not be followed blindly, but it nevertheless forms a useful aid to judgement both for the Engineer and the Contractor.

DR LAKE
This reference to the selection of particular plant for chalk excavation is probably a special case due to the historical development of problems with this material in the UK. However, with rare exceptions, this situation should not be the generality because the Specification should define the objectives and the required result or product, giving the minimum of constraints and allowing the Contractor the maximum flexibility in execution. The distinction between these two approaches is very important.

MR THREADGOLD
The work by Ingoldby (pp 137-142) which relates saturation water content to chalk crushing value as a means of identifying the types of plant and restrictions on winter working should provide a useful guide at the site investigation stage for road projects and should allow more economical programming and design of construction works.

Figure 27 shows a plot derived from fig.8 of Ingoldby (p.140). In this plot the chalk tested, which is from the Upper Chalk formation, has also been identified by use of the scale of grading which was used for the Mundford site using both visual description and the correlation of Wakeling[16] which uses SPT values.

The higher quality chalk, classified as grade 4 and better, forms a population of points generally within class B; grades 5 and 6 are more frequently in the more sensitive classes C and D. However, there is considerable overlap. This suggests that difficulties could well be experienced if methods of excavation and their timing are related purely to grades of chalk, when the sensitivity to remoulding would appear to be the critical factor.

MR CHILD
Soil Mechanics Ltd, in conjunction with the Transport and Road Research Laboratory, have investigated two proposed chalk cuttings for a motorway using the chalk crushing value (CCV). The material is Upper Chalk, above the water table and saturated with porosities of 36-48%, equivalent to grades 4-6 on the Mundford scale. The CCVs generally lie between 3.5 and 4.3 and cover zones A-D of the classification for chalk fill (see fig.8 of Ingoldby, p.140). The results are shown in Fig.28, from which a mean saturation moisture content (and porosity) against CCV relationship was assessed. As this is relatively insensitive to CCV it was possible to include in the assessment for suitability a large number of porosity tests. The conclusions are that for what on a log may appear to be, say, all grade 4 in situ chalk there may be types A-D chalk for fill and thus the two classifications are not closely related. Results from grades 5 and 6 do not necessarily indicate worse construction problems but may not be representative if these grades contain a significant matrix.

In fig.7 of Ingoldby (p.140) the placement moisture content has been assessed as on average 0.85 times the saturation moisture content although most of the samples used in the assessment were for chalks with a saturation moisture content of less than about 20% where construction should not be a problem.

MR PARSONS
Although a wide range of values of saturation moisture content was encountered during these studies, in the range above 20% only three points out of 17 lie above the line representing a slope of 0.85. Thus the results indicate that in the critical area (i.e. where chalk has a saturation moisture content in excess of about 23%) there is a fairly low probability of the fill moisture content exceeding 0.85 of the saturation moisture content.

MR BODEN
Ingoldby's classification for chalk (pp 137-142) is intended for use in the prediction of the behaviour of chalk as a freshly placed fill material in relation to the method of excavation, transport, placement and compaction. The Mundford classification[17] is intended primarily for use in the prediction of the behaviour of chalk as an in situ foundation material. The systems are based on the measurement of different parameters and therefore I would not necessarily expect to find correlation between them. Further, I would not be surprised to find that a chalk which falls within a single grade on the Mundford scale could belong to more than one class in Ingoldby's classification.

MR H. GRACE, *Scott Wilson Kirkpatrick and Partners*
In the early years of the motorway programme the upper moisture content for wet fill was fixed at the plastic limit + 2%, but it was doubtful if anyone at that time realized the financial implications of this requirement. It resulted in large quantities of material, which were perfectly satisfactory in all respects other than their moisture content, being tipped to spoil and drier material being hauled in from nearby borrow pits. Millions of pounds were spent in this way. An instrumented trial embankment on the M6 in Westmorland (see pp 113-118) demonstrated beyond doubt that fill appreciably wetter than that normally permitted by specifications then current could be used if drainage layers and settlement periods of up to three months with a suitable surcharge were provided. Appreciable volumes of wet fill were used in selected locations on this section of the M6. Looking back, it is difficult to understand why this method of construction was not used sooner. The principles governing consolidation and settlement were understood by Terzaghi[18] in 1925 and since the Second World War dam engineers had used drainage layers in embankment dams to dissipate pore pressures and so increase shear strength. Fig.1 on page 117 was prepared from a theoretical study of the problem and shows the relationship between permeability, spacing of drainage layers and the time in days required for 90% consolidation. Providing drainage layers are provided at 2 m spacing and a settlement period of 100 days can be allowed, then all soils having a permeability greater than about 10^{-8} cm/s can be used in this manner. This

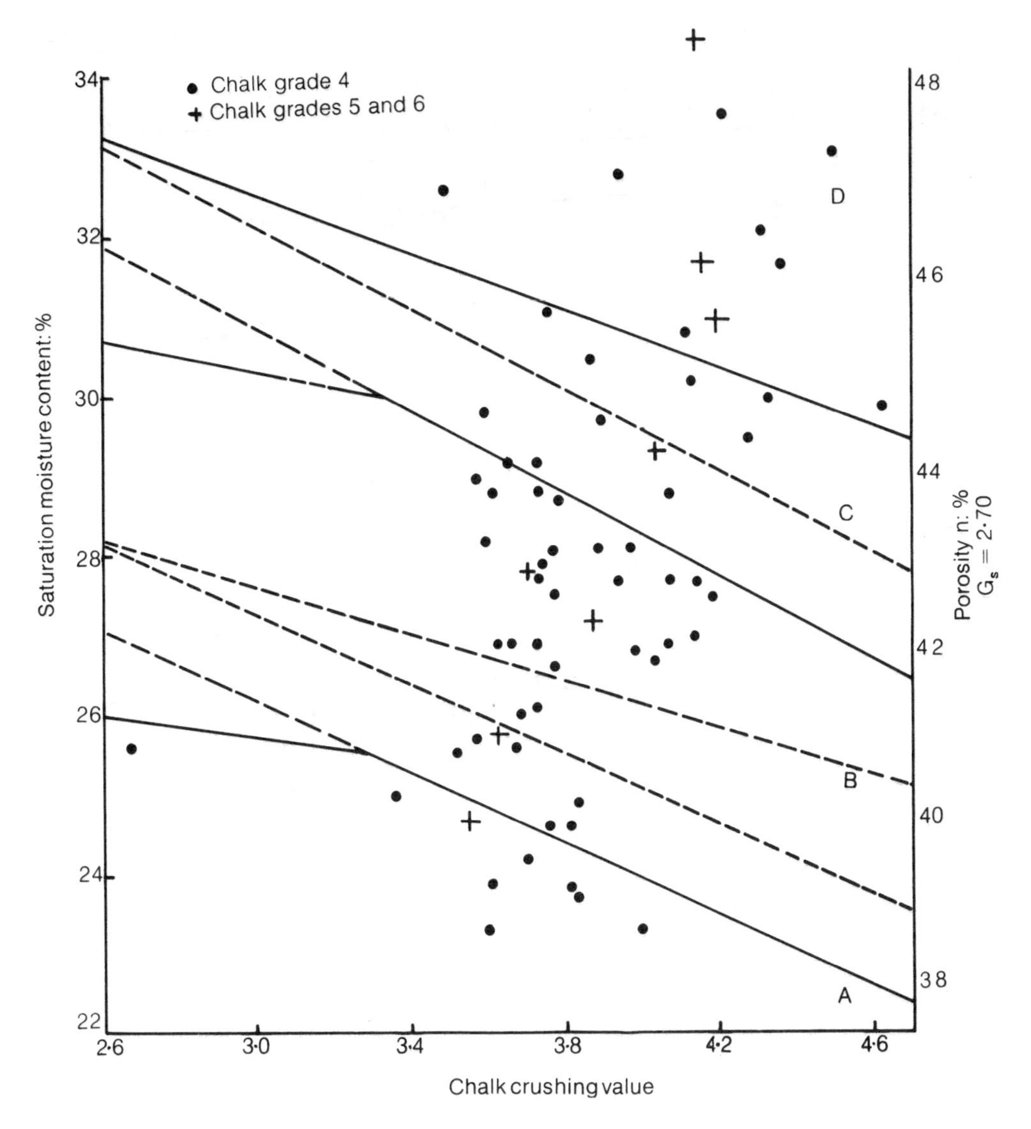

Fig. 28. Chalk crushing value compared with chalk grade

Fig. 29

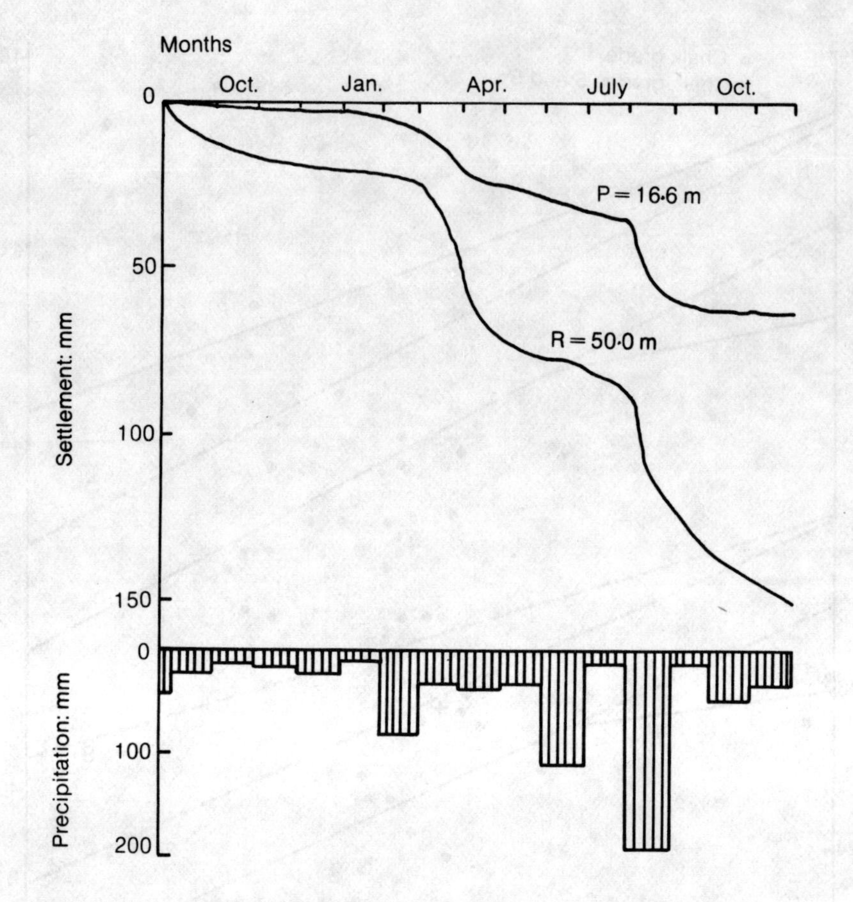

Fig. 30. Influence of infiltration on settlement

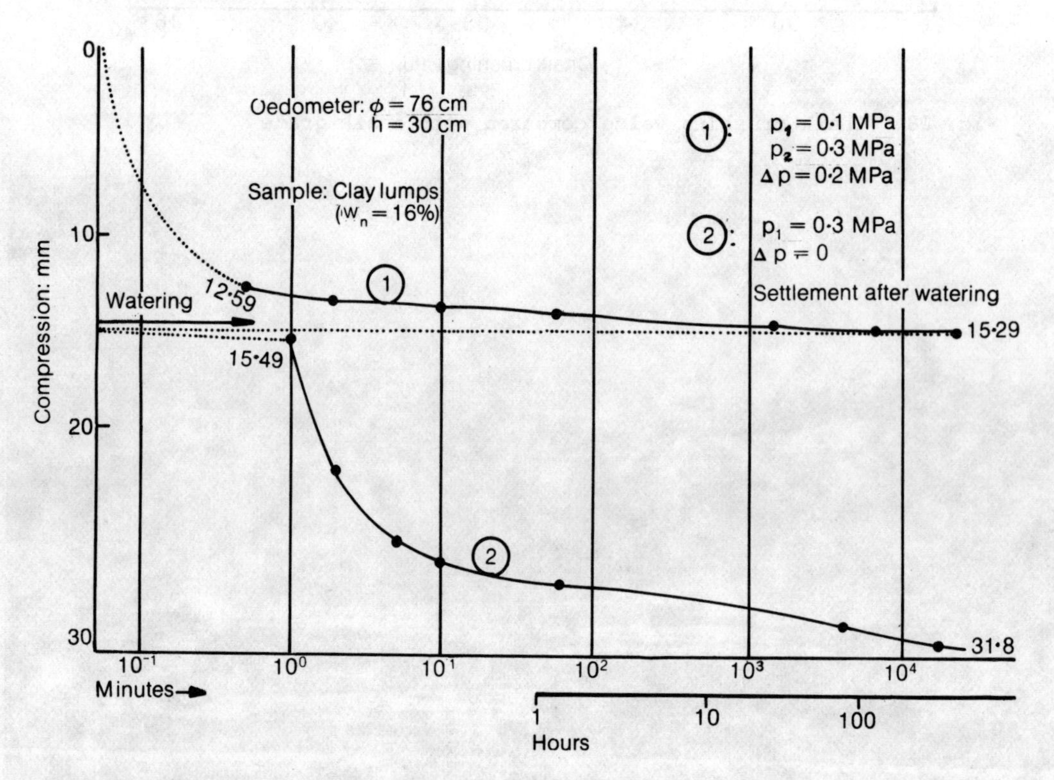

Fig. 31. Time-consolidation curves

means that nearly all British soils would function satisfactorily if used in this manner.

MR S. VARAKSIN, *Techniques Louis Menard*
Dynamic consolidation has been used on three sections of expressway in France. The treatment has restored the in situ soil parameters of the clay from borrow pits when used for embankments. On the A2 and A13, 95% of modified Proctor values were obtained in 3-10 m embankments where the material was placed in one continuous operation and then treated by dynamic consolidation. The lime stabilization used on other sections was not necessary. On the A4 expressway an old gravel quarry was backfilled with marl lumps. The depth was 6 m of which 5 m was under water. Shear and deformation characteristics of the undisturbed marl were obtained by two phases of dynamic consolidation.

MR V. BUREŠ, *Stavební geologie, Prague*
Dumped fill from clay stripping in an open-cast coal mine in Northern Bohemia will be used as an embankment for an express railway, a motorway and a river. This section of the fill is 5 km long and its height varies between 60 m and 130 m (Fig.29). The stripped material consists of lumps of tertiary clay up to 30 cm, of stiff to hard consistency. The average values of plasticity are PL = 25-30%, LL = 50-75% and natural moisture content w_n = 15-25%.

The material has been dumped in layers 30-50 m thick by overburden dumping machines. The age of the fill when the railway traffic starts in 1980 will be 1-15 years.

Using the results of field and laboratory geotechnical investigations from the completed sections of the fill, probable settlements of the railway embankment were calculated. According to the age and height of the fill the final total settlement was evaluated as 5-12 m. For the first year of traffic the expected value of settlement is 50-100 cm. In order to reduce the settlement and especially to eliminate sudden irregular slumps along the line, dynamic consolidation of the fill is proposed. For the same purpose, a sand/gravel cushioning several metres thick will be built on the crest of the fill.

The railway line will be doubled in both directions for the first ten years to enable realignment of the rails without interruption to traffic. Also the river will be transferred to steel pipelines to prevent leakage into the fill.

One of the important characteristics of the fill is its permeability and influence of percolation on settlement. According to field tests, the upper layers of the fill have a coefficient of permeability of 10^{-2} to 10^{-4} cm/s. (These values were measured to a depth of about 50 m.)

High permeability of the fill is also evident from long-term measurements of settlement which is related to precipitation in this area, as shown in Fig.30. Settlement greater than 50 mm follows immediately after rainfall or sudden melting of snow.

Rapid collapse of the soil structure caused by watering of a soil sample is shown in Fig.31, which presents two consolidation curves from the test performed in an oedometer 76 cm in diameter. The sample was prepared from stiff clay lumps. Curve 1 is for a consolidation load of 0.3 MPa (300 kN/m^2). Immediately after allowing water into the consolidated loaded sample, further compression appeared (curve 2).

A field experiment, based on the experience gained from oedometer tests, was carried out. The surface of the fill was watered for two weeks to imitate daily precipitation of 25-50 mm. This experiment was not successful as water did not penetrate the weathered or remoulded layer of the fill, so that surface heave instead of settlement was measured.

MR R.W. PEARCE, *Cooper MacDonald and Partners*
It is my contention that the success of dynamic consolidation treatment on clay fill sites depends to a great extent on the use of granular fill. First, there is the surface problem connected with tracking heavy plant on a clay site in wet weather conditions. In addition, as the weight hits the ground, clay lumps may be ejected 50-60 m in all directions. Various forms of skirt around the weight have been adopted with limited success. Thus a carpet of granular material at least 500 mm deep is necessary if law suits are to be avoided from bombarded neighbours.

Secondly, to achieve any significant compaction at depth the deep initial compaction points must be back-filled with granular material. Dynamic consolidation is carried out in a number of phases starting with deep widely spaced compaction points and gradually moving in subsequent phases to overlapping low energy compaction. Treatment is given first to the deepest zones of fill and then to higher layers (Fig.32).

The compaction of loose granular soil requires plastic deformation of the soil and an absorption of energy. Following the plastic deformations the soil becomes increasingly elastic. At a certain stage of compaction the top layer behaves almost purely elastically and the stresses are propagated to deeper layers where plastic deformation, compaction, can again take place. Hence the compacted zone can gradually be extended. However, cohesive soil retains its plastic properties even in a compacted state and therefore propagation of stresses is severely limited (Fig.33).

It seems likely that any improvement in the clay fill at depth is caused by lateral displacement of the clay as the granular fill is punched down. There will be severely remoulded zones around each compaction position leaving the clay between largely undisturbed. The sensitivity of the clay is therefore critical. The treatment is likely to extend only slightly beyond the depth to which the stone piers have penetrated.

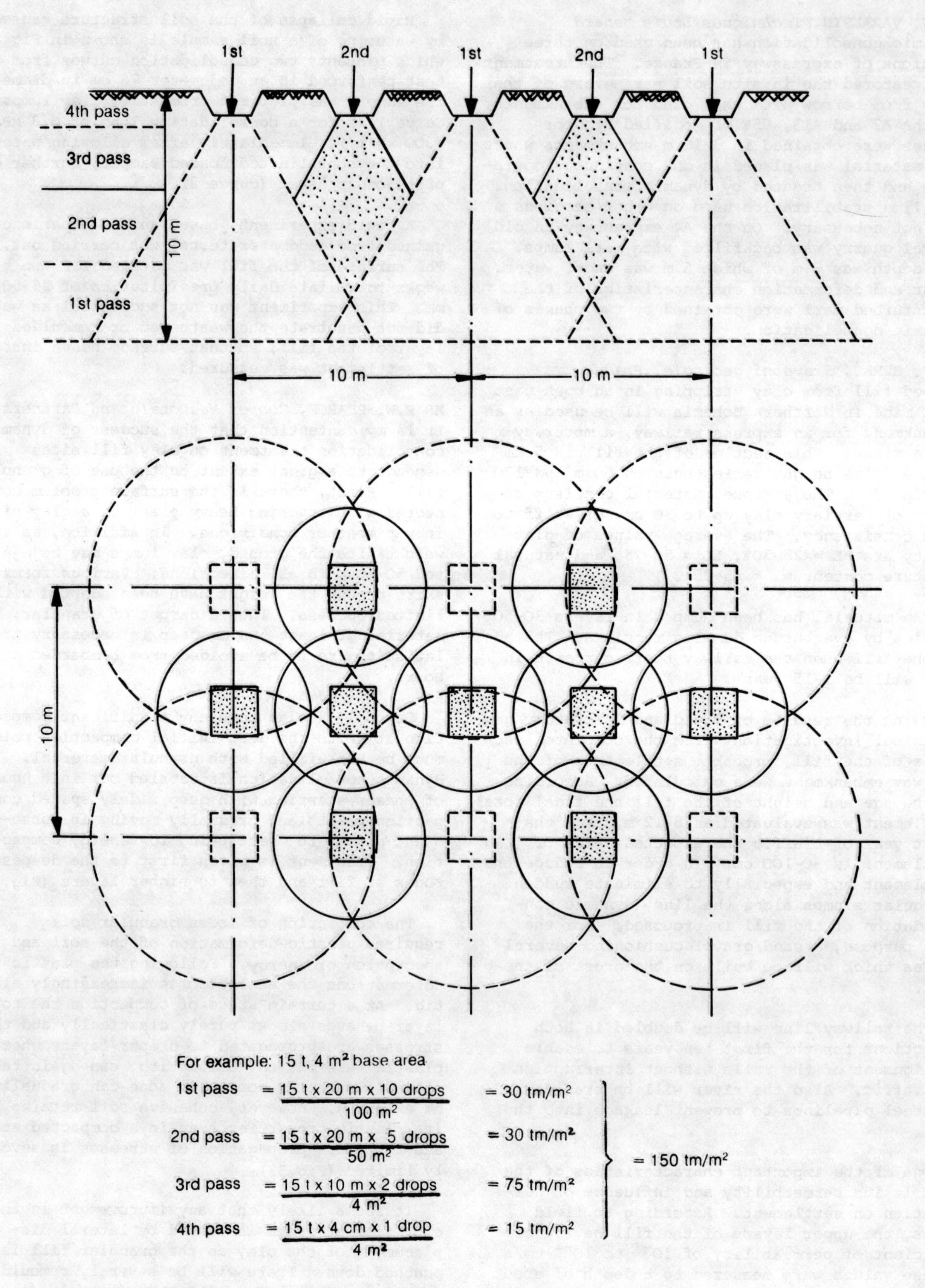

Fig. 32. Application of dynamic consolidation energy

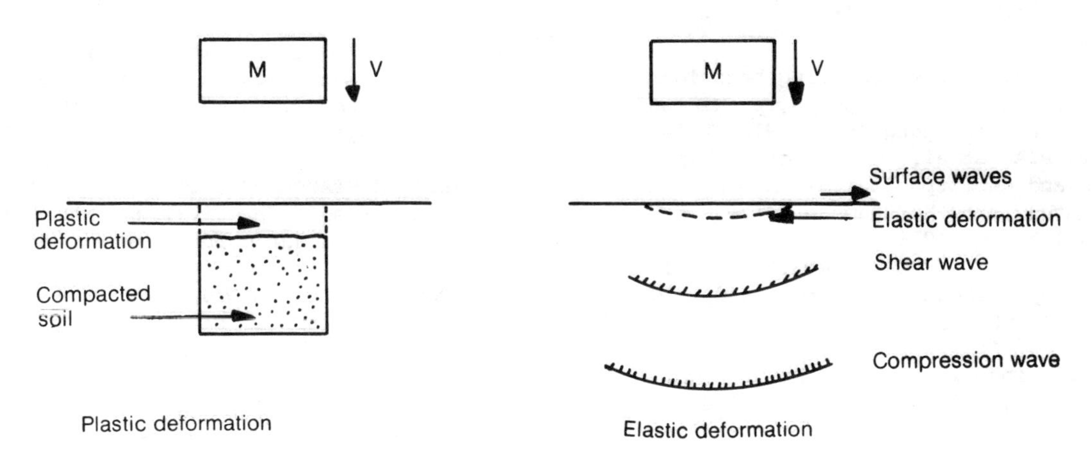

Fig. 33

Unlike the stone column technique of vibroflotation, the major shearing action of dynamic consolidation will cause some closure of voids and compaction of the fill. Evidence of this is, as Thompson and Herbert (pp 197-204) point out, the depression of the site surface during the treatment. Nevertheless, the final treated ground must behave as only slightly compacted clay reinforced with stone piers with a substantial granular surface raft helping to spread the loads on to the piers.

The economics of treating a clay filled site must therefore be controlled largely by the cost of importing the granular fill.

I should be interested to know what proportion of the total contract sum the granular fill represents.

MR D. JOHNSON, *Cementation Piling and Foundation Ltd*
With reference to the dynamic consolidation of clay fill currently being undertaken at Surrey Docks, during the early stages of this contract substantial volumes of granular material were used to form a firm working surface and to fill imprints formed during the tamping process. The total volume used was equivalent to approximately 1.0 m^3/m^2 of the treatment area. Recently the character of the fill and the general weather conditions have been such that it was not necessary to import granular material to fill imprints.

MR D.J. AYRES, *British Railways Board*
I have noted in discussion with French railway colleagues that they have specified the treatment of the top 200 mm of clay or silt subgrades with lime or cement according to a soil classification method. This was applied both on embankments and in cuttings. It does not appear to be the practice elsewhere. I wonder if the advantage of the application is that it provides a zone rich in calcium to counteract the weathering or cationic base exchange due to rainfall.

DR SMART
I wonder if it would be possible to combine the functions of a horizontal drainage blanket and a horizontal reinforcement in a suitable type of plastic blanket.

MR M.S. ATKINSON, *Soil Mechanics Ltd*
The case described by Boman and Broms (pp 49-56) is one of the few where a true control section without drains has been incorporated into the works. In many published case histories the field results are compared with predictions based on laboratory test results. These are of less value as inaccuracies of predicting field drainage parameters from laboratory testing may nullify the conclusions drawn regarding the effectiveness of the drains.

In fig.11 on page 53 the settlements after approximately eight months are greatest in the section with vertical drains, whereas in fig.12 on page 54 the greatest settlements after the same time interval are in the undrained section. I wonder why this is so.

Recent work by my company indicates that sometimes use of the Barron theory[20] approach leads to very inaccurate determinations of drain spacings. We have recently developed a three-dimensional computer model simulating the function of artificial drains. It has been successfully tested on a number of case histories and shows that where drains are installed in thick, very compressible strata, or very wet soils (i.e. wherever large volumes of water are to be expelled from the soil), the overall resistance to flow of the drain can be appreciable. This can lead to a serious overestimate of the drain effectiveness and an incorrect design spacing. The accuracy of a design using Barron's simple theory is dependent on the nature and thickness of the soils to be drained and the type of drain used.

Frequently the requirement is for a given percentage consolidation in a specified time. If the required consolidation is defined in terms of settlement then two problems arise: the necessity of waiting a sufficient time to determine what is 100% primary consolidation, and the effect of secondary settlements. If dissipation of excess pore water pressure is used as the criterion the problem becomes one of deciding at what position in the consolidation soil excess pore pressures are to be measured or of installing sufficient piezometers make it possible to determine the volume under the excess pore pressure curve.

REFERENCES

1. JONES, R.H. The engineering behaviour of glacial materials. Proc. Symp. Midlands Soil Mech. Fdn Engng Soc., 1975, 253-254.
2. KNOTT, P.A. et al. The relationship between soils and geomorphological mapping units: a case study from Northamptonshire. Private communication, 1978.
3. MINISTRY OF TRANSPORT. Notes for guidance on the specification for road and bridge works. HMSO, London, 1978.
4. RODIN, S. Ability of clay fill to support construction plant. Civ. Engng Publ. Wks Rev., 1965, Vol. 60, 197-202 and 343-345.
5. SNEDKER, E.A. Choice of an upper limit of moisture content for highway earthworks. Highw. Des. Constr., 1973, Vol. 41, Jan., 2-5.
6. FORDE, M.C. The tyre-terrain system. Proc. 5th Int. Conf. Int. Soc. Terrain-Vehicle Systems, Detroit, 1975.
7. SKEMPTON, A.W. and NORTHEY, R.D. The sensitivity of clays. Géotechnique, 1952, Vol. 3, 30-52.
8. MINISTRY OF TRANSPORT. Specification for road and bridge works. HMSO, London, 1969.
9. TERZAGHI, K. and PECK, R.B. Soil mechanics in engineering practice. Wiley, New York, 1948.
10. BLACK, W.P.M. A method of estimating the California bearing ratio of cohesive soils from plasticity data. Géotechnique, 1962, Vol. 12, no. 4, 271-282.
11. SHERWOOD, P.T. The reproducibility of the results of soil classification and compaction tests. Transport and Road Research Laboratory, Crowthorne, 1970, LR 339.
12. GOLDER, H.Q. Relationship of runway thickness and under-carriage design to the properties of the sub-grade soil. Institution of Civil Engineers, London, 1946, Airport paper 4.
13. BRITISH STANDARDS INSTITUTION. Methods of test for soil for civil engineering purposes. British Standards Institution, London, 1975, BS 1377.
14. WOOD, D.M. and WROTH, C.P. The use of the cone penetrometer to determine the plastic limit of soils. Ground Engng, 1978, Vol. 11, no. 3, 37.
15. DENNEHY, J.P. A new method for assessing the suitability of cohesive soil for use as embankment fill. MPhil thesis, University of Surrey, 1976.
16. WAKELING, T.R.M. A comparison of the results of standard site investigation methods against the results of a detailed geotechnical investigation in Middle Chalk at Mundford, Norfolk. *In* In situ investigations in soils and rocks. British Geotechnical Society, London, 1970, 17-22.
17. BURLAND, J.B. and LORD, J.A. The load-deformation behaviour of Middle Chalk at Mundford, Norfolk. Building Research Station, Garston, 1970, current paper 6/70.
18. TERZAGHI, K. Erdbaumechanic. Deuticke, Leipzig, 1925.
19. LABORATOIRE CENTRAL DES PONTS ET CHAUSSEES. Catalogue des structures types de chaussées. Laboratoire Central des Ponts et Chaussées, Paris, 1971.
20. BARRON, R.A. Consolidation of fine-grained soils by drain wells. Proc. Am. Soc. Civ. Engrs, 1947, Vol. 73, no. 6, 811-835.

P.R. VAUGHAN, BSc, PhD, MICE, Imperial College of Science and Technology

General report: Engineering properties of clay fills

The behaviour of compacted clay fills can generally be predicted using simple and conventional methods of analysis. These require various soil parameters to be evaluated, such as drained and undrained strength, stiffness and compressibility and short-term and long-term pore pressures. The trafficability of fills may be expected to depend on undrained strength, although other factors may be involved. There is a need to evaluate the appropriate material parameters for fills which are being constructed, and, more particularly, at the preliminary site investigation stage. This report considers the nature and evaluation of material parameters for compacted fills of clay and other fine-grained soils.

Compacted fills offer, in principle, one of the simpler problems of geomechanics - the simplicity lies in the fact that material properties can be studied in the laboratory without the problems of sampling and sample disturbance. However, other problems concerned with admixture, treatment and workmanship create countervailing complications which make the determination of engineering properties uncertain. A particular problem is the differences in soil structure produced by different types of compaction in the field and in the laboratory, and the importance of structure in controlling fill behaviour.

It is convenient first to consider the definition of soils which behave in a clay-like manner. Permeability is the most useful parameter in determining the type of behaviour which is likely to occur. The use of other than undrained soil properties involves the assumption of the principle of effective stress. In partly-saturated soils with a two-phase pore fluid, this principle is complicated and of uncertain predictive value.

Special problems include the effect of mixing different soils, the effect of sensitivity and loss of strength during placing, the degradation of granular materials to yield fine-grained fills after placement and the effect of freezing on subsequent fill properties.

THE PRINCIPLE OF EFFECTIVE STRESS

The use of parameters expressed in terms of effective stress presumes that this principle is valid. In partly-saturated fills the existence of two separate fluid phases at different pressures due to capillarity causes problems. In general, it is suggested that some simplifications can be made, based on whether the air is present in occluded or continuous form. When the air is continuous the permeability of a soil to air is relatively high compared with its permeability to water. This arises due to differences in viscosity. Thus, in a field situation, the pore air pressure equilibrates relatively rapidly to atmospheric pressure. In these circumstances the pore water pressure must be negative, due to capillarity.

When the air is occluded the permeability to air is very low as air travels only by diffusion. The pore water pressure may be negative or positive. Langfelder et al.[1] show that air permeability decreases dramatically when the compaction water content is increased to a value just above optimum. It can be presumed that compaction to refusal produces the occluded air state. A soil which initially has continuous air will change to one with occluded air if it is compressed by a confining stress, and this process will occur more readily if air drainage can occur while the air remains continuous, as is likely in the field. The air is therefore likely to be occluded once the pore water pressure becomes positive.

The presence of continuous air implies quite high pore water suctions in a fine-grained soil, and in this condition the principle of effective stress cannot be defined using the pore water pressure alone.[2] Moreover, strains may occur which are opposite in sign to those predicted by changes in apparent effective stress. Cox (pp 79-86) demonstrates collapse and decrease in volume while the effective stress is decreasing. Fortunately, the engineering problems occurring in fills in this state are generally restricted to those of heave and collapse. When the air is occluded, and certainly when the pore water pressures are positive, experience shows that the evaluation of effective stress using the pore water pressure alone in the conventional manner is adequate for engineering purposes.

This view is supported by the test data shown in Fig.1 for both field and laboratory compacted Weald clay. The volumetric compressibility in terms of conventional effective stress is shown. Systematic compression occurs when the pore pressures in the test specimens become positive. Systematic compression as a function of effective stress is not displayed

Table 1. Permeability of fills related to consolidation, drainage, wetting and drying

CONSOLIDATION DURING CONSTRUCTION

$$k = \frac{H^2 . m_v . \gamma_w . T}{t}$$

H = embankment height
t = construction time

$m_v = 0.1 m^2/MN$, H = 20m, t = 1 year

	no underdrain T	no underdrain k m/s	full underdrain T	full underdrain k m/s
Full consolidation	200	$5x10^{-9}$	50	$1x10^{-9}$
No consolidation	1.0	$2x10^{-10}$	0.2	$6x10^{-11}$

DRAINAGE - FALL OF FREE BOUNDARY dh_b/dt

$dh_b/dt = k. i_b/n_e$: $k = n_e.dh_b/i_b.dt$

n_e = effective porosity = 0.2

i_b = hydraulic gradient at boundary

H = 20m, t = 1 year

		Permeability k m/s no underdrain $i_b = 0.1$	full underdrain $i_b = 1.0$
No drainage	$dh_b/dt = 0.05H/t$	$6x10^{-8}$	$6x10^{-9}$
Full "	$dh_b/dt = H/t$	$1x10^{-6}$	$1x10^{-7}$

WATER RETENTION (Fig.2)

Substantial - suction to desaturate 10m $k = 10^{-9}$
Negligeable - " " " 1m $k = 10^{-6}$ m/s

SWELLING

Significant $du/u_o = 0.25$ at 0.5m depth in 0.02yr.
$C_s = 5m^2/yr$ $m_v = 0.1m^2/MN$ $k = 1x10^{-10}$ m/s

DIRECT INFILTRATION Rate V m/s

$V = k. i/n_e$ V = 20mm/hour, i = 1.0, $n_e = 0.2$
$k = 1x10^{-6}$ m/s

AIR DRYING FROM FILL SURFACE

Substantial drying Suction (S_o = 100%) $k = 1x10^{-9}$
Early desaturation " " $k = 5x10^{-7}$ m/s

CONSOLIDATION UNDER WHEEL LOAD

$T = c_v.t/D^2 = 2.0$ D = 0.5m, t = 0.1s, $c_v = 5m^2/s$
$m_v = 0.1m^2/MN$ $k = 5x10^{-3}$ m/s

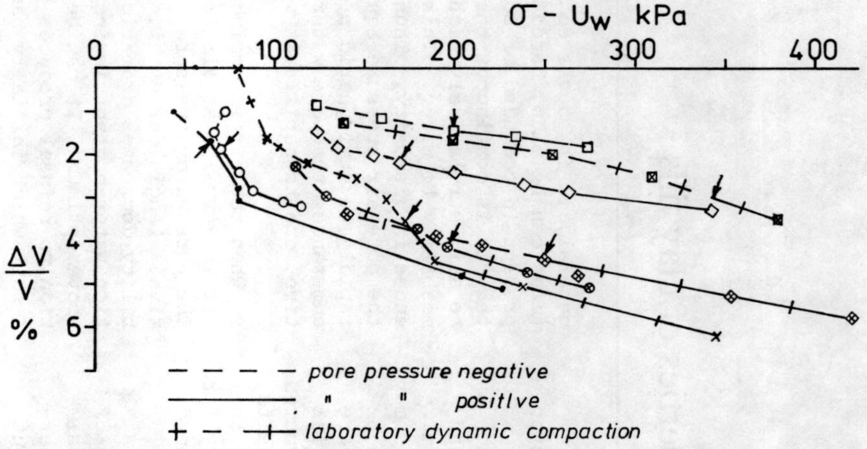

	SAMPLE NO.	COMPAC-TION	W_P %	W_L %	W_i %	γ_i t/m³	W_f %	γ_f t/m³
○	12A	F	27	62	22·6	2·08	22·6	2·15
⊗		L			22·9	2·05	22·9	2·16
•	12B	F	25	61	23·6	2·06	22·5	2·16
×		L			23·5	2·04	22·8	2·17
□	13A	F			21·4	2·07	21·4	2·10
⊠		L			21·6	2·01	21·6	2·09
◇	13B	F			21·2	2·07	20·8	2·14
◈		L			21·0	1·98	20·7	2·11

12A & 13A Undrained throughout test.
12B & 13B Air drainage allowed untill u_w+ve. Then undrained until final consolidation stage.

Fig.1. Isotropic compression: Weald clay compacted in the field and in the laboratory

while the pore water pressures remain negative.

CLAY-LIKE BEHAVIOUR AND PERMEABILITY

Clays as fill exhibit the well-known characteristics of poor drainage, water retention, low undrained strength when they are wet, and susceptibility to deterioration if placing occurs during wet weather. A number of mechanisms are involved and they can be quantified in terms of permeability.

Consider first the degree of consolidation which occurs during construction. The coefficient of consolidation, $c_v = k/\gamma_w m_v$, depends primarily on the permeability k and presuming a coefficient of volume compressibility $m_v = 0.1$ m^2/MN, then k (m/s) = 3.1×10^{-11} c_v (m^2/year).

Gibson[3] considered the rate of consolidation of a fill being constructed at a uniform rate $\Delta H/\Delta t$. His results give the bounding values for c_v and k in Table 1. If full consolidation occurs during construction, then the pore pressures are given by the equations of unsteady seepage. Full consolidation implies zero pore pressure only if the fill is underdrained and of uniform permeability.

The rate at which the free boundary drops during unsteady seepage is given by $\Delta h_b/\Delta t = ki_b/n_e$, where i_b is the vertical hydraulic gradient at the boundary and n_e is the effective porosity. The use of this equation implies that the effects of capillarity can be neglected. The value of i_b ranges from unity with uniform permeability and full underdrainage to about 0.1 with no underdrainage. Values of k bounding negligible and substantial drainage are summarized in Table 1. From Table 1, movements of the boundary during consolidation are clearly negligible. With underdrainage, pore pressures are approximately zero or negative throughout drainage. Without underdrainage, the pore pressures decrease from positive values during drainage.

Figure 2 shows the capillary rise in some typical soils, taken from the data of Lane and Washburn,[4] plotted against permeability, which is a good measure of effective pore size. The values of permeability for substantial and negligible water retention are shown in Table 1. The results show that water retention is typically significant only where drainage is negligible.

The next factor to be considered is the penetration of water into the compacted fill during wet weather. A typical swelling calculation is given in Table 1. It shows that, in the true clays ($k < 10^{-10}$ m/s), swelling effects are of limited significance unless the exposed surface is exposed to free water for considerable periods. Direct penetration of rainfall into a fill of low saturation becomes of significance when $k > 10^{-7}$ m/s, and full penetration occurs when $k \simeq 5 \times 10^{-6}$ m/s. At permeabilities greater than about 5×10^{-6} m/s, rapid drainage compensates for ease of penetration, and rainfall is unlikely to cause problems.

The factor to be considered next is air drying from the fill surface. Drying causes substantial negative pore water pressures, and if these suctions cause desaturation of the fill such that continuity of the pore water is lost, then hydraulic conductivity drops dramatically and the drying process is interrupted. Fig.2 shows that this is likely to become significant as permeability increases above about 10^{-9} m/s. With permeabilities below this value, drying is effective, and consolidation and swelling in response to wetting and drying of the surface are of approximately the same magnitude. With permeabilities greater than this the fill can wet up more readily than it can dry from its surface. Drainage must be relied on to remove the water which penetrates, and reference to Table 1 shows that the permeability at which drainage is fully effective is considerably higher than that at which water penetration is significant.

One further factor can be mentioned for fill with a high degree of saturation. Trafficking will be controlled by the undrained behaviour of the fill unless it can consolidate during transient wheel loading. The sample calculation of Table 1 indicates that a permeability greater than 5×10^{-3} m/s is typically required for such consolidation to occur. A critical condition arises if a water table is maintained close to the fill surface, such as when a fill is dumped into water from a working level close to the water surface. Desaturation due to drainage is then prevented. In the range of permeabilities where drainage can be effective in desaturating the fill ($k > 5 \times 10^{-7}$ m/s), the degree of saturation in the absence of a high maintained water table can only be high enough to cause compaction and traffic problems during and shortly after heavy rain.

Three ranges of behaviour can be identified based on permeability.

(a) True clays with permeabilities less than about 5×10^{-10} m/s. They do not consolidate significantly during construction unless closely spaced drains are used. Once compacted, they do not take up much water unless their surfaces are exposed to free water for considerable periods, or unless free water is worked into them by the rutting and remoulding action of plant running on a surface of inadequate undrained strength. The latter factor suggests that wetting up in wet weather is likely to be worse in fills that are already wet than in fills placed drier and stronger. Air drying is effective in this range.

(b) Soils of permeability between about 5×10^{-10} m/s and 10^{-7} m/s. They consolidate readily during construction, so that construction pore pressures do not cause a stability problem, particularly if underdrainage is provided. However, water can readily enter these soils during rainfall, and neither air drying nor drainage is effective in removing it. Thus, poor trafficking conditions can

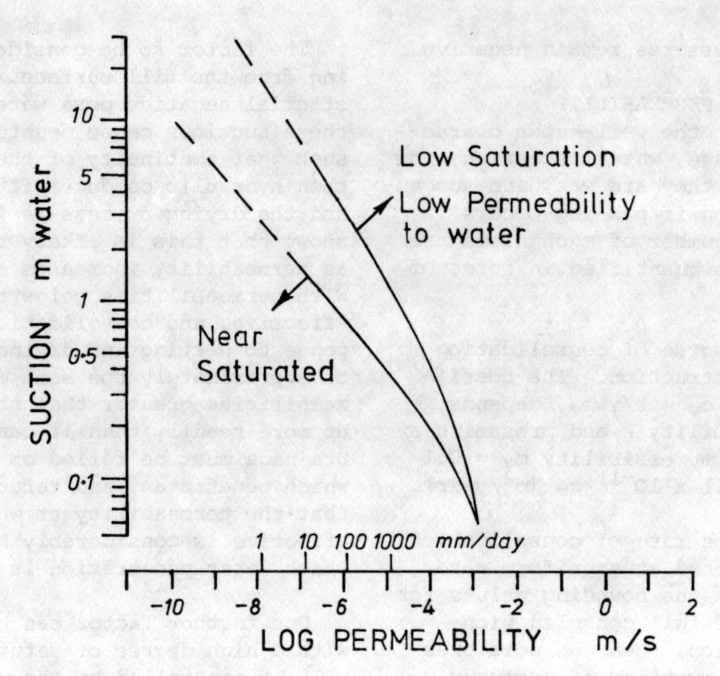

Fig.2. Capilliary rise in fine grained soil (after Lane and Washburn[4])

Table 2. The effect of compacted structure on soil parameters

Parameter	Effect
Drained strength c', ϕ'	Slight
Permeability k	Slight (unless macro voids are present)
Consolidation and swelling. C_v, C_s	Slight
Volumetric stiffness $\frac{\Delta V}{V}$ $\Delta\sigma'$	Slight
Undrained strength C_u, c, ϕ	Large in plastic clays - probably decreasing with increasing strain rate. Slight in sandy clays of low to medium plasticity
Undrained Pore Pressure response	Large
Shear stiffness E', E_u	Large

develop easily and can persist long after surface infiltration has ceased. These soils can present worse placement problems than the true clays.

(c) Soils of higher permeability. Drainage is effective in these soils and maintains a low degree of saturation unless a high water table is artificially maintained. Neither construction stability problems nor trafficking problems are likely.

It is suggested that permeability is a valuable classification parameter which can predict the type of behaviour which is likely during placement and embankment construction. It can be examined readily at the site investigation stage.

COMPACTED STRUCTURE

The effect of type of compaction on the structure of the soil produced in the laboratory has been studied extensively, as summarized, for instance, by Lambe.[5] Vaughan et al. (pp 205-217) and Vaughan[6] suggest that there is a further significant difference between the structure of laboratory and field-compacted clays. In field compaction, large lumps of clay are incorporated into the fill in a distorted but unbroken-down state, and the amount of fully broken-down matrix material is limited. If the borrow material is saturated then the lumps remain saturated. There will be much more finely broken-down material in a typical specimen compacted in the laboratory, and the structure will be closer to that due to complete remoulding. The probable effects of the influence of this difference in structure on material properties are summarized in Table 2.

Changes in material index properties between borrow pit and placed fill are discussed by Harpster (pp 119-125). Changes of this nature, as well as differences in compacted structure, suggest that caution is needed in predicting field behaviour from tests on specimens compacted in the laboratory at the site investigation stage.

SHEAR STRENGTH PARAMETERS IN TERMS OF EFFECTIVE STRESS

Vaughan et al. (pp 205-217) find that this strength is not dependent on compacted structure, at least at high normal effective stresses. Thus it can be determined by standard laboratory tests. A particular problem arises with the strength at very low normal effective stresses, which operates in the shallow embankment slope failures which are common. Expressed in conventional terms, the cohesion intercept c' then controls the strength. In principle, this component of strength increases with placement density. Chandler and Pachakis[7] show that c' may be zero at modest effective stresses in an old, loose-dumped clay fill. It is clear that a significant cohesion intercept operates in a modern compacted fill (Vaughan et al., pp 205-217), and that construction methods are desirable which ensure full compaction of the fill placed in the outside of embankment slopes.

The precise amount of compaction does not seem to be important. The results of Vaughan et al. (pp 205-217) and Abeyesekera et al. (pp 1-14) do not show significant changes in strength with small changes in initial sample density or compaction effort. The swelling data presented by Cox (pp 79-86) show that, after swelling, differences in initial density are reduced.

Wrightman (pp 227-229) presents data which indicate that compaction can cause shear surfaces in fill, and similar instances are reported and discussed by Vaughan et al. (pp 205-217), Pavlakis (pp 266-269), Chaplin (p.241) and Parsons (p.269). It seems that both horizontal surfaces can be formed by smooth rollers and more random surfaces by rutting under scraper wheels. The influence of this type of feature on both drained and undrained strength can be considerable, the continuous plane being potentially more serious than random surfaces formed during rutting. A critical reduction in strength can occur only in the more plastic clay which shows a large drop from peak to residual strength (Vaughan et al., pp 205-217). A single surface can be readily destroyed by scarification before placing of the next layer, and may be destroyed naturally by the wheels of placing plant. In embankments and parts of embankments where drained shear strength is critical, inspection to see if shear surfaces are being formed seems prudent. If they cannot be avoided or destroyed by the appropriate placing methods, then a reduction of the assumed strength for the fill may be necessary.

UNDRAINED STRENGTH AND BEHAVIOUR

Three factors are of importance in the design and construction of fills

(a) undrained deformation characteristics, which control embankment displacements during undrained construction

(b) pore pressure response during undrained loading, which is of importance when drainage measures are planned and when pore pressure measurements are used to monitor stability

(c) undrained shear strength, which controls short-term stability and traffickability.

These factors are discussed by Symons (pp 297-302) and Parsons (pp 307-314).

Deformation characteristics and pore pressure response

Vaughan et al. (pp 205-217) and Vaughan[6] show that the compacted structure of a typical plastic clay controls both stiffness in shear and pore pressure response when slow loading rates allowing pore pressure equilibrium are used. They also show that pore pressure responds to average stress change and not significantly to shear stress. Prediction of construction pore pressures can be based on laboratory tests provided test specimens with the correct compacted structure are used.

Pore pressure does not respond solely to average stress in the typical sandy clay fills. Compacted structure may well control pore pressure generation at working shear stresses. Fig.3 shows the pore pressure generated in a sandy clay till after dynamic compaction in the laboratory when subject undrained to an increasing isotropic stress. The response of the same clay in the field is also shown. The large difference may be due in part to the imposition of cyclic stresses in the early stages of field placing. Whatever the reason, it is clear that field pore pressures cannot be predicted reliably on the basis of laboratory tests. Stiffness in shear of undisturbed and fully remoulded specimens of sandy clay of the same strength is different, and it may be inferred that compacted structure will influence stiffness.

Undrained strength

There has been a tendency to presume that undrained strength is a unique soil property. The undrained strength of a particular soil is not unique and is influenced by many factors. Four of these are of potential importance in the study of fills

(a) initial compacted structure
(b) stress path during loading
(c) degree of saturation and average confining stress
(d) rate of loading.

Two aspects of the performance of a clay embankment fill are controlled by undrained strength. The first is short-term, undrained slope stability. The rate of loading is slow, allowing local redistribution of pore pressure, and is much slower than that in conventional laboratory quick tests. Stress paths during embankment construction approximate to that of the triaxial compression test. Average confining stress varies within the embankment. The second aspect is performance during trafficking, and it seems to be agreed that undrained strength is the dominant factor controlling trafficking, even if it is not the only one. Wheel loads will increase the average stress in the fill and, typically, they will be applied quickly.

Estimation of undrained strength may be made by a variety of means. The first is by direct test in which reasonably uniform and directly calculable stresses are applied at a known rate, such as the triaxial compression test. The second is by direct test in which the average shear stress is calculable, but in which the stresses are not uniform and the whole stress path is unknown, such as the shear box and more particularly, the vane test. The third is by a direct test similar to the second type. but in which the average shear stress cannot be calculated directly, such as the cone and plate penetration tests. Such tests can be used to determine strength only if they are calibrated against soils for which a strength has been determined by some other means. The fourth method is to relate the water content of the soil to index properties. This method is clearly likely to succeed best if the index tests used determine the water content of the soil at a particular strength, and if they measure the strength in an appropriate way and at an appropriate rate.

If a fill is non-homogeneous, then the size of the specimen tested will influence the scatter of results obtained and, perhaps, the average result. Subject to this effect, the accuracy with which strength is determined is likely to decrease from the first category of tests to the fourth.

Perhaps the most important factor in determining the undrained strength of a fill is rate of loading. Fig.4 shows rate of shear effects of a typical clay as summarized by Skempton and Bishop.[8] Rates typical of embankment construction, quick and slow laboratory tests, in situ tests and wheel loading are indicated. It is convenient to consider the slower rates of loading first. At rates slower than the conventional quick laboratory test the dominant mechanism is probably pore pressure redistribution. Table 1 of Vaughan et al. (p.210) shows that compacted structure has a strong influence on the undrained strength of a typical plastic clay when significant pore pressure redistribution occurs, and a smaller influence when a conventional quick loading rate is used. The effect of compacted structure, varying from full remoulding to intact, on the undrained strength of typical sandy clay tills determined in conventional quick undrained tests has been found to be negligible.[8,9] This view is supported by Lieszkowsky (pp 236-237) and Threadgold (pp 251-253). It seems safe to assume that laboratory compacted specimens can be used to determine the ultimate undrained strengths of field compacted fills of this type of clay.

The strength of a fill at the slow loading rates appropriate for end of construction stability might be estimated from quick tests if a reliable correction for rate effect could be applied. However, there is evidence that this varies with the type of clay. Table 1 of Vaughan et al. (p.210) shows a substantial reduction in strength with decreasing loading rate. Vaughan et al.[9] found the opposite effect for remoulded specimens of a sandy clay fill. In the absence of a more general picture concerning rate effects, it is clear that the undrained strength relevant to end of construction stability should be determined using slow undrained tests in which at least partial local pore pressure equilibrium is allowed. The effects of partial saturation and confining pressure can be studied readily using such tests. However, the total error in using quick tests on laboratory compacted specimens of any type of soil does not involve a ratio greater than 1.7, and it should be much less for a sandy clay. Thus quick tests may be used to examine whether or not a stability problem is likely to exist.

The undrained strength relevant to trafficking problems and its estimation by direct and indirect means has been the subject of much discussion. The stress paths involved in wheel

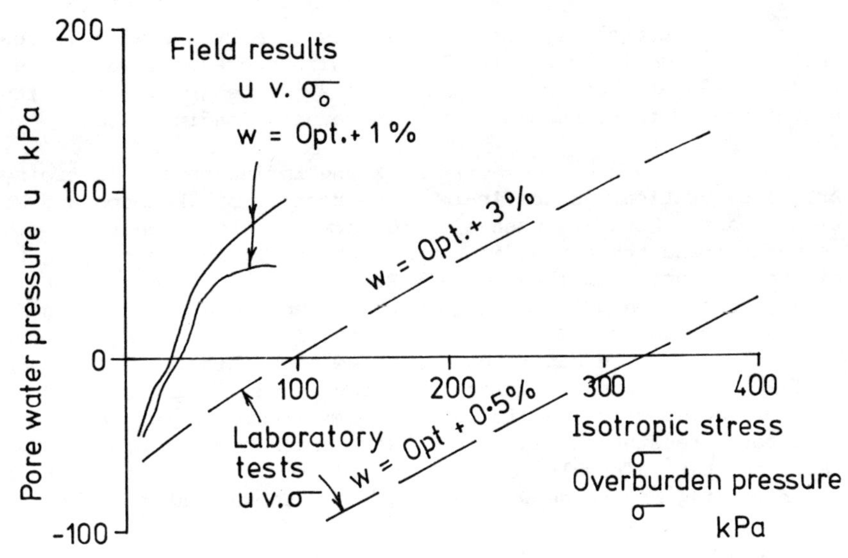

Fig.3. Selset boulder clay: pore pressure response in specimens compacted in the laboratory and measured in the field

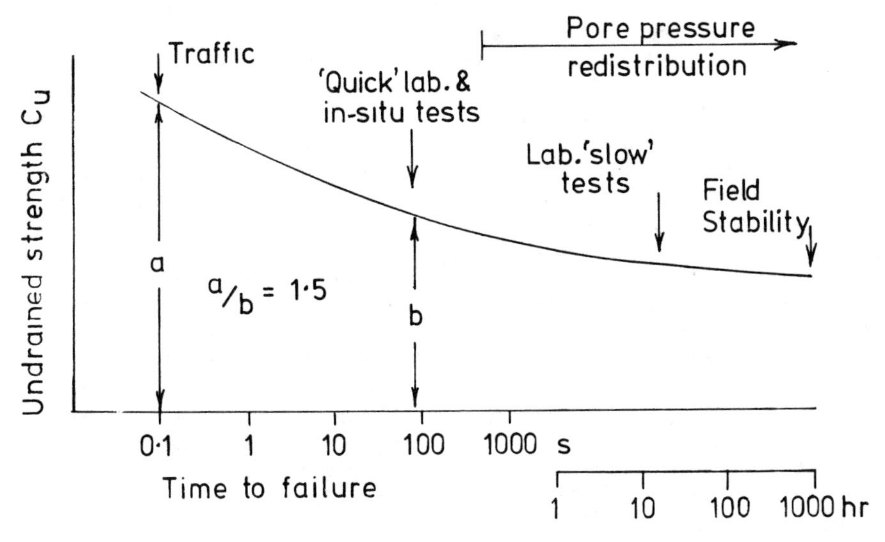

Fig.4. Typical effect of loading rate on the undrained strength of clay (after Skempton and Bishop[8])

Table 3. Undrained strength at the liquid limit (after Skempton and Bishop[8])

Clay	w_L %	I_p %	C_u at w_L* kPa	Decrease x
Horton	30	14	1.45	1.00
London	73	48	1.03	0.71
Gosport	80	50	0.88	0.61
Shellhaven	97	65	0.74	0.51

* Determined by laboratory vane

loading are clearly complex, but probably the triaxial compression test provides an adequate simulation. Dennehy (pp 87-94) indicates that similar results are obtained by the vane and triaxial test.

The effect of partial saturation and confining stress may be significant. However, when a clay is compacted to refusal and the air in it is occluded, the gain in strength from the unconfined state to full saturation by confining pressure is not more than, typically, 50%, and the increase is achieved largely by a confining stress of the order of 200 kPa. Thus the effect is of less importance for wetter clays and it is probably insignificant for strengths of less than 30 kPa. Increased scatter in results and uncertainty as to their meaning may arise when tests in which confining pressure is uncontrolled, such as the vane and cone, are used to evaluate strengths higher than this value. The unconfined compression test may give a large scatter of results as it will fully reflect differences in initial density.

Figure 4 indicates that the strength relevant to trafficking should be measured at high loading rates, or corrected for rate effect if slower loading rates are used. Pore pressure effects are not significant at high loading rates. The effect of compacted structure at these loading rates is uncertain. Discussion indicates that at conventional quick loading rates it is reduced in the plastic clays and negligible in the sandy clay tills. It seems reasonable to presume that strength at the faster strain rates is dominated by the remoulded matrix behaviour where the matrix is weaker than the lumps retained in the fill, as will usually be the case, particularly when the fill has become wetter during excavation and placing. Thus the estimation of the quick strength of the fill from that of the fully remoulded soil seems justified. This requires further study.

A distinction must be made between prediction at the site investigation stage, and measurement as a fill is being placed. In the latter case large laboratory specimens or in situ tests can be used, so that the structure is adequately represented in the test. However, if this is done then different strength/water content relationships may be expected from fully remoulded specimens and from field compacted specimens, due to differences in lump and matrix water contents in the field-compacted material. Tests on fully remoulded specimens may predict the strength of the fill better than they predict the strength/water content relationship. Laboratory tests on remoulded specimens offer a particular advantage in the field control of sandy clay tills, as the effect of structure is least in these materials and the problems of stones which influence both undisturbed sampling and in situ testing (Rocke, pp 265-266) are avoided.

The effect of very high loading rates on undrained strength has received little study. Mitchell[10] quotes data which indicate that it can be substantial, and deduces that it is likely to increase with increasing plasticity and with decreasing water content. There is also evidence (Fig.1) that it increases with increasing loading rate.

The influence of plasticity is illustrated by the results of Skempton and Northey.[11] They determined the strength of four remoulded clays at their liquid limit using the laboratory vane. Recent use of the cone test (the results of which depend on the undrained strength at fast strain rates) to determine the liquid limit indicates that this limit is the water content at which the dynamic undrained shear strength has a common but unknown value. If rate effects were unaffected by plasticity then the vane would show a common strength at the liquid limit. Table 3 shows that the vane strength decreases with increasing plasticity, indicating that the rate of shear effect increases significantly with increasing plasticity. The correction factor proposed by Bjerrum[12] for the application of the vane test to foundation problems in soft clays, which he attributed mainly to rate effects, shows a trend similar to that in Table 3.

Measurement of fast rate effects may be relevant to trafficking problems, and to the classification of clays generally. It presents difficult technical problems. A possible approach may be to use the cone test, which has the advantage that it depends on the dynamic shear strength, and the disadvantage that an actual strength can be deduced only by correlation with other tests inevitably done at much slower loading rates. Wood and Wroth[13] as discussed by Jones (p.274) suggest using the cone test for the indirect determination of a water content at which the dynamic shear strength is one hundred times greater than that at the liquid limit. They propose this as an alternative to the existing plastic limit, and use the principle that the work done by the falling cone is a function of the dynamic shear strength.

If trafficking problems are controlled by dynamic undrained strength, then the use of this alternative index test, together with the liquid limit, should give a much better correlation based on water content than most other current index test methods. A simple extension of the method could give absolute information concerning the rate effect. If the cone were used in a constant rate of penetration test in which the same penetration was achieved in about the same time as is taken for failure in conventional laboratory shear tests, the work done can be deduced from the load-penetration curve. This can be compared with the work done by the cone in free-fall and the rate effect deduced. By comparison of the slow cone penetration result with the strength of the soil determined at the same loading rate, by the triaxial or vane test, an absolute calibration of the cone could be produced, and the dynamic strength could be measured in absolute terms.

The effects of very high strain rates, and their relevance to the trafficking problem and to the use of other, slower methods of strength

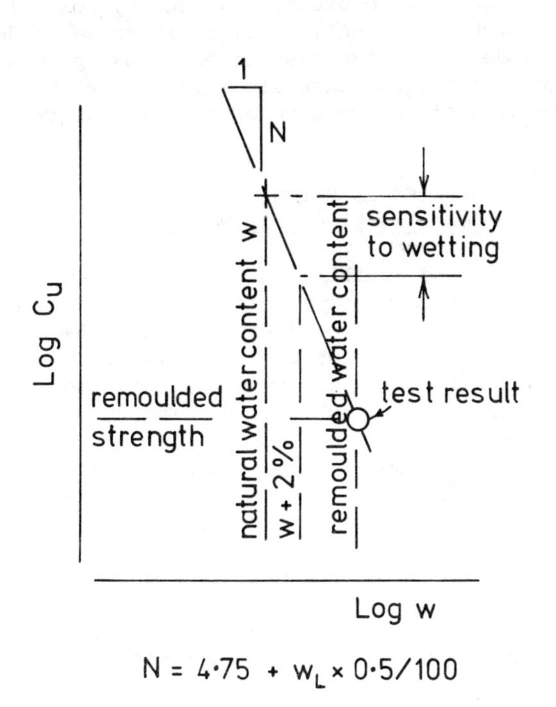

Fig.5. Construction of a remoulded strength/water content relationship from a single strength determination

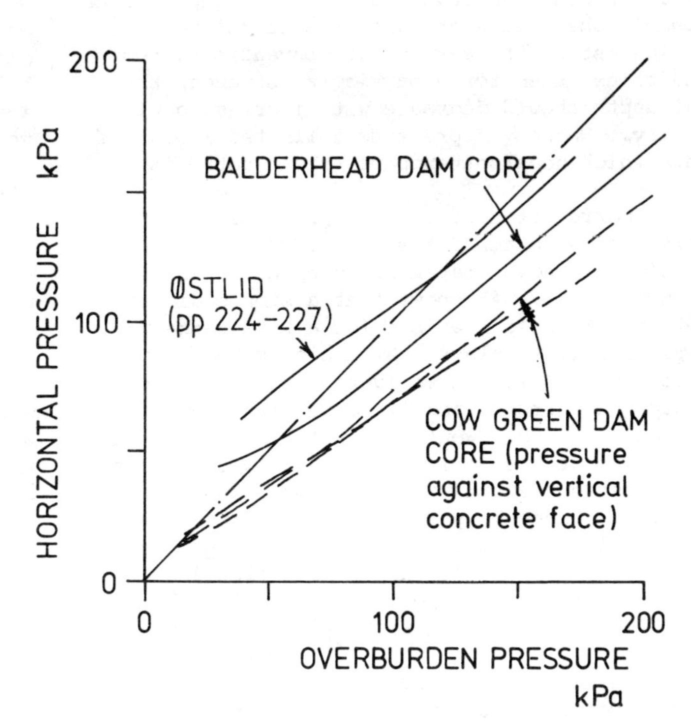

Fig.6. Measurements of lateral stress in compacted fills

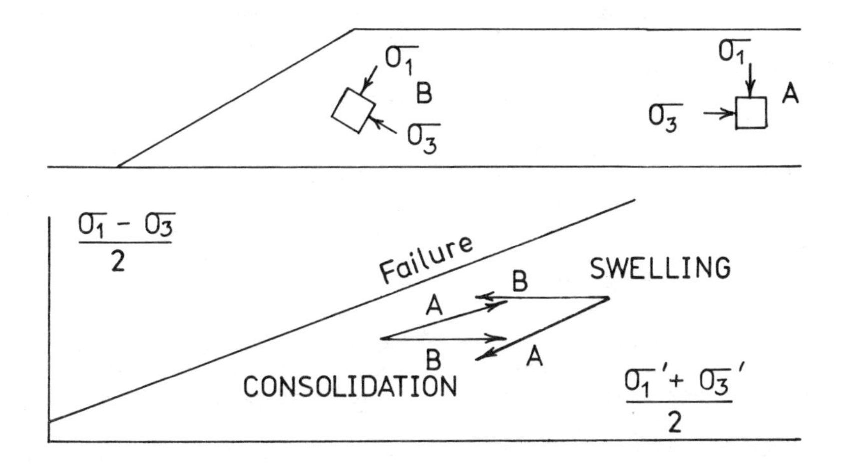

CONSOLIDATION σ_1'/σ_3' constant or reducing - vertical compression & small horizontal strains

SWELLING A No horizontal strain - σ_1'/σ_3' reducing vertical expansion & heave

B σ_1'/σ_3' increasing - expansion & shear strain - horizontal spreading & vertical heave or settlement

Fig.7. Pore pressure equilibration in embankments: stress paths in consolidation and swelling

measurement, clearly require further study. An increase in rate effect with increasing plasticity would mean that, if rut depth depended on dynamic shear strength and were correlated against strength measured at conventional testing rates, then for a particular strength the rut depth should decrease with increasing plasticity. Dennehy[14] presents a limited amount of data which does not support this hypothesis.

If correlation of traffic problems with strengths determined for remoulded soil at conventional loading rates is accepted, then Dennehy (pp 87-94) shows that a simple direct determination is feasible for a wide range of typical British clays. A simple method, suitable for the site investigation stage, largely replacing current identification test methods and costing less, would be as follows. The best linear relationship is obtained between strength and water content if a double log plot is used (fig.3 of Dennehy, p.91). Further, the gradient of the plot is almost independent of the type of clay tested for those clays which plot just above the A line. $\Delta \log C_u / \Delta \log w$ increases from 4.75 to 5.25 with increasing plasticity.

Suppose that a disturbed sample of clay is taken together with its natural water content. A remoulded test specimen is then prepared either for the laboratory vane - in which case it should have a strength less than 25 kPa to avoid uncertainties due to partial saturation - or for the triaxial test - in which case the sample can be saturated by a high confining pressure irrespective of its strength. A second water content is taken after the strength has been measured and, even if the specimen has been remoulded at the natural water content, the effect of drying out during handling is known and controlled. The water content/strength relationship can then be established as shown in Fig.5. Possible trafficking problems and the sensitivity of the soil to wetting up are both established.

IN SITU STRESSES IN FILLS DUE TO COMPACTION

Østlid (pp 224-227) presents results which indicate that placement and compaction of a fill results initially in horizontal stresses which are considerably higher than vertical stresses. Some other results are shown in Fig.6. Such stresses will influence construction pore pressures and will cause additional horizontal strains and displacements during construction. The development of large horizontal strains probably due to this cause is reported by Vaughan et al.[15]

EQUILIBRIUM PORE PRESSURES

Ample evidence has been presented to indicate that quite high positive pore pressures typically develop in the long term in clay embankments, given that net infiltration into the slopes and crest occurs. Vaughan et al. (pp 205-217) suggest that positive pore pressures can develop in a fill even when there is completely effective underdrainage, due to a decrease in permeability with depth and increasing stress and density. This effect needs to be incorporated in studies of long-term seepage pressures.[16] It did not occur in the embankment reported by Inada et al. (pp 127-132), but the fill in that case was a high water content, low density volcanic soil which showed only a slight increase in density under load. Matyas[17] showed that permeability did not increase significantly with increasing effective stress in a similar tropical soil.

PORE PRESSURE EQUILIBRIUM

Rates of consolidation and swelling

When the pore air phase is occluded then consolidation and swelling should follow mechanisms which are broadly compatible with the assumption of the Terzaghi consolidation theory. As the pore fluid is compressible, consolidation and swelling will be slower than for the same soil when saturated. Vaughan et al. (pp 205-217) summarize values of c_r and c_s inferred from field rates of pore pressure change and from direct measurement in the laboratory. Farrar (pp 101-106) presents additional data. It may be concluded that compacted structure has only a slight influence on consolidation and swelling rates, and that these coefficients can be predicted from tests on specimens compacted in the laboratory with an accuracy sufficient for engineering purposes.

Deformation during consolidation and swelling

The stress paths involved in the consolidation and swelling of embankment fills after the end of construction are illustrated qualitatively in Fig.7. Consolidation generally involves either a constant or decreasing effective stress ratio, and small horizontal strains are likely. This is supported by experience which indicates that there are insignificant horizontal deformations of an embankment during consolidation, as reported by Inada et al. (pp 127-132).

The settlement of an embankment during consolidation can be predicted with adequate accuracy using the results of conventional oedometer tests, provided that the pore pressure changes are known, the total stresses do not change significantly by arching, and samples of field compacted fill are used. The results reported by Farrar (pp 101-106), Coumoulos and Koryalos (pp 73-78) and Boman and Broms (pp 49-56) broadly support this view.

The influence of compacted structure on compressibility is uncertain. Fig.1 shows the compressibility of field and laboratory compacted Weald clay under increasing isotropic effective stress. Structure has no significant influence on compressibility once pore pressures are positive, although the stiffness of this clay in undrained shear is strongly influenced by structure.[6] The results of Farrar (pp 101-106) also support the view that structure is not important. Coumoulos (pp 221-224) presents data which indicate that structure may have an effect but, as pointed out by Penman (pp 219-221), compression results from narrow dam cores are subject to the possible influence of arching on the assumed total stress. Other evidence[18,19] suggests that structure does not affect one-dimensional compression, and that laboratory compacted test

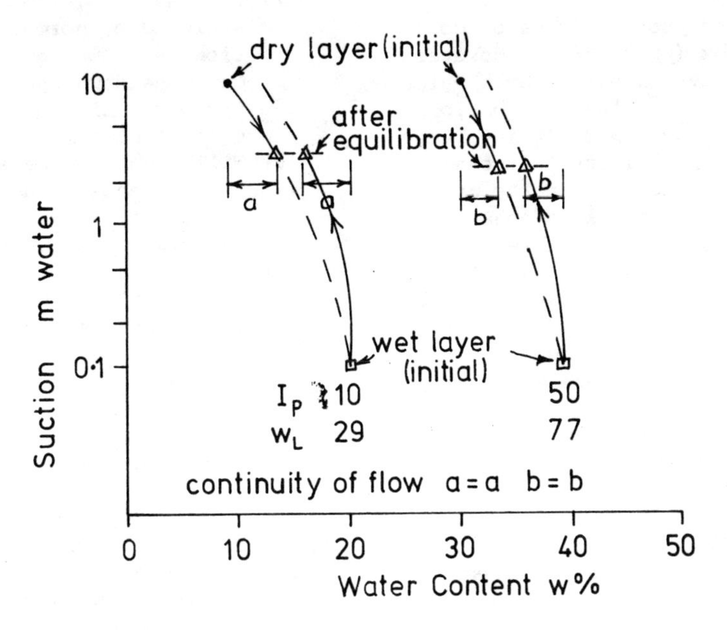

Fig.8. Effects of placing wet and dry clay together in equal quantities

Table 4. Breakdown of fills during normal placing: a possible correlation with the crushing strength of the as-dug material

	crushing strength MPa
Complete breakdown, homogeneous fill formed. Suitable for dam covers.	< 0.15
Some macro voids after compaction	0.15 - 0.5
Substantial breakdown	0.5 - 5.0
Breakdown slight and can be avoided. Can be achieved if required.	5.0 - 10.0
Rockfill	> 30.0

specimens can be used for post-construction settlement estimates without a major loss in accuracy.

Swelling involves the possibilities of collapse and heave (see Cox (pp 79-86)). However, the oedometer test involves a stress path which is inappropriate for studying the deformations of an embankment during swelling. It involves a significant increase in horizontal stress (Fig.7) and this is possible only in the centre of a wide embankment. In other locations (Fig.7) the fill and foundations will be unable to resist the horizontal stresses, lateral strain and deformation will result, and heave will be reduced. In the extreme, as a low factor of safety is approached during swelling, shear strains will become large and there will be vertical compression and settlement. Edwards (p.236) shows an example where significant lateral movements occurred during swelling of an otherwise stable embankment.

THE EFFECT OF MIXING FILLS

There have been several reports on unsatisfactory results obtained when two different fills are mixed during placement. Three possible mechanisms for this can be deduced. The first is contamination and a reduction in permeability, which may convert a free draining fill into a non-free draining one. Perry (pp 189-196) reports that chalk fill contaminated by clay became a relatively uniform mixture.

The second mechanism will operate if a clay is mixed with a wet sand. The clay will have substantial suctions as excavated. The sand will be unable to sustain these suctions without substantial loss of water, which will be available to allow swelling and softening of the clay.

A third mechanism will operated when wet and dry clay are mixed, and can be illustrated from fig.3 of Black and Lister (p.40). Suppose that a layer of wet clay is placed together with a drier layer of the same clay. Fig.8 shows that, because of hysteresis in wetting and drying, the wet layer remains wetter than the dry layer after pore pressure equilibration, and the effect is not the same as full mixing and remoulding together. The effect described persists even when the volume of dry clay considerably exceeds the volume of wet clay. The equilibration of a wet layer of fill within a dryer fill is discussed by Vaughan et al. (pp 205-217) who suggest that the result in a sandy clay may be worse than that predicted from the results of Black and Lister (pp 37-48).

These mechanisms confirm that the mixing of a poor material with a better one is seldom likely to be effective.

BREAKDOWN OF FILLS

A number of contributors discuss the breakdown of material structure during placing of fills, and both soft rocks and sensitive clays are considered.

In soft rocks the questions to be asked are as follows.

(a) Will an open, free-draining material be achieved by normal placement and compaction?
(b) Can breakdown and the development of a low permeability fill be avoided by special placement methods?
(c) Will breakdown be accompanied by undesirable effects in addition to a reduction in permeability, such as the release of water when low density chalk is crushed?
(d) If an open, free-draining structure is likely will it be permanent, or will undesirable effects occur after consolidation, such as collapse on loss of high suctions, break up of bonding of the rock fragments by swelling and chemical deterioration? Such effects are considered by Abeyesekera et al. (pp 1-14), Cedrun and Simik (pp 57-61), Cox (pp 79-86), Ingoldby (pp 137-142), Perry (pp 189-196) and Sherill (p241).
(e) If an open, free-draining fill is likely, but is unsatisfactory in the long term, what amount of breakdown is necessary to avoid long-term problems, and can it be achieved by special placing and compaction techniques? Cedrun and Simik (pp 57-61) give an example of such techniques.

Ingoldby (pp 137-142) presents a test using a special compaction apparatus for evaluating the degradeability of chalk during placement. The results are related empirically with field experience. Chaplin (p.241) suggests that porosity after compression, which is a measure of breakdown, should be a function of the crushing strength of the lumps. This seems logical, and it suggests that breakdown can be related to unconfined compression strength of the lumps of as-dug material. Such an approach may be valuable in allowing experience with a wide range of materials to be placed on the same basis. Some tentative figures relating crushing strength to breakdown are suggested in Table 4. Breakdown must also depend on the size of material likely to arise from excavation and thus on the spacing of bedding planes, joints and fissures, and so on.

A second problem involves soils which lose strength when reworked, but have acceptable properties when placed. Volcanic clay soils from Japan which have this behaviour are discussed by Inada et al. (pp 127-132) and Kuno et al. (pp 149-156). They describe the cone index test as a method of correlating the properties of these soils with placement problems in the field. This test has similarities to that described by Ingoldby (pp 137-142) in that a dynamic compaction test is used to simulate the effect of placing and compacting plant. The undrained strength after compaction is measured, rather than the rate of compression of the material during compaction, as is done by Ingoldby (pp 137-142). The amount of work done on the sample during compaction to simulate field working appears to be similar in the two otherwise widely dissimilar tests. This suggests that it may be possible to quantify the action of plant on the general basis of work done per unit volume of soil. If so, it might be possible to put experience of the breakdown of widely different soils on a common basis, and so to reinforce the methods of predicting the behaviour of particular soils.

CONCLUDING REMARKS

The contributions to this Conference have described a wide range of practical experience with clays and other fine-grained soils. There have been significantly fewer contributions discussing the fundamental engineering properties of the materials involved. A number of empirical and semi-empirical methods have been proposed which relate field experience with various types of index test. Such approaches have many practical attractions of simplicity and convenience but they suffer from a fundamental drawback: they apply to restricted and ill-defined ranges of conditions, and given a new material, not covered by the range in which experience has been obtained, there is no way of deciding whether or not the method will be valid and useful. More fundamental soil parameters, properly defined and measured, can be determined for all types of soil, and they can usually be directly and theoretically related to behaviour and performance. The use of fundamental methods, and their correlation with field experience and with the simpler and more convenient empirical methods, is essential if the power of the geotechnical engineer to predict unsatisfactory performance is to be improved.

REFERENCES

1. LANGFELDER, L.J. ET AL. Air permeability of compacted cohesive soils. J.Soil Mech. Fdn Engng Div. Am. Soc. Civ. Engrs, 1968, vol. 94, SM4, 981-1001.
2. BISHOP, A.W. ET AL. Factors controlling the strength of partly-saturated cohesive soils. Proceedings of conference on shear strength of cohesive soils. American Society of Civil Engineers, New York, 1960, 503-532.
3. GIBSON, R.E. The progress of consolidation in a clay layer increasing in thickness with time. Géotechnique, 1958, vol.8, no.4, 171-182.
4. LANE, K.S. and WASHBURN, D.E. Capillarity tests by capillarimeter and by soil-filled tubes. 26th Annual Meeting Highways Research Board, 1948, 460-473.
5. LAMBE, T.W. Soil stabilisation. *In* Foundation engineering, G.A. Leonard (ed.) McGraw-Hill, New York, 1962, 351-437.
6. VAUGHAN, P.R. Panel Discussion: Session 3. Proc. 9th Int. Conf. Soil Mech., Tokyo, 1977, vol.3, 411-413.
7. CHANDLER, R.J. and PACHAKIS, M. Long-term failure of a bank on a solifluction sheet. Proc. 8th Int. Conf. Soil Mech., Moscow, 1973, vol. 2.2, 45-51.
8. SKEMPTON, A.W. and BISHOP, A.W. Soils. *In* Building materials: their elasticity and inelasticity, M. Reiner (ed.). North Holland, Amsterdam, 1954, 417-482.
9. VAUGHAN, P.R. ET AL. The design, construction and performance of the Cow Green embankment dam. Géotechnique, 1975, vol.25, no.3, 555-580.
10. MITCHELL, J.K. Fundamentals of soil behaviour. Wiley, New York, 1976, ch.14.
11. SKEMPTON, A.W. and NORTHEY, R.D. The sensitivity of clays. Géotechnique, 1952, vol.3, no.3, 30-53.
12. BJERRUM, L. Problems of soil mechanics and construction on soft clays: state-of-the-art report. Proc. 8th Int. Conf. Soil Mech., Moscow, 1973, vol.3, 109-159.
13. WOOD, D.M. and WROTH, C.P. The use of the cone penetrometer to determine the plastic limit of soils. Ground Engng, 1978, Apr., 37.
14. DENNEHY, J.P. A new method for assessing the suitability of cohesive soil for use as embankment fill. MPhil thesis, University of Surrey, 1976.
15. VAUGHAN, P.R. ET AL. Discussion on application. Field instrumentation in geotechnical engineering. Butterworths, London, 1974, 617-623.
16. DE MELLO, V.F.B. Reflections on design decisions of practical significance to embankment dams. Géotechnique, 1977, vol.27, no.3, 279-356.
17. MATYAS, E.L. Air and water permeability of compacted soils. Permeability and capillarity of soils. American Society Testing Materials, 1966, STP 417, 160-175.
18. BISHOP, A.W. and VAUGHAN, P.R. Selset Reservoir: design and performance of the embankment. Proc. Instn. Civ. Engrs, 1962, vol.21, 305-346.
19. BISHOP, A.W. and AL-DHAHIR, Z.A. Some comparisons between laboratory tests, in-situ tests and full-scale performance with special reference to permeability and coefficient of consolidation. *In* In situ investigations of soils and rocks. British Geotechnical Society, London, 1969, 251-264.

I.F. SYMONS, MSc, MICE, Transport and Road Research Laboratory

General report: Performance of clay fills

This report considers problems in design, site practice and conditions experienced in service with emphasis on stability, deformation and use of instrumentation to monitor performance. The types of problem considered reflect real differences in the nature and scale of the operations between highways and dams. The construction of long lengths of shallow embankments for roads means that different materials from local sources are frequently incorporated within the fill and the properties of the completed work can be greatly influenced by both the methods used and the conditions prevailing during construction. As these factors can be defined only in general terms before construction they restrict the accuracy with which performance can be predicted. For road embankments a particularly important requirement is the limitation of differential movement from swelling or consolidation in service which can affect pavement performance or riding quality.

For earth dams the much larger scale and more restricted area of operation combined with the extended time scale and greater consequences of failure mean that more detailed assessments of materials and predictions of performance are necessary and possible. There is also greater scope for field monitoring of behaviour and control of construction which enable the design phase to be extended into the construction period. As well as considering overall stability and settlement the design and construction process must aim at minimizing local discontinuities and reductions in stress between zones which may lead to cracking of the core and leakage in service.

STABILITY DURING CONSTRUCTION

Stability during construction is of major concern for earth dams but for road embankments in the UK it is only likely to be critical for higher fills and those made of material too wet for normal placement methods or for embankments constructed on soft foundations where a combined fill/foundation failure can occur. This can be demonstrated by the use of the simple stability chart of the type given in fig.1 of Arrowsmith (p.27). A survey of major roads has indicated that about 95% of the total length of embankments have heights of less than 8 m. For this height with side slopes of 1 in 2 an undrained strength of about 30 kN/m^2 should ensure stability and this is similar to the values quoted as the lower limit for operation of normal plant (see Parsons, pp 307-314).

Method of assessment

Stability during construction can be assessed using total or effective stress methods of analysis; the relative merits of these methods are considered by Lieszkowszky (pp 236-237), Cole (pp 237-239) and Vaughan et al. (pp 205-217). The total stress method is valid only for conditions of undrained loading but has the advantage of speed and simplicity. The undrained strength of the fill used in this method is relatively easy to specify and check during construction but may prove difficult to determine in advance of construction. The effective stress method is less restrictive and allows the improvement in stability resulting from pore pressure dissipation during construction to be assessed. The strength parameters can be more accurately defined but the initial pore pressures and rates of dissipation are difficult to predict accurately prior to construction.

Penman (pp 177-187) refers to the use of effective stress methods of analysis linked with pore pressure measurements to assess stability during the construction of embankment dams of boulder clay. Inada et al. (pp 127-132) describe similar assessments of stability during and after construction of a high road embankment composed of soft volcanic soil. These authors refer to the use of drainage blankets within the fill for improving stability during construction. Lieszkowszky (pp 157-164) mentions total stress analyses for design and effective stress analyses for control of construction of a zoned earth dam founded on, and with a core constructed of high plasticity glacial clay. High pore pressures measured in the foundation necessitated pauses in construction.

Comparison of the methods

In both the total and effective stress methods of analysis it is assumed that deformations will be acceptably small at conventional factors of safety. For valid comparison between the methods a distinction is therefore necessary between fills which exhibit strain hardening and those which show strain softening behaviour. Vaughan et al. (pp 205-217) make such a distinction by considering separately the undrained behaviour of sandy clay and plastic clay fills.

For plastic clay fills exhibiting brittle behaviour the peak strength is mobilized at relatively small strains, but breakdown by field compaction is likely to reduce the rate of loss in strength with increasing strain and thereby inhibit the type of progressive failure normally associated with these materials in their natural state. For stable embankments similar values of the factor of safety are likely from total and effective stress analyses. However, field compacted specimens are likely to be required for reliable estimates of undrained strength and for prediction of construction pore pressures which will generally be low unless placed above the natural moisture content. A conservative approach is necessary for design in assessing strength from conventional tests on remoulded specimens and for major fills pre-construction trials may be justified. In such fills instability is likely to be manifested by displacements contained within a narrow zone which may restrict the value of general observations of deformation and pore pressure for purposes of construction control (Wrightman, pp 227-229).

For sandy clay fills exhibiting strain hardening behaviour the peak strength is mobilized at large strains and can be assessed from laboratory tests on remoulded specimens. Construction pore pressures are a function of placement conditions and therefore method and cannot be reliably predicted from laboratory tests. Moreover, the pore pressures are governed by the level of shear stress applied and this inhibits the prediction of pore pressures during the later stages of construction from earlier field measurements. For stable embankments effective stress analyses based on field measurements are likely to indicate significantly lower factors of safety than total stress analyses. Lieszkowszky (pp 236-237) suggests that the latter method overestimates short-term stability and this may be compatible with the need to limit deformations in these materials. During undrained construction of fills of sandy clay a stage is likely to be reached where further increase in height is accompanied by generalized and increasing deformations which continue until a limiting final height is attained. Field observations[1,2] generally confirm this pattern of behaviour but there are insufficient data to establish how the onset of deformations and the limiting heights compare with the critical heights obtained from total and effective stress analyses. It seems probable that these methods provide upper and lower bound estimates of stability near to failure and observations of incremental deformation are then likely to be essential for stability control.

Vaughan et al. (pp 205-217) present data which indicate that for a wide range of British clays the boundary in terms of material behaviour between plastic clays and sandy clays may coincide with the change in classification from CH to CI (PI $\simeq$ 27). Such a distinction in terms of plasticity is likely to provide only a general guide to material behaviour and in any particular case a detailed knowledge of the stress-strain pore pressure relations for the material is required.[3]

Pore pressures during construction

The difficulty of predicting pore pressures for use in effective stress analyses before construction is referred to by a number of contributors. Pavlakis (pp 266-269) mentions a dam project where centrifugal model testing provided useful design data on pore pressures during construction. Such centrifuge studies, particularly if linked with finite element analyses, are likely to provide a useful design tool for examining sensitivity and behaviour mechanisms, but cannot take into account the effects of layer construction and placement method on the stresses and pore pressures in fill. The more traditional approach for major fills has been to measure pore pressures during construction.

Penman (pp 177-187) describes early work at the Building Research Station on the development of piezometers and monitoring systems for use in embankment dams and gives records which illustrate the effect of moisture content and grading on the pore pressure response in boulder clay fills. Field data from a number of dam projects are summarized by Vaughan et al. (pp 205-217) to show the influence of plasticity of the fill on the pore pressures developed. These authors mention the effect of construction plant on the development of high initial pore pressures in sandy clay fill. High construction pore pressures caused by large horizontal stresses in fill due to heavy compaction are also referred to by Al-Shaikh-Ali (pp 15-23). Lieszkowszky (pp 157-164) gives measurements of pore pressure which show a different pattern of response to loading between the dam core and the foundation which he attributes to the effect of shear deformation in the non-saturated core material. The pore pressure observations in the foundation outlined (Lieszkowszky, pp 236-237) suggest that appreciable lateral dissipation may have taken place in the varved clay. Kenney[4] has shown that this may have an adverse effect on stability in soils where the coefficient of consolidation is appreciably greater than the coefficient of swelling. The data of Inada et al. (pp 127-132) on pore pressures indicate the additional benefit of drainage layers in reducing the pore pressure response with increase in applied loading.[5]

Instability during construction

When slips in road embankments occur during construction they are often dealt with on the spot and largely go unrecorded so that the opportunity is lost to examine common factors. A number of examples of such problems are given in the discussion which, in particular, indicate the detrimental effect on stability of discontinuities in the fill created by the method of placement.

Problems experienced during the construction of road embankments containing layers of granular material within clay, are outlined by Arrowsmith (pp 25-36) who suggests that where they are likely to occur their effects may be minimized by placing the material of higher permeability in level layers and by adopting protective measures to prevent ingress of water when the presence of erodible material at the top of the fill is unavoidable. Wrightman (pp 227-229) describes major instability which occurred during the construction of two road embankments of London clay where the probable cause was the presence within the fill of smooth outward sloping planes which had been formed, using smooth-wheeled rollers, to promote discharge of surface water during breaks in construction. Chaplin (p.241) also refers to the formation of discontinuities in fill by the movement of surface layers ahead of heavy rollers. Pavlakis (pp 266-269) describes highly polished laminations in and between lifts, caused by construction traffic and compaction equipment, which were discovered during placement of the clay core of an earth dam. Cole (pp 237-239) gives a case history of instability in the outer slopes of a road embankment on sidelong ground where the likely cause was water transmission through more permeable zones created in the fill by surface desiccation during construction.

STABILITY IN SERVICE

For dams and water-retaining structures the consequences of failure in service are generally such that any risk of instability during or after reservoir filling must be eliminated. As discussed by de Mello,[6] the design must ensure stability even under conditions of limiting pore pressures. The slopes adopted for shallower road and railway embankments are generally much steeper because superficial instability which does not affect operation is tolerable. For such fills the designer must aim at achieving an economic balance between construction and maintenance costs and the type, frequency and magnitude of instability in service is then of prime concern.

Stability assessment

Stability in the long term is governed by the strength parameters in terms of effective stress and the equilibrium pore pressures developed within the fill. These equilibrium pore pressures depend on the flow pattern established by infiltration and seepage of water through the surface boundaries. The changes in stability following completion of construction are controlled by the changes in effective stress as pore pressures adjust to their equilibrium values and by any alteration in strength resulting from the consolidation and swelling process. The magnitude and direction of the changes in stability and effective stress therefore depend on the total stresses and condition of the fill on completion of construction. For high embankments placed appreciably wet of optimum, as described by Inada et al. (pp 127-132), consolidation following completion of construction results in an improvement in overall stability with time; for shallower fills placed in an initially dry condition swelling will lead to a decrease in stability with time as described by Chandler (p.234). Many embankments will fall between these limits with consolidation occurring at depth and swelling taking place nearer the surface boundary as indicated by Farrar's (pp 101-106) measurements of pore pressure within a road embankment of London clay.

For homogeneous clay embankments complete pore pressure equalization may take many decades with the rate dependent on the consolidation and swelling properties of the fill and the proximity of internal drainage and surface boundaries. Vaughan et al. (pp 205-217) point out that decrease in permeability with stress and depth can cause more rapid equilibrium of pore pressures in the near surface layers and this can lead to the formation of perched water tables. The pore pressures within these layers will respond rapidly to climatic variations and the maximum values will exercise a controlling influence on stability against superficial movements. Pore pressure observations are given by a number of contributors using standpipe piezometers but the response time of this equipment is too slow to detect rapid fluctuations in the surface layers of the fill.

The protection afforded by vegetation cover on the slopes of railway embankments (Ayres, p.239) is likely to have a doubly beneficial effect on stability at shallow depths, by reducing net infiltration and through the reinforcing action of the root structures. The formation of shrinkage cracks in a road embankment during the abnormally dry summer of 1976 is referred to by Al-Shaikh-Ali (pp 15-23) and there are indications that on other roads extensive cracking of a similar type occurred in the embankment slopes. Ingress of water into the cracks during the subsequent winter would induce increased pore pressures in the near surface layers and an increase in frequency of superficial movements could then be expected.

The strength of fill in terms of effective stress required for the assessment of stability in service is considered by Vaughan (pp 283-295). The results of laboratory studies given by Cox (pp 79-86), Abeyesekera et al. (pp 1-14) and Vaughan et al. (pp 205-217) suggest that the strength parameters are unlikely to be greatly affected by compacted structure, consolidation and swelling, and can therefore be determined from tests on laboratory prepared specimens. For shallow fills stability is largely governed by the cohesion component C' which results from the precompression effect of compaction and may therefore be absent in loosely tipped material.

Vaughan et al. (pp 205-217) suggest that progressive failure is likely to have little influence on stability within fills and peak strength parameters can therefore be used for design. In contrast Dohaney and Forde (pp 95-100) refer to the need to use residual strength parameters for assessing long-term stability in fills where loss of strength occurs with time. This would lead to much flatter slopes for road and railway embankments and there would seem to be little evidence to justify this approach for most fill materials encountered in the UK. In this context Ayres (p.239) points out that a great proportion of the length of railway embankments over 100 years old shows no signs of previous instability. This does not necessarily mean that the performance of these embankments would be acceptable for modern roads because routine adjustment of railway tracks can compensate for deformations which may be detrimental to road pavements. Whyte (p.237) also considers stability in service solely in terms of the effective angle of friction ϕ' of the fill and suggests that with the water table at the slope surface the maximum stable slope angle would be reduced to $\phi/2$. Such extreme conditions of pore pressure are only likely close to the slope surface where stability would be greatly influenced by even a small cohesion component of strength.

Instability in service

The performance in service of homogeneous clay fills is considered by Vaughan et al. (pp 205-217) who predict a high frequency of local failures in steep-sided shallow fills typical of road and railway embankments, a few years after construction when equilibrium pore pressures are achieved in the near surface layers of fill, with a reduction in frequency thereafter. The data given also indicate that for equivalent heights and slopes, fills of sandy clay are likely to be more stable than those constructed from the more plastic clays. For highways in the UK large scale earthworks effectively started with the advent of motorway construction in the late 1950s and a survey in the mid 1960s[7] indicated that the frequency and type of instability then occurring was not a serious problem.

Instability in the side slopes of a road embankment composed of clay and gravel layers due to softening by water seepage from the median drain is described by Ingold and Clayton (pp 133-136). Normal drainage practice for road embankments in the UK is outlined by Wood (p.244). Snedker (pp 229-232) suggests that more positive drainage systems are needed to prevent entry of water into clay fills containing lenses and layers of higher permeability material and gives examples of slope failures and subsidence affecting road pavements which occurred in fills up to eight years old after a period of intense rainfall in early 1977. Threadgold (pp 232-234) also attributes instability in a 16 year old embankment of Lias Clay to increase in pore pressure resulting from entry of water from the drainage system and through the road pavement. The data which he gives from back analyses of these slips indicate that the average strength in operation along the shear surfaces was below the peak value. Hill (p.234) describes instability in a Lias Clay embankment of similar age where the failure surface was found to extend into the natural ground and to be bounded by the base of a solifluction sheet. Embankment failures associated with the development of perched water tables in fills founded on soliflucted material have also been reported by Chandler et al.[8] Ayres (p.239) mentions that a significant number of more recent failures in railway embankments over 100 years have involved circular slips extending below natural ground. These embankments are generally of similar height and slope to the more recent highway embankments but because of their greater age are likely to have been constructed to a lower initial density and to have achieved equilibrium.

The principal types of failure encountered in clay tips from mines and quarries are outlined by Miller (p.236) whose observations indicate that slicken sided surfaces formed by the movement of poorly compacted soil down the slope face can impede drainage from the body of the fill and lead to the development of shallow surface movements. This mechanism is less likely in recently constructed highway embankments where overfilling and trimming back and the operation of plant on the slope face for final profiling will prevent the formation of loose uncompacted zones.

These field data indicate that although delayed failures in relatively steeply sided embankments are initiated by adverse weather conditions, there are also factors which adversely affect the flow pattern within the fill and increase the rate of swelling. These include the presence of lenses or layers of more permeable material, density gradients caused by layering and compaction, and ingress of water through road pavements and from unlined or defective drainage systems. There is also some indication from the older embankments that depth of movement may increase with time as pore pressure equilibrium extends within the fill. A wider survey of the type and frequency of instability in motorway embankments would be needed to establish whether or not any changes in current slope design or drainage practice are required.

SETTLEMENT AND DEFORMATION

In the design of embankments of cohesive fill a main objective must be to limit the magnitude and effect of differential movements. For zoned earth dams differential movements occurring during construction can cause reduction in stress which increases the risk of cracking in the core and leakage on impounding; for road embankments differential settlements occurring after construction can affect riding quality and pavement performance. Data on the deformation, stresses and volume changes within cohesive fills are now considered in terms of the performance of dams and embankments.

Performance of dams

The excessive leakage which occurred from the Balderhead and Hyttejuvet dams on first filling[9,10] can be explained in terms of differential settlement between the core and shoulders causing the total stresses within the core to be reduced to a value equal to the reservoir head, thereby reducing the effective stresses to zero and allowing cracks to form. The mechanism of hydraulic cracking is complex and any reduction in stress is likely to be dependent on the dimensions and shape of the core as well as the relative compressibility and strength of the core material. The two opposing arguments for placement of the core material wet and dry of optimum to minimize load transfer are referred to by Penman (pp 177-187, 219-221).[11] The field measurements presented indicate that in wetter and softer cores, higher total stresses are likely to occur and that the additional thrusts reduce movements on first impounding. Flexibility was a prime requirement in the dam (Harpster, pp 119-125) which crosses a potentially active fault and the specification called for the material to be placed 1-2% above optimum. Kennard et al. (pp 143-147) describe the use of a shear strength specification for core materials placed wet of optimum in a number of recent dams. The strength range commonly used was 55-100 kN/m^2 with the upper bound set by the need to ensure a plastic core and the lower bound usually governed by the assumed limits of trafficability and not by stability considerations.

When arching and load transfer occurs during construction appreciable changes in deformation and stress are likely during reservoir filling[12] and these may reduce the risk of hydraulic cracking. If impounding is slow and a rise in pore pressure occurs from seepage then shearing may develop within the core loading to an increase in total stress. Swelling of the core under reduced effective stress is also likely to increase stresses and field evidence indicates that hydraulic cracking is more likely in low plasticity materials having low coefficients of swelling.[13] Coumoulos and Koryalos (pp 73-78) give measurements of stress and deformation during the construction of a 90 m high zoned earthfill dam with a core of low plasticity clay placed at about optimum moisture content. The measurements show reasonably good agreement between field compressibility and laboratory measurements on specimens from the core. At another site, where compessibility was determined from tests on laboratory compacted specimens, the agreement with field data was poor (Coumoulos, pp 221-224). The earth pressure measurements suggest that the vertical stresses were appreciably below overburden during the later stages of construction and the effect of this on the settlement comparisons is referred to by Penman (pp 219-221). Coumoulos (pp 221-224) reports that no lateral movement of the core on first filling was detected by the inclinometers, which indicates that pressures within the core may have been higher than the recorded values.

Cracking in dams associated with changes in geometry in the longitudinal and transverse direction has been described by Lowe.[14] Chaplin (p.241) points out that because of geometry only a small proportion of the weight of the fill placed near the top of the dam contributes to the vertical stress at depth in the core and that this effect is particularly apparent with granular cores.

Performance of road embankments

For road embankments of cohesive fill the magnitude of post-construction settlements must be limited, particularly in the vicinity of rigid structures, to ensure an acceptable riding quality. Studies of the performance of backfill to structures have shown that quite small settlements occur within well compacted cohesive soils placed at moisture contents close to the plastic limit. For such conditions settlements within the embankment may be comparable to those occurring in the subsoil. This is also the case in the study reported by Farrar (pp 101-106) where measured deformations in a London clay embankment were broadly in agreement with predictions based on tests on undisturbed and laboratory compacted specimens.

For fills of wet cohesive material the use of drainage layers to accelerate consolidation and minimize post-construction settlements is described by Grace and Green (pp 113-118), Inada et al. (pp 127-132), Boman and Broms (pp 49-56) and Østlid (pp 224-227). Grace and Green (pp 113-118) also refer to the difficulty of defining acceptable deformations and draw attention to the need for more information on the vertical accelerations which can be tolerated by passengers and vehicles. Østlid's (pp 224-227) measurements indicate high all-round stresses within a soft clay fill during the early stages of construction with a decrease in lateral stress towards earth pressure at rest during consolidation. A similar pattern of behaviour is shown by the earth pressure measurements against a timber wall given by Boman and Broms (pp 49-56) and these demonstrate the effect of drainage conditions on the rate of stress change. Their observations of settlement indicate compression of about 7%, which agrees with predictions from oedometer tests on samples from the fill although the measured rates of consolidation were faster than the laboratory tests predicted. The data from extension compression gauges report by Inada et al. (pp 127-132) show the effect of closer drain spacing in reducing lateral strains within a 20 m high road embankment of soft volcanic soil. The settlement data show a total compression of only about 5% with residual settlements after construction of about 1%. Similar values are quoted by Kuno et al. (pp 149-156) for another volcanic soil with similar characteristics.

Cox (pp 79-86) gives the results of a laboratory study of the volume changes occurring in fills placed appreciably dry of optimum and outlines the type of problem which may be encountered with such materials in road embankments and dams. In dry materials the use of a method specification for compaction could lead to a less than adequate state of compaction and the resulting increase in bulk permeability may reduce the time required to achieve equilibrium conditions. In this context Edwards (p.236) describes a pavement failure on an approach embankment which occurred within two years of construction, where swelling within the upper layers of the fill as a result of ingress of water is the most probable cause. The embankment had been constructed from overconsolidated clay obtained from considerable depth in cutting and compaction had produced relatively high air voids. For narrow embankments of this type and in the outer regions of wider embankments, lateral deformation may be of significance as a result of increased shear deformations accompanying swelling. Threadgold (pp 232-234) gives data on the consolidation behaviour of samples from a five-year old embankment of London clay which indicate a potential for further swelling in the middle of the fill. Sherrell (p.241) refers to the large volume increase involved in the conversion of pyrite to gypsum and suggests that swelling problems in mudstone and clay fill may be caused when iron sulphide contents are greater than 1%.

SUMMARY

The papers and discussion have provided useful information concerning practical aspects of the design and performance of clay fills and have highlighted particular problems and areas where more data are needed.

The relative merits of effective and total stress methods for assessing stability during construction are considered and the limitations imposed by material behaviour are pointed out. Field data are given which indicate that high initial pore pressures and stresses can be developed, and discontinuities formed within fill, by the action of construction plant.

A number of cases are presented of delayed failures and problems of subsidence and volume change in shallow road and railway embankments associated with a decrease in effective stress in service. These indicate the adverse effect on the flow pattern and rate of softening within the fill caused by ingress of water from road pavements and drainage systems, by the inclusion within cohesive fills of more permeable materials and by inadequate compaction of dry materials. A wider survey of the type and frequency of instability in road embankments would be required to assess whether or not any changes in current slope design and drainage practice are necessary. More information is needed on the equilibrium pore pressures attained in embankment slopes and on the effect of compaction and swelling on the cohesion component of strength which governs shallow movements.

The effect of placement moisture content on the performance of dam cores is considered and data are presented which indicate that higher stresses are likely in wetter materials, with a consequent reduction in movement and in risk of hydraulic cracking on first filling of the reservoir.

Post-construction settlements in road embankments may affect riding quality and case records are presented which demonstrate the advantages to be gained from the use of drainage layers in wet cohesive fill to accelerate consolidation during construction. More information is needed on acceptable deformation for roads.

REFERENCES

1. BANKS, J.A. Construction of Muirhead Reservoir Scotland. Proc. 2nd Int. Conf. Soil. Mech., Rotterdam, 1948, vol.2, 24-31.
2. LINELL, K.A. and SHEA, H.F. Strength and deformation characteristics of various glacial tills from New England. Proceedings of research conference on shear strength of cohesive soils. American Society of Civil Engineers, New York, 1960, 275-314.
3. BISHOP, A.W. Contribution to discussion. Field instrumentation in geotechnical engineering. Butterworths, London, 1974, 666-671.
4. KENNEY, T.C. Pore pressures and bearing capacity of layered clays. Norwegian Geotechnical Institute, Oslo, 1965, Publication 63.
5. BISHOP, A.W. Some factors controlling pore pressures set up during the construction of earth dams. Proc. 4th Int. Conf. Soil Mech., London, 1957, vol.2, 294-300.
6. de MELLO, V.F.B. Reflections on design decisions of practical influence to embankment dams. Géotechnique, 1977, vol.27, no.3, 279-355.
7. SYMONS, I.F. The magnitude and cost of minor instability in the side slopes of earthworks on major roads. Road Research Laboratory, Crowthorne, 1970, Report LR 331.
8. CHANDLER, R.J. et al. Four long term failures of embankments founded on areas of landslip. Q.Jl Engng Geol., 1973, vol.6, no. no.4, 405-422.
9. VAUGHAN, P.R. et al. Cracking and erosion of the rolled clay core of Balderhead Dam and the remedial works adopted for its repair. Trans. 10th Int. Congr. Large Dams, 1970, vol.1, 73-93.
10. KJAERNSLI, B. and TORBLAA, I. Leakage through horizontal cracks in the core of Hyttejuvet Dam. Norwegian Geotechnical Institute, Oslo, 1968, Publication 80, 39-47.
11. PENMAN, A.D.M. The failure of Teton Dam. Ground Engng, 1977, vol.10, no.6, 17-27.
12. SQUIER, L.R. Load transfer in earth and rockfill dams. J. Soil Mech. Fdns Div. Am. Soc. Civ. Engrs, 1970, vol.96, SM 11, 213-233.
13. VAUGHAN, P.R. Cracking of clay cores of dams. Report by A.D.M. Penman. Proc. Instn Civ. Engrs, 1970, vol.46, 115-117.
14. LOWE, J. III. Recent developments in the design and construction of earth and rockfill dams. Trans. 10th Int. Congr. Large Dams, 1970, vol.5, 1-28.

N.W. LISTER, BSc(Eng), MICE, Transport and Road Research Laboratory

General report: Road subgrades

Over the years the importance assigned to soil subgrade in determining road performance, and hence design, has fluctuated widely. After the Second World War it was considered to be of such importance that road design was carried out mainly by staff with soil mechanics training. At a later stage only the design of the sub-base of the road was directly affected by the subgrade strength and at present the picture is changing again.

The reasons for these changes are not hard to find. In 1952 it was possible to say that 'in the past no design method was used and much the same thickness of road construction was sometimes placed on weak clays and stony gravels, the range of soil strengths being at least 25 to 1; premature failures on weak subgrades were naturally not uncommon.'[1] The problem appeared to be one in the province of the soils engineers, namely to design a road so that the maximum shear stress generated by traffic in the subgrade did not exceed its shear strength. Although this approach led to different thicknesses of roads on subgrades of different strengths it did not prevent the development of unacceptable subgrade deformation under pavements, the construction of which was generally of granular material under a relatively thin bituminous surfacing. At that time it was not possible to predict the subgrade stresses required in the design process or to treat the influence of repetitional traffic-loading; the design methods subsequently adopted were not in the main stream of soil mechanics. They were based increasingly on the empirical correlation of road performance with some form of penetration test on the subgrade, such as the California bearing ratio (CBR) and North Dakota cone tests or with a soil classification test, typified by the group index method. Increasing traffic led to the inclusion as design parameters of wheel load and intensity of commercial traffic; in this phase the toll of the soil, although important, was somewhat diminished.

In the 1960s full-scale road experiments had demonstrated that the heavier traffic that was using the new motorway and arterial trunk road network was best catered for by thick bitumen or cement-bound road bases. These stiff materials so reduced traffic stresses transmitted to the subgrade that it was unlikely that the soil could be of major importance in pavement design. Soil strength influenced only the design of sub-base, the thickness of which is determined as much by its ability to serve as a stable platform for constructing the road as by its influence on the in-service behaviour of the completed pavement.

More recently the pendulum has begun to swing back. Under a given pavement the soil stiffness (as indicated by CBR) has been shown to be related to the elastic deflexion of the pavement surface; in turn, deflexion has been strongly correlated to structural performance under traffic. Also, increasing knowledge of the mechanism governing pavement deterioration has indicated that, under the most adverse in-service conditions (i.e. heavy wheel loads applied at high road temperatures and possibly associated with a high seasonal water table), the stresses generated by traffic are sufficient to bring about appreciable deformation in the subgrade. Both developments indicate the importance of the soil in determining road performance, as does the recent Department of Transport specification[2] of capping layers designed to ensure an improved subgrade immediately beneath the pavement.

There is now considerable activity world-wide aimed at producing structural methods of pavement design based on an understanding of the physical behaviour of pavement materials and soil; considerable emphasis is therefore placed on the stress/strain behaviour of soils under repeated loading. This comes at a time when the mainstream of soil mechanics is showing a similar interest in stress/strain rather than in strength and stability; in the study of offshore problems the repeated loading aspects are also present. The coming together of geotechnics and the subgrade aspects of pavement design is gratifying.

Black and Lister (pp 37-48) show that the suction method developed at the Transport and Road Research Laboratory can be used to predict changes in the CBR or undrained shear strength of remoulded soils and therefore is applicable to clay fill materials. The method relies on correlations between plasticity data, soil suction and the bearing capacity factor Nq. It has been used to quantify the changes in subgrade strength resulting from various moisture conditions which may occur during construction and subsequently under the completed pavement. The effect of hysteresis observed in the relationships between soil suction and moisture content is such as to ensure that, under equilibrium conditions of moisture under the completed pavement, the strength of the subgrades that become very wet during construction remains considerably less than that of subgrades that are protected during the construction phase. Experimental relations between pavement deflexion and CBR strength, and between deflexion and pavement performance under traffic, demonstrate that wet weather during construction can result in a pavement life which is only about half that expected for a road constructed on fill during average weather conditions.

Roads are of low cost per unit area and only a limited amount of strength testing of the subgrade is therefore acceptable economically. Workable correlative methods between strength tests and the cheaper soil classification tests are therefore attractive as a possible means of extending the available design information. The envelope of results of CBR measurements made by Al-Shaikh-Ali (pp 15-23) on Cheshire lodgement tills can be reproduced reasonably well by the suction method, indicating that these materials behave in the same predictable manner as other British cohesive soils. The large scatter of results about the mean is to be expected, given the known variability often obtained in measuring the plastic and liquid limits and the practice which fails to measure the moisture content on the same sample of soil used to determine the plasticity data. Measurements of shear strength reported by Dennehy (pp 87-94) can also be satisfactorily predicted. However, CBR values measured by Perry (pp 189-196) and shear strengths measured by Parsons (pp 169-175) are inaccurately estimated; in each case predicted changes of strength are much greater than measured changes. The explanation may be found in the effect of different methods of compaction on strength, as mentioned by Vaughan et al. (pp 205-217), and in the effect on the strength obtained of the elapsed time between compaction and the measurement of strength. The latter effect has been noted on a range of soils in addition to the chalk-clay mixtures studied by Perry (pp 189-196).

Arrowsmith (pp 25-36) presents subgrade strengths obtained on several contracts where control of clay fills earthworks was by moisture content measurements related to the plastic limit. Fig.2 on p.28 is typical and indicates the inadequacy of control by this approach. The suction method when applied to the results predicts satisfactorily the envelope of results when likely differences in measured moisture contents and plastic limits, not made on identical samples, are taken into account.

The validity of some of the assumptions made in the derivation of the suction method has been considered in detail. There are theoretical shortcomings of the method that can result in considerable errors in forecasting the absolute strengths of some soils. However, the method has been shown to predict with good precision changes in strength resulting from changes in moisture content of a wide range of soils. Its effectiveness is improved if excess pore pressures generated during compaction are allowed to dissipate before strength testing is carried out.

There are a number of approaches to improving soil moisture conditions during the construction of a road and hence the performance of the road under traffic. Solutions vary from waterproofing the site to blanketing the site in a deep sand layer on the assumption that, in the UK, the soil will always be liable to be very weak. The latter solution will no doubt work, but at a cost. An intermediate solution is that of the capping layer of granular material, or improved subgrade.[2] The presence of this additional thickness of material should increase the effectiveness of the sub-base in carrying construction traffic, reduce the level of pavement deflexion and its variability and therefore improve the performance of the completed pavement.

A more economical solution may be found in cement stabilization of, at least, the upper subbase layers; alternatively, lime could be used to improve the strength of existing subgrade. Machinery capable of mixing effectively materials to greater depths is now available; site trials would increase confidence. Doubts are expressed as to the long-term effectiveness of the process but there have not been reports of deterioration from countries where the process is widely used.

The capping layer concept, introduced to take account of the growing awareness of the importance of soil in determining road performance, is only at an interim stage of development and can be changed and widened to reflect increasing knowledge and experience. Whatever the final form it is fairly certain that the designs, at least for heavy traffic, will be more conservative in relation to the soil; this will correctly reflect the economic reality that the cost of traffic delays on such roads, arising from the need for premature repair, far outweighs any savings from reduced first costs associated with designs inadequate in relation to the subgrade.

The question arises as to whether more of different soils testing at the design stage would be beneficial. On level ground soil conditions probably change more rapidly than on cut and fill and more testing would give more information as to soil variability along the line of the road, and better allowance could possibly then be made for this variability in the design. In deep cut extra information could only be obtained at considerable extra effort and cost but there is, in any case, some doubt at present about the strength of undisturbed formations exposed in deep cuts: it is not certain that CBR tests on remoulded soils adequately represent their strength. On fills direct testing is obviously not possible at the design stage because the fill is not in position to test; estimation of the strength of soils for pavement design is an obvious weakness in the present design procedure.

A possible alternative approach to bringing

about improvement would be for the design of the pavement to be modified during construction in the light of circumstances prevailing; the site engineer might be given the power to vary the thickness of sub-base to take care of likely changes in design strength as the result of moisture conditions during construction. Strength testing during construction would be necessary; testing with the penetrometer or by the moisture condition value method could be used to assess present strengths and a guide based on the suction approach would then indicate changes in subgrade strength likely to occur in attaining equilibrium conditions under the completed road. This approach would be particularly valuable on embankments where information on the strength of the subgrade at the design stage is, of necessity, relatively speculative; testing of the fill is in any case required for control of the construction of the embankment. The principle of varying sub-base thickness according to soil conditions on site has been successfully used on some past contracts in Scotland.

When considering possible improvements to site practice, the obvious most important single consideration is keeping the subgrade dry. Sealing the formation when still dry is superficially attractive but it is difficult to make such a seal sufficiently robust to be able to withstand compaction of sub-base materials and has the disadvantage of restricting drying out of a subgrade wetted as the result of a break in the seal. A better alternative would appear to be a high quality seal on top of a granular sub-base, the sub-base to be laid as quickly as possible after the formation had been prepared; this arrangement would allow drainage of the subgrade to take place in case of leakage through the seal. However, it can prove difficult to make and maintain a seal on an open-textured granular sub-base. Relatively impermeable shale sub-bases, whenever available, or sub-bases stabilized with cement will provide the necessary seal.

Consideration should also be given to providing more comprehensive weather protection in the form of a movable tent - not to cover the whole site but only the final trimming of the formation before covering with sub-base and sealing. On concrete roads laid by conventional concreting trains the necessary rails to carry a cover are already in position.

Another option is to leave sufficient clearance under overbridges and sufficient weight allowance on underbridges for extra pavement thickness in the form of additional surfacing to be added. This would be done if deflexion tests on the completed road made some months after it was opened to traffic, when equilibrium subgrade conditions had been reached, showed the design to be inadequate.

In building roads to deadlines in the uncertain climate of the UK the necessary rate of progress is inevitably obtained at the price of reduced performance of the in-service; the penalty is often the result of wet subgrade conditions during construction. There are a number of ways in which this undesirable effect can be reduced. Experience on actual jobs is the best way to evaluate them.

REFERENCES

1. DEPARTMENT OF SCIENTIFIC AND INDUSTRIAL RESEARCH. Soil mechanics for road engineers. Transport and Road Research Laboratory, Department of Scientific and Industrial Research, Crowthorne, 1952.
2. DEPARTMENT OF TRANSPORT. Specification for roads and bridgeworks. HMSO, London, 1976.

A.W. PARSONS, Transport and Road Research Laboratory

General report: Construction and placement of clay fills

Embankment construction involves the employment of highly expensive equipment capable of achieving high rates of output in the earth-moving and compaction processes. Costs of construction are related to the output of the plant, and in any given situation the achievement of maximum potential output should result in the lowest costs of construction. The Engineer, however, has to ensure that the soil being placed in the embankment is capable of achieving the designer's requirements and that it is placed to an adequate state of compaction; such control as is necessary to achieve this must be carried out at a rate compatible with the output of the construction equipment if construction costs are not to be adversely affected. Other important issues affecting the efficiency and economics of construction are those associated with the ability of the construction equipment to operate effectively in the prevailing soil conditions and the effect of plant operations on the quality of certain types of fill material.

SPECIFICATION AND CONTROL OF QUALITY OF MATERIAL

The specification with regard to the quality of the material has to reflect the needs of the designer to ensure that the performance he requires from the fill can be achieved. The limits set by the specification may be dictated by the stability and settlement requirements (usually in the case of high embankments), by the needs for impermeability and flexibility (in the case of cores for dams), or by the potential difficulties of plant operations (for shallow embankments such as those usually encountered in road construction).

It is essential that any measurement of quality reflects the relevant engineering properties of the soil and to this end a direct measurement of such a property (e.g. undrained shear strength) might be appropriate or alternatively an indirect type of measurement which correlates with the engineering property might be used. The latter could be preferable when it has the advantages of increased practicability and rapidity at the time of construction, especially as the method of control must be such that effective control is obtained at a rate compatible with the rate of embankment construction. In addition, the method and criterion adopted should be applicable over the range of soil variations encountered, in particular over those variations in plasticity and in stone content which are difficult to identify without additional laboratory testing.

The criteria by which to judge the effectiveness of a method of control of quality of the material could be listed as

(a) direct measurement of or relation with the engineering property of interest
(b) accuracy and reproducibility
(c) quickness in providing a result
(d) ability to test a representative sample
(e) applicability to the range of soil types likely to be encountered
(f) capability of use at any convenient location, e.g. in the site laboratory or on site at the point of excavation or placement.

In the hurly-burly of major earth-moving operations it is not possible to use complex control tests. Testing at the construction stage should be as simple as possible yet still complying with the criteria listed.

Alternative methods of specification and control are now presented.

Moisture content

The limit of acceptability of a material may be specified in terms of moisture content. The method would not be expected to satisfy the need to relate to an engineering property unless the soil were extremely consistent. The use of the method for highway embankments is described by Al-Shaikh-Ali (pp 15-23) but the main purpose in this particular application was to provide a means of rejecting the more difficult clay soils with high plastic limits. Lieszkowszky (pp 157-164) describes an application of the method in the control of the quality of material used in the clay core of a dam. In this case the consistency of the material appeared to be a major factor contributing to the success of the method. Undrained shear strength was found to correlate well with the moisture content and provided an additional means of control.

Ratio of moisture content to plastic limit

The comparison of moisture content with plastic limit has been used as a means of assessment of the suitability of clay fill in highway embankments for a number of years. The development of the use of the ratio of moisture content to plastic limit is described by Arrowsmith (pp 25-36) and applications of the method are mentioned by Al-Shaikh-Ali (pp 15-23) and Threadgold (pp 251-253). However, when undrained shear strength is the engineering property of major interest in deciding the criterion of suitability, then the evidence indicates that there are substantial difficulties associated with the use of the ratio of moisture content to plastic limit as an on-site method of control.

Evidence of poor correlation between the ratio and the undrained shear strength is provided by Arrowsmith (pp 25-36), Dennehy (pp 87-94), Black (p.260) and Parsons (p.263). One reason suggested for the poor correlation is the low accuracy of the plastic limit test and the ensuing variability of the results (Black, p.260; Powell, pp 260-263; Parsons, p.263). Another reason for poor correlation with strength could be the discrepancy between the samples selected for the determination of moisture content and of plastic limit. Methods of correction for such discrepancies are described by Arrowsmith (pp 25-36, 265) and Forde (pp 263-265).

Moisture content and liquid limit

Although also susceptible to discrepancies between samples selected for the two separate measurements involved, the comparison of the moisture content with the liquid limit has been suggested as a preferable alternative to the use of the ratio of moisture content to plastic limit as a measure of the quality of the soil. Dennehy (pp 87-94) shows that liquid limit is a more accurate index of strength than plastic limit, and other data are provided by Webb (p.263). In general, the data indicate that to achieve a given value of undrained shear strength, the ratio of moisture content to liquid limit must be increased as the liquid limit decreases. Other comments in favour of such a method are made by Powell (pp 260-263) and Lake (p.273). However, Black (p.260) quotes results that indicate that the method is not universally applicable.

Wetness index or consistency index

The use of a wetness index, as proposed by Dohaney and Forde (pp 95-100), or a consistency index, involves the measurement of two parameters as well as moisture content. In the case of the wetness index these parameters are liquid limit and the optimum moisture content determined in a prescribed laboratory compaction test, and for the consistency index they are the liquid limit and the plastic limit. Disregarding the problems of sampling discrepancies and possible variability of test results, the amount of effort required to determine either the wetness index or the consistency index would appear to inhibit their use as methods of on-site control. They could be used at the design stage, but where variable soil conditions occur such methods appear to have little practical application during construction. Powell (pp 260-263) refers to an early attempt to use consistency index which was foiled by difficulties with the measurement of liquid limit by the test method then in use.

Moisture condition test

A new form of test, the moisture condition test, is proposed by Parsons (pp 169-175). It is comparatively simple to perform and is claimed to satisfy to a large extent the criteria for judging the effectiveness of a method of assessing the suitability of soil. The test is relatively rapid and is reported to be applicable over a wide range of soil types. The resulting moisture condition value (MCV) is shown to correlate well with the undrained shear strength. Further data on the correlation of the MCV with moisture content, undrained shear strength or CBR are provided by Threadgold (pp 251-253), Edwards (pp 253-258) and Webb (p.263). Forde (p.253) questions the application of the test to low plasticity stony clays, but Edwards (pp 253-258) shows the application of the test over a wide range of soils including such materials. The restriction of the method to material passing a 20 mm sieve could be overcome, if thought appropriate, by increasing the scale of the test apparatus (Parsons, pp 258-260). The analysis of the relation between MCV and undrained shear strength by Charles (pp 315-321) shows that the undrained shear strength of the fully compacted specimen is almost directly proportional to the energy required to compact it.

Parsons (pp 169-175) also demonstrates the possible application of the relation between MCV and moisture content to the classification of soil, and this is also referred to by Webb (p.258).

Direct measurements of strength

The direct measurement of undrained shear strength on a number of dam construction sites using triaxial tests on 100 mm dia. specimens is described by Kennard et al. (pp 143-147). Depending on the site the samples were either taken directly from the compacted fill or remoulded in the laboratory; triaxial cell pressures varied considerably between sites. On one site triaxial testing was superseded by in situ field vane tesing. Lieszkowszky (pp 157-164) also describes the use of field vane tests in the compacted clay core of a dam as a form of control of both quality of the soil and of the compaction. Arrowsmith (pp 25-36) describes how, in a future road construction project, it is intended to determine undrained shear strength using unconfined compression tests on 38 mm dia. specimens taken from material recompacted in the laboratory. Black (p.260) recommends that where a clay is sufficiently homogeneous and of low sensitivity, a penetrometer method can be used to determine undrained shear strength.

Evidence is also available on the potential difficulties in performing tests involving the direct measurement of soil strength and in interpreting the results. The type of soil and the methods of sampling and testing can all have an influence and give rise to radically different results. Both Arrowsmith (pp 25-36) and Lieszkowszky (pp 157-164) describe tests made with the field vane where the shear strength values were shown to be about 1½ times the shear strength determined by unconfined compression testing. Vaughan (pp 283-295) draws attention to the influence of the rate of strain during strength testing. He advocates a high rate, such as is achieved in the fall-cone test, if strength is to be related to trafficability problems. Rocke (pp 265-266) describes the difficulty in sampling for undrained shear strength tests where stony clay is encountered. He achieved rates as low as one successful test in six sampling attempts, and he states that the method was inefficient and expensive in manpower. Child (pp 269-271) considers that the use of vane tests and 38 mm dia. specimens does not incorporate sufficiently large samples to take account of soil structure, although this factor is clearly of greater importance where undisturbed tests are concerned. The need to sample larger volumes of fill material in certain instances can be overcome by use of in situ plate-bearing tests or large scale tests in the laboratory. Applications of such tests are described by Cox (p.271) and Rocke (pp 265-266).

It appears that the direct measurement of strength could work well and is generally preferable in homogeneous clay soils. The control of the quality of a clay with little or no stone content for use in the clay core of a dam would be a particularly appropriate application. Where soils are variable or stony larger scale tests have to be applied and the practicability of the methods and the potential rate of testing have to be carefully considered.

Visual control

Although the quality of the soil cannot be specified in terms of appearance, it is clear that testing effort can often be reduced considerably by the intelligent application of visual and other forms of inspection. Thus Lieszkowszky (pp 157-164) describes visual and tactile examination of the material and Coumoulos (p.269) refers to the visual identification of suspect areas before testing and the subsequent reduction in costs of field quality control. Chartres (p.269) found that the operation of dump trucks on a clay core was a practical indicator of the strength of the fill material.

Factors affecting the criteria for soil suitability

The quality of the soil being excavated can vary from day to day as a result of changes in ground conditions, in soil type or in the weather. The influence of weather on the quality of soil demands particular attention and Parsons (pp 169-175) describes how MCV calibration data may lead to a means of predicting the effects of wet weather on a soil. The potential changes in soil conditions between site investigation and construction are considered by Jones (pp 247-251) to be worthy of research and he makes a plea for the careful recording of site data for subsequent scientific evaluation. Dennehy (pp 274-275) considers that an additional factor that influences the comparison of conditions at site investigation with those during construction is the potential error resulting from delayed testing during site investigation. He recommends that immediate measurements of moisture content should be made on duplicate samples during the site investigation.

With heterogeneous soils involving mixtures of wet and dry lumps of clay, such as are created when water is added to clay core material, results can be produced that are difficult to interpret. Often the strength of the weak lumps of clay controls the overall behaviour of the samples of such materials.

The control of the quality of the material should logically be made at the excavation location so that, if judged unsuitable, the soil may be immediately transported off the site, so avoiding the considerable wasted effort that would result if such a judgement were made subsequent to the placement of the unsuitable material in the fill area. Allowance has to be made, however, for the effects of weather conditions on the quality of the soil during the construction process. Because the strength of the material may be affected by the remoulding process, the test of quality made in the excavation location should ideally be a remoulded test and not a measurement of the in situ strength. In situ tests (plate bearing and field vane tests) and undisturbed shear strength tests (triaxial or unconfined compression tests) would appear to be more appropriate in the fill area, but here the interpretation of a poor result could be more difficult because of the combined influence of the quality of the soil and its state of compaction. The advantage, however, is that such testing provides an indication of the as-constructed quality of the fill with a single method of test (Vaughan, pp 271-273).

SPECIFICATION AND CONTROL OF COMPACTION

In the UK the specification for compaction of clay fills in highways differs significantly from specifications used in large dam embankments. In the latter case an end-result specification is normally used, frequently in terms of relative compaction (in situ dry density expressed as a percentage of that achieved in the British Standard laboratory compaction test using the 2.5 kg rammer). In highway earthworks compaction is to a method specification, where the thickness of layer and number of passes are specified for various types of compaction plant for three different classes of soil. In most other countries embankment fills are generally placed with an end-result specification. Steger (p.241) commented that the uniformity of soil in the clay cores of dams is likely to allow the practical application of precise moisture content and compaction specifications, whereas in highways the variability of soils dictates the use of less precise specifications.

End-result specification

In end-result specifications the required state of compaction is specified; for clay fills this is usually in terms of either relative compaction or air voids. In relative compaction the British Standard 2.5 kg rammer test (the normal Proctor test) is most commonly used as the standard against which to judge the in situ density. During construction the control to an end-result specification requires measurements of in situ dry density and the subsequent translation to relative compaction or air voids. The testing necessary to produce the results can be laborious and variable soil conditions can necessitate many tests so that it is possible to make a confident appraisal of the state of compaction achieved. Cole (pp 237-239) warns of the delays caused by testing to determine the in situ dry density. These delays during road construction in Africa gave rise to excessive drying of the surface of the compacted layers resulting in seepage planes and subsequent slips. Coumoulos (p.269) describes how the frequency of testing for a relative compaction specification was reduced by visually identifying suspect areas.

Kennard et al. (pp 143-147) describe construction procedures on several earth fill dams where an end-result specification for compaction was used. This varied in form between sites, the in situ dry density being either 95% of the maximum dry density achieved in the normal Proctor test or 97% or 98% of the dry density achieved in the normal Proctor test at the same moisture content. Where a percentage of the laboratory maximum dry density is required there is a degree of moisture content control incorporated (i.e. only within a certain range of moisture contents can the required relative compaction be achieved by a given compactive effort), whereas with laboratory compaction at the field moisture content only, an additional moisture content control is required. This latter method could lend itself, therefore, to appplication in clay core construction where the need for flexibility is reflected in the use of shear strength criteria in addition to a compaction requirement.

The laboratory compaction test, using the 2.5 kg rammer, if carried out strictly to British Standard procedure, entails the removal of material coarser than 20 mm. Where the natural soil contains a significant percentage of material retained on the 20 mm sieve there are two courses open: either to increase the scale of the laboratory compaction test so as to incorporate larger particles, or to correct the dry density to allow for the excluded material in the laboratory test. Kennard et al. (pp 143-147) refer to the determination of relative compaction after correction for stone content. Lieszkowszky (pp 157-164) also describes the application of a relative compaction specification but during construction he found that the results of field vane tests correlated well with relative compaction and could be used as a means of compaction control.

Kuno et al (pp 149-156) describe how the use of relative compaction specifications is difficult with a highly sensitive clay, the reason being that the maximum dry density achieved in the laboratory test on any given sample varies depending on the initial moisture content. They have therefore adopted a degree of saturation or air void criterion. Coumoulos (p.269) shows how in the laboratory compaction test the density was severely reduced by organic matter content; to avoid a high organic matter content in the fill material a loss on ignition test was applied.

A scheme using an air voids criterion is described by Rocke (pp 265-266) and reference is made to trials of a large diameter in situ density test for use in a stony glacial till.

A further means of control of compaction that could be classified as an end-result specification is that incorporating proof rolling, i.e. passing over the compacted layer with a heavy machine such as a pneumatic-tyred roller or a loaded scraper and noting the amount of deformation produced. The application of this method to a sand formation is described by Grace (p.245).

Method specification

The application of a method specification for compaction requires control by inspection. This entails monitoring of the number of passes of the compactors and the thickness of layer being compacted. Because of the difficulty in counting passes in an embankment construction situation some Engineers compile tables which include the potential output of relevant compaction plant when working according to the Specification. The potential output must exceed the rate of earth-moving in the particular fill area if the Specification is to be achieved. Control of compliance with the Specification then consists of checking the thickness of layer and ensuring that the compactors work systematically over the area. Arrowsmith (p.34) gives an example of such a table. Results of measurements of acceptable states of compaction achieved under the terms of the Department of Transport's method specification for compaction are given by Al-Shaikh-Ali (pp 15-23).

Over-compaction

In general, the wetter and weaker the clay fill, the easier it is to compact to its lowest potential void content at its existing moisture content. Thus the final passes of the compactor and the passage of earth-moving plant can occur over already fully compacted soil, leading to the generation of positive pore pressures with the well-known symptoms of large elastic and plastic movements under the plant. Clearly some reduction in compactive effort is required to avoid continued trafficking when the maximum potential density is approached. Al-Shaikh-Ali (pp 15-23) and Arrowsmith (pp 25-36) describe this problem, and the latter suggests that it is associated with the heavier plant now used in road construction. Ingoldby (pp 137-142) describes how compaction of unstable chalk fills can become impractical at the construction stage due to the generation of positive pore pressures.

It may not always be easy to alleviate the problem of over-compaction by reduction in compactive effort. With an end-result specification the effort can he reduced and the state of compaction measured to ensure compliance with the Specification, but with a method specification the resulting state of compaction is not normally checked and the required reduction in compactive effort would be unknown. In the past there have been instances of the application of large increases in the thickness of the compacted layer, as in the relaxed specification for chalk described by Ingoldby (pp 137-142). Although this action has the effect of eliminating the positive pore pressures, large voids could be created in the bottom of such thick layers, to the detriment of the subsequent performance of the fill. Such voids could be created even when the soil is very wet, producing a more permeable fill with a considerable increase in the risk of subsequent settlement.

Under-compaction

As soil becomes drier it is more difficult to compact. With an end-result specification this difficulty becomes apparent in the results obtained during the compliance testing and the necessary action can be taken to increase the compactive effort and raise the state of compaction to be within specification. With a method specification, where the requirements provide an adequate state of compaction at some prescribed level of moisture content or shear strength, inadequate states of compaction would be achieved if the material was in a drier (i.e. stronger) condition than the prescribed level on which the specification was designed. Edwards (pp 253-258) mentions the difficulty of achieving states of compaction equivalent to 5% air voids in materials with MCVs exceeding 16.

Prediction of over-compaction or under-compaction

Given that problems of over-compaction or under-compaction are greater with the use of a method specification than in the case of an end-result specification, consideration should be given to the possibility of using a control test to indicate the amount of compactive effort necessary to achieve an adequate state of compaction. If the test used to control the quality of the soil is a form of shear strength test or an empirical test that correlates well with shear strength, then the results could provide such an indication. Parsons (pp 169-175) suggests that MCV, if adopted as a routine test for suitability of fill material, could provide the necessary scale by which the method specification could be varied to avoid over-compaction or under-compaction.

Discontinuities

The effect on stability of discontinuities created by construction plant is considered by Symons (pp 297-302). Such discontinuities can often by produced by heavy smooth-wheeled rollers. Vibrating smooth-wheeled rollers can also produce discontinuities by creating polished surfaces when compacting clay at moisture contents drier than the plastic limit. Some compaction plant (e.g sheep's-foot rollers and tamping rollers), by reason of their design, could be expected to key together layers of compacted soil. However, Pavlakis (pp 266-269) describes how a tamping roller clogged-up during compaction of a clay core where slip planes were found to have been formed. Parsons (p.269) and Chartres (p.269) describe how such clogging of tamping and sheep's-foot rollers arises when mixtures of wet and dry soils are encountered. The conclusion must be that special measures may have to be taken to avoid the formation, in critical situations, of discontinuities within or between compacted layers. These measures could comprise the effective use of certain types of compactor to key the layers together or some other type of operation. For example, Pavlakis (pp 266-269) found that the scarifying of the surface of the compacted layers destroyed the potential slip planes.

TRAFFICABILITY

In general, soil from cuttings and other areas of excavation that is deemed to be unsuitable for use in embankments still has to be excavated, loaded into vehicles, transported and deposited in a spoil area, where it will almost certainly have to be trafficked by the plant. In many cases it is likely to involve less work and so be more economical to place that material in the embankment than to take it off site and import other soil to replace it. Assuming that at all times the most economical earthworks should be constructed within the constraints imposed by a satisfactory performace of the embankment, it appears vital that the limits of operation of various types of earth-moving plant should be known. This would allow the economics of plant operations when building embankments with more difficult wet material to be compared with the economics of treating that material as unsuitable and replacing it by some better quality material from further afield.

Limits of operation

Several approaches have been adopted towards the assessment of the limits of operation of earth-moving plant. Dennehy (pp 87-94) has determined the relation between rut depth and undrained shear strength and from this has deduced the values of undrained shear strength below which rubber-tyred scrapers, divided into two categories depending on tyre pressure, would be unable to operate. Parsons (pp 169-175) has related rut depth, speed of travel and limits of operation to MCV for medium-sized scrapers and MCV has been related to undrained shear strength. Kuno et al. (pp 149-156) used the cone index value to assess the potential performance of plant on Kanto loam - a difficult material with high sensitivity to remoulding; they give a factor for converting the cone index to an undrained shear strength. Arrowsmith (pp 25-36) and Threadgold (pp 251-253) also provide limiting values of undrained shear strength for operation of certain types of earth-moving plant. The various values quoted, converted to undrained shear strength where necessary, are given in Table 1.

The results provided by Parsons (pp 169-175) and by Kuno et al. (pp 149-156) show the importance of power to weight ratio by the clear improvement in performance of twin-engined scrapers over the single-engined or self-loading types. It appears, therefore, that the limiting condition for the operation of plant cannot be assessed from a consideration of rut depth alone, although in certain circumstances rut depth may be an appropriate criterion, e.g. to avoid damage to a prepared formation. The results quoted for scrapers that do not specify the number of engines can be only of marginal usefulness to the design engineer who is considering the ability of plant to handle material of a given quality. The data in Table 1 show discrepancies and, taken with the ranges of minimum values quoted, confirm

Table 1. Minimum values of undrained shear strength for operation of various types of earth-moving plant (kN/m^2)

Type of plant	Method of measuring C_u				
	Derived from MCV/hand vane relation (Parsons, pp 169-175)	Derived from cone index/unconfined compression test relation (Kuno et al., pp 149-156)	Unspecified (Arrowsmith, pp 25-36)	Field vane and tri-axial tests (Dennehy, pp 87-94)	Triaxial tests (Threadgold, pp 251-253)
Bulldozer with extra low contact pressure		15-20			
Bulldozer with low contact pressure		20-40			
Small tracked plant					30
Ordinary bulldozer		40-70			
Towed scraper		50-70	35		40
Medium twin-engined scraper	25-40				
Twin-engined scraper		40-50			
Medium single-engined scraper	70-110				
Self-loading scraper		At least 100			
Large rubber-tyred scraper			50		
Scrapers with tyre pressure of 340-380 kN/m^2				60-80	
Scrapers with tyre pressure of 240-310 kN/m^2				40-60	
Medium/heavy plant					60
Heavy motorized scrapers					100
Dump truck		At least 100			

the need for further research into this important aspect of earthwork construction.

Forde and Davis (pp 107-112) show the differences in behaviour of high and low plasticity clays at approximately the same undrained shear strength when trafficked in a mobility rig. The results show that greater traction was obtained with the high plasticity clay, thus raising the question of the validity of undrained shear strength, taken in isolation, as a method of assessing trafficability. The evidence indicates that adhesion between soil and tyre - a factor that may play an important part - depends on the type of clay.

Additional factors affecting plant selection

Factors additional to the soil conditions that affect the selection of earth-moving plant are suggested by Dengate (p.273) and by Quinion (pp 273-274). Among such factors, the more important are the availability of particular types of plant, including facilities for hire of such plant, the topography of the site, the length of haul, and the cost of spoil and borrow areas. In certain instances the selection of plant may be based solely on the requirement to preserve the quality of the fill material (Ingoldby, pp 137-142; Boden, pp 275-276).

Cost of earthwork construction using wet fill

The cost of earthworks can be affected by the way that the soil is handled. Al-Shaikh-Ali (p.274) recommends the Contractor to employ an experienced materials engineer to advise on such aspects and clearly this could be of prime importance when marginal soil conditions are encountered. Grace and Green (pp 113-118) describe the use of wet fill on a road scheme and estimate that a saving of at least £½ million was achieved in 35 miles of motorway construction. However, the use of wet fill may not always effect savings. Given two materials - one wet and one at a more acceptable moisture content, and both to be hauled over the same distance - the material at the more acceptable moisture content would be expected to be less expensive to handle. Only if the use of wet fill involves shorter haul distances will such use normally be justified, and then only if more detailed cost estimates confirm the economic case.

EFFECT OF PLANT OPERATIONS ON SOIL QUALITY

The operation of earth-moving plant can lead to deterioration of the quality of the soil being excavated. This can come about by the intermixing of two otherwise suitable soils to produce an unsuitable mixture or from breakdown of the material. Occasionally the breakdown of material can have a beneficial effect in providing a better graded material for the achievement of a good state of compaction.

Mixing of suitable soils to produce unsuitable mixtures

The intermixing of diffferent soil types could well be a major cause of variations in quality of material from that predicted from the site investigation findings (Jones pp 247-251). Al-Shaikh-Ali (pp 15-23) and Arrowsmith (pp 25-36) describe the effect of mixing suitable sand and clay to produce an unsuitable mixture. Adding sand to clay causes a reduction in plastic limit without the same relative reduction in moisture content. Perry (pp 189-196) illustrates how the gain in strength with time associated with compacted soft chalk is reduced when clay becomes intermixed with the chalk. Construction operations should, if possible, be designed to avoid such mixing of materials if they are to be used as embankment fill. The proper selection of either vertical face or horizontal face excavation methods (face shovel or scraper respectively), depending on the configuration of the interface between the soil types, could be expected to assist in the achievement of this objective.

Breakdown of material

Cedrun and Simic (pp 57-61) describe the construction of highway embankments with a marl rock where the breakdown of the material by the excavation and compaction processes is essential to the achievement of a satisfactory embankment. The objective was to produce a relatively impervious embankment with voids between remaining rock fragments filled with a compacted clayey soil. In this way the effects of weathering after construction would be minimized. Satisfactory results were obtained during construction of the highway embankments using a tamping roller (impact roller) which simultaneously compacted the fill and produced the necessary final breakdown of the material. Rocke (pp 265-266) has highlighted the need to break down large lumps of glacial till material excavated by large earth-moving machines so that satisfactory states of compaction can be achieved in layers of acceptable thickness.

Ingoldby (pp 137-142) describes how plant operations can lead to deterioration of the quality of chalk. He provides a classification method to assist in the selection of methods of excavation which are likely to maintain a satisfactory quality of material so that plant operations can continue effectively. He shows the variations in degree of degradation that can result from different methods of excavation, with a face shovel providing minimal breakdown and scrapers the maximum breakdown of chalk, over a wide range of chalk qualities.

Addition of water and removal of stones

The addition of water and removal of stones are normally restricted, in the UK at least, to the construction of clay cores of dams. A number of methods of increasing moisture content are described by Kennard et al. (pp 143-147). In all cases the methods of adding water and mixing the soil were incapable of producing a homogeneous material. The effects of time and of compaction were also necessary to achieve some measure of homogeneity. Vaughan (pp 271-273) provides data showing the increased variation in moisture content after the addition of water to clay core material; in his case mixing included the use of rotavators which would have provided a more positive form of mixing than the methods mentioned by Kennard et al. (pp 143-147). Rocke (pp 265-266) also mentions the problems of adding water and highlights the difficulty of achieving a uniform distribution of water with high-pressure water-distribution equipment.

The removal of stones from glacial till is mentioned as a difficulty by Rocke (pp 265-266). In this instance boulders were removed by machine and by hand.

Chalk classification

To provide a method of identifying chalk which may give rise to problems during embankment construction, either in terms of difficulties of trafficability in the freshly placed fill or in terms of difficulties with excavation and compaction in very hard chalk, Ingoldby (pp 137-142) suggests a classification based on a test of hardness of the chalk lumps and the maximum amount of water that the lumps can hold (saturation moisture content). In the classification, methods of excavation are proposed which minimize the difficulties with the various classes of chalk. Threadgold (p.276) and Child (p.276) give results of such classification tests and make comparisons with the Mundford classification.[1] In general, the two forms of classification would not be expected to agree (Boden, p.276) as they relate respectively to the properties of excavated chalk for use as fill and of in situ chalk for use as a structural foundation.

CONCLUSIONS

For the control of quality of the soil the use of a measurement of undrained shear strength seems to be preferred provided it can be carried out at a reasonable rate. The construction of the clay cores of dams, where the strength is of prime interest and often the soil is more amenable to shear strength testing, can be a particularly appropriate application. Considerable evidence has been put forward regarding the inappropriateness of moisture content and plastic limit tests as a valid means of control of quality, and even when allowance is made for variations in plasticity index they are still shown to be misleading in relation to engineering properties. Empirical strength tests, such as the moisture condition test, which correlate well with undrained shear strength appear to offer the solution where soils are not amenable to easy direct measurements of strength. In addition the causes of variations in quality of excavated material from that predicted do not appear to be fully understood.

The criteria on which to base the level of suitability of a soil will vary depending on the application. Stability will be the criterion in high embankments, whereas for more shallow embankments it is generally agreed that trafficability of plant controls the limits of quality of the soil. However, from the conflicting evidence submitted it is clear that the limits of trafficability cannot yet be accurately stated. They will depend on the type of plant involved. It has also been stated that the rate of strain in the method of test used to assess the limits of trafficability should ideally be similar to that experienced by the soil under the moving wheels of the plant. More information is required on the limit of trafficability before such limits can be confidently used as criteria of quality of the soil for the construction of shallow embankments. Where such information is provided the test procedure and rates of strain should be fully detailed.

The addition of water to achieve the necessary strength limits in soil (e.g. in clay cores) is a further problem. No totally satisfactory method appears to exist for the process. Although the required average levels of strength can be achieved, the variation of values usually exceeds the specified levels. Further developments may well be required in methods of distributing water in a uniform manner and in the subsequent mixing of the soil. The requirements appear little different from those in the soil stabilization field and possibly soil stabilization methods could be adapted to the needs of the dam construction industry.

REFERENCE

1. BURLAND, J.B. and LORD, J.A. The load-deformation behaviour of Middle Chalk at Mundford, Norfolk. Building Research Station, Garston, 1970, current paper 6/70.

J.A. CHARLES, PhD, MSc(Eng), MICE, Building Research Establishment

General report: Methods of treatment of clay fills

Two distinct sets of circumstances have been described in which some form of treatment is required to obtain a satisfactory performance from a clay fill. The first situation arises when the borrow material for a clay fill is wet and the resulting compacted clay fill has a low undrained shear strength and needs to be treated so that its strength is increased. The other situation occurs where some form of construction has to take place over an existing uncompacted clay fill and treatment is needed to reduce air voids and increase the density of the loose fill. The relationship between these two situations is shown in Fig. 1.

TREATMENT OF WET CLAY FILL

Introduction

The wetness of a clay fill and its corresponding low undrained shear strength may present difficulties in both the design of an embankment and its construction. There are a variety of solutions to these problems. Where stability is a problem, an embankment can be redesigned with flatter slopes. Grace and Green (pp 113-118) and Kuno et al. (pp 149-156) describe how construction problems can be overcome by special earthmoving techniques. However, in many cases it may be economic to reduce the moisture content of the clay fill and hence increase its undrained shear strength (Fig. 1).

Aeration

Kuno et al. describe the successful drying in Japan of a volcanic clay soil of high plasticity - Kanto loam. Drying of the spread loam was achieved using a disc harrow and a reduction in moisture content of about 20% is reported. However, the method was practical only in the best weather conditions. Lieszkowszky (pp 157-164) reports the drying of a highly plastic, sensitive, varved clay which was used for the clay core of a dam in Northern Ontario. The contractor used two tractors pulling a farmer's disc and harrow to rip and break up the clay for drying in the borrow area. During a warm sunny day, a decrease in moisture content of up to 10% was measured. However, there was a good deal of wet weather during the construction period of the dam and in general the drying was difficult to achieve. Grace and Green cite the successful use of aeration at Sasumua Dam in Kenya but conclude that climatic conditions are such that the method is not generally applicable in the British Isles. This is in agreement with Rodin[1] who considered that the drying of clayey or silty soils in the UK by aerating the soil by disking or harrowing was usually impracticable as the prevailing weather conditions do not normally contain prolonged dry periods.

The effectiveness of an aeration technique depends not only on the climatic conditions but also on the change in moisture content w that is required to produce the desired increase in undrained shear strength. Cooling and Smith[2] presented data on the undrained shear strength versus moisture content relationship for a number of remoulded clays. Fig. 1 shows the relationship between moisture content and undrained shear strength for a cohesive fill of intermediate plasticity. It is seen that to increase the undrained shear strength from 30 kN/m^2 to 50 kN/m^2 would require about 2% reduction in moisture content. Skempton and Northey[3] have demonstrated that many remoulded clays show a similar relationship between undrained shear strength C_u and liquidity index LI, where LI = (w - PL)/PI where PL is the plastic limit and PI is the plasticity index. Over a restricted range of undrained shear strength the relationship approximates to

$$\log_{10} C_u = A - B\,(LI) \tag{1}$$

where A and B are constants. Hence

$$\Delta w = -\frac{PI}{2.30B}\frac{\Delta C_u}{C_u} \tag{2}$$

For the soil shown in Fig. 1, with 20 $kN/m^2 < C_u < 100\ kN/m^2$, equation (2) becomes

$$\Delta w = -\frac{PI}{5.8}\frac{\Delta C_u}{C_u} \tag{3}$$

Consequently the change in moisture content required to produce a certain percentage change in undrained shear strength of a clay fill is a linear function of the plasticity index of the soil. With high plasticity soils much larger reductions in moisture content are required than with low plasticity soils to produce the same improvement in undrained shear strength. Rodin[1] comments that with fill of low clay or silt content (i.e. low plasticity) some drying may be practical on occasions, even in the UK.

Treatment with lime

Kuno et al. describe the use of quicklime to reduce the moisture content of Kanto loam. Either in-place mixing or pre-mixing is said to be usual with a central mixing method rarely used. Using a pre-mixing technique, in which trenches were dug in the loam and filled with quicklime before excavation, the moisture content of the loam was reduced from 108% to 70% and the plasticity index from 50% to 14%. The lime content was approximately 20% of the dry weight of the soil.

Van Ganse[4] stated that over 11 x 10^6 m^3 wet loess and loam soils were treated with quicklime for the construction of embankments in Belgium in 1968-71, mainly on motorway construction sites. The usual procedure was to spread lime in the cutting. In 1968, when treatment with lime began in Belgium, additions of 3% of quicklime were used, but this had been subsequently reduced to only about 1%. It was found that the major part of the improvement of the soil occurred within one hour of mixing. Van Ganse commented that in western Europe powdered quicklime was used in preference to slaked lime as it was cheaper and, in addition to the effects produced by slaked lime, quicklime reduced the water content by 0.65% per 1% of quicklime used.

Ayres (p.281) draws attention to the French practice of treatment of the top 200 mm of clay or silt subgrades with lime. In some parts of the USA hydrated lime and quicklime have been used extensively for many years to treat clay subgrade soils.

In the UK little use appears to have been made of lime as a method of improving the properties of wet clay fill. Dumbleton[5] carried out a laboratory investigation into the use of hydrated lime to stabilize soils for road construction in Great Britain. He commented that it might be possible to treat wet subgrades to assist construction. However, Snedker[6] considered that the practical and economic aspects of stabilization in small quantities on roadworks needed more consideration before it was likely to appeal to highway engineers in the UK. The use of quicklime, primarily as a construction expedient, to dry out waterlogged sites has been described by Dunn.[7] Lake (p.273) refers to a recent dewatering, excavation and compaction trial in Essex which included lime treatment of the fill. The soils are described as very loamy clays and clayey fine sands and silts. It is affirmed that the lime treatment was effective only with the more clayey material. It may be that safety and environmental problems will be a barrier to the widespread use of quicklime for drying wet clay fills in the UK.

Yamanouchi et al.[8] have described embankment construction in Japan in which 5 cm thick layers of quicklime, sandwiched between cardboard, were placed in soft clay fill at about one metre vertical intervals. As water is absorbed from the clay the quicklime-cardboard sandwich acts both as a drain and as a reinforcing element in the embankment. The mixing of lime with clay to form columns which will act as both drains and reinforcement under foundations has been described by Broms and Boman[9].

Drainage layers

The use of drainage layers to increase the rate at which excess construction pore pressures dissipate, and hence increase the shear strength of wet cohesive fills during embankment construction, is well established. Penman (pp 177-187) describes an early use of this method to maintain stability during the construction of the Usk earth dam in 1953. The drainage layers consisted of an inner element of crushed rock with a sand and gravel mixture above and below. Two layers were placed in the 35 m high clay embankment. Following the construction of the Usk dam, drainage layers have been a frequent feature in the design of clay embankment dams in the UK and their behaviour is well understood. Gibson[10] analysed the consolidation of a clay fill with drainage blankets during construction. Gibson and Shefford[11] have examined the efficiency of horizontal drainage layers in clay fills.

Grace and Green (pp 113-118) describe the use of drainage layers in motorway embankments of wet clay fill. Road embankments are generally comparatively small in the UK and, if built on a strong foundation, settlement rather than stability is likely to be the controlling consideration in design. On this assumption Grace and Green prepared their fig. 1 showing the required spacing of drainage layers for soils of different permeability (assuming a particular fixed value for compressibility). Grace (pp 276-279) comments that if drainage layers are provided at 2 m centres and if a settlement period of 100 days can be allowed, then all soils having a permeability greater than 10^{-8} cm/s can be used in this way, and that this includes nearly all British soils. He says that in view of the expense involved in rejecting cohesive fills simply because they are too wet, it is surprising that this type of construction was not used at an earlier stage of the UK motorway construction programme.

Inada et al. (pp 127-132) used drainage layers in a 20 m high motorway embankment in Japan, when stability was a major problem. The embankment was constructed using a sensitive volcanic cohesive fill of very low undrained shear strength (14 kN/m^2). In the test embankment, drainage layers consisted of 20 cm thick layers of sand and were placed at 5 m or 15 m intervals. The slopes of the embankment were 1 vertical in 2 horizontal. In the 32 months following placing of the fill, the moisture content decreased by 10%. The pore water pressures at the end of construction were much lower in the section of the embankment with drainage layers at 5 m centres.

Boman and Broms (pp 49-56) describe placing a clay fill with an undrained shear strength of only about 3 kN/m^2. Not surprisingly the fill was of very limited height (1.8 m). Plastic-paper drain strips were placed in the fill at 50 cm spacings. The clay fill was placed by end tipping and the drains were inserted afterwards.

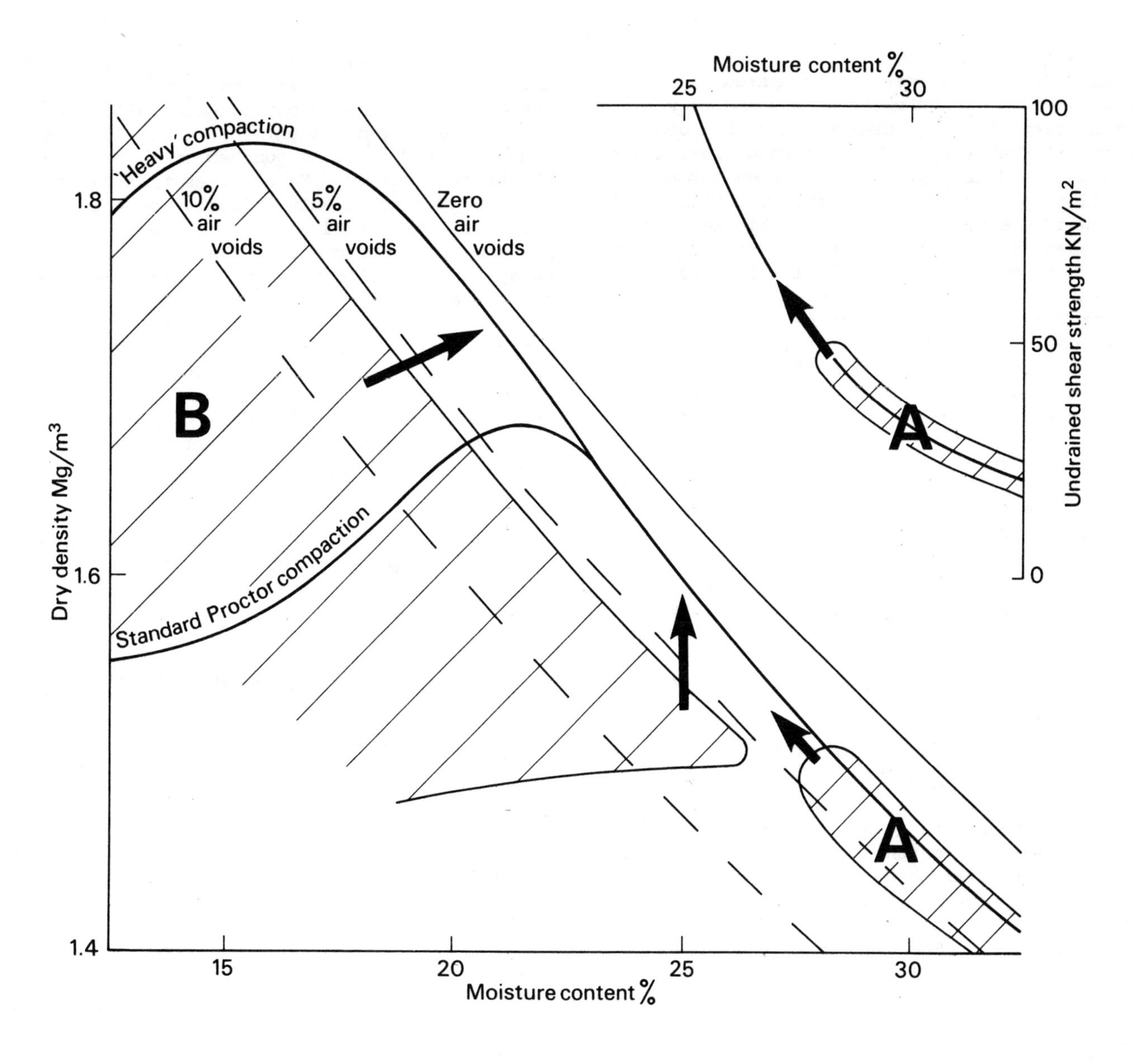

Fig. 1. Treatment of clay fills; A wet clay fill, B existing uncompacted clay fill

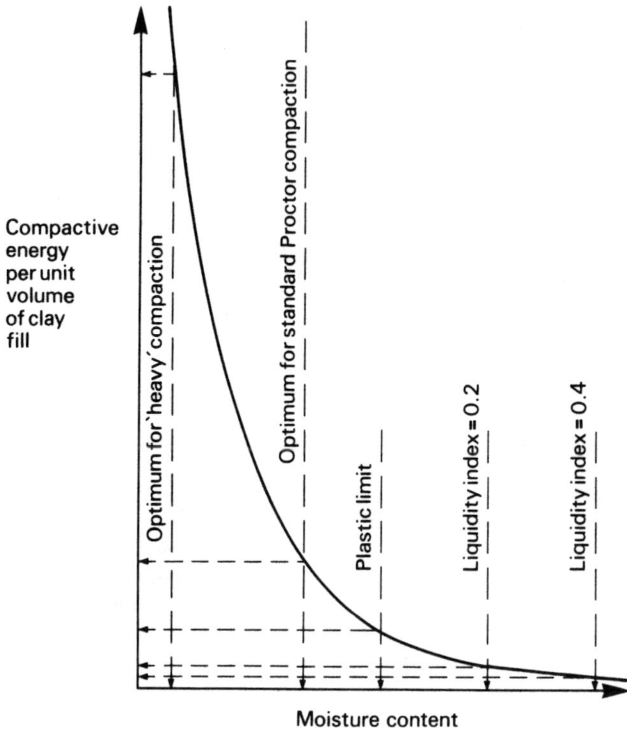

Fig. 2. Energy required to compact a clay fill as a function of moisture content

There was an undrained control section of the fill. The fill was well instrumented with piezometers and settlement gauges and it was demonstrated that the time required for consolidation of the clay fill was reduced where drains were used. Smart (p.281) questioned whether the functions of drainage and reinforcement could be combined in some form of plastic blanket. Many non-woven fabrics can convey water in their own plane but would be relatively ineffective as reinforcing elements. Loudiere[12] has described an earth dam in which a vertical drain was made of a non-woven synthetic fabric. He has also described the use of fabrics above and below horizontal drainage layers of granular fill in earth embankments.

The temporary surcharge loading of a clay fill with drainage layers could accelerate consolidation. However, a surcharge of fill on top of an embankment of wet clay fill might well cause instability. The application of a vacuum to drainage layers sealed into the fill could produce a similar beneficial increase in rate of consolidation but, as pore pressures are reduced, without any tendency to cause instability. This type of vacuum technique was suggested for natural soft clay soils by Kjellman.[13]

TREATMENT OF EXISTING UNCOMPACTED CLAY FILL

Introduction

With increasing frequency it is necessary to build structures and roads on sites where there is a considerable depth of uncompacted clay fill. Some of these fills have resulted from opencast coal and ironstone mining, others from uncontrolled back filling of old clay and gravel pits, docks and so on. The principal defect of such fills as foundation materials is their loose condition with large voids, and the basic requirement of ground treatment is to compact them. Obviously there are other solutions. It might be economic with a fill of shallow depth to remove the fill and replace it in thin layers with adequate compaction. Alternatively structural loads can be transferred down to the underlying natural ground by piles. The insertion of stone columns into an uncompacted clay fill by a vibro technique, as described by Moseley and Slocombe (pp 165-168), is basically a method of reinforcing rather than compacting the fill. However, in many situations the best approach to the problems of an uncompacted clay fill is to adopt some form of ground treatment to reduce the air voids and densify the fill as shown in Fig. 1. Case records of the use of several different methods have been presented and discussed.

Dynamic consolidation

In dynamic consolidation, deep compaction of soils is attempted by repeated impacts of a heavy weight on to the ground surface.[14] The Menard system of dynamic consolidation, as it has been applied at clay fill sites in the UK, has typically involved dropping a 15 t weight from heights of up to 20 m. Primary tamping has usually consisted of repeated impacts at a number of points on a fairly widely spaced grid. Pearce (pp 279-281) describes a typical sequence of operations and also points out that it is usually necessary to import granular fill on to the site to form a working platform. He asserts that the success of dynamic consolidation on clay fill is heavily dependent on the use of this granular fill which is punched into the clay fill at the grid points. He believes that the fill then behaves as an only slightly compacted clay fill reinforced with stone piers. However, the enforced settlement of the ground surface due to dynamic consolidation at the sites described by Charles et al. (pp 63-72), Moseley and Slocombe (pp 165-168), and Thompson and Herbert (pp 197-204) appears to show that the treatment causes significant compaction of a clay fill.

The effectiveness of dynamic consolidation, and the economics of using it on a clay fill, depend very much on the amount of compactive effort required to reduce the air voids in the fill to close to the minimum, which is usually 2-3%. Parsons (pp 169-175) describes a laboratory test programme in which soils were compacted into a cylindrical container with a rammer of almost the same diameter as the container. For several different types of cohesive soil he found a relationship between undrained shear strength (measured by a vane) and moisture condition value MCV of the form

$$\log C_u = a + b(\mathrm{MCV}) \qquad (4)$$

where a and b are constants.

From Parsons' definition of moisture condition value it follows that

$$\log C_u = \log d + b\ (10 \log N) \qquad (5)$$

where N is the number of blows required to compact the soil and where

$$\log d = a$$

Equation (5) can be re-written as

$$C_u = dN^{10b} \qquad (6)$$

In Parsons' test, the energy per blow per unit volume of soil is about 22 kNm/m^3 and typical values for a and b are a = 0.8 and b = 0.11. Consequently the relationship between compactive energy required per unit volume of soil E/v and the undrained shear strength is (in kN/m^2)

$$E/V = 4\ C_u^{0.9} \qquad (7)$$

Parsons' tests are limited to the range 10 kN/m^2 < C_u < 100 kN/m^2 but Threadgold (pp 251-253) presents data from tests of a London Clay fill confirming the relationship between undrained shear strength and moisture condition value for undrained shear strengths up to 700 kN/m^2. Edwards (pp 253-258) shows similar relationships for a number of different types of cohesive fill. However, Edwards also demonstrates that with a moisture condition value of 16 or above samples may not be compacted in the test to an air voids of 5% or less. Thus at this stage the moisture condition test may be underestimating the compactive effort required to compact the soil fully.

The total amount of compactive effort needed to compact a clay fill depends not only on the undrained shear strength of the fill but also on the actual type of compaction applied (including such factors as the energy applied per blow and the diameter of the rammer). Proctor[15] indicated that in his type of laboratory compaction test, the size of the mould affected the result significantly. It is clear therefore that equation (7), derived from a particular laboratory test, could not be directly applied to other types of compaction. The importance of the relation is that it suggests that, for a particular form of impact compaction on a clay fill, the compactive effort required is almost a linear function of the undrained shear strength of the fill. The implication of this is illustrated in Fig. 2 which shows the type of relationship, which will be obtained for a cohesive fill, between required compactive effort and moisture content. It is seen that the amount of energy needed to compact a clay fill at a moisture content corresponding to optimum in the heavy (4.5 kg rammer) compaction test (test 13, BS 1377[16]) might typically be ten times the energy required to compact the same fill at a moisture content equal to the plastic limit. It is interesting that all the case records of dynamic consolidation of clay fills described by Charles et al., Moseley and Slocombe, and Thompson and Herbert appear to relate to fills at moisture contents close to the plastic limit. Bures (p.279) describes a site where the clay fill has a moisture content significantly smaller than the plastic limit and where dynamic consolidation is proposed. It may be that to be effective either a much larger input of energy will be required at this site or, before dynamic consolidation, the fill will have to be inundated to increase its moisture content.

The depth within the clay fill to which dynamic consolidation is effective is also of interest. Menard and Broise[14] have suggested that this depth is primarily a function of the energy delivered per blow of the falling weight. Obviously the total energy input, the diameter of the weight and the spacing of the compaction grid points also have some effect on the depth to which the treatment is effective. At the clay fill sites described by Charles et al., Moseley and Slocombe, and Thompson and Herbert dynamic consolidation seems to have had a significant effect on the fill down to depths of 5-6 m.

Varaksin (p.279) describes the use of dynamic consolidation in constructing expressway embankments up to 10 m high in France. The clay fill was placed in one operation and then treated by dynamic consolidation. Clearly in considering this as an alternative to placing and compacting in thin layers, the economics of the two alternative approaches should be examined.

Pre-loading with surcharge of fill

The treatment of an uncompacted clay fill with a temporary surcharge of 9 m of fill is described by Charles et al. (pp 63-72). Fig. 8 of their paper shows that settlement of the clay fill occurred mainly as the surcharge was being placed and can be attributed to reduction of air voids in the fill rather than the squeezing out of water from the clay lumps. This immediate settlement is important, as surcharging is sometimes rejected as a means of ground treatment because it is considered to be a lengthy process and this is not the case for an uncompacted fill material. Where a supply of extra fill is locally available, surcharging of an existing uncompacted clay fill would seem an attractive solution. As the surcharge need not be left in position for a long period, treatment of a large site can be achieved by continuously moving a heap of fill around the site. Tomlinson and Wilson[17] describe how an uncompacted colliery waste, which had been end-tipped into a flooded clay pit, was treated by pushing a 5 m high shale fill surcharge across the site with a large bulldozer.

The height of surcharge required to treat an uncompacted clay fill might be calculated on the basis that at every point within the clay fill the soil should be subjected to a greater stress by the surcharge than the stress that will result from the actual construction on the site. However, damaging movements in the clay fill may be due to causes other than structural load, e.g. creep settlement under self-weight of the fill and collapse settlement on inundation of an unsaturated fill. Consequently ground treatment should aim at eliminating or reducing these other types of ground movement. In this context, the depth to which a surcharge will effectively compress the clay fill is of interest. Where a surcharged area is small compared with the depth of the clay fill, the depth to which treatment is effective will be controlled by the minimum dimension, in plan, of the surcharge. Otherwise the depth of effectiveness is likely to be principally a function of the height of the surcharge. The clay fill described by Charles et al. was compressed down to a depth of about 12 m, e.g. about 1.3 times the surcharge height.

In dynamic consolidation, as the moisture content of a clay fill is reduced, the amount of compactive effort required to compact the soil rapidly increases. In a similar manner the amount of compression produced by a given surcharge load becomes smaller as the moisture content of the clay fill is reduced and a larger surcharge is required to reduce air voids to a minimum. This means that with dry clay fills surcharging may be effective only if inundation takes place first.

In the situation described by Tomlinson and Wilson[17] fill had been end-tipped into a flooded pit. Where there is a high water table in an uncompacted clay fill, lowering the water table by pumping may be an alternative way of effectively surcharging the fill. As the water table is lowered the effective stress in the fill is increased. This approach was considered by Tomlinson and Wilson and, although not adopted, some pumping was carried out at one stage and an increased rate of settlement was observed.

Inundation

Many loose unsaturated fill materials undergo collapse settlement on inundation. If such movements occur within a clay fill during or after construction on the fill, then serious problems may arise. A corollary of inundation after construction being a potential hazard on a clay fill site is that inundation before construction may be an effective method of ground treatment.

A number of laboratory investigations of collapse compression of cohesive fills have been described. All the tests were one-dimensional compression tests carried out in oedometers. Abeyesekera et al. (pp 1-14) report a laboratory test in which a collapse strain of 16% was measured after one day when a uniformly graded shale aggregate (10 mm maximum particle size) under a vertical stress of 140 kN/m^2 was flooded. Of this compression, 70% occurred within 7 minutes of introducing water into the sample. The collapse was attributed primarily to slaking degradation of the shale. Cox's fig. 3 (p.80) suggests that collapse compressions greater than 8% were measured when mudstone samples placed at low density and moisture content and under a vertical stress of 400 kN/m^2 were inundated. Bures (p.279) reports that when a sample formed of stiff clay lumps and under a vertical stress of 300 kN/m^2 was watered a compression of 6% occurred over 17 days. Of this compression, 70% occurred within 10 minutes of watering the sample. Fig. 2 of Charles et al. (p.64) shows the relationship between vertical compression on inundation and the air voids in the sample before inundation for some clay fill samples under a vertical stress of 100 kN/m^2. It was necessary to reduce the air voids to 5% to avoid collapse compression on inundation in these particular circumstances.

Bures presents measurements of the settlement of a deep uncompacted fill of stiff clay in northern Bohemia. He considers that the rate of settlement was related to precipitation in the area. However, when a field inundation test was attempted by watering the surface of the fill, it was found that the water did not penetrate the surface layer of the fill. It was because of the likelihood of this difficulty that in the inundation test at Corby described by Charles et al. trenches were dug through the surface layer of the clay fill. The collapse strain produced in the top 5 m of fill at Corby was generally between 1-2%, although a compression strain of 6% was measured at one location. The attempt at this site to pretreat an uncompacted clay fill by inundation was not very successful. However, the inundation caused sufficient settlement to show that movements due to wetting of the clay fill subsequent to construction could be a problem. An obvious difficulty with inundation as a ground treatment method is that water put into the fill at ground surface will tend to run down the largest fissures and voids and will not necessarily produce a uniform treatment throughout the fill. Where a ground water table rises and saturates a clay fill, this problem does not occur. The clay fill at Corby was significantly wet of standard Proctor optimum moisture content; the obvious way to densify such a fill is by surcharge loading or dynamic consolidation (Fig. 1). Where a clay fill is fairly dry, as the clay fill described by Bures appears to be, it may be that inundation is the most appropriate form of ground treatment. In such a situation, it might be necessary to follow inundation by some other form of ground treatment: surcharging or dynamic consolidation.

Assessment of treatment of uncompacted clay fill

When an uncompacted clay fill has been subjected to some form of ground treatment, it is desirable to carry out a programme of testing to confirm that the treatment has been effective and that the clay fill will be an adequate foundation for the proposed development of the site. As a major problem with uncompacted fills is the possibility of extreme variability, it is clear that large-scale field tests are preferable to laboratory tests on small, possibly unrepresentative, samples. A test programme is most likely to be useful if it is closely related to those aspects of the behaviour of the clay fill giving most cause for concern. These could include creep settlement of the fill under its self-weight, compression under structural loading or collapse settlement on inundation. Creep movements can be investigated by monitoring the movements of surface settlement stations by precise levelling over an adequate period of time. Plate loading tests would seem a practical way of assessing the behaviour of the clay fill under structural loading. In most instances the settlement under working load over a lengthy period is of more interest than a hurried bearing capacity test. Cox (p.271) had described a simple and practical form of test in which a 1 m square reinforced concrete base is loaded to the required vertical stress using a rubbish skip as kentledge. With this arrangement there is no difficulty in carrying out a long-term test. In some situations it may be more appropriate to load a larger area of the treated ground by placing a small surcharge of fill directly on it. Where there is the possibility of collapse settlement, field inundation tests may be carried out.

REFERENCES

1. RODIN, S. Earthworks - some practical aspects in the United Kingdom - Part 2. Muck Shift. Bulk Handler, 1964, May, 34-41.
2. COOLING, L.F. and SMITH, D.B. The shearing resistance of soils. J. Instn. Civ. Engrs, 1935, vol. 3, 333-343.
3. SKEMPTON, A.W. and NORTHEY, R.D. The sensitivity of clays. Géotechnique, 1953, vol. 3, no. 1, 30-53.
4. VAN GANSE, R. Immediate stabilisation of wet soils with lime. Proc. 8th Int. Conf. Soil Mech., 1973, vol. 2.2, 233-237.
5. DUMBLETON, M.J. Lime stabilised soil for road construction in Great Britain. Rds. Rd. Constr., 1962, vol. 40, 321-325.
6. SNEDKER, E.A. Choice of an upper limit of moisture content for highway earthworks. Highws Des. and Constr., 1973, Jan., 2-5.
7. DUNN, C.S. Drying out waterlogged construction sites. Construction, 1975, No. 16, 34-35.

8. YAMANOUCHI, T. et al. In-site experiments on soft clay banking by means of multiple-sandwich method using cardboard wicks and quicklime. Proc. 4th Asian Reg. Conf. Soil Mech., Bangkok, 1971, vol. 1, 341-345.
9. BROMS, B.B. and BOMAN, P. Lime columns - a new type of vertical drains. Proc. 9th Int. Conf. Soil Mech., 1977, vol. 1, 427-432.
10. GIBSON, R.E. The progress of consolidation in a clay layer increasing in thickness with time. Géotechnique, 1958, vol. 8, no. 4, 171-182.
11. GIBSON, R.E. and SHEFFORD, G.C. The efficiency of horizontal drainage layers for accelerating consolidation of clay embankments. Géotechnique, 1968, vol. 18, no. 3, 327-335.
12. LOUDIERE, D. The use of synthetic fabrics in earth dams (in French). Proceedings of international conference on use of fabrics in geotechnics, Paris, 1977, vol. 2, 219-223.
13. KJELLMAN, W. Consolidation of clay soil by means of atmospheric pressure. Proceedings of conference on soil stabilisation, Institute of Technology, Massachusetts, 1952, 258-263.
14. MENARD, L. and BROISE, Y. Theoretical and practical aspects of dynamic consolidation. Géotechnique, 1975, vol. 25, 3-18.
15. PROCTOR, R.R. Relationship between foot pounds per cubic foot of compactive effort to soil density and subsequent consolidation under various loadings. Proc. 2nd Int. Conf. Soil Mech., 1948, vol. 5, 223-227.
16. BRITISH STANDARDS INSTITUTION. Methods of test for soils for civil engineering purposes. British Standards Institution, London, 1975, BS 1377.
17. TOMLINSON, M.J. and WILSON, D.M. Preloading of foundations by surcharge on filled ground. Géotechnique, 1973, vol. 23, no. 1, 117-120.

Index

227–229, 229–232, 232–234, 234, 236, 236–237, 237, 237–239, 239, 239–241, 241, 243, 243–244, 244, 244–245, 245, 245–246, 247, 247–251, 251–253, 253, 253–258, 258, 258–260, 260, 260–263, 263, 263–265, 265, 265–266, 266, 266–269, 269, 269–271, 271, 271–273, 273, 273–274, 274, 274–275, 275–276, 276, 276–279, 279, 279–281, 281

Errata

Page 17. Replace Fig. 3 by

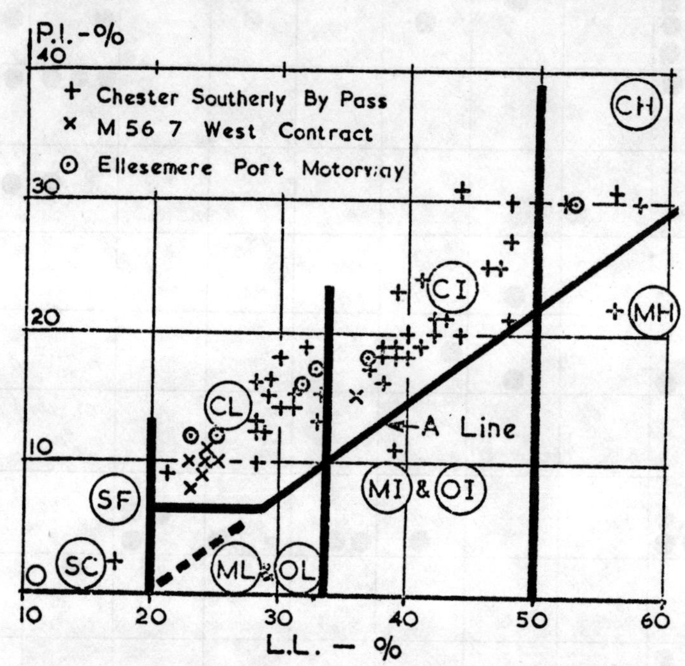

Fig. 3. Plasticity index vs liquid limit

Page 18. Line 30 should read
$k = 10^{-6} - 10^{-8}$ cm/sec

Page 146. Line 15 of right-hand column: for 'not' read 'or'